LANDMARK PAPERS:
GRANITE PETROGENESIS

Selected by J. D. Clemens
University of Stellenbosch, South Africa

and F. Bea
University of Granada, Spain

2012
Published by the Mineralogical Society of Great Britain and Ireland

THE MINERALOGICAL SOCIETY

The Mineralogical Society of Great Britain and Ireland was instituted in 1876. The prime aim of the Society is to advance the knowledge not only of Mineralogy but also of Crystallography, Geochemistry and Petrology, together with kindred subjects. This is done principally through the publication of scientific journals, books and monographs, and through arranging or supporting scientific meetings. The Society speaks for Mineralogy in Great Britain, linking with British science through the Royal Society and cooperating closely with the Geological Society. It maintains close liaison with other European mineralogists as a member society of the European Mineralogical Union, and is the body that nominates British representatives to the International Mineralogical Association. There are some 900 members of the Society, of whom approximately half reside abroad.

The Society's contact details are:
Address: 12 Baylis Mews, Amyand Park Road, Twickenham TW1 3HQ, UK
Tel. +44 (0)20 8891 6600
Fax +44 (0)20 8891 6599
E-mail: info@minersoc.org
Website: www.minersoc.org

Copies of this book are available from the Mineralogical Society office and from the Society's online bookshop (www.minersoc.org).

First edition 20012.

Typeset in 10.5/12 pt Times by Almaroca Ltd., West Kirby, Wirral, UK
Printed and bound by Lightning Source, UK

ISBN: 978-0-903056-30-4

Publishers of original papers, reproduced with permission in this volume, together with links to their websites or publications:

American Geophysical Union (http://www.agu.org/)
Deutsche Gesellschaft für Geowissenschaften (http://www.dgg.de/cms/front_content.php)
Elsevier (http://www.journals.elsevier.com/tectonophysics/)
Geological Society of America (http://www.geosociety.org/)
Geological Society (www.geolsoc.org.uk/)
Nature Publishing Group (www.nature.com)
Oxford University Press (www.oup.co.uk)
Royal Society of Edinburgh (http://www.royalsoced.org.uk/555_LearnedJournals.html)
Springer (http://www.springer.com)
The Society of Resource Geology (http://www.kt.rim.or.jp/~srg/english.html)
University of Chicago Press (http://press.uchicago.edu/index.html)
Yale University (http://www.geology.yale.edu/journals/Ajs.html)

Preface

A compilation of landmark papers on granite petrogenesis is overdue. Over the previous century, petrological thought revolved mainly around the origins of mafic and ultramafic magmas, with strong emphasis on processes such as assimilation, fractional crystallization and magma mixing. Work on the origins of felsic rocks had been confined mainly to considerations of the explosive volcanic magmas erupted at subduction margins through the sorts of processes that occur in relation to the evolution of mafic magma. For a long period, consideration of granitic rocks was mired in sterile controversies over their supposed origin by metasomatic replacement of pre-existing rocks or by forceful emplacement of bulbous and buoyant diapirs. Both of these erroneous concepts have now been laid to rest, although the corpses twitch occasionally.

The experimental work of Tuttle and Bowen (1958) helped to provide petrologists with the firm concept of granitic rocks as magmatic products, and various subsequent studies extended the concept. However, it was not until the 1970s that a kind of petrological renaissance occurred, with the realization that granitic rocks are dominantly the products of crustal melting during high-grade metamorphic events, and that they could provide a valuable window into the commonly unexposed depths of the continental crust. This was sparked mainly by the work of Australian scientists, principally Bruce Chappell and Alan White, as initially aired in Chappell and White (1974). Following this impetus, numerous studies used geochemical, isotopic and experimental methods to appreciate, for the first time, the range of physicochemical conditions involved in the formation of granitic magmas and their consequences for the compositions and behaviour of the magmas. This, in turn, allowed for the recognition of the place held by granitic magmatism within the system of continental growth and differentiation.

The most recent phase of discoveries regarding granite petrogenesis revolves around the rates and mechanisms of crustal melting, the segregation and migration of felsic magmas, the shapes of batholiths, the creation of space for granitic intrusions and the roles of chemical disequilibrium and entrainment of solids in the origin of chemical variations in suites of granitic rocks. In compiling this set of landmark papers, we have tried to single out the most significant advances that have taken place and then to trace these to what appear to be their origins in particular publications. As with all such compilations, readers will disagree on what has been significant, so we beg your indulgence. This is our choice. Of course, there is an acceleration of knowledge as time progresses and work going on now, or recently published, may well turn out to be ground-breaking research that deserves a place among the landmarks. Anyway, this is how it seems to us who observe the present. There is a good chance that this perception is incorrect and only the passage of time can really settle the question. For this perceptual reason, we have not included any work published after 1996.

This volume is organized into a number of sections that represent what we regard as the main research fields in which progress has been made in understanding granites and their genesis. The origin of granitic magmas is part of the cycle that unites tectonic processes with the generation of mantle magma, crustal growth, the transfer of mantle heat to the crust, high-grade regional metamorphism, crustal melting and crustal differentiation. Thus, the extended subject area is far too wide to deal with in a single volume. We have therefore limited our considerations to papers dealing with the formation, physical behaviour and chemical evolution of granitic magmas, and just touching on the matter of associated ore deposits. The papers are grouped into the following sections:

- The Origins of Granitic Magmas

- Field Relations, Magma Transport and Emplacement

- Typology and Magmatic Evolution

- Melt compositions, Experimental Petrology and Economic Potential

Apart from betraying our personal prejudices, we hope that our selection will serve to bring to a 21st century audience, many of whom were not yet born, an appreciation of how thought on granitic magma genesis and evolution developed during the previous century, to shape what we think we know today and to guide our paths to further enlightenment (or possibly delusion – time will tell).

John Clemens
Stellenbosch
Fernando Bea
Granada
2011

Volume 4 Editors

Prof. John D. Clemens FRMIT, PhD, FMinSoc

John Douglas Clemens was born in 1953, in the now non-existent coal-mining town of Yallourn, in Victoria, Australia. In 1973 he gained his Fellowship Diploma of Geology from the then Royal Melbourne Institute of Technology (now RMIT University). In 1982, he received his PhD in Geology from Monash University, for work on granitic and silicic volcanic rocks, and experimental petrology. From then until 1986 he was a postdoctoral fellow in the Department of Chemistry of Arizona State University, involved in research on nuclear waste disposal, experimental petrology and high-temperature calorimetry of geomaterials. In 1987 he moved to the Department of Geology at the Blaise Pascal University of Clermont-Ferrand (France), as an Associate Director of Research. There he continued in experimental petrology and modelling of crustal melting. In 1989 he joined Manchester University as lecturer in Geology. From 1994 to 2007, he held the Chair of Geosciences at Kingston University (London) and, from 2001, was also Associate Dean in the Faculty of Science. His present appointment is Professor in Geology and Executive Head of Earth Sciences at Stellenbosch, South Africa. He has published >60 papers, mostly in international journals, and several book chapters. The papers have accumulated >3000 citations. His most important work has been on the origin and evolution of granites and silicic volcanic magmas, experimental studies and modelling of partial melting and the ascent and chemical evolution of granitic magmas. These continue to be his main interests.

Prof. Fernando Bea, Catedrático de Petrología y Geoquímica

Fernando Bea took a grade in Geology at the University of Oviedo, Spain, in 1972 and a doctorate in Geological Sciences at the University of Salamanca, in Spain, with a thesis on Variscan granites of western Spain, supervised by the late Prof. García de Figuerola. In Salamanca he was was responsible of the geochemical laboratory and lectures in geochemistry. In 1992 he moved, as full professor of Igneous Petrology and Geochemistry, to the University of Granada where, jointly with Prof. P. Montero, he started the ICP-MS (1993) and TIMS laboratories. Recently, he led a project to install a SHRIMP II ion microprobe, which has been operational since September 2011. His research interests have been in the petrology and geochemistry of granite rocks and the mechanisms of generation of granite magmas, having studied in detail the role of heat-producing elements. He has worked in Spain, the Urals, Kola, Egypt and Morocco.

LANDMARK PAPERS: GRANITE PETROGENESIS

Foreword

The story of granite is inextricably tied to the story of geology itself.

It was in Glen Tilt, in 1785 that James Hutton first found the compelling evidence he sought to prove that granite, a crystalline rock, was formed from molten material and that this melt had intruded into pre-existing and therefore older strata. His 'plutonism' ideas were vindicated, his rationale for the Theory of the Earth, presented to the Royal Society of Edinburgh that year in outline form as "Concerning the system of the Earth, its duration, and stability", supported, and the fate of the neptunist view of the Earth sealed.

James Hutton's proof of the hot, magmatic origin of granite was published in 1794 as "Observations on Granite". (Hutton, J., 1794, *Transactions of the Royal Society of Edinburgh*, **3**, 77). Over two centuries later we are still making 'observations on granite', some of them surprising and revolutionary, many of them of great significance for evaluating the processes that control the formation and chemical differentiation of the crust, constraining the rates of growth of the continents, and realising the economic potential of granite systems themselves.

This Landmark volume introduces and explains some of the most important works that have led to our current understanding. It quite rightly does not present the Hutton (James!) contribution to granite petrogenesis: the major contributions to this had to await the introduction of the petrological microscope and use of experiments to determine melting conditions and equilibria, amongst other things. The rationale for the selection of the landmark papers is in their contributions to understanding or stimulating discourse and debate on the formation, physical behaviour and chemical evolution of granitic magmas. John Clemens and Fernando Bea, each of whom brings unique experiences, approaches, insights and passion to the subject area, are highly successful in doing this – all readers can learn from the papers assembled in this compilation and from the concise and incisive commentaries provided on them by John and Fernando.

As Read emphasised, there are 'granites and granites', and as John Clemens and Fernando Bea note here, "our personal field experience conditions, to a great extent, our ideas about granite petrogenesis". Despite its ubiquity, to some extent granite is in the eye of the beholder. My own perspective on granite genesis is conditioned by growing up in east Australia and being exposed early in my education to the ideas of Alan White and Bruce Chappell. I still vividly recall attending and being stimulated by their talks on S- and I-type granites at the 25th IGC in Sydney in 1976. I also recall subsequently (in 1977) mapping for weeks in a veritable 'sea' of peraluminous but ultimately monotonous S-type granite in northern Victoria, relieved and excited by the nearby presence of cordierite-sillimanite migmatites. No doubt this initiated my metamorphic petrologist's perspective on granite genesis via migmatites, and interest in the roles played by accessory minerals such as zircon and monazite at and near-source. It is to the editors' credit that their selection of papers for this Landmark volume embraces a broad spectrum of approaches and perspectives that will be of interest not only to igneous petrologists and geochemists but also to those of us at the 'migma', thermal modelling and tectono-metamorphic ends of the vast range of geoscience research areas to which granite studies contribute.

I return to Hutton. The 200th anniversary of the publication of his first version of the Theory of the Earth (Hutton, J., 1788, *Transactions of the Royal Society of Edinburgh*, **1**, 209), provided the inspiration for convening a conference to celebrate his work, in particular his insights on granite. Thus we saw the first *Hutton Symposium on Granites and Related Rocks*, held in Edinburgh in 1988. Not only was this a timely celebration of Hutton, but also something of a watershed in granite research. The Hutton Symposia since 1988 have brought together researchers with diverse perspectives on the origins, transport and emplacment of granites, broadening the scope of granite research and integrating ideas and concepts from migmatites, isotope geochemistry, geophysics and deformation studies. Whilst this has been a catalyst for new directions in granite research, it also bears witness to the lively and progressive development of approaches to using and understanding granites that has been stimulated by the contributions so well represented by the excellent papers included in this Landmarks volume.

Simon Harley

LANDMARK PAPERS: GRANITE PETROGENESIS

by John Clemens and Fernando Bea

Table of Contents

CHAPTER 1

Bunsen, R.W. (1861) Uber die bildung des granites. *Zeitschrift der Deutchen geologischen Gesellschaft*, **13**, 61−63.

Opinions vary on whether the Bunsen burner was invented by Robert Wilhelm Bunsen (1811−1899), but it was certainly named after him. Most remember him as a prominent chemist who worked in Göttingen, Kassel, Marburg, and finally Heidelberg. He was interested in isolating pure compounds from their mixtures and was one of the founders of spectroscopy. He certainly invented a number of other pieces of apparatus used in physical chemistry and was involved in the joint discovery of caesium and rubidium. A true experimentalist, he commonly performed dangerous tasks, and was nearly killed by arsenic poisoning and was also blinded in one eye by a laboratory explosion.

What is perhaps less known is that Bunsen harboured a lifelong passion for geology, stemming from a visit to Iceland with mineralogist colleagues. At his time, the great geological debates were carried on in two opposing camps – the Neptunists (who favoured a sedimentary origin for all rocks) and the Plutonists (who favoured a variety of deep origins involving high temperatures). Plutonists corresponded roughly to people whose ideas about petrogenesis would today be grouped as igneous and metamorphic petrology.

In the middle of the nineteenth century, petrologists were examining rocks, among them granites, in thin section. The main minerals observed were, of course, quartz and feldspars, with alkali feldspar usually dominating over plagioclase. There were no techniques for *in situ* analysis of these mineral grains and so much attention was paid to the textures. Quartz and alkali feldspar were seen to be intergrown in a variety of ways, but one important observation was that these textures decreed that the quartz must usually have crystallized after the feldspar. This seemingly mundane fact was the cause of a huge debate between the Neptunists and the Plutonists.

The Plutonists were sure that granite is an igneous rock, solidified from a magma that had intruded into the surrounding wall rocks and gradually crystallized. The Neptunists maintained that granite was the initial precipitate from the primordial ocean, and around which the other rocks were later deposited. Most field geological facts seemed to support the Plutonists' view. However, amongst other observations, Neptunists pointed out that the textures observed in granites presented the Plutonists with an apparently insurmountable problem.

The reasoning went that it had been shown in laboratory experiments that quartz melts at a very high temperature indeed (>1700°C), while orthoclase melts at a much lower temperature (~1100°C). Thus, they reasoned, if the textures of granites showed that quartz usually crystallized after the feldspars, this disagreed with the experiments and therefore the granites cannot have solidified from a molten state. Geologists, at this point in history, were field-based observationalists and armchair theorists who commonly had little training in or knowledge of physics, chemistry or biology.

Enter Bunsen, the physical chemist! In the following almost scathing letter, published in 1861 in the *Journal of the German Geological Society*, he pointed out that chemists had long known that, in mixtures of compounds in solution, the order of crystallization of the compounds depends on the bulk composition of the system and the pressure. He expressed surprise that this fact seemed to have escaped the geologists who were busy arguing about the origin of granites. He provided the analogue example of the system $CaCl_2 - H_2O$ and discussed how this behaves. He then dealt with the orthoclase–quartz system and pointed out that the order of crystallization here is similarly dependent on the ratio of feldspar to quartz in the melt. Bunsen also made three further highly significant points for petrologists. The first is that a silicate melt should be viewed as a kind of solution, which must obey the same rules as the lower-temperature solutions that chemists commonly deal with – the temperature difference being of no fundamental importance. The second is that the final crystallization temperature of a quartz-feldspar mixture will be very much lower than the melting points of either pure mineral – in other words, eutectoid behaviour. The third additional point he makes (though somewhat obtusely) is that there may well be a compositional continuum between a silicate melt with dissolved H_2O and a high-temperature aqueous solution with dissolved silicate. Bunsen coined the word 'pneumatolytic'.

Thus, Bunsen's short article represents the first insertion of the principles of hard physical chemistry into the understanding of granites (and all igneous systems), and provides a foretaste of issues surrounding the different types of melting reactions and the origins of hydrothermal fluids. Thus, we have no doubt that Bunsen (1861) deserves a prominent position in this collection of landmark papers. The translation given here was produced by the University of Stellenbosch Language Service, with minor contributions by JDC. To make the paper accessible to a modern audience, it omits the flowery and circumlocutive elements of writing that were common at the time, while retaining the spirit of the original.

Journal

of the

German Geological Society

Volume XIII
1861.

including six illustrations

Berlin, 1861
Wilhelm Herts (Bessersche Buchandlung)
Behren-Strasse Nr. 7

61

2. On the formation of granite

by Mr R. Bunsen in Heidelberg

(from a letter to Mr A. Streng in Clausthal)

"....It is about time that you critically discuss the origin of quartz in granite rocks. In this matter, I want to draw your attention to a curious error that for a long time has played an important part in geological hypotheses about the formation of granite. To correct it will probably somewhat reassure geologists who see the results of careful and well founded observation of nature threatened by the deductions of experimenting analytical chemists.

Quartz solidifies at a higher temperature than orthoclase. Orthoclase solidifies at a higher temperature than mica. Thus, according to the opponents of the plutonic origin of this type of rock, if granite originally consisted of a molten mixture of these three minerals, on cooling, first the quartz, after that the orthoclase and finally the mica would have solidified. However, as the petrographic structure of granite rock normally reveals a different sequence of solidification, it is asserted that such rocks cannot be of molten origin. Indeed, it is difficult to understand how for years geologists have accepted such a fallacy and even harder to understand how, even today, it has repeatedly been cited in support of geological hypotheses. No-one seems to have considered that the temperature at which a substance solidifies in its pure state is never that at which it solidifies in solutions with other substances. The point of precipitation of a chemically pure compound depends solely on its composition and on pressure. In contrast, the point of precipitation of a substance in solution with other substances depends mainly on the relative proportions of the substances that are held in the solution.

Certainly, no analytical chemist would get the absurd idea of assuming that a solution would cease to be a solution if it were heated to 200, 300, 400 degrees, or up to such a temperature at which it begins to be self-luminous, i.e. molten. That would be saying that a mixture of ice and crystallized calcium chloride which has liquefied certainly is a solution, whereas a mixture of quartz and feldspar, because it only liquefies at red heat, is not a solution.

In fact, no-one can doubt that what applies to solutions at low temperatures is also valid for solutions at high temperatures. If one observes the processes taking place in any solution, for instance one of ice and crystallized calcium chloride, the following becomes evident. With a certain content of crystallized calcium chloride, the liquid only begins to solidify at $-10°C$. Then, if the temperature is lowered only slightly, it will freeze to more or less pure ice surrounding crystals of calcium chloride. If one successively increases the amount of calcium chloride in such a solution, it can retain its liquidity or solidify at -20 degrees, -30 degrees, -40 degrees, -50 degrees, etc., and the processes of solidification are similarly repeated at these temperatures. Thus, the temperature at which the water and the calcium chloride solidify changes, depending on their ratio in the mixture. Evidently, the freezing point of water can sink here by more than -59 degrees C. The solidification of calcium chloride, which on its own takes place at $+26$ degrees, can even be lowered by nearly 100 degrees. Potassium sulphate, saltpeter, etc. can solidify in solutions at temperatures that lie between 600 to 800 degrees below their melting points.

Everyone knows that, depending on its concentration, first water and then salt, or first salt and then water will crystallize. Staying with the example of a solution of water and calcium chloride, water will not start to solidify at its melting point of 0 degrees and calcium chloride at its melting point of $+26$ degrees C.

Calcium chloride would not always solidify before the water, therefore one cannot assume that quartz and feldspar would have to solidify in their molten solution at their respective melting points. On the contrary, in agreement with empirical knowledge about solutions generally, we find that, in graphic granite rich in feldspar, the quartz was precipitated before the feldspar. In other granites this happened at the same time, and in yet others, after it.

Now, as Rose has shown in his latest, interesting and important research paper, quartz changes to an amorphous soluble modification with a density of 2.2 near to its melting point. If this mineral crystallized in the molten mass of granite, at very different temperatures and always only below its melting point, the only conclusion is this: that quartz crystallizing from a molten granite mass below its melting point, like that crystallizing further below this melting point from aqueous solutions, in all probability would not show the specific gravity 2.2 but a density of 2.6 and have the characteristics associated with quartz."

Zeitschrift

der

Deutschen geologischen Gesellschaft.

—◦—

XIII. Band.
1861.

Mit achtzehn Tafeln.

———————————

^c Berlin, 1861.

Bei Wilhelm Hertz (Bessersche Buchhandlung).

Behren-Strasse No. 7.

61

2. Ueber die Bildung des Granites.

Von Herrn R. Bunsen in Heidelberg.

(Aus einem Schreiben an Herrn A. Streng in Clausthal.)

„ — Dass Sie beabsichtigen, die Entstehungsweise des Quarzes in den granitischen Gesteinen einer kritischen Besprechung zu unterwerfen, ist gewiss ganz an der Zeit. Ich möchte Sie dabei auf einen seltsamen Irrthum aufmerksam machen, der in den geologischen Hypothesen über die Granitbildung seit geraumer Zeit eine grosse Rolle gespielt hat und dessen Berichtigung den Geologen, welche die Ergebnisse sorgsamer und wohlbegründeter Naturbeobachtungen von den Schlussfolgerungen der experimentirenden Chemiker bedroht sehen, wohl zu einiger Beruhigung gereichen wird. Der Quarz erstarrt bei einer höheren Temperatur als der Orthoklas, der Orthoklas bei einer höheren als der Glimmer. - Bestand daher der Granit ursprünglich aus einem feuerflüssigen Gemenge dieser drei Fossilien, so muss — behaupten die Gegner der plutonischen Entstehung dieser Gebirgsart — bei dem Abkühlen eines solchen Gemenges der Quarz zuerst, der Orthoklas darauf und der Glimmer zuletzt fest werden. Da nun die petrographische Structur der granitischen Gesteine gewöhnlich eine andere Reihenfolge der Erstarrung erkennen lässt, so können — behauptet man weiter — jene Gebirgsarten nicht feuerflüssigen Ursprungs sein. Es ist in der That schwer begreiflich, wie ein solcher Fehlschluss sich jahrelang hat bei den Geologen in Geltung erhalten können und schwerer noch begreiflich, wie derselbe selbst heute noch immer wieder zur Stütze geologischer Hypothesen reproducirt zu werden pflegt. Niemand scheint daran gedacht zu haben, dass die Temperatur, bei welcher ein Körper für sich erstarrt, niemals diejenige ist, bei welcher er aus seinen Lösungen in anderen Körpern fest wird. Der Erstarrungspunkt einer chemisch reinen Verbindung hängt allein von ihrer stofflichen Natur und dem Drucke ab, wogegen der Erstarrungs-

62

punkt eines mit anderen Substanzen zu einer Lösung verbundenen Körpers ausserdem noch und zwar hauptsächlich von dem relativen Verhältniss der sich gelöst haltenden Substanzen bedingt wird. Es wird gewiss kein Chemiker auf die widersinnige Idee verfallen, anzunehmen, dass eine Lösung aufhöre eine Lösung zu sein, wenn sie bis auf 200, 300, 400 Grad oder bis zu einer Temperatur erhitzt wird, bei welcher sie anfängt selbstleuchtend zu werden, d. h. feuerflüssig zu sein, also z. B. anzunehmen, dass ein Gemenge von Eis und krystallisirtem Chlorcalcium, welches flüssig geworden ist, wohl eine Lösung sei, ein flüssiges Gemenge von Quarz und Feldspath dagegen nicht, weil es erst in der Glühhitze flüssig wird. Niemand kann vielmehr den leisesten Zweifel darüber hegen, dass was für Lösungen in niederen Temperaturen gilt auch für Lösungen in höheren Temperaturen gültig sein muss. Betrachtet man nun irgend eine Lösung, z. B. eine Lösung von Eis und krystallisirtem Chlorcalcium in Beziehung auf die Vorgänge, welche bei dem Festwerden derselben eintreten, so zeigt sich Folgendes: Bei einem gewissen Gehalt an krystallisirtem Chlorcalcium wird die Flüssigkeit erst bei — 10 Grad C. anfangen fest zu werden, dann bei nur wenig sinkender Temperatur bis zum letzten Tropfen zu mehr oder weniger reinem Eis erstarren, in welchem Chlorcalciumkrystalle eingebettet sind. Vermehrt man successive den Chlorcalciumgehalt einer solchen Lösung, so kann man sie beliebig bis — 20 Grad — 30 Grad — 40 Grad — 50 Grad etc. flüssig erhalten oder erstarren lassen, wo sich dann bei diesen Temperaturen jene Vorgänge des Erstarrens in ähnlicher Weise wiederholen. Es wechselt also die Temperatur, bei welcher das Wasser und das Chlorcalcium fest wird, je nach den Mischungsverhältnissen. Der Erstarrungspunkt des Wassers kann hier, wie man sieht, um mehr als 59 Grad C. unter seinen Gefrierpunkt sinken, der Erstarrungspunkt des Chlorcalciums, welcher für sich bei + 26 Grad liegt, sogar um nahezu 100 Grad erniedrigt werden. Schwefelsaures Kali, Salpeter etc. können aus ihren Lösungen bei Temperaturen fest werden, die 600 bis 800 Grad unter ihrem Schmelzpunkt liegen. Jedermann weiss ferner, dass aus Lösungen je nach der Concentration derselben zuerst Wasser und dann Salz oder zuerst Salz und später Wasser krystallisirt zu erhalten ist. So wenig daher — um bei demselben Beispiel stehen zu bleiben — aus einer Chlorcalcium-Lösung das Wasser

63

bei seinem Schmelzpunkt 0 Grad und das wasserhaltige Chlorcalcium bei seinem Schmelzpunkt + 26 Grad C., so wenig ferner das Chlorcalcium immer vor dem Wasser erstarrt, eben so
wenig ist die Voraussetzung zulässig, dass Quarz und Feldspath
aus ihrer feuerflüssigen Lösung bei ihren respectiven Schmelz-
punkten fest werden müssten. Wir finden vielmehr in völliger
Uebereinstimmung mit den Erfahrungen, die wir bei allen Lösungen machen können, dass in dem an Feldspath reichen Schrift-
granit der Quarz vor dem Feldspath, in anderen Graniten gleich-
zeitig mit demselben und wieder in anderen nach demselben
ausgeschieden wurde. Wenn nun der Quarz, wie ROSE in seiner
neuesten interessanten und wichtigen Arbeit gezeigt hat, nicht
einmal weit von seinem Schmelzpunkt in die amorphe lösliche
Modification von der Dichtigkeit 2,2 übergeht und wenn dies
Mineral aus dem geschmolzenen Granitgemenge bei der aller-
verschiedensten Temperatur auskrystallisiren konnte und zwar
stets nur unter seinem Schmelzpunkte, so wird man daraus wieder in völliger Uebereinstimmung mit der Erfahrung nur schlies-
sen können, dass der unterhalb seines Schmelzpunktes aus dem
feuerflüssigen Granitgemenge krystallisirende Quarz, gerade so
wie der noch weiter unterhalb dieses Schmelzpunktes aus wässri-
gen Lösungen krystallisirende, aller Voraussicht nach nicht das
spezifische Gewicht 2,2, sondern die Dichtigkeit 2,6 und die
damit verbundenen Eigenschaften zeigen werde."

CHAPTERS 2 AND 3

Read, H.H. (1948) Granites and granites. *Geological Society of America Memoir*, **28**, 1–19.
Bowen, N.L. (1948) The granite problem and the method of multiple prejudices. *Geological Society of America Memoir*, **28**, 79–90.

These two papers are, in fact, the opening (Read) and the closing (Bowen) talks given in the Meeting of the Geological Society of America held in Ottawa on December 30, 1947, and devoted to the 'Granite Problem'. They were published in the famous Memoir 28 of the Geological Society of America. The former also was published in Read's book *The Granite Controversy* (Read, 1956).

Granite scientists have sustained lively (and recurrent) controversies over the years. One of the most famous was between the Transformists, those believing that granites arose by granitization, and the Magmatists, those believing that granites arose by crystallization of a totally fluid magma – led at that time by H. H. Read and N. L. Bowen, respectively. Whereas Bowen, mostly based on experiments using physical-chemical considerations, proposed that granites were the products of differentiation of basaltic magmas, Read, mostly based on field work and petrography, considered that granites are closely connected to migmatites and regional metamorphism, being in most cases produced by "granitisation" – the process by which "solid rocks are converted to rocks of granitic character without passing trough (sic) a magmatic stage". It must be remembered that magma, at that time, did mean a "totally fluid rock substance" (sic).

These two papers represent the zenith of that discussion. They are little gems, with plenty of humour and acerbic arguments, that ought be read by anybody wishing to understand the evolution of granite science. Besides, they teach us how gentlemen must behave during scientific discussions, a virtue which is, unfortunately, often forgotten nowadays.

Read was not an extreme Transformist. The main point of friction with Bowen was Read's "freeing the granites from their bondage to a parental basaltic magma" (sic). For Read, at the heart of the granite problem was the room problem, which can be considerably softened if granites were derived from metamorphic rocks; accordingly, he favoured granitization. This was not a single unitary doctrine but included a wide spectrum of ideas, from those involving magma (as it is currently understood) and partial melting of crustal rocks, to others involving soaking and transforming fluids (ichors), to the most extreme variants that involved just solid-state diffusion.

Read favoured the idea that intrusive granites were formed by movements of mushes composed of a mixture of relict solid pieces in a fluid base of molten rock, what we today would call a magmatic mush and, adopted for this the term migma coined by Reinhard (1935). Read believed that if the process that gave rise to migma intensifies, the migma will become a magma capable of moving into higher levels of the crust to produce discordant granite intrusions with contact metamorphic aureoles. The frontispiece illustration of his book *The Granite Controversy* with the leading sentence "per migma ad magma" reflects precisely his idea about the genesis of granites, which is at the core of what many granite petrologists implicitly or explicitly accept today.

A modern granite petrologist can learn several lessons from Read's paper. First, that there is no unique solution to the granite problem; there are granites and granites (see Pitcher, 1987). Second that our personal field experience conditions, to a great extent, our ideas about granite petrogenesis. As the diversity of granitic rocks and the environments in which granites are produced is so large, we must be aware of broad generalizations extrapolated from small-scale observations, however accurate these may be. The road to geological hell is paved with such generalizations.

In contrast with the soft-core position of Read as a Transformist, Bowen was a hard-core Magmatist, at least at that time. He defended the idea that most granites were generated through crystallization differentiation of mafic magmas. In this paper he made a devastating revision of the most extreme theories of granitization, those involving only solid-state diffusion or small fractions of "ichors". More recent data on diffusion rates and heat production fully confirmed his criticism. Bowen said that there are four views among the Magmatists. Those who claim that granitic magma is formed by melting of geosynclinal sediments, those who claim that granitic magma is formed by melting at the base of the granite layer of the Earth, those who propose that granitic magma is the result of differentiation of a syntectic magma formed by the solution of felsic material in basaltic magma, and those, like himself, who adhere to the view that granite is the product of differentiation of basaltic magma.

Most of us today will agree that these four mechanisms are possible but probably disagree on the relative importance of each. In my personal opinion (F. Bea), the two first mechanisms are the most common, followed by the third and, far distant in terms of volume of granitic magma produced, the fourth. Bowen was absolutely right that the first granites produced on Earth must have resulted from fractional crystallization of basalt, but this tenet cannot be sustained for all Earth's history. Once a granitic crust was established, probably during Hadean times or even before (Watson and Harrison, 2005), recycling of crustal material with variable addition of mantle juvenile components seems to have been the main mechanism for generating granitic magmas.

The discussion between Read and Bowen was a frontal collision between two massive trains. Both defended some ideas that the years have proven incorrect and others that have proven correct. As often occurs, especially in geology, the intersection between the ideas of both sides, that is, that most granites resulted from the crystallisation of magmas generated by partial melting of crustal materials, is probably the position closest to what most granite scientists would nowadays agree.

GRANITES AND GRANITES

BY H. H. READ

(Royal School of Mines, London, S.W. 7)

INTRODUCTION

There can have been few occasions in the past when the President of the Geological Society of London—the mother of all geological societies—has been able to accept the hospitality of that most splendid of daughters, The Geological Society of America. I come bearing most hearty greetings and good wishes from my Society to yours.

This meeting celebrates, I believe, the Centenary of the Geological Survey of Canada, and I am honoured to be present at the inauguration of a second century of fruitful geological investigation. Our countries are bound together by the cords—gossamer yet stronger than steel—of a common blood, a common creed, and a common history—we spring from the same roots. But we geologists are bound still closer to one another by devotion to our common science—nothing can unloose that tie. As these proceedings continue, we shall find maybe that we differ on a great many aspects of geological interpretation, but this is a sign of vitality and can in no way affect our regard for one another as fellow-seekers after the truth concerning the crust of Mother Earth.

As President of the Geological Society of London, I salute The Geological Society of America and, contemplating your glorious past, I look forward to your yet more glorious future.

In the foregoing paragraphs I have spoken in my official presidential capacity; from now on, I speak as a private individual—the personal pronoun will occur, I expect, with astounding frequency. I am not certain in what role I appear before this assembly, whether as a prophet blessed with a new revelation or, more in keeping with Professor Bowen's (1947, p. 264) valuation of me, as an old soak seeking the penitent's bench. Maybe, like most men, I fit best into some intermediate category. This question of the origin of granite is perhaps the most lively of geological topics today—but we should remember that it always has been. About every twenty years or so, the problem has been firmly settled, and a sort of uneasy peace has broken out. This indicates to my mind that there is no unique solution of the problem—there are granites and granites. Assuming the character of a trimmer (in the ancient and best sense), I protest that there is no need for an *either-or* attitude. Bigots or, if you like, enthusiasts, on both sides do a deal of harm, and pontiffs, I suggest to Professor Bowen, while capable of a greater number of good deeds, are also capable of a greater number of bad deeds than the village drunk. If we keep our tempers, whilst not pulling our punches, we shall receive great profit and pleasure from these debates. I quote from Hutton, the founder of our science:

"While man has to learn, mankind must have different opinions. It is the prerogative of man to form opinions; these indeed are often, commonly I may say, erroneous; but they are commonly corrected and it is thus that truth in general is made to appear."

1

ORIGIN OF GRANITE

What I propose to do on this occasion is to present the results of my consideration
of the granite problem that I have reached during the last half dozen years. This
is not the time or place to describe new work, either mine or my students! I have,
as it were, to draw up a trial balance sheet of my personal faith in this matter, leaving
it to the future to change the two sides of the account. During those magnificent
but episodic war years in London, when one went to earth at dusk, I kept myself
geologically fit by pondering on granites and granitisation (Read, 1943; 1944; summarized in 1946). Today I summarize these meditations and bring them up to date;
in doing so, I must needs repeat a great deal that I have already said in other places,
and, as time presses, often in the same words.

THE NATURE OF THE GRANITE PROBLEM

I see the granite problem as essentially one of field geology—it is not primarily
one of petrography, mineralogy, physical chemistry, or of any other ancillary discipline. It certainly does not fall within Gignoux's category of *la pétrographie de
tiroirs*. Granites are very big things, not hand specimens. Now every geologist
must judge a geological question against the background of his own field experience.
This self-evident fact determined the divergences between the French masters like
Michel-Lévy and Lacroix on the one hand and the German petrographers like Rosen-
busch on the other. A great part of my own field excursions has come from areas
of granites, migmatites, and metamorphic rocks of regional development. If, by
viewing granites in a certain way, I find that my field experience can be interpreted
more comfortably, then I propose to adopt that view no matter how uncomfortable
it may be to another geologist with a different field background. This is not a matter
of spiritual pride but just plain common sense.

To me, the making of granite, migmatisation, and regional metamorphism are all
parts of one process—they result from the agency that Lyell a century ago called
plutonic. In his classic metamorphic theory of the origin of granite, the master
placed the granitic and metamorphic rocks in the closest genetic association:

"granite itself, as well as the altered strata, have derived their crystalline texture from plutonic
agency" (1838, p. 19).
  "The transmutation has been effected by the influence of subterranean heat acting under great
pressure, and aided by thermal water or steam and other gases permeating the porous rocks, and
giving rise to various chemical decompositions and new combinations, the whole of which action
has been termed 'plutonic' as expressing in one word all the modifying causes brought into play at
great depths and under conditions never exemplified at the surface. To this plutonic action the
fusion of granite itself in the bowels of the earth as well as the development of the metamorphic
texture in sedimentary strata may be attributed." (1875, p. 139).

Before he was borne down, like the rest of his British and American contemporaries,
by the onset of magmatism from Germany, Lyell clearly separated the Plutonic from
the Volcanic rocks, the latter being "the products of igneous action" (1838, p. 4–5,
11–13). The existence of two fundamentally opposed groups of eruptive rocks, the
one granitic and plutonic, the other basic and dominantly volcanic, had likewise
been realized by Elie de Beaumont (1847, p. 1253, 1254, 1288, 1297), and this distinction was later emphasized by Michel-Lévy and others of the French school.
The French saw fundamental differences in the mode of eruption of granites and of
other eruptive rocks. A modern presentation of these differences is provided by

W. Q. Kennedy (1938) in his division of igneous bodies into volcanic and plutonic associations, the first association being basic, dominantly effusive, truly magmatic and non-orogenic, the second being granitic and granodioritic, associated with processes of assimilation, and with orogenic movement. Drawing from Lyell and the French masters and fortified by Kennedy, I have recently (1944; 1946) advanced a threefold classification of rocks that suits my field experience best. It is:

Neptunic—the sedimentary rocks, dominantly marine;

Volcanic—the magmatic, igneous, rocks, dominantly effusive and basic;

Plutonic—the metamorphic, migmatic, and granitic rocks.

The granite problem is essentially a plutonic problem—the granitic rocks are tied up at one stage or other of their formation with the rest of the plutonic rocks and only rarely and almost fortuitously are they genetically connected with the volcanic rocks of my classification. The contrast in the geological setting of the Volcanic and the Plutonic rocks is of so fundamental a character that I am unable to accompany Niggli, Bowen, and many others in their nimble leaps horizontally or vertically from one pigeon hole to any other pigeon hole in the usual classification of rock specimens. If the classification were truly genetic, such leaps would be justified, but I cannot follow, for example, Niggli (1942, p. 6) in the vertical leap from rhyolite to granite which he performs because of the undeniable existence of lavas of granitic composition, nor Bowen (1947, p. 263, 264) in a similar leap from andesite to diorite because vast volumes of andesitic lava undoubtedly existed. Even the giant Lyell (1838, p. 217) took two jumps to do such feats. The volcanic rocks can be interpreted along uniformitarian lines; the plutonic rocks are too deep a matter.

Perhaps it was Edward Suess in his *Das Antlitz der Erde* who persuaded us that the transition from volcanic to plutonic was a reality by his masterly reconstruction of a denudation series from a modern cinder cone to a deep-seated mass of granite. As Raguin (1946, p. 127) has recently recalled, such a transition is nowhere displayed as a whole but is the projection of a theoretical view based upon the interpretation of fragmentary observations. Perfect chemical and petrographical transitions may be demonstrable, but the much more important geological transition has to be argued. These are matters discussed in detail by Raguin (1946, p. 127–141) and W. Klüpfel (1941). The geological factor involves time, a dimension lacking in the *petrographie de tiroirs* and in Suess' denudation series. Vulcanism and plutonism closely associated on the ground may be separated by respectable stretches of time; their sequence appears to be irregular, and one does not demand the other; their association may be accidental. A conversation (*Unterhaltung*) between Hans Cloos (1939) of Bonn and A. Rittmann of Basle indicates how international this view of the independence of *Vulkanismus* and *Plutonismus* now is: to Rittmann's remark that these two processes have magmatically nothing to do with one another, Cloos replies *Zugegeben*, and later proceeds to reinforce this opinion from his vast stores of knowledge of granite tectonics and the regional settings of plutonic bodies. Time again enters into the conversation—the granitic and plutonic rocks belong to the orogeny, the basic and volcanic rocks belong to the geosynclinal phase—a view of general application as I suggested long ago (Read, 1927).

I feel much more comfortable, therefore, in seeing the granite problem as a plutonic problem not directly attached to basic rocks of the volcanic association. I have before welcomed this "freeing the granites from their bondage to a parental basaltic magma" (1943, p. 85)—conduct that has not met with the approval of Bowen (1947, p. 274) who reminds us that since the earth is dominantly basic, the original granitic layer at least must have been derived from basaltic magma. I agree, of course, but protest that this gigantic operation, like a great many others that went on at that distant date, is not necessarily going on now, nor need it ever have gone on by crystal settling. Here is that dimension of time turning up again—and we should remind ourselves that uniformitarianism cannot be true for all time.

It will be convenient at this stage of my argument to recall other expressions of the volcanic:plutonic duality. From the earliest days of our science, the dominance of granitic rocks among the intrusives and of basaltic rocks among the extrusives has been recognized. Daly (1933, p. 32–41) and Barth (1939, p. 113) have made quantitative estimates that confirm the early suggestions—granites and basalts are the commonest rocks of the crust, and, for a geologist, the most important rocks are the commonest. These remarkable facts early led to the proposals of Bunsen in 1851 and Durocher in 1857 that two magmas, one acid and one basic, were available in the crust from which all the variety of igneous rocks could be produced, and this two-magma view has been argued along modern lines by F. Loewinson-Lessing (1911). It was the belief of Durocher (1857) that the two magmas had their determined positions in the crust, the lighter viscous acid magma resting on the denser fluid basic magma. From geophysical evidence it is considered that Durocher's liquid shells do not exist as such, but the same evidence suggests that a light granitic sialic layer overlies a dense basaltic simatic layer. The very crust itself appears to have a volcanic:plutonic duality.

Before going further, I must give a few more details concerning the contents of my Volcanic and Plutonic classes only outlined in the foregoing. The Volcanic rocks include the intrusive diabases and gabbros which are considered simply as portions of basaltic magma that have failed to become extrusive. Most of the effusive rocks may be interpreted as produced by crystal fractionation of primary basaltic magma; the formidable question of the position of the voluminous andesites must be left till we have a better synoptic view of them, both in time and space. Turning to the Plutonics, the granitic members with which this debate is primarily concerned include, of course, the granites and granodiorites. The important syenites and diorites must each be considered as a special case, and its origin determined on the field evidence, but in many examples they appear to me to be of hybrid or contamination origin closely attached to granite. Parenthetically, I express the hope that I shall be able to take back to Britain a chip of the normal igneous diorite known to Professor Bowen (1947, p. 263)—it will be a pleasant memento of this visit.

To summarize this first section of my remarks, then, I present the two classes of rocks—Volcanic and Plutonic—differing fundamentally in their nature and geological setting. The granitic rocks of this debate are held to be plutonic.

MAGMATIC DISPLACEMENT VERSUS NONMAGMATIC REPLACEMENT

If we have to select a unique solution for the granite problem then the choices are two:

(1) Granites are igneous rocks resulting from the consolidation of an intrusive magma that structurally displaces the country rocks.

(2) Granites are rocks resulting from a granitisation during which the country rocks have been replaced (or, better, transformed).

It will be an advantage to state the senses in which I use the technical terms. The *New English Dictionary* defines *granite* as "a granular crystalline rock, consisting essentially of quartz, orthoclase-feldspar, much used in building." With the extensions requisite to cover granodiorite and the like, and with the added information that can be gathered from a piece of granitic rock—namely, that the constituent minerals are large enough to be recognized by the naked eye and that, under the microscope, they are seen to have interfered with one another's free growth—we have a perfect definition of granite as a rock specimen. The *igneous* rocks are those produced by consolidation of *magma*, which is completely fluid rock substance. If a rock cannot be demonstrated to have been formed by consolidation of magma then it cannot be demonstrated to be an igneous rock—a reasonable requirement often disregarded. Lastly, *granitisation* means the process by which solid rocks are converted to rocks of granitic character without passing through a magmatic stage.

THE GRANITIC MAGMA PRIMARY?

If granites are igneous rocks then we have to consider the proposals concerning the origin of granitic magma. These proposals fall into two groups—in the first the granitic magma is primary, in the second it is derived.

We have already noted the classic view of Durocher and Bunsen that there are two independent primordial magmas, one providing the granitic rocks, the other the basaltic. If these two magmas must be made, it appears to me reasonable to derive them from the two outer earth shells postulated from earthquake investigation. As I have said before, two shells, two magmas seems sound. From the sialic layer there might be produced granitic magma by pure melting or by partial melting or solution. Pure melting might be brought about either by depressing the sial into the hot sima or by raising hot simatic material into the sial. E. M. Anderson (1938, p. 78) sees no difficulty in the suggestion that along the geosynclinal belts the cover may become thick enough to cause the sial to melt and the resulting granitic magma to rise up into the fold roots. Many of the characters of batholiths appear consonant with this suggestion. Pure melting as a means of producing magma has been favored by many eminent petrologists, as recorded by Daly (1933, p. 289-290). "Many granitic magmas may have their immediate origin in the remelting, say by deep burial, of a granite," states Bowen (1928, p. 319). On the other hand, the association of granitic and basaltic rocks in certain igneous complexes might be explained, as A. Holmes (1931; 1932) has proposed, by hot basaltic magma rising through and melting the overlying sial. I would add here that bodies of granite of this origin would be smallish and of restricted association.

Many geologists who have dealt with regions of metamorphic and migmatic rocks have preferred selective refusion to pure melting for the production of granitic magma. In this regard, the early suggestions of Holmquist, Van Hise, Lane, Daly, and others have been especially elaborated by P. Eskola (1932; 1933) who considers that such magma may arise in connection with orogenic movements by the squeezing out of the lowest-melting quartzo-felspathic materials, either from basic rocks or from the deep portions of the geosynclinal piles.   The granitic magma so formed moves into higher levels of the crust where it may perform metasomatic operations akin to what I call granitisation.   Eskola believes in crystal fractionation in the production of Cambrian and later granites but sees no evidence for crystal settling in the great granitic terranes of the Archean.

I find it provocative that Daly (1933) too is quite prepared to accept anatexis of this kind for the Archean granite magmas but feels compelled to put forward half a dozen or so alternatives for magmas of later date.   I find it so because the Archean supplies the prime examples of my Plutonic class of granitic, migmatitic, and meta-morphic rocks.   Are there granites and granites after all?

Certain authorities consider therefore that a primary granitic magma could arise by melting, pure or partial, of the sialic crust, once it has been separated from the simatic layer.   This magma is independent and not derived from a primary basaltic magma.   This classic two-magma view that agrees with the distribution of rock types in the crust would be more acceptable to me—if I had to accept either—than the one-magma view, popular today in certain seminaries, that degrades granitic magma to the status of the dregs of the primary basaltic.   This opens a discussion that is entitled, especially in this den, to a fresh section of these remarks.

THE GRANITIC MAGMA SECONDARY?

On many geological grounds it is reasonable to admit basaltic magma as primary in its own right.   It is the source of my Volcanic class of true and undoubted igneous rocks.   As we have seen, vast volumes of basaltic magma have been delivered to the earth's surface or near it.   Sinking of early-formed heavy iron-magnesium sili-cates from such a magma would leave a residual enriched in silica and alkalies, a residual tending toward a granitic magma.   By the squeezing out at appropriate stages of the fluid residuum from the early-formed crystals, either through pressure during orogeny or just by the weight of the mesh of crystals, there could be collected partial magmas of various qualities; the extreme term is held to be granitic magma giving the granitic rocks.   I do not propose to attempt in this assembly to discuss the validity of the general theory of crystallisation-differentiation as I am sure that that will be well looked after by Professor Bowen; I confine my remarks to certain criticisms concerned with the production of *granitic* magma *in quantity* by this proc-ess.   These criticisms have already been detailed (Read, 1943, p. 81–84).

A first criticism is concerned with the nature of the residual liquid arising by the crystal fractionation of basaltic magma.   Fenner (1929; 1931) has argued that such liquids should be rich in iron.   Wager and Deer (1939) have provided a field demom-stration on a grand scale that this is the case and conclude that crystal fractionation of basaltic magma leads to ferro-gabbros and not to intermediate rocks of the calc-alkaline series.   Niggli (1942, p. 72) dismisses this conclusion as a mere generalisation from a special case, and Bowen (1947, p. 273) damns it with faint praise as being only one possible result of the crystal fractionation of basic magma.   It seems to me, however, to be a test of the one-magma view on a very noble scale—amounting to a great many "little crucibles."   The second criticism concerns the quantity of the residuals produced during the fractional crystallisation of basaltic magma. This criticism must be considered with the dominance of basalt *and* granite in the crust in mind. Many petrologists—Grout, Holmes, Daly, Krokström, Fenner, and others —have pointed out that only a small quantity of granitic magma can be formed in this way. Daly (1933, p. 401, 425) has remarked on "the stupendous quantity of basalt that would have to crystallize in order to make a batholith of granite by [this] process. Is such volume credible in the case of any post-Archean batholith?" Remembering the granitic cast of the Archean, I propose for one to answer Daly's query, generalised, in the negative. The third major criticism deals with the mechanism of the expulsion of the residual flow from the crystal mesh. Holmes (1936) has dealt with this matter. This operation is of a kind and of a magnitude that it should be capable of ready geological demonstration—and that is certainly not the case.

Bowen (1947, p. 272) has pointed to those geologists who "claim that the proportion of granitic residual liquid is so small that it could never be squeezed out of the crystal mesh to form a separate mass" as being the same people who find no difficulty in squeezing out small quantities of interstitial granitic liquid formed by anatexis; this is probably true, but that does not make the residual liquids of Bowen's mechanism any the more abundant. This is, I feel, another example of what Doris Reynolds (1947b, p. 212) has recently called authoritarian bluff. Crystal settling can be observed to operate in the laboratory, and it can be proposed to have operated on a gigantic scale in the formation of the simatic and sialic layers of the early earth— there is enough basaltic material for that, at least. But whether it operates or not in the crust is a question that can be decided mainly on field evidence. Petrology is not a restricted branch of physical chemistry—to quote a protest of Reynolds (1947b, p. 222), but is a branch of geology. On geological grounds, I maintain the independence of the granitic Plutonic rocks and the basaltic Volcanic rocks. At my time of life I do not propose, in Hutton's prophetic words, to "judge of the great operations of the mineral kingdom, from having kindled a fire, and looked into the bottom of a little crucible."

THE ROOM PROBLEM

It is the problem of room that lies at the heart of the granite problem, and the room problem is a matter to be dealt with by field geology. Let us for the moment concede with the magmatists that granitic magma is available in great quantities in the crust. We have to enquire how that magma, in small and large draughts, becomes emplaced at higher levels. Two types of mechanism have been proposed, the first by structural readjustments of the country rocks and the second by piecemeal stoping of the cover by the magma.

Small bodies of magma could undoubtedly become emplaced to give igneous masses of sill, laccolith, or dyke form. Nobody can quarrel with that. Even gigantic bodies of magma have been intruded as sheets as, for example, in the Sudbury and Bushveld masses—but these are of basic material. The great granitic masses do not exhibit a sheet form, and igneous bodies of this form are not customary in the fold zones. However, I see no objection to the emplacement of small bodies by doming or other structural adjustment of allied type. Rather larger bodies may be emplaced by subsidence of a segment of the crust along straight or curved fractures. By such a mechanism, as displayed by Billings (1945), it is considered by Bowen (1947, p. 274) that the Room Problem is solved, but this seems to me to be beyond the warrant. In New Hampshire, Billings has described a province characterised by ring intrusion controlled by subsidence of blocks bounded by circular fractures and has extended this mechanism to account for the emplacement of the White Mountain batholith of that region by the coalescence of many such ring-dikes. It is important to get the size of ring-dikes in proper focus: the average radius of the New Hampshire examples is 1900 feet, and the largest in the world, to quote Billings, has a radius of just over 9 miles. A survey of other ring-dikes in Britain and in Nigeria shows that they are never large bodies, and, further, observations on the West African province by a student of mine, Dr. R. R. E. Jacobson, suggest that they are of specialised petrographical types. (So noteworthy is the identity of the Nigerian and New Hampshire rocks and structures that I suggest you have another look for tin in the Eastern States.) In any event, ring intrusions seem to me to be of a small and specialised character—they are not the common rocks we are interested in today. To pass the ball back to Billings and Bowen, I invite them to contemplate the emplacement of the enormous magma reservoirs underlying the ring-structures and shown in many of the diagrams illustrating Billings' work. They seem to me to have lowered the Room Problem by a few thousand feet, not to have solved it.

Turning to the stoping hypothesis, one would be foolish to deny its efficacy over a limited range. In some of the Cornish granites, for example, its operation can be inferred over tens of feet—the country-rock blocks look as if they could be fitted back into the roof. But this is a roof or margin phenomenon, and it is questionable whether blocks of ordinary country rocks could sink far in a magma of granitic composition expecially if, as Bowen (1947, p. 274) would have us believe, the large granite masses are merely cappings to still larger basic masses below. Further, as soon as the stoped blocks sank out of the relatively impotent marginal regions, they should become incorporated in the active magma and give rise to derivatives that should be revealed in the subsequent history of the province. But in many granite masses it appears extremely unlikely that stoping can have occurred. To begin with, the stratigraphy and structure of the country rock can be traced through the granite; blocks in the granite have not moved. Further, granite tectonics shows that in many plutons there has been upward movement of a viscous material. As Mayo (1941, p. 1069) has remarked for certain Californian masses, "it is not easy to imagine much downward stoping where highly viscous masses moved upward" and he clinches this valuation of stoping by recording the "fact that throughout the area, with a few local exceptions, the internal structures of granitic and metamorphic rocks alike are conformable parts of the regional pattern" (p. 1070). Similarly, Ernst Cloos (1936, p. 436–438) has noted many features of the Sierra Nevada pluton in the Yosemite region that tell against the stoping hypothesis.

Except for small bodies, Mayo rejects the hypothesis of forceful intrusion on account of the disharmony between the enormous force required for such a process and the existence in many cases of relatively thin roofs, and on account of the nature of the wall-rock structures. He falls back upon "permissive" intrusion, tectonically controlled; it is of interest to me to find that even in such a process Mayo (1941, p. 1074) considers that "wherever the tension counteracted somewhat the local pressure, the deep-seated rocks are melted, or the travel of heat and fluids was somehow facilitated. As a result, melts, or softened masses of rock might rise and fill any potential cavity as it formed." But the roof problem is still not solved, and Mayo ends his detailed 80-page paper with a pregnant sentence—"if it can be shown that some of the so-called intrusive rocks were derived from metamorphic rocks, the demand for space for the emplacement of the intrusions will be correspondingly reduced." In other words, he would be more comfortable about it all, and a similar desire in others has led to the development of the granitisation theory, involving little or no magma, that we now proceed to examine.

GRANITISATION

Granitisation means different things to different people—to me it means what I have said on an earlier page: the process by which solid rocks are converted to rocks of granitic character without passing through a magmatic stage. The Transformists, as Reynolds (1947b, p. 209) has happily named them, propose that no great volumes of granitic magma are involved in making the great granitic masses—these are the product of piecemeal transformation on the spot of the country rocks, and the granites are thus only incidentally igneous. I have given a detailed analysis elsewhere (Read, 1944) of the development of ideas on granitisation, but I feel compelled to give also a summary here—since the ideas are largely European.

During the last century, French field geologists have developed views on granitisation and felspathisation that are now coming into their own. I give my summary of their achievements (Read, 1946, p. 66):

"The French emphasized that there are many kinds of granite contacts; in some the country-rocks are sharply delimited from the granite, in others the country-rocks gradually pass into granite through a transitional zone of metamorphic and felspathised rocks. In this latter type, abundant transfer of material from the granitic zone into the metamorphic zone could be determined. They correlated style of contact with depth of granite formation, clean and definite in the upper levels, hazy in the lower. They related this to the operation in depth of a granitisation-process involving either an intensification of metamorphism or the activity of a granitic melt which itself produces the metamorphism. In either operation, they invoked the action of energetic fluids and gases—mineralisers, volatiles or emanations—that made the country rock as it stood over into granite or prepared it for an advance of the granitic melt. They envisaged conditions under which metamorphic rocks could become igneous rocks."

For the most part, the French geologists maintained, and still do, that a "granitic magma was involved in granitisation—its intrusion was preceded by a cortège of emanations that transformed the wall-rocks into granite and so made possible their assimilation by the advancing magma."

During his life work in Finland, the great master Sederholm demonstrated a wide extension of the granitic phenomena made familiar—to anyone who would take the trouble to read about them—by the French. He proposed a regional refusion of deeply buried portions of the crust—a refusion connected with granitic magma. This magma mixed with the country rocks to produce mixed rocks, the migmatites.

As his investigations proceeded, Sederholm laid more stress on the activity, meta-somatic or otherwise, of granitic juices or ichors in the production of migmatites—but still granitic magma was regarded as their prime source. Under the influence either of ichors or of magma, the country rocks could be converted to a new anatectic magma that, on consolidation, gave igneous rocks (as I define them).

It is fair to state that neither the classic French school nor Sederholm proposed thorough-going granitisation in the modern sense—a granitic magma of one prove-nance or another was always in evidence; around this igneous rock was a felspathised and granitised aureole. I think it is true that today a respectable but not particu-larly vocal party would adopt this moderate position—we shall assess the strength of its field evidence in a moment, but first have to display the developments leading to the pure granitisation or nonmagmatic theory advanced by other Fennoskandians such as Wegmann, Kranck, and Backlund.

Wegmann and Kranck (1931) re-examined some of Sederholm's classic ground in southwest Finland and reached conclusions of great importance for the granitisation theory. The celebrated Hangë granite was considered by Wegmann not to be derived from a molten silicate mass but to arise by granitic solutions soaking into the country rock and rendering portions of the transformed rocks mobile enough to behave as intrusions; the pegmatite dikes were interpreted as parts of the granitising solutions that crystallised in open spaces in the soaked mass. Around this central core was formed a mantle showing different kinds of enrichment in different parts—internally one of Si-K enrichment leading to the production of quartz and feldspar, medially as Mg enrichment revealed by cordierite, and externally an epidote-rich zone of unakites. Since material was brought in at one place it must disappear at another place, and a whole series of chemical readjustments was set in motion, con-trolled by concentration, permeability, temperature gradient, and deposition. The petrological record of these operations is presented by Kranck. In the granitisation zone he finds the typical minerals to be cordierite and anthophyllite; farther out andalusite, kyanite, and garnet appear, and at the exterior staurolite and tourmaline are formed at the lowest temperatures. The process is one of exchange of minerals rich in MgFe(Ca) against KAlSi—a magnesia metasomatism associated with the formation of a potash granite. Kranck concludes that granitisation is a metasoma-tism rather than a migmatitic intrusion (Kranck, 1937, p. 85).

In his general discussion of migmatisation, Wegmann (1935a, 1935b, 1936) devel-oped the doctrine of fronts, hinted at above, that has now become one of the main features of the granitisation proposals. Migmatisation passes like a wave through the country rocks leaving a granitic and gneiss zone behind. Fronts of different qualities move at different speeds with different results through the framework of the diverse country rocks. Wegmann emphasized the importance of an intergranular film or pore-liquid in this process. The relation between supply and cooling deter-mines the movement of the front; if the front is stationary it may give a sharp contact. As one front advances it may cause the advance of another; for example, the forma-tion of a soda or potash front as witnessed by felspathisation may push out the unwanted magnesia and give rise to a cordierite front. A pint pot holds not more than a pint, as I have remarked elsewhere. Wegmann confesses that the source of the emanations—to use a nice noncommittal word—is not known; they must come from the depths in great quantity, concentrated perhaps by crystallisation-differentiation and squeezed out. Applications of Wegmann's views to well-exposed ground are given in his Greenland work (Wegmann, 1935c; 1938).

Backlund, writing about the same time as Wegmann, advanced very similar proposals which he has since developed (Backlund, 1936; 1937; 1938a; 1938b; 1938c; 1946). As the migmatite front advances, different country rocks react differently; pelites need only a small influx of soda and silica to change them into granite, but quartzites, limestones and greenstones are converted into granitic and other types by more selective and complex metasomatisms. For Backlund, all the so-called magmatic rocks—except basalts and their differentiates—are the results of granitisation, this including such processes as migmatisation and anatexis. Evidence includes the preservation of sedimentation and folding structures in the granites, the regional orientation of xenoliths and their relations to dyke-intrusions. The Finnish granite masses are heterogeneous since heterogeneous sediments have been granitised to give them. Backlund concludes that granitisation is the only way in which the room problem can be solved for the emplacement of granite in the folded and compressed Archean.

Viewing these later developments in Fennoskandia, we see that Sederholm's requirement of a magma in migmatisation has been largely abandoned. Migmatisation and granitisation have become almost synonymous operations.

Chemical dissection of the granitisation process has been largely the work of Doris L. Reynolds (1943; 1944; 1946; 1947a, 1947b) in her investigations of the Newry granodiorite in Ireland. In 1943, she described "replacement" bodies of granodioritic rock rimmed by biotite-enriched hornfels zones and showed that NaCaSi were introduced into the original hornfels to give trondhjemitic types, whilst AlFeMgKHTiPMn were removed to provide the biotite-rich rims. By a wider investigation, furnished with much chemical and petrographical detail, she (1944) demonstrated that various

"granodiorites have been developed from hornfelsed sedimentary rocks, and that their evolution depended on introduction and precipitation of certain constituents, together with internal migration, fixation and removal of certain constituents. The minimum introductions necessary for granitisation were Na, Ca and Si, and the material removed from the system at the end of the process was rich in Al, Fe, Mg, and Ca" (Reynolds, 1944, p. 234).

This outgoing basic material is discovered in a biotite-enriched contact aureole and in basic and ultrabasic roof rocks with an "igneous-look"—to use Grout's term. The zones of biotite enrichment represent an MgFe front, and the main granodiorites of the complex an NaCa front; in the basic and ultrabasic roof rocks a whole series of fronts, primary and secondary, are recognized, and "each specific rock now represents a subsidiary zone of maximum concentration of some constituents" (Reynolds, 1944, p. 236).

The FeMg zone of Wegmann, characterised by cordierite, has thus become a real *basic front* considered capable of leading to the formation of igneous-looking basic and ultrabasic rocks. Dunn (1942) applied the term *diabrochite* to the basic complements of granitisation, not wishing to use the term migmatite for these nongranitic

ORIGIN OF GRANITE

products. This basic front is a proposal of far-reaching importance in that it may go to solve another room problem—that of certain basic and ultrabasic bodies occurring on the margins of granitic masses or as multitudes of small masses in migmatitic complexes. In the great Sutherland (Scotland) migmatite region (Read, 1931), for example, some hundreds of very small bosses of basic and ultrabasic material (hybrids of Ach'uaine type) are exposed over an area of a thousand square miles; magmatic emplacement of such bodies over such an area seems to me impossible—but a replacement through a basic front associated with the regional granitisation may be found to be workable.

Recently Reynolds (1946) has discussed the chemical changes that take place in various types of rock when they react with granitic "magma"—gathering her data from a score or more occurrences. She finds that the initial change in rocks of all types includes enrichment in mafic constituents and alkalis and that only subsequently are the rocks granitised in the strict sense of the term. Thus, pelitic rocks in contact with granite "magma" become molecularly desilicated and attain the composition of syenite or basic or ultrabasic igneous rocks with the culmination of alkalis and/or CaFeMg; during the second stage, the desilicated rock receives Si and alkali and loses AlCaFeMg and approaches a granitic composition. A somewhat similar twofold operation is shown to have occurred in quartzites, limestones, and basic igneous rocks subjected to "granitic" reaction.

It seems to me that the granitisation theory, as a result of the development of the notion of basic fronts, is now in a yet more interesting condition than normally. If it is demonstrated that there is a zone of FeMg enrichment around any granite mass, its explanation may be an entertaining matter. In Miss Reynolds' own words (1947a, p. 212–213):—"it has to be assumed either (a) that considerable proportions of Fe and Mg remain in the residual solutions of granite magma, which is contrary to experimental evidence; or (b) that gaseous or volatile compounds of Fe and Mg are given off from granite magma, i.e., that the process is one of pneumatolysis; or (c) that the aureole of Fe and Mg enrichment is a basic front, to be explained by the fixation of material that migrated from a central locus, now occupied by granite, at the time the granite was emplaced; the granite itself representing, not a crystallisation product of magma, but the granitisation product of the country rocks within which it occurs." There is plenty of work still to be done.

DENTS DE CHEVAL

In the previous section of these remarks I have given what may appear a theoretical armchair interpretation of the granite problem and, accordingly, I hasten to remark that this is essentially a deduction from vast amounts of detailed field work in well-exposed ground. I join with Miss Reynolds (1947b, p. 209–210) in protesting against suggestions by magmatists like Grout (1941, p. 1565) that transformists "state that there is evidence but do not give the evidence to their readers." The main purpose of any paper surely is to give the evidence to the reader.

The field evidence for granitisation comes from all over the world and especially from those regions where the Pleistocene glaciation has laid bare exceptionally good exposures. Scores of sections could be listed that show transitions from country rock to granitic rock through an intermediate zone of migmatisation and felspathisation. I have already given details of a picked half dozen of such investigations selected because they were embellished with many plates and figures (Read, 1944, p. 81–83)—the list is Barrois' Rostrenon contact, Goldschmidt's Stavanger investigation, G. H. Anderson's work in the Northern Inyo Range, Wegmann's two areas in South and Northeast Greenland, Fernando's felspathised schists in the Shetland Islands, mine and Cheng's work on the Sutherland migmatite complex, and Doris Reynolds' study of the Newry transformations to which reference has just been made. The transition country rock—migmatite—granite is without doubt one of the most firmly established facts. But this transition is interpreted in two ways: *either* it shows a relatively unimportant migmatised and felspathised zone between the country rock and an intrusive magmatic granite, *or* it shows the production of granite from country rock by the completion of those granitisation processes that are seen in a half-way stage in the intermediate migmatite zone. We have just seen how Miss Reynolds' elaboration of the basic front may affect our choice of interpretation. It may also be affected by what Raguin (1946, p. 183) has styled *"le phéno-mène de la double enclave"*—the occurrence of felspars, identical with those of the granite, in the country rock. I have dealt with this elsewhere at some length (Read, 1944, p. 74–87).

Such felspars were commented upon by the French a century ago, and the phenomenon became so familiar to them that felspathisation seemed a reasonable process. Harker (1932, p. 250) and Niggli (1942, p. 29) have objected that in metamorphism felspars can arise without addition of material; but such rocks are not like the felspathised schists of the migmatite zones (Read, 1944, p. 81) as a glance at any of the figures illustrating the selected samples already mentioned would show. Thus, Barrois (1884, p. 14) gives a figure showing irregular veins of porphyritic granite in micaceous greywacke in which are developed numerous large crystals of orthoclase, *"diseminées au hasard, à la façon des andalousites dans un schiste, au voisinage du granite."* Michel-Lévy (1893–1894) made famous similar felspars from the Flamanville granite under the local name of *dents de cheval*. They are recorded from innumerable granite contacts and are displayed with advantage in almost every polished slab of granite used for building. Chemical and mineralogical investigations of many examples show that the large felspars occurring in enclaves and country rock are identical even in small detail with the large felspars occurring in the "magmatic" granite. The most remarkable example of "the phenomenon of the double enclave" is provided by the rapakivi granite of Finland which we may now examine.

The rapakivi granite is characterised by large ovoids of potash felspar mantled by a ring of small oligoclase crystals. An extraordinary variety of explanations has been advanced by magmatists to account for these mantled ovoids (Read, 1944, p. 75)—by crystallisation of drops of magma, repeated sinkings and coatings of early crystals, eutectic crystallisation of orthoclase and oligoclase, alternate supersaturation for these two felspars, reduction in gas pressure, crystallisation in a very viscous magma, and so forth. As Backlund (1938, p. 367) has remarked, "no routine explanation based on the usual concepts of 'magmatic' behavior satisfies all the implications disclosed by close examination of the ovoids." And this remark becomes explosive when it is realised that exactly similar ovoids occur in the country rock adjacent to the rapakivi.   Further, ovoid felspars in granite and country rock are not peculiar to rapakivi—they have been described for example by Wegmann, from Greenland, by Grantham from Shap in the English Lake District, by Fernando from Unst in the Shetlands.   I repeat a question I have asked before: are we to believe that the complicated physico-chemical controls required in variety by the magmatists operate also in the country-rock environment in exactly the same way as they did in the consolidating magma?

Some observers have suggested that the large felspars were mechanically forced into the country rocks; examination of the field exposures shows that this is impossible—for example, in the Skaw granite of Unst enclaves with felspars can be seen in three dimensions with their delicate sedimentary structures perfectly undisturbed (Read, 1942).   Sederholm relies on permeation of the country rock by the granitic magma.   Raguin (1946, p. 184) has argued that the physical states of the aureole rocks and the granitic magma at the time of formation of the identical felspars were the same—the aureole rocks being permeated with solutions and the magma having innumerable particles in suspension.   Fenner (1931, p. 81) has suggested that impregnation of the aureole by gases gives rise simultaneously to the same minerals that are being formed from the liquid.   Recently, Walker and Mathias (1946) for the famous Sea Point contact at Cape Town have regarded both the porphyroblastic and porphyritic felspars as potash replacements.   It must be recalled that the *dents du cheval* are only the most noticeable of the granitic minerals growing in the aureole rocks—other alkali felspars and appropriate ferromagnesian minerals likewise occur, and there is no doubt that thorough granitic rocks are produced from country rocks by this process.   If the country-rock environment and the granitic magmatic environment are considered to be physico-chemically identical then what has been deduced from the "little crucibles" seems to me to be without application.   I would prefer to consider that because the enclaves containing the felspars are sedimentary, then what encloses the enclaves is also sedimentary—the porphyritic granites have been produced from sediments by an intensification of the felspathisation process.

The *double enclave* has been used with great vigor by Perrin and Roubault (1937; 1939) in their objection to the formation of granites from magma.   They conclude that granites are made in the solid as the result of large-scale chemical diffusion—a mechanism that we now briefly consider.

DIFFUSION IN THE SOLID

Among transformists, opinion appears to be moving away from the view that granitisation is connected with some kind of soaking by juice or ichor and toward the invocation of ionic migration.   A dozen years ago, Wegmann discussed atomic migration through stationary and moving frameworks, and Backlund correlated the details of the rise of the migmatite front with the differing atomic radii of the participating elements.   Two more recent statements are: granitisation is "a migration of ions within solids by way of structural faults, deformations and crystal discontinuities and by means of potential differences of lattice energies, the result being a re-modelling and substitution" (Backlund, 1946, p. 114); and, "It is, therefore, necessary to stress the fact that granitisation depends, not on the bodily introduction of magma, juice or ichor, but on a complex series of ionic migrations, with balanced additions and subtractions, so that as the rocks undergoing granitisation approach the composition of granite, others receive additions including Fe and Mg, and become more basic" (Reynolds, 1947b, p. 210–211).

The whole question of diffusion in the solid state and its bearing on the granitisation problem has been considered in detail by Bugge (1945) and by Perrin and Roubault (1937; 1939).   The latter deal with a mass of metallurgical and petrographical material of great variety and conclude that granite is formed by reactions in the solid state; they are certain of this since they have not met "*aucun fait qui prouvât, dans le cas d'un granite quelconque, indiscutablement son origine par refroidissement d'un magma fondu*" (1939, p. 147).   I have commented on enthusiasts in an early page and shall do so again immediately.   Bugge likewise considers that granites and granodiorites are due to metasomatism in connection with long distance migration; "the rocks are, so to speak, immersed in a molecular and ionic disperse system of particles which moves through the interstices of the minerals and through the minerals themselves, altering metasomatically every part of the rocks" (1939, p. 58).   Reynolds (1947b, p. 220–222) has summarised the ways in which solid diffusion can take place—through spaces in the crystal lattice, from one lattice point to another within the crystal mesh, and along the boundaries of closely packed crystal grains.   We have to begin to look at minerals not simply as chemical compounds but as chemical compounds with architectures, as J. de Lapparent (1941) has put it.

These new methods of attack on the problem of metasomatism of country rocks to produce "igneous looking" rocks appear to me to be of the greatest promise.   It is premature to treat them flippantly as Bowen (1947) has recently done.   On the other side, it will be of no advantage to the transformists' cause to go too fast.   For example, I wish to suspend judgment on some applications of the solid diffusion thesis: I have yet to be convinced, as Perrin and Roubault (1941; 1945) are convinced, that sections hitherto considered to show Mesozoic sandstones and conglomerates resting unconformably on metamorphosed older rocks are really to be interpreted as revealing the metamorphic front which, propagated by solid diffusion, has been stopped on meeting porous layers.   I will wait and see; in your idiom, I come from Missouri.

THE PLUTONIC ROCKS

Early in these lengthening remarks, I stated that I saw the granite problem as one closely tied up with migmatisation and metamorphism—as a plutonic problem. I have discussed this matter at length (Read, 1939).   In many areas of regional metamorphism an orderly sequence of metamorphic grades can be established culminating in a central core of migmatisation.   I have interpreted this by an application of the doctrine of fronts.

"The process of granitisation taking place in the kernel leads to the expulsion of waves of metasomatism which, passing through the surrounding country-rocks under changing chemical and physical controls, promote the formation of the series of metamorphic zones about it.   Regional metamorphism, therefore, has little to do with stress or folding or deformation—it is largely an affair of permeation by metasomatising solutions.   It is only by this kind of operation that the delicate original textures of the sediments can be preserved when the rocks are thoroughly recrystallised" (Read, 1946, p. 668).

A great deal of detailed work is to be done in tracing out the movements of the fronts: in Antrim and southwest Scotland Miss Reynolds (1942) has begun by establishing a KFeMgAl front, represented by biotite—rich schists, which moved ahead of an NaSi front, represented by albite schists—both fronts moving in the direction of axial planes of recumbent folds as advance guards of a migmatisation front.

The proposition that zones of regional metamorphism are related to granitisation cores may thus receive support when enough chemical data are available. A first and necessarily generalised survey by Lapadu-Hargues (1945) indicates that the distribution of elements in the series, phyllites: chlorite-sericite schists: biotite-muscovite schists: two-mica gneisses and felspathised schists: granite gneisses: granites, varies with the grade of metamorphism. He interprets the variations in terms of atomic mobility, the smallest ions having migrated farthest; the order of increasing mobility is K, Ca, Na, Mg, Fe. I am of the opinion that the results of Lapadu-Hargues need to be tested statistically and very much expanded before they can be accepted, but they show a path that has to be followed.

The close genetic connection between granitisation and regional metamorphism has, of course, long been the belief of many workers, especially the French and Fenno-skandians, on the plutonic terranes. I have to recall, however, that some modern French geologists, notably Jung and Roques, consider that the migmatite front moves independently of the metamorphic zones (or metamorphic front)—but this is a matter that I propose to deal with in another place. For myself, I find comfort in my view of the plutonic rocks; regional metamorphism seems inexplicable when it is detached from granitisation.

MOBILISATION

Some granites appear to be intrusive, just as some rock salt, clay, and other mobile materials are intrusive. I interject that, obviously, intrusive does not imply igneous or magmatic. We have to enquire how it is proposed to get the products of granitisation to become intrusive—how they are mobilised, to use the current term. We have left chemical mobility and have to consider mechanical mobility.

Early it was realised in France and in England that increase in the degree of metamorphism, ultrametamorphism or granitisation would render the products capable of movement. Later, Wegmann and Backlund in Fennoskandia, and Mac-Gregor and Wilson (1939) and Reynolds in England have discussed the mechanism. Mobility could be induced in granitisation products by pore magma, introduced fluids, increased temperatures, and increased pressures. A mash of mixed material— Reinhard's *migma*—would move with less than a quarter of its volume liquid. I hold with MacGregor and Wilson (1939, p. 210) that "the complete fluidity of any major mass (of granite) has never been conclusively demonstrated. In most cases a mobile mash of crystals and liquid would equally well account for the observed facts." There are a great many textures in the crystalline rocks that can be interpreted in a variety of ways, and in any case, having in mind the nature of the mobile mashes, there will be plenty of opportunities for homogenisation.

As a pendant to a great deal of the foregoing, I must remind you that one of the best examples of mobilisation is that described by Bowen (1910) from Gowganda,

Ontario, where granophyre and aplite veins arise by transfusion of shale at a diabase contact: *Forsan et haec olim juvabit meminisse*, or, to translate "those were the days!"

GRANITES AND GRANITES

Whether as migma or, if the process goes far enough, as magma, granitic material may move into higher levels of the crust and produce there the discordant granite bodies with aureoles of thermal metamorphism. There might arise a whole series of types of granite masses depending upon the mobility of the granitic material—there might be granites and granites. This *granite series* has always been a tenet of the French geologists—from Élie de Beaumont a century ago through Delesse, Michel-Lévy, Lacroix, to Termier and the moderns. Termier maintained that if a granite is surrounded by a narrow aureole it is certain that the granite has come from somewhere else, ready made; if a granite is surrounded by a vast aureole of metamorphic rocks, it has certainly been formed in place whilst the neighboring rocks were regionally metamorphosed. And coming down to last year, we find Raguin continuing the tradition in his *Géologie du granite* with the two categories *granites d'anatexie* and *granites en massifs circonscrits* with transitional types between.

Though there may be granites and granites, most of them are of one kind, and all of them may likely be of one connected origin. Their geological setting will decide where they come in the series. I see vast and inexhaustible fields for steady study and speculation amongst my Plutonic Rocks. Let us address ourselves to the task with a good heart.

BIBLIOGRAPHY

Note: Many additional references are given in Read, 1943: 1944, and in Holmes, 1945.

Anderson, E. M. (1938) *Crustal layers and the origin of magmas*, Bull. Volc. Series II, Tome III, p. 24–82.

Backlund, H. G. (1936) *Der "Magmaaufstieg" in Faltengebirgen*, Soc. Géol. Finlande, C. R., no. IX, Bull. Com. Géol. Finl., no. 115, p. 293–347.

———— (1937) *Die Umgrenzung der Svecofenniden*, Geol. Inst. Upsala, Bull., vol. 27, p. 239.

———— (1938a) *The Rapakiwi puzzle, A reply*, Geol. Fören. Förhandl., vol. 60, p. 105–112.

———— (1938b) *Zur "Granitisationstheorie": eine Verdeutlichung*, Geol. Fören. Förhandl., vol. 60, p. 177–200.

———— (1938c) *The problems of the Rapakiwi granites*, Jour. Geol., vol. 46, p. 347.

———— (1946) *The granitisation problem*, Geol. Mag., vol. 83, p. 105–117.

Barrois, C. (1884) *Mémoire sur le granite de Rostrenon*, Soc. Géol. Nord., Ann., vol. 12, p. 14.

Barth, T. F. W. (1939) *in* Barth, Correns and Eskola, *Die Entstehung der Gesteine*, Ferdinand Enke, Stuttgart.

Beaumont, Élie de (1847) *Note sur les émanations volcaniques et métallifères*, Soc. Géol. France, Bull., vol. 12, 2d ser., p. 1247–1333.

Billings, M. P. (1945) *Mechanics of igneous intrusion in New Hampshire*, Am. Jour. Sci., vol. 243A, p. 40–66.

Bowen, N. L. (1910) *Diabase and granophyre of the Gowganda Lake District*, Jour. Geol., vol. 18, p. 658–674.

———— (1928) *The evolution of the igneous rocks*, Princeton Univ. Press.

———— (1947) *Magmas*, Geol. Soc. Am., Bull., vol. 58, p. 263–277.

Bugge, J. A. W. (1945) *The geological importance of diffusion in the solid state*, Norsk. Videns.-Akad. i Oslo, I Math. Naturv. Kl. No. 13.

Cloos, E. (1936) *Der Sierra Nevada Pluton in Californian*, Neues Jahrb. Min. Geol. Palaont., Beil. Bd. 76, Abt. B., p. 355–450.

Cloos, H., and A. Rittmann (1939) *Zur Einteilung und Benennung der Plutone*, Geol. Rundschau. Bd. XXX, p. 600–608.

Daly, R. A. (1933) *Igneous rocks and the depths of the earth*, McGraw-Hill Book Co., New York.

Dunn, J. A. (1942) *Granite and magmation and metamorphism*, Econ. Geol., vol. 37, p. 231–238.

Durocher (1857) Annales des Mines, vol. XI, p. 217.

Eskola, P. (1932) *On the origin of granitic magmas*, Tscher. Min. Pet. Mitt., vol. 42, p. 455.

———— (1933) *On the differential anatexis of rocks*, Soc. Géol. Finl., C. R., no. 7, p. 12.

Fenner, C. N. (1929) *The crystallization of basalts*, Am. Jour. Sci., 5th ser., vol. 18, p. 225.

———— (1931) *The residual liquids of crystallizing magma*, Mineral. Mag., p. 539.

Grout, F. F. (1941) *The formation of igneous-looking rocks by metasomatism*, Geol. Soc. Am., Bull., vol. 52, p. 1525–1576.

Harker, A. (1932) *Metamorphism*, London.

Holmes, A. (1931) *The problem of the association of acid and basic rocks in central complexes*, Geol. Mag., p. 241.

———— (1932) *The origin of igneous rocks*, Geol. Mag., p. 543.

———— (1936) *The idea of contrasted differentiation*, Geol. Mag., p. 228.

———— (1945) *Natural history of granite*, Nature, vol. 155, p. 412.

Kennedy, W. Q. (1938) *Crustal layers and the origin of magmas*, Bull. Volc. Series II, Tome III, p 24–82.

Klüpfel, W. (1941) *Ueber die Altvulkane und die Neuvulkane und ihre Abstammung*, Zentralb. Mineral., Abt. B, p. 230, 249, 281, 313.

Kranck, E. H. (1937) *Uber Intrusion und Tektonik im Kustengebiete zwischen Helsingfors und Porkola*, Soc. Géol. Finlande, C. R., no. X.

Lapadu-Hargues (1945) *Sur l'existence et la nature de l'Apport chimique dans certaines séries cristallo-phyllienes*, Soc. Géol. France, Bull., 5th ser., T. 15, p. 255–307.

Lapparent, J. de (1941) *Logique des minéraux du granite*, Rev. Scientifique, p. 285–292.

Loewinson-Lessing, F. (1911) Geol. Mag., p. 248–257.

Lyell, Sir Charles (1838) *Elements of geology*, 1st ed.

———— (1875) **Principles of geology**, 12th ed.

MacGregor, M., and Wilson, G. (1939) *On granitization and associated processes*, Geol. Mag., vol. 76, p. 210–212.

Mayo, E. B. (1941) *Deformation in the interval Mt. Lyell-Mt. Whitney, California*, Geol. Soc. Am., Bull., vol. 52, p. 1001–1084.

Michel-Lévy, A. (1893–1894) *Contribution à l'étude du granite de Flamanville et des granites francais en général*, Carte Géol. France, Bull., vol. V.

Niggli, P. (1942) *Das Problem der Granitbildung*, Schweiz. Min. Petr. Mitt., vol. XXII, p. 1.

Perrin, R., and Roubault, M. (1937) *Les réactions à l'état solide et la géologie*, Bull. Serv. Carte géol. l'algerie, 5th ser., Petrographie, no. 1.

———— (1939) *Le granite et les réactions à l'état solide*, Bull. Serv. Carte géol. l'algerie, 5th. ser., Pétrographie, no. 4.

———— (1941) *Observation d'un "Front" de métamorphisme régional*, Soc. Géol. France, Bull., 5th ser., T. II, p. 183.

———— (1945) *Observations de métamorphisme du Trias dans les Alpes autochtones au Lac de la Girotte (Savoie)*, Soc. Geol. France, 5th ser., T. 15, p. 171.

Raguin, E. (1946) *Géologie du granite*, Paris.

Read, H. H. (1927) *The igneous and metamorphic history of Cromar, Deeside, Aberdeenshire*, Royal Soc. Edinburgh, Tr., vol. 55, p. 317–353.

———— (1931) *Central Sutherland*. Geol. Soc. Scotland, Mem.

———— (1939) *Metamorphism and igneous action*, British Assoc., Dundee, Sect. Cx, Presidential Address.

———— (1942) Geol. Assoc., Pr., vol. 53, p. 107.

———— (1943) *Meditations on Granite: Part One*, Geol. Assoc., Pr., vol. 54, p. 64–85.

Read, H. H. (1944) *Meditations on granite, Part Two*, Geol. Assoc., Pr., vol. 55, p. 45–93.
———— (1946) *This subject of granite.* Sci. Prog., London, vol. 34, p. 659–669.
Reynolds, D. L. (1942) *The albite-schists of Antrim and their petrogenetic relationship to Caledonian orogenesis*, Royal Irish Acad., Pr., vol. 48 B, p. 43–66.
———— (1943) *Granitization of hornfelsed sediments in the Newry granodiorite of Goraghwood Quarry, Co. Armagh*, Royal Irish Acad., Pr., vol. 48 B, p. 231.
———— (1944) *The south-western end of the Newry igneous complex*, Geol. Soc. London, Quart. Jour., vol. 99, p. 205–246.
———— (1946) *The sequence of geochemical changes leading to granitization*, Geol. Soc. London, Quart. Jour., vol. 102, p. 389–446.
———— (1947a) *The association of basic "fronts" with granitization*, Sci. Prog., vol. 35, p. 205–219.
———— (1947b) *The granite controversy*, Geol. Mag., vol. 84, p. 209–223.
Wager, L. R., and Deer, W. A. (1939) *The petrology of the Skaergaard intrusion, Kangerdlugussuaq, East Greenland*, Medd. Grönland, vol. 105, no. 4.
Walker, F., and Mathias, M. (1946) *The petrology of two granite-slate contacts at Cape Town, South Africa*, Geol. Soc. London, Quart. Jour., vol. 102, p. 499–518.
Wegmann, C. E. (1935a) *Zur Deutung der Migmatite*, Geol. Rund., vol. 26, p. 305.
———— (1935b) *Über einige Fragen der Tiefentektonik*, Geol. Rund., vol. 26, p. 449.
———— (1935c) *Caledonian orogeny in Christian XI's land*, Medd. Grönland, Bd. 103, no. 3.
———— (1936) *Geologische Merkmale der Unterkruste*, Geol. Rund., vol. 27, p. 43.
———— (1938) *Geological investigations in southern Greenland*, Medd. Grönland, Bd. 113, no. 2.
———— **and Kranck, E. H.** (1931) *Beiträge zür Kenntnis der Svecofenniden in Finland. I, II*, Com. Géol. Finlande, Bull. 89.

THE GRANITE PROBLEM AND THE METHOD OF MULTIPLE PREJUDICES

BY NORMAN L. BOWEN

(Geophysical Laboratory, Washington, D. C.)

INTRODUCTION

In his well-known essay, *The method of multiple working hypotheses,* Chamberlin (1897) discussed various modes of attack upon scientific problems. Of a theory hastily advanced in explanation of any group of observations he said, "From an unduly favored child it readily grows to be a master and leads its author whithersoever it will." The theory thus becomes a "ruling theory." It might be said that it becomes a prejudice. As we may readily suppose, he regarded the ruling theory as having little to commend it, though he did point out that in the method of the ruling theory the investigator's "very errors may indeed stimulate investigation on the part of others." Chamberlin then passed on to consider the method of the working hypothesis, which he naturally regarded as having notable advantages, but added "... the distinction is not such as to prevent a working hypothesis from gliding with the utmost ease into a ruling theory...", and further, "To avoid this grave danger the method of multiple working hypotheses is urged" where "the mutual conflicts of hypotheses whet the discriminative edge of each." Chamberlin thus had in mind a happy situation where an individual investigator diligently sought all reasonable processes that might lead to an observed relation, carefully considered the full consequences of each process envisioned, impartially compared these deduced consequences with the re-examined facts and thus reached a conclusion as to the probable process or group of processes that were operative. 'Tis a consummation devoutly to be wished; yet when one views the present state of petrology he cannot fail to wonder whether any individual is capable of such detachment. Each of us is, of course, quite sure that he himself has done just what Chamberlin recommends, but he is equally sure that the other fellow has done no such thing. We hurl epithets, we hurl them back. We are most thankful that we are not as other men are. That fictional character, the impartial observer, would probably say that we are all nursing pet prejudices. The situation is deplorable, but may not be hopeless. It may be that the mutual conflicts of prejudices will dull the rabid edge of each, and that the communal petrologic mind can eventually reach the truth by this sorry method as surely, if not as directly, as Chamberlin's ideal individual mind. Some such hope must, presumably, have animated those who conceived this symposium. So let us proceed, shall we say, *en toute objectivité.*

THE CONFLICT

It may be true that the details of the story of granites "yet survive, stamp'd on these lifeless things", but it is clearly not true that he who runs may read. "The best geologist is he who has seen the most rocks", says Read (1939). To a degree this position is tenable, yet Read himself makes certain unspecified reservations which

79

may be imagined and which lead me to suggest that it is less a question of sight than of insight.   Look as one will at the rocks he cannot see them in process of formation.

As a result of field observation we ordinarily succeed only in setting up a number of rival hypotheses.   To choose between them we should not only have more field observations but we should bring to bear every resource of the fundamental sciences. This some petrologists attempt to do, but in spite of their efforts there remains much divergence of views.   The ebb and flow of opinion on the granite question has been repeatedly portrayed in recent writings, and this aspect of our subject may therefore be briefly dismissed here.   On the one hand we have the view that granite is the product of consolidation of a magma, on the other the view that it is formed by the replacement of pre-existing rocks through the agency of some diffuse medium which induces "granitization."   Since the latter action is one that is accepted as operative in certain metamorphic processes, and the former is a typical igneous phenomenon, we may for brevity state that the question is whether granites are igneous or metamorphic.

Although it was Hutton's igneous views that dealt Wernerism its death blow, it may nevertheless be said, at this distance, that the view that granites are metamorphic is the older.   Recent devotees of this view like to dwell on this fact, not in modest disclaimer—modesty is unfortunately not one of the outstanding characters of controversial writings—but apparently because they regard the support of the old masters as indicative of the verity of the process.   Yet among the old masters are numbered those who sternly rejected organic evolution, continental glaciation, and other established processes.   The modern magmatist, too, can find some comfort in the early adverse views.   He had these views before him; on the evidence, so he believes, he rejected them as inadequate and thus cannot be accused of making a snap judgment. Thus the old masters please everyone, indeed an unusual situation.   But, as has already been said, the historical and, we may add, the nationalistic and even the "school" aspects of the controversy have already been well treated in recent writings, and we may proceed to the consideration of matters of detail.

It is possible that a reader would gain from the foregoing an exaggerated impression of the cleavage between the two camps.   There is probably no magmatist who does not believe that igneous and especially granitic material has occasionally been introduced into and replaced other rocks to such an extent that they acquire the composition of granite.   He does reject the contention that most granitic rock has been formed by this process of granitization.   Similarly there is probably no upholder of the metamorphic view who does not accept the possibility of a little granitic magma on occasion, and of the formation of some granite by direct consolidation of this magma.   The real question is, then: How much granite is magmatic and how much metamorphic?

The various views now current on the question of the origin of granite will be treated in the order of decreasing respectability today.   This is also in my opinion the order of increasing probability, for granites by and large.

DRY GRANITIZATION

The extreme proponents of granitization would develop all great batholithic masses by processes of replacement.   In the older view the replacing substances were transported in solution in a pervasive fluid medium (variously denominated mineralizers, emanations, ichor, etc.) and were deposited where conditions were appropriate. Recently it has been proposed that the migration of ions through crystals, a solid diffusion, has been the process in operation. This is on first thought, an astounding proposal, but it has many supporters, and will require our careful consideration. At first, solid diffusion was advanced as a hypothesis in opposition to that of transport by a pervasive fluid. Perrin and Roubault (1937) rejected liquids or fluids entirely. They even went so far as to call upon solid diffusion to explain certain phenomena occurring in silica brick used in the lining of metallurgical furnaces, although Greig (1927) had studied the factors involved in detail, and had excellent justification in his experimental studies for stating that interstitial liquid works its way in from the edge of the brick, corroding the crystalline grains. Perrin and Roubault called Greig's liquid "hypothetical" although his studies show that liquid must form in the existing compositions at the temperatures concerned, and anyone can assure himself of the existence of interstitial liquid (glass) by microscopic examination of the brick. Such extreme advocacy of solid diffusion for cases where existence of a liquid is demonstrable is bound to defeat its purpose.

Later proponents of the view that solid diffusion is an important if not a dominant factor in petrology make no such tactical errors, indeed, there is a disposition for the ichor camp and the solid diffusion camp to join forces against the magmatists. Transport by solid diffusion may be the dominant or even the sole process in certain instances, in others it may have been assisted by fluid transfer, according to this consolidated view. It is necessary to examine more closely these views in their modern dress.

A crystal has been described as "an empty space with small charged particles distributed at enormous distances from each other" and occupying "no more than 10^{-15} part of the space" (Joffé, 1928). As such, it would seem to present an ideal medium for the play of diffusion. Actually these particles are, as a result of the electrical and magnetic fields, arranged in a definite manner which is the characteristic lattice of the individual substance, and there must be definite expenditure of energy if this arrangement is to be disturbed in any manner. According to theory, the ideal lattice is attainable only at the absolute zero of temperature. At all other temperatures thermal vibration of the particles induces certain departures from the ideal. These defects, as they are called, are believed to be of two kinds. The one consists of holes in the lattice—i.e., certain lattice positions are unoccupied; in the other some atoms occupy interstitial positions—i.e., positions that do not correspond with regular lattice positions. With increased amplitude of the thermal vibrations, consequent upon increased temperature, the proportion of these defects increases, but the proportion remains small until the temperature approaches the melting point of the crystal, where they rapidly multiply and the lattice is destroyed. These relations have been determined for ionic crystals with the aid of certain assumptions, by measuring the electrolytic conductivity of the crystal. Thus for AgCl it is found that 0.1 per cent of the ions are interstitial at 350°C. but only about 0.02 per cent at 250°C. (Mott and Gurney, 1940).

It is the presence of defects in the lattice that enables ordinary solid diffusion to occur, whether it be auto-diffusion or the entry of foreign ions into the crystal. As a result of the rapid decrease of the thermal vibrational energy and of the proportion of defects, with falling temperature, they are soon reduced to a value below which diffusion is ineffective, indeed there is apparently a value below which diffusion cannot occur. This threshold value is said to obtain at a temperature which is different for different types of substances but is given as 0.8 to 0.9 T for silicates, where T is the absolute melting temperature of the individual silicate (Eitel, 1929). If this condition were rigidly true, it would be possible to rule out solid diffusion in silicate assemblages entirely, for the melting temperature of the assemblage is usually so far below the melting temperatures of the individual phases that crystals would not yet have formed until the temperature had fallen to a value lower than that at which solid diffusion of ions in the individual phase could occur.

Actually there are additional factors. At the border of a crystal the fields of force are different from those in the interior and even in the interior of the crystal there may be defects other than those of the thermal equilibrium type just described. These may be inherited from a condition originating during early rapid growth of crystal nuclei (lineage) (Buerger, 1934) or may be induced by strain. For present purposes it is sufficient to regard the crystal as divisible into subparallel blocks. At the outer borders and at the interior borders between blocks the activation energy necessary to enable an ion to move out of its cell is less than that in the main lattice. Diffusion will thus begin at a lower temperature.

Buerger and Washken (1947) have investigated the conditions for recrystallization of a plastically deformed (compressed) mass of powdered individual crystalline substances. They worked with anhydrite, fluorite, and periclase and for each they found a critical temperature below which recrystallization (diffusion) did not take place. The relation had been observed for deformed metals (Van Liempt, 1935). A certain minimum activation energy is thus necessary for each substance.

Not merely for the thermal equilibrium types of defects but for the defects connected with grain boundaries and with strain there is thus a minimum temperature at which diffusion begins. It would seem certain, therefore, that sharply zoned plagioclase crystals must have formed at a temperature below the threshold value at which diffusion can take place in plagioclase; otherwise they could not retain the hair-line borders between zones that frequently prevail.

Since they do retain such sharp borders, how, then, did they grow? Certainly not by solid diffusion of ions through hundreds or even thousands of meters to feed their growth, for in such case diffusion within the crystals themselves, *pari passu* with their growth, would obviously have brought them to complete uniformity. Appeal to stress as an aid to the long-distance diffusion does not help, for the plagioclase crystals growing in such surroundings would themselves be subject to stress. Yet solid diffusion is said to be the method by which the great granite batholiths have been formed. They represent the breat bulk of all our granites, and granites characteristically contain zoned plagioclase (Johannsen, 1932). It matters not that the plagioclase may not be sharply zoned, for if fed in the manner postulated they would be of uniform composition. That zoned mix-crystals can diffuse to uniformity when conditions are appropriate is an established fact. I have watched the process going forward under the microscope in solid solutions of NH_4Cl in NH_4NO_3, but the temperature was only 3°C. below the melting point (Bowen, 1926).

Perrin and Roubault (1939) attempt to laugh magmas out of court by saying that Bowen literally makes the feldspars dance in his efforts to explain oscillatory zoning. It is true that something other than continuous, uniform cooling of the magma is necessary to explain the phenomenon, but think of the persistence of any kind of zoning in a crystal and the problem which it poses for men who would have solid diffusion take place through thousands of meters and literally move mountains!

The most elaborate discussions of the geological significance of solid diffusion are those of Ramberg (1944) and of Bugge (1945). Each attempts, in his own way, to build up what might be termed a unified theory of petrogenesis. They would, as it were, write a single equation to encompass all petrologic phenomena. Such attempts have a natural attraction. According to Ramberg, solid diffusion takes place wherever there is an activity gradient, according to Bugge, a gradient of chemical potential. These gradients will exist wherever there is a concentration, a temperature, or a pressure gradient. Thus there is a tendency for atoms to migrate down the normal temperature-pressure gradient of the earth, since high temperature and high pressure increase the activity or the potential. But all the atoms are so affected, and the significant factor is the differential effect on the various atoms. What will migrate? Without giving too convincing reasons Ramberg and Bugge choose just what they need in order to get granite composition. Ramberg (1944, p. 103) does say that "the variation, with mechanical pressure, of the partial activities of minerals depends on the specific gravity of the minerals and on the atomic weight of the elements." He does not specify at this point what substances are the most migrant but after analysis of the effect of the temperature gradient he then sums up the situation by saying (p. 108):

"According to my opinion the granitic minerals: potash feldspar, quartz, albite and perhaps some minerals rich in Fe in relation to Mg, have the greatest chemical activities (under the P. T. conditions existing in the geologically known levels of the crust). These minerals are those which are able to migrate long ways."

Thus there seems to be envisaged an upward flow, not merely of light ions but also of light minerals formed from these ions, all in a completely solid material, and although it is not specifically mentioned, there is of necessity a return flow of heavy ions and of heavy minerals formed from them. In a later communication Ramberg (1946) makes specific provision for the return flow.

We may note that Mg has a lower atomic weight than Fe and also forms lighter minerals. Ramberg's assumed greater facility of upward diffusion of Fe is thus contradictory to his own statement of the controlling factors. Apparently realizing this difficulty Ramberg makes Fe migrate downward in his later communication but in so doing leaves unexplained the relatively high Fe:Mg ratio in granites, although he recognizes this as a fact, witness the above quotation. When such difficulties are encountered appeal is made to the disturbing effect of the chemical forces, and, since these are not known, it is assumed that they are just right to give the results seen in the rocks.

ORIGIN OF GRANITE

To the magmatist the controlling factor is concentration in residual liquids during fractional crystallization, and the nature of the residual liquid, including the Fe:Mg relation, falls naturally into the picture.

The advocates of solid diffusion, except perhaps a few extremists, do not, indeed, reject magmas nor yet the concept of fractional crystallization in magmas. Bugge (1945, p. 59) says,

"In addition to these (O^{2-} and OH^-) the chief components of the ichor are Si, Al, Na, K. These four elements are concentrated in the mother liquor by crystallization of a magma, they are the first elements mobilized by differential anatexis and they are also the elements showing the highest speed of diffusion both through the crystals and through the intergranular film."

In this version the effects of all the processes assumed to be operative would be so nearly identical that there would be no chemical criteria by which their effects could be distinguished and their relative importance evaluated. Yet it may be doubted that in the case of solid diffusion the chemical effects could so simulate impregnation or actual magma consolidation. There are too many stumbling blocks. The K atom for example is by no means a very small or a very light one, yet it must be assumed to be among the most susceptible to solid diffusion if granites are to be formed by that process. Moreover Mg, whose relation to Fe has already been discussed, is both lighter and smaller than K. Why does Mg lag behind in this process of granitization by solid diffusion? It is a natural laggard in any process of transfer as a constituent of residual liquids. They are nearly free from Mg, on account of the refractory character of its compounds and their consequent early separation from magmas.

In attempting to assess the relative migratory powers of the ions some very curious adjustments of views have been necessary. It was early concluded that Si and Al would rise because they were small, but this led to difficulty with O and OH which are large and by the same reasoning should lag behind. Bugge solved this difficulty for OH by assuming that the H could migrate, but this only enhanced the difficulty in the case of O. There was grave danger of developing an earth layer consisting entirely of O, which we might regard as a sort of companion for the core of solid hydrogen of one theory of the earth. Apparently thinking that this O layer was a bit unrealistic, Ramberg came to the rescue by assuming, *not* that O *could migrate in spite of* its great size, but that it *must migrate* upward *because of* its great size. The O ions are too big to remain where the pressure is great and up they come (Ramberg, 1946). The whole process of solid diffusion is nothing, if not versatile.

Another aspect of the hypothesis calls for comment. In the supposed migration of ions from a locus of high pressure and their coming to rest in a lattice at a locus of low pressure they set free a heat of dispersion at the latter point, and the gradual accumulation of such heat is said to be a possible means of conversion into magma. When we have regard for rates it can readily be seen that this heating process could have no importance. The coefficient of diffusion in crystal silicates is not known. It has been measured in some liquid silicates and found to be of the order 10^{-7} to 10^{-6} (Bowen 1921). In crystal silicates it is not likely to be greater than 10^{-10} to 10^{-9} even at temperatures of 1400°–1500°C, if the contrast that exists between liquid and crystal in the known case of metals is indicative. Now the coefficient of diffusivity of temperature in rocks is of the order 10^{-2}, enormously greater than the diffusivity for ions.

Heat liberated by "condensation" of ions arriving by the supposed method would thus be rapidly dissipated down the geothermal gradient as an utterly negligible addition to that heat flow.   If the earth is a cooling body and if the final cooling by conduction alone has yet penetrated to only a relatively moderate depth, then below that depth there would be only a small temperature gradient whose value would depend on the mode of evolution of the earth.   At such depth this heat transfer due to transfer of ions down the pressure gradient might be of importance.

Solid diffusion probably explains some synantetic phenomena and its present popularity will lead to fruitful investigations of order-disorder in silicates, but in the present state of knowledge it seems doubtful that it can be regarded as a factor of major importance in geology.

WET GRANITIZATION

Magmatists have generally supposed that in the crystallization of granite there may be produced in the late stages a liquid that has a high concentration of water and other very fusible materials, in which remain dissolved the more soluble constituents of the granite itself.   It is thus what may be called a very thin granitelike magma, though most magmatists, myself included, would ordinarily object to the extension of the term "magma" to such a liquid.   The term "liquid" has been used but will now be withdrawn, and the noncommittal "fluid" substituted for it in order to avoid possible controversy as to the state of the material, which does not matter for immediate purposes.   Such a tenuous fluid, magmatists suppose, sometimes escapes into the country rocks of a granitic intrusive and introduces granitic material into them, often by a process of replacement.   In short it effects granitization.

Now in the cooling of the whole earth and the crystallization of that part of it which may be crystalline there might be produced a somewhat similar residual fluid.   Such a fluid is said to be still in part entrapped in the deeper crust and to be ever rising into higher levels, though its rise may be facilitated locally.   A fluid of some such character and origin is what I suppose the granitizationists have in mind when they speak of an ichor (the word signifies "plasma", but, as frequently used, "miasma" might be a more appropriate rendering) which has produced, so it is said, granitization of colossal masses of the earth's crust.   No rock or structure is immune from attack, but the ichor preys especially upon the great prisms of geosynclinal sediments that have from time to time formed as elongate belts on the earth.   Not only does the ichor produce granite from other rocks by metasomatic replacement, but upon occasion the intensity of the action may be such as to cause fusion and the production of true granitic magma, which may be injected as an independent intrusive, according to the hypothesis.

The great barrier to the acceptance of such large-scale granitization is the fact that it requires a huge supply of energy and no supply is in sight.   The rocks of geosynclinal prisms, the principal candidates for granitization, are dominantly shales, characterized by an abundance of low-grade minerals—clays, hydromicas, chlorite. The conversion of these into the largely anhydrous, high-grade minerals of granite is a strongly endothermic reaction.   An ichor amounting to only a minute fraction of the mass of the rock which it pervades would first need to raise the temperature of the whole mass to a value such that these low-grade minerals would no longer be stable, in itself a colossal task, but it would then require to maintain this temperature in the mass in the face of the endothermic transformation effects. The ichor, travelling excessively slowly through the minute pores of rocks immediately beneath those suffering granitization, could not reach the current seat of granitization with the temperature of *l'écorce terrestre*. It could have only the slightly higher temperature of the immediately subjacent rocks. It is difficult to conceive how the ichor, even if continuously replaced by a fresh supply, necessarily an excessively slow process, could bring about the colossal effects attributed to it.

The prospect is even bleaker than I have painted it, for the spent ichor must take into solution (again an endothermic process) and carry on with it all the material not needed for granite. This will be largely mafic or cafemic in character, average shale being much more mafic than average granite. Yet the ichor has, *ex hypothesi*, come from deep levels in the earth where, by all geophysical indications, the material is largely mafic, and if mafic material is more soluble in it than granitic, as the metathesis postulated in granitization would require, why did it not bring along dominantly mafic material in the first place?

The energy requirements of granitization have usually been ignored by its wet protagonists, but a few have evidently realized the difficulties. Holmes says and says again "emanations plus energy." Some energy is indeed required, for the thermodynamic picture leaves us with a negative supply. Is there an extra-thermodynamic supply, such as that afforded by atomic disintegration? Holmes and Harwood (1937, p. 253) do mention radioactivity as a possible source of energy in the emanations but make no specific suggestions, nor do they point out any evidence of exceptional concentration of radioactive substances. We may feel sure that Holmes has exhausted the possibilities, for here he is playing on his own home grounds.

It has been suggested that the radioactive isotope of potassium was sufficiently abundant in early pre-Cambrian times to constitute an important source of heat for such phenomena. This does not help much for the late Mesozoic Coast Range batholith of British Columbia, which was formed, so it is said, by granitization.

Other problems raised by the hypothesis of wet granitization will be discussed at a later point, especially the question of the "basic front."

GRANITE

In the opinion of many, at the moment not very respectable, petrologists most of the granitic rock that we see was formed by crystallization of an intrusive molten magma closely resembling in character the extrusive molten material of rhyolitic lavas, though containing a greater proportion of water and other volatile constituents than most of the extrusives. Those holding such views are termed magmatists. Within this disreputable group there is wide divergence of opinion as to how granitic magma came into being.

There is the view that it is formed by fusion of geosynclinal sediments, or that it is produced by the refusion of the base of the granitic layer of the earth, that it results from the differentiation of a syntectic magma formed by the solution of granitic or other salic material in basaltic magma, or, finally, that it is the product of differentiation of basaltic magma as such.   These are again arranged in the order of decreasing respectability today.

No magmatist, perhaps, can be regarded as adhering exclusively to any one of these views to the exclusion of the others.   The opinion that geosynclinal sediments are fused is not always clearly distinguished from the metasomatic (wet granitization) view, though fundamentally they are distinct, for in the metasomatic process only a small part of the mass would be liquid at any one time, and that minutely disseminated, whereas in the magmatist view there is complete conversion of a significant unitary volume of sediment into liquid.   There might, perhaps, be a passage of one process into the other.   An eminently sane discussion of the various processes is the Eskola essay of 1932.

In the present state of our knowledge we must accept all these methods of formation of granitic magma as possible.   Nevertheless the disgraceful representative of the lowest group who now speaks to you cannot refrain from asking the question: "Whence came the first granite that suffered refusion to granitic magma?" and again, since salic sediments can be produced only by erosion of a salic terrane, "Whence came the first salic sediments that suffered refusion to granitic magma?".   Seeking a satisfactory answer to these questions and finding none, the reprobate is constrained to believe that, initially at least, granitic magma and the first granites were formed by pure differentiation of the basic (mafic) material of the earth.   The method of differentiation was most probably that which is widely demonstrated in many floored intrusives of later times, and finds ample sanction in experimental studies of silicates, namely fractional crystallization (Bowen, 1937).   Some solution of granite and other salic material in basaltic magma is not an unlikely means of augmenting the granitic differentiates of the later basaltic intrusives, and some granitic magma may be the direct result of refusion of granite or salic sediments, but even here there is a suspicion that this is not due to secular causes but to the more catastrophic effect of the intrusion of hot basaltic magma from the depths which, though not itself entering the zone of remelting, contributed the necessary additional heat to that zone.

On the whole these purely regenerative methods of production of granite do not seem to meet all the requirements, and we must still entertain the hypothesis that most granites have been produced throughout geologic time by differentiation of basic (basaltic) magma, in part somewhat modified in the course of its rise from the depths.   The granite is the normal uppermost differentiate produced by repeated gravitative and tectonic (filterpressing) separation of a series of crystals from an ever-changing mother liquor.

On this hypothesis there should be basic and ultrabasic masses of aggregate bulk several times that of the granite.   In a recent polemic with Nockolds, McIntyre (1947) points out this fact and thunders "Where are these?", at the same time knowing the answer full well.   On the hypothesis they should be down below, and geophysical data for the whole earth indicate this condition.   It is not necessary to suppose that every granite is in simple direct downward continuity with its congeners. In its long history the changing liquid may have migrated laterally as well as vertically.   The local existence of a vertical section of granite a mile thick is no guarantee that the same granite extends downward 10 or 15 miles, and the statement that it does is based on assumption just as definitely as the statement that it gradually changes.

This answer may not be accepted as satisfactory, presumably will not, else the question would not have been asked, but the question itself tempts one to put a question to the wet granitizationists which can be phrased in the very same words.   In wet granitization the emanations are said to carry upward and outward from the seat of granite formation all the materials not needed for granite.   These are deposited in a frontal zone (Reynolds, 1947).   If this were true for the great masses of granite, there should be associated with each of them colossal masses of basic or ultrabasic rock.   "Where are these?", I falter.   They should be readily accessible to view, more so than the granite itself.   If the Coast Range batholith of British Columbia were formed by this method, as has been claimed, much of the rest of British Columbia would be occupied by rocks of the basic frontal zone.   No such condition exists. Pushed to its logical consequences for a large mass, that corollary of wet granitization, the "basic front," is thus a basic affront to the intelligence of the geologic fraternity.

There are small-scale occurrences of basic rocks in association with granitic intrusives that may really have been formed in the manner postulated for the *"much neglected* basic front" (McIntyre, 1947),* though, with little doubt, many masses that have been so interpreted were originally of basic character (limestones, basic extrusives) and have been modified by addition of salic material from the granite magma. The magmatist expects the emanations from magma to accomplish metasomatic replacement, including granitization of surrounding rocks, but he expects it to be of moderate extent as compared with the truly magmatic body.   Moreover he finds no difficulty in supplying the relatively modest energy demands, which he refers to the bodily and catastrophic transfer of heat from the depths in the form of hot magma. He denies, too, that it is proper to extrapolate to the whole body the metasomatic replacement relation that may be patent for part of its marginal facies.

Though, locally, volume for volume replacement of individual beds may be demonstrated, the claim that wet granitization offers a solution of the "room problem" is altogether spurious.   Material pushed up chemically occupies just as much room as the same material pushed up mechanically.   Even though repeated equivolume replacement is postulated, it cannot be repeated *ad infinitum*.   There would necessarily be a large net addition of material to the upper crust.   Wet granitization does not solve, it merely evades the room problem.   The magmatist has, however, in some instances, found not unreasonable real solutions of the problem.

The "wets" would do well to learn from the "drys" and let basic material go down, as salic material goes up.   The "drys" do not, indeed, appear to have a satisfactory method, but perhaps a better co-ordinated union of the "wet" and "dry" forces would have tangible results.

In the meantime the really efficient method of having salic material rise and mafic material sink remains the method of crystallization differentiation of magma.   That the course of crystallization in the multicomponent system of the rock-forming oxides is appropriate to give a late liquid of granitic (or related) character is well supported

* Italics are mine.

H.H. Read and N.L. Bowen: Granites and granites

by a series of experimental investigations collated elsewhere (Bowen, 1937). Later studies extend the evidence to a wider range of compositions, but there would seem to be no point in giving further details here.   Instead the discussion will be closed by offering a sonnet that reflects the international petrologic situation today, and is suggested by the presence of our two distinguished visitors from abroad, Professor Read and Professor Niggli.

THOUGHTS OF A PETROLOGIST ON THE LIQUIDATION OF MAGMA
or
ENGLAND AND SWITZERLAND, 1947

(With apologies to Wordsworth for changing three words)

Two Voices are there; one is of the Sea
One of the Mountains; each a mighty voice:
In both from age to age thou didst rejoice,
They were thy chosen music, *Verity*!

There came a tyrant, and with holy glee
Thou fought'st against him,—but hast vainly striven:
Thou from thy *Alban* holds at length art driven,
Where not a *comber* murmurs heard by Thee.

—Of one deep bliss thine ear hath been bereft;
Then cleave, O cleave to that which still is left—
For, high-soul'd Maid, what sorrow would it be

That Mountain floods should thunder as before,
An Ocean bellow from his rocky shore,
And neither awful Voice be heard by Thee!

REFERENCES CITED

Bowen, N. L. (1921) *Diffusion in silicate melts*, Jour. Geol., vol. 29, p. 295–317.

———— (1926) *Properties of ammonium nitrate*, Jour. Phys. Chem., vol. 30, p. 732.

———— (1937) *Recent high-temperature research on silicates and its significance in igneous geology*, Am. Jour. Sci., 5th ser. vol. 33, p. 1–21.

Buerger, M. J. (1934) *The lineage structure of crystals*, Zeitschr. Krist., vol. 89, p. 195–200.

———— and Washken, E. (1947) *Metamorphism of minerals*, Am. Mineral., vol. 32, p. 296–308.

Bugge, J. A. W. (1945) *The geological importance of diffusion in the solid state*, Norske Vidensk-—Akad. I, Mat. Naturv. Kl. no. 13, p. 5 59.

Chamberlin, T. C. (1897) *The method of multiple working hypotheses*, Jour. Geol., vol. 5, p. 155–165.

Eitel, W. (1929) *Physikalische Chemie der Silikate*, Leopold Voss, Leipzig, p. 436.

Eskola, P. (1932) *On the origin of granitic magmas*, Mineral. Petrol. Mitteil, vol. 42, p. 455–481.

Greig, J. W. (1927) *Immiscibility in silicate melts*, Am. Jour. Sci., 5th ser., vol. 13, p. 483.

Holmes, A., and Harwood, H. F. (1937) *The volcanic area of Bufumbira*, Geol. Survey Uganda, Mem. III, pt. 2, p. 253.

Joffé, A. F. (1928) *The physics of crystals*, McGraw-Hill, New York, p. 1.

Johannsen, A. (1932) *Petrography, Vol. II*, Univ. Chicago Press, p. 148.

McIntyre, D. B. (1947) *Crystallization of plutonic and hypabyssal rocks*, Geol. Mag., vol. 84, p. 119–123.

Mott, N. F., and Gurney, R. W. (1940) *Electronic processes in ionic crystals*, Oxford Univ. Press, p. 48.

Perrin, R., and Roubault, M. (1937) *Les reactions à l'état solide et la géologie*, Bull. Serv. Carte, Géol. Algerie, ser. 5, no. 1, p. 45.

Perrin, R., and Roubault, M. (1939) *Le granite et les réactions à l'état solide*, Bull. Serv. Carte, Géol. Algerie, ser. 5, no. 4, p. 64.

Ramberg, H. (1944) *The thermodynamics of the earth's crust, I: Preliminary survey of the principal forces and reactions in the earth's crust*, Norsk. Geol. Tidsk., Bd. 24, p. 98–111.

————— (1946) *Chemical equilibrium in the gravitational field, and some geological implications*, Medd. fra Dansk Geol. Forening., Bd. 11, p. 27.

Read, H. H. (1939) *Metamorphism and igneous action*, British Assoc., Dundee, Sect. C, Presidential Address, p. 2.

Reynolds, D. (1947) *The Hercynian Fe-Mg metasomatism in Cornwall: A reinterpretation*, Geol. Mag., vol. 84, p. 48.

Van Liempt, J. (1935) Zeitschr. Phys., vol. 96, p. 534.

CHAPTER 4

Fyfe, W.S. (1973) The generation of batholiths. *Tectonophysics*, **17**, 273−283.

This short paper is an essay that tries to link experimental petrology with geophysics, thereby proposing a model for batholith generation that has been highly influential among latter generations of granite petrologists.

The paper begins with a summary of the experimental evidence concerning the generation of granitic melts from common crustal rocks. It concludes that whenever a thick, low-density crust is formed, melting and melt migration from lower to upper crustal levels will occur on a timescale of <10 My, leading to the formation of a new crustal architecture that is closer to gravitational equilibrium. After reviewing the geological evidence for partial melting in high-grade metamorphic terranes, the paper explores the physics of melt accumulation and motion through the crust.

For melt accumulation in partially molten zones, Fyfe used a simple Stokesian approximation and concluded that, if the viscosity of the partially molten rock mass is $<10^{13}$ poise, gravitational buoyancy could lead to the accumulation of a 'clean' molten layer, beneath the melting interface, in ~1 My. Following this, gravitational instabilities in the top interface of that low-density molten layer would generate a wavy surface out of which buoyant, bubble-like diapirs of granitic magma would detach and ascend to upper-crustal levels. It should be remembered here that Fyfe was working at a time when the diapir model for granite ascent and emplacement, with its roots in the early 20th century work of Mrazec (1915) and Nicolesco (1929), was a ruling paradigm that was virtually unchallenged. Thus, using equations developed by Elsasser (1963) for the formation of Earth's core from molten iron and those of Ramberg (1967) for diapirism, Fyfe found that a bubble-like pattern of rising plutons with a diameter of 10 km, and a wavelength on the order of tens of kilometres, would form if the lower part of a thickened crust underwent extensive melting. He considered that, to a first approximation, the upward motion of the plutons might be Stokesian, at least in the warmer and less viscous lower crust, but that it must be far more complex in the cooler and more brittle upper crust.

Fyfe also used the Elsasser–Ramberg equations to explain the distribution of granitic, basaltic and andesitic igneous activity (and related metamorphic belts) in Archaean terranes, emphasizing the close relationships between crustal thickness, heat production and the generation of granitic magmas. He stated that, after the Archaean, the generation of large granitic batholiths would only be possible in a tectonically thickened crust, where this magma formation would represent the product of the expected structural relaxation. Heat for partial melting of the crust was envisaged to have been supplied through radioactive decay of the important heat-producing elements (U, Th, K) and that it could be accelerated by heat derived from magma rising from the mantle, through a process of mixing with granitic crustal melts to produce intermediate, andesitic magmas. Some of the details these processes have been disputed and some of the proposed mechanisms are no longer thought to be valid, in the light of new physical and geochemical information. Nevertheless, Fyfe's ideas and general approaches showed the way forward in the integration of physics into geological thought. They provide the basis for many of the kinds of models for petrogenesis of felsic and intermediate magmas that are still influential among present-day petrologists. For this reason this paper deserves its place as a landmark.

Tectonophysics, 17 (1973) 273–283

© Elsevier Scientific Publishing Company, Amsterdam – Printed in The Netherlands

THE GENERATION OF BATHOLITHS

W.S. FYFE

Geology Department, Manchester University, Manchester (Great Britain)

(Accepted for publication November 7, 1972)

ABSTRACT

Fyfe, W.S., 1973. The generation of batholiths. In: P.J. Wyllie (Editor), *Experimental Petrology and Global Tectonics. Tectonophysics*, 17(3): 273–283.

The formation of the melts which produce intrusions of the granite family are considered to result from the partial fusion of high-grade metamorphic rocks. The melting behaviour of such materials is considered. Such melts will rarely be water-saturated and the degree of water-saturation must set limits on the ability of the melt to rise. The natural residue of fractional fusion will be metamorphic rocks of the granulite facies. The motion of the melts seems reasonable in relation to the theory of Taylor instability and Stokesian rise of the materials. Differences between Archaean batholith patterns and those in modern belts of plate subduction will be considered. It is suggested that such differences could result from fusion processes in the upper mantle occurring at higher levels than at the present time and "ocean ridges" being more closely spaced.

INTRODUCTION

I think one always has a certain advantage when one is asked to contribute to symposia of this kind, and in these few pages I would like to try to briefly summarize those things which appear to me to be the main areas requiring attention at this time. I think today we can accept as fact that the vast volume of materials of the granite clan which are intruded into moderately high crustal levels, attain their position in the form of silicate melts. Thus perhaps the first problem we face is that of the site of generation of such liquids.

CRUSTAL FUSION

There is very good evidence to suggest that lower crustal levels may be subject to partial fusion. The first lines of evidence come from the relations between the composition of the continental crust and internal heat production, crustal heat flow and thermal gradient and experimental data on the melting behaviour of metamorphic rocks which must be present at lower crustal levels. The continental crust on average is about 30 km thick but becomes much thicker, 50–70 km, in mobile regions. Heat flow in the shield areas averages around $1\ \mu cal.cm^{-2}\ sec^{-1}$ which would correspond to a geothermal gradient of about $20°C\ km^{-1}$; in mobile regions the heat flow is perhaps twice this value. We might then expect temperatures

Landmark papers: Granite Petrogenesis

at a depth of 30 km to be of the order of 600°C or perhaps greater and if for any reason crust with a large granitic, and hence radioactive, component is thickened we might expect temperatures to approach 1000°C. Next we must compare these generalizations with laboratory study of the melting of appropriate materials.

Quite clearly the materials which must be studied are metamorphic rocks for geological evidence shows that before any rock reaches a temperature where significant fusion is likely to occur it will be a metamorphic rock in the amphibolite facies of metamorphism. This conclusion taken with data from rock mechanics, indicates that in this region, water, which so profoundly influences silicate melting phenomena, will not be present in vast amounts and what is present will be almost entirely confined to three minerals; muscovite, biotite and hornblende. Moreover, as these hydrates themselves play a part in the crystallization of the granite clan, we can suspect that their limits of stability may well overlap with the regions of fusion.

Numerous studies (see for example Winkler, 1967; Wyllie, 1971; Brown and Fyfe, 1970) have shown that at moderate depths, we would expect crustal fusion to commence at about 700°C and to increase in amount as the temperature is raised. Basically, the melting will occur when we have reactions of the type:

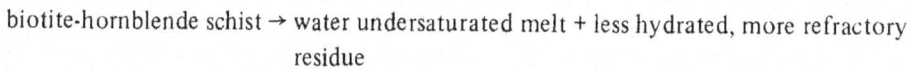

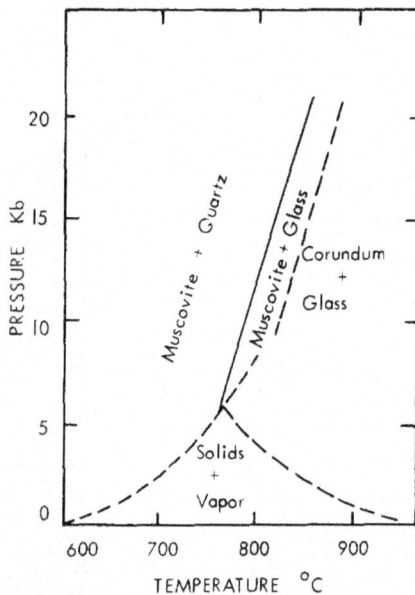

biotite-hornblende schist → water undersaturated melt + less hydrated, more refractory residue

Fig. 1. Melting diagram for the system muscovite-quartz after Segnit and Kennedy, 1961. Note that the melting curve is positive in slope.

We can illustrate the situation with respect to Segnit and Kennedy's (1961) study of the melting of muscovite, one of the first studies of this type of behaviour. In this case (Fig. 1) melting commences where the vapour-pressure curve of the hydrate intersects the curve for water-saturated melting. The melting reaction can be summarised by:

$$muscovite \rightarrow melt + corundum$$
$$or: \quad muscovite + quartz \rightarrow melt + sillimanite$$

In the crust we would expect a complicated series of melting reactions involving muscovite, biotite and eventually amphibole to produce melts as illustrated in Fig. 2. It is clear that how much melt is produced depends critically on how much hydrated phase is present and how much easily fusible material is present in the volume of rock so effected. Numerous analytical studies have indicated that the liquids formed are of the granite family and show trends with P and T as shown in Fig. 3 from Brown (1970).

Thus, the data so far discussed, make it seem highly probable that whenever a thick, low-density crust is formed, melting may occur leading to a new crustal configuration approaching the equilibrium gravitational configuration. Such processes must limit the temperatures in the lower crust (just as basalt production must limit upper-mantle temperatures) and must limit the thickness of this more radioactive part of the earth for the heating up time of any great thickness is rather short compared to related dynamic processes.

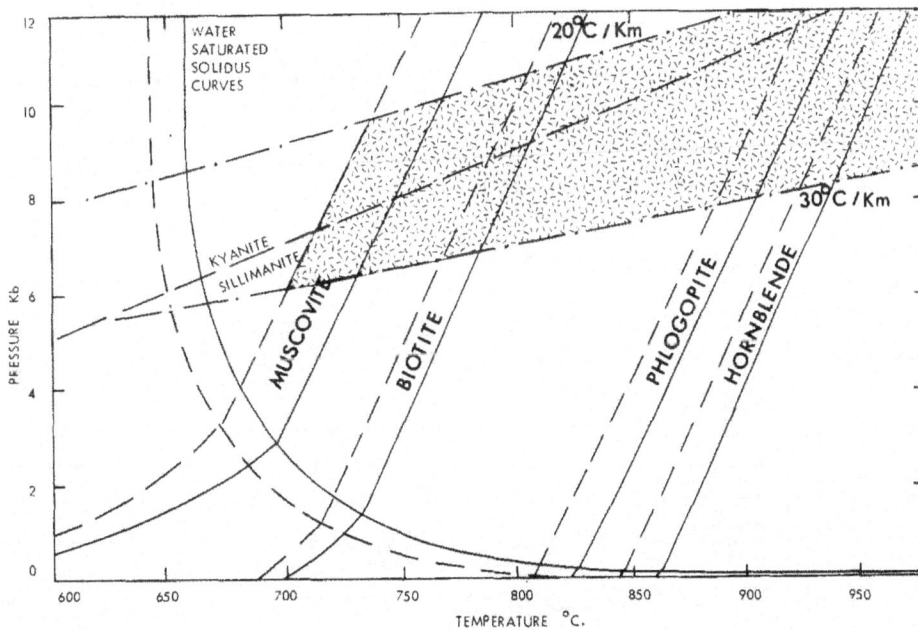

Fig. 2. Beginning of melting in granitic (dashed lines) and dioritic (solid lines) compositions (after Brown and Fyfe, 1970). The shaded area is a probable region for maximum magma generation under modern thermal gradients.

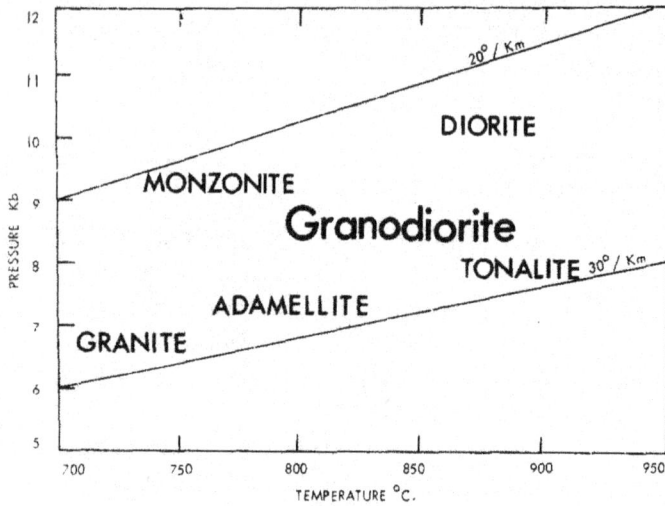

Fig. 3. Compositional trends of liquids formed by the melting process of Fig. 2 (after Brown, 1970).

It should be noted that on our model the melting curves on which granites are formed have dP/dT positive. Such liquids can rise without crystallizing and can assimilate country rocks to a limited extent. One might expect that only rather dry melts are likely to be extrusive (rhyolites and ignimbrites) for only these will cut through the negative water-saturated minimum curves near the surface (see Fig. 2). The water content of the melts will be greatest for those formed near muscovite breakdown but for higher temperature melts the water content may be quite low (say 2%) which agrees with the remarkable lack of large hydrothermal aureoles around many large, high-level batholiths.

GEOLOGICAL EVIDENCE FOR FUSION OF METAMORPHIC ROCKS

In crustal regions where metamorphic rocks of the highest grades are exposed, evidence for the sweating out of granitic components is common. It is rare to find rocks tending towards the pyroxene-bearing granulite facies which do not show such effects. Some workers term such rocks migmatites. Experimental work would indicate that basic amphibolites and granulites would be the natural residues of partial fusion and that these could be expected in the crustal roots beneath large batholiths. Seismic evidence (see Bateman and Eaton, 1967; Wyllie, 1971) beneath such regions is in accord with such an observation. A model based on partial fusion seems necessary to explain the formation of granulites (see Fyfe, 1972) and to account for their extremely low radioactivity (Heier, 1972).

Mineral assemblages in migmatite regions also reflect conditions where water is not present in excess; reaction relations between the phases present and water are common. This situation contrasts sharply with the common observations from assemblages in the lower grades of metamorphism. Assemblages illustrating such effects might include some of the minerals: biotite-garnet-magnetite-pyroxene-amphibole-feldspars-sillimanite etc. A simple way of rationalizing such mixtures is to assume that $P_{total} \simeq P_{melt} > P_{H_2O}$. The silicate phases are buffering P_{H_2O} by producing melts.

Limited data from oxygen isotope thermometry suggest that the temperatures of the highest grades of metamorphism approach melting temperatures (Epstein and Taylor, 1967) as do many observations based on the phase relations of metamorphic minerals. It should be noted that limited data available on self-diffusion coefficients for oxygen in silicates make it highly probable that such thermometry will not be reliable in very high temperature rocks. In summary, it appears difficult to escape the conclusion that the crust does melt.

THE ACCUMULATION OF CRUSTAL MELTS

Let us assume that partial melting commences in a crustal segment at depth. Clearly at all greater depths more melting may be expected to occur. The amount of melt which can form depends on the quantity of water available (the amount of biotite, etc., in the rocks) and the composition of the region as well as on the P–T regime. The most dramatic effect melting will have on the region will be to lower the overall viscosity. As lower crust has a viscosity of about 10^{22} poises and granitic liquids have viscosities in the order of 10^6 poises, the melting region will have much lowered viscosities. The simple Einstein relation between the amount of liquid and viscosity, while being clearly too simple for the present case, may provide a guide (see Scarfe, 1972). The viscosity in this region will have a great influence on determining if, and at what rate, the liquid can accumulate at the melting interface.

Imagine that at a depth of 30–40 km, average crust (50% granite, 50% basalt) is melting. The latent heat of fusion of granite is about 50 cal.gm^{-1}. Heat production in granites is about $2.6 \cdot 10^{-13}$ cal.g^{-1}sec^{-1} and for basalt, $3.8 \cdot 10^{-14}$ cal.g^{-1}sec^{-1}. If we then assume the average heat production is about $1.4 \cdot 10^{-13}$ cal.g^{-1} sec^{-1}, and we assume that half the material melts, we will need 25 cal.g^{-1}. This process will absorb all the heat production for about 6 million years. Hence the time constant for producing plutons is in the order of 10^7 years, unless extra heat is added from below (a possibility not to be dismissed).

Next, consider a drop of liquid 5 cm in radius which has a density around 2.4 g cm^{-3} moving upwards in a zone of partial fusion. Let us assume that for the liquid to collect we must consider motions in the order of 1 km. What viscosity will allow such motions in times of 10^6 years, appropriate to the generation of the melts? From Stoke's law, such motions would be possible if the viscosity of the system was less than 10^{13} poises, and such a viscosity in a rock 10% molten is highly probable (Scarfe, 1972). It thus seems that there is plenty of time for liquid to move (and residue to sink) so that a rather clean molten layer can accumulate beneath the melting interface.

MOTION OF THE MELT

The melting process generates what is known as Taylor instability where a low-density layer is formed beneath a layer of higher density. Examples of the motions of such materials have been considered by Elsasser (1963) and are beautifully illustrated in Ramberg (1967). In essence, a wavy surface will develop on the interface, and certain wavelengths will tend to grow more rapidly than others. Analysis of the problem (Ramberg, 1967) indicates that these preferred wavelengths are a function of viscosity contrasts, and the thickness parameters of the layers.

For a magma, we must also note that small intrusions of the liquid layer into the overburden will be cooled rapidly and lose the density contrast necessary for motion. Thus thermal parameters set limits on the size of "bubbles" which may rise by flow into the crust. Some time ago Grout (1945) and Larsen (1945) considered such problems in relation to rising plutons.

Elsasser (1963) simplified some of the dynamic equations in his discussion of core formation by molten iron drops falling into the core. What is striking about Elsasser's analysis is the long wavelengths that would be predicted for earth phenomena of this type; these arising from the times constants and viscosities of the solid parts of the earth. In particular I would stress his equation limited to the case of a very large viscosity contrast between liquid and solid:

$$n = g/2K\eta \cdot (\rho_2 - \rho_1)/\rho_1$$

relating the time for bulge growth ($n^{-1} \alpha t$), the density contrast, the viscosity of the overburden, and the wave number K. This equation indicates that magmatic intrusions will have very long wavelengths. The same conclusions could be drawn from Ramberg's analysis.

Thus for the case of lower crust, assuming a viscosity of 10^{22} poises, and allowing a time constant of 10^6 years, we would predict a wavelength pattern of 30 km for the spacing of plutons. Clearly it is of the correct order of magnitude! Further, the equation makes it clear that as processes of magma generation become slower, λ becomes smaller, and as η becomes larger, λ increases in proportion. These results could be of immense significance in considering the differences between basalt and granite accumulation.

I think these relations set the stage for our argument. If crustal fusion is occurring we would anticipate the formation of a low-viscosity basal layer to the crust. Magma and granulite-amphibolite residue will tend to separate forming a layer of low-density silicate liquid with a more basic refractory basement. This layer will form bulges on a wavelength scale in the order of tens of kilometres invading the upper crust.

Perhaps we should next turn to geology itself and look at the structure of batholiths as shown in more recent mobile belts such as the Andean or Sierra Nevada batholiths occurring in regions of thickened crust. The volume of granitic material in the Sierra Nevada batholith is about $5 \cdot 10^5$ km^3 and the number of separate plutons is large — several hundred The volume per pluton, while variable, is in the order of $2 \cdot 10^3$ km^3 which would corre-

spond to an average bubble radius in the order of 8 km. The phenomenon is beautifully
displayed on the Saudi Arabian plateau where circular plutons invade crust in the greenschist
facies (Fig. 4). A very average diameter of plutons is around 10 km.

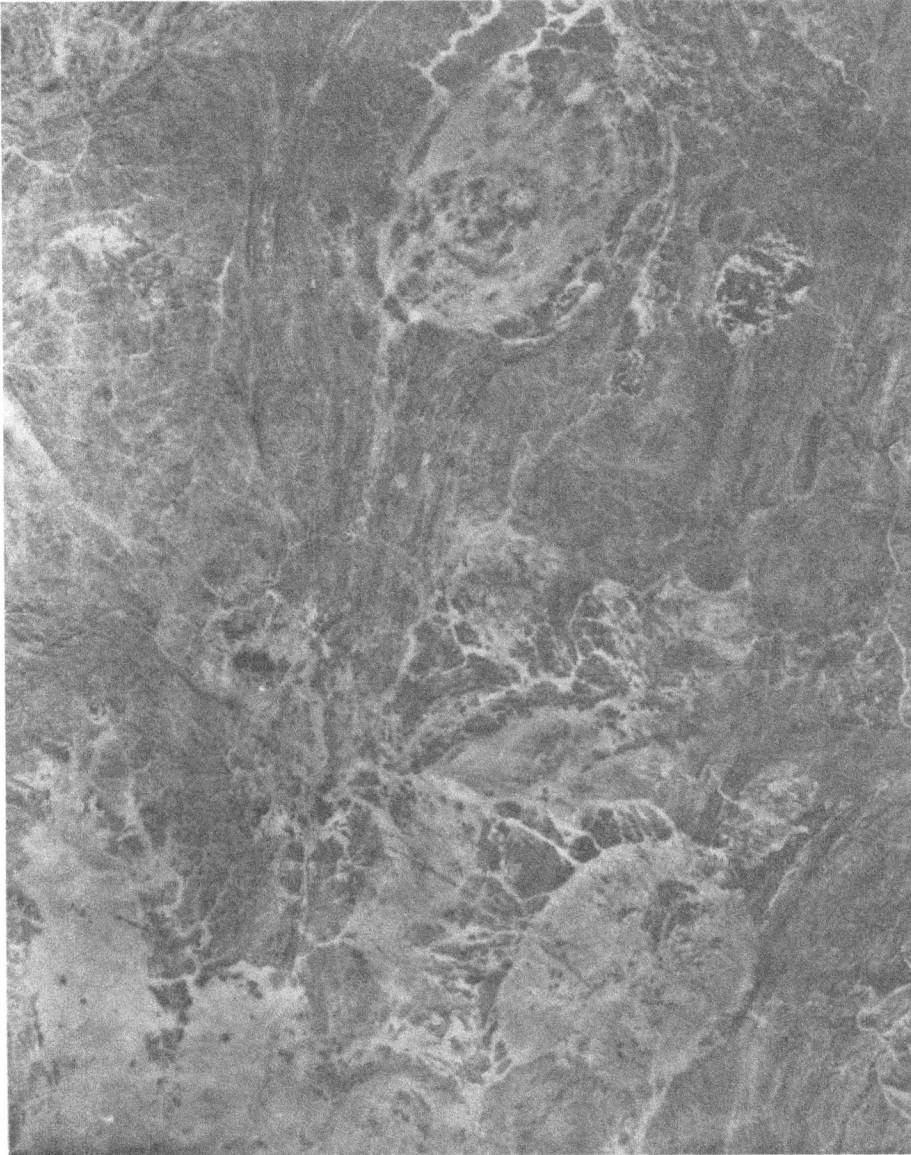

Fig. 4. Air photograph of plutons in the Saudi Arabian crust. The two large circular masses have a diameter of about 10 km. (Reproduced by courtesy of the Directorate of Mineral resources, Saudi Arabia.)

If large masses of granite break off bulges at the fusion interface, it is not unreasonable that their motion might be Stokesian to a first approximation. On account of the large viscosity of the medium and the slow velocities involved, the Reynolds number must be very small. For example, a bubble of 5 km radius rising 10 km in 10^6 years (a velocity required to prevent cooling the mass to complete solidification) would be consistent with a viscosity of the lower crust of 10^{21} poises. In this way, as I pointed out in 1971, we could use plutons to measure the viscosity of the lower crust. But the mechanics of such motion must be complicated for a rising pluton must heat up and generate a gaseous envelope in the marginal rocks. Oxygen-isotope studies (see Turi and Taylor, 1971) should soon elucidate such features around plutons. The bubble pattern of granite intrusion seems to fit many examples over the past 10^9 years or so. Obviously, any such model is far too simple for, particularly at higher levels, crustal structure must influence the form of intrusion and extrusion.

ARCHAEAN GRANITES

When we compare the present distribution of granitic—basaltic—andesitic igneous activity and related metamorphic belts with the Archaean, certain differences seem to occur. I would like to use the Rhodesian craton to illustrate such a different pattern. Ages on this craton exceed $3 \cdot 10^9$ years. MacGregor (1951) first drew attention to the spacing of huge granitic domes and schist belts. Unlike the present pattern the volcanic schist-belt separation is much smaller (say, 100 km) and the belts tend to form closed elipses. Between these belts are huge, composite granite domes. In the Chindamora batholith the outer zone appears to be tonalitic, this dome is then spread by later granodioritic intrusions and later spread again by more granitic types. Also plutonism and volcanism seem closely associated; the crust was thin. Further, deep burial metamorphic rocks are lacking. The bubble picture does not seem to apply; at least not in such simple fashion.

In thinking of Archaean magmatism one must consider substantially greater rates of heat production and hence faster melting at much shallower levels. This is in accord with the large areas of low-pressure cordierite-sillimanite granulites in the Archaean. Such rocks would not be expected on any large scale under present geothermal gradients. One might also consider that the upper mantle could be melting at higher levels, and if this was the case, it is perhaps not so surprising that ultrabasic liquids (komatiites) pass to the surface without extensive modification (Viljoen and Viljoen, 1970). If one considers such a situation in relation to Taylor instability in the crust, if molten layers are thinner, one would anticipate a smaller wavelength pattern to develop (Ramberg, 1967) but shorter time constants of magma production would lead to a larger pattern. If heat production was perhaps 3–4 times the present we would expect bulge spacing to be proportionally larger. Further, if the crust was melting at, say, 10–15 km, it is rather difficult to imagine separate large bubbles breaking off as can occur in a 50-km crust and a model based on doming on a long wavelength pattern seems appropriate.

I think, too, that a second point must be considered in thinking of Archaean magmatism.

If the more acidic crust is melting at higher levels on account of greater radioactivity, mafic liquids coming from below will tend not to penetrate shallow, low-viscosity crust. Injection and flow under such a crust could be a widespread phenomenon leading to underplating by less radioactive material. Today, we use andesitic volcanism and blueschist metamorphism to indicate belts of plate subduction of down-flowing convection cells. Do the Archaean schist belts represent the same phenomena but on a much smaller wavelength pattern? The blueschists are not surprisingly lacking; but do the andesites represent zones of crust–mantle mixing? If so it could be this, plus an underplating mechanism that could control the behaviour of the base of the acidic crust which might become partially molten over its entire dimensions. The intrusive patterns of the melts could then be a result simply of the thermal motions in the upper mantle with the cells rising under the dome centres.

MANTLE WAVELENGTH PATTERNS

I would like to return to the Elsasser–Ramberg equations and their implications in terms of melting in the upper mantle and particularly in connection with the suggestion that patterns in the Archaean had a shorter wavelength. Changes in the bulge pattern (ocean ridges) of basic liquid collection in the upper mantle could result from changes in the viscosity of the region and changes in the time constants of liquid production. Because of the enormous increase in pressure with depth one might expect viscosity to increase substantially. This would be expected if thermal gradients were low and if there were no drastic changes in the state of matter. The question does not seem to be settled (Verhoogen et al., 1970) but viscosity increases of a few powers of ten do not seem to be excluded. From the Elsasser equation any such changes alter the wavelength by the same magnitude. From Ramberg's data, it is clear that if the low-velocity layer represents a small degree of melting and if this layer has thicknesses of the order of 100 km (Wyllie, 1971) then the wavelength of bulging will be in the order of a few thousand kilometres. Any significant analysis of this point requires more exact knowledge of viscosity in this region and the time constants of melt production. A further point which should be stressed is that because of the boundary conditions in a sphere, the number of bulges would be quantized and if thermal gradients or depth of melting or thickness of the melting zone changed with time, the number of cells would also be expected to change.

Such quantization has been discussed by Runcorn (1965) based on the very suggestive data from age distributions of peak magmatic–metamorphic–tectonic events. Runcorn discussed this problem on the basis of a model which assumes changing dimensions of the core and he concluded that cell size was diminishing with time. He then used this conclusion to explain continental distribution with large cells sweeping the crust together at an early stage. But there are other ways of explaining this phenomenon as suggested by Elsasser. On his model of assymetric and rapid core formation, the early continental distribution would reflect asymmetry of core formation. Clearly, key information in the elucidation of such arguments is to be found in the structures of ancient cratonic regions; I find it difficult to escape evidence for small cells in the upper mantle.

A further point certainly needs amplification. There is still considerable doubt about whether or not the amount of continental crust has increased through time (see Verhoogen et al., 1970, pp. 664–666). Ocean-floor spreading processes tend to sweep up the evidence. But if the primitive earth was hotter it is difficult to see how there could have been more hold up of more fusible and more radioactive material in a hotter, less viscous, upper mantle (Birch, 1965). If this is reasonable, then the thinner Archaean crust could have also been more widespread (3 ×) than at present. It is quite remarkable that present crustal thickness and heat production and melting behaviour fit together quite reasonably. Igneous phenomena in the crust constitute a spreading process.

SUMMARY

In conclusion I would stress the following points:

(1) Fusion processes must limit the thickness of more acid crust. Such crust must have been much thinner and perhaps more extensive in area in the Archaean.

(2) Perturbations leading to thickened crust and large-scale batholith production at the present time should only be possible in a tectonically thickened crust, the thickening being driven by upper-mantle convective processes. Batholith generation will be the expected relaxation phenomenon. Rising magma from a subduction zone may provide energy for accelerated crustal fusion.

(3) The intrusive patterns of granites seem rational in terms of the mathematics of Taylor instability and Stokesian rise of plutons.

(4) The same theories may be critical in explaining the spacing of convection cells and sites of basalt production and the variations with time.

(5) Archaean structural patterns are considered to indicate smaller size of convection cells in the past.

REFERENCES

Bateman, P.C. and Eaton, J.B., 1967. Sierra Nevada batholith. *Science*, 158: 1407–1417.

Birch, F., 1965. Speculations on the Earth's thermal history. *Geol. Soc. Am. Bull.*, 76: 133–154.

Brown, G.C., 1970. *Experimental Studies on Granites and Related Rocks Under Varying Conditions of Water Vapour Pressure and Hydrated Mineral Content.* Dissertation, Manchester University, 221 pp.

Brown, G.C. and Fyfe, W.S., 1970. The production of granitic melts during ultrametamorphism. *Contrib. Mineral. Petrol.*, 28: 310–318.

Elsasser, W.M., 1963. Early history of the Earth. In: J. Geiss and E.D. Goldberg (Editors), *Earth Science and Meteoritics.* North-Holland, Amsterdam, pp. 1–29.

Epstein, S. and Taylor, H.P., 1967. Variation of $^{18}O/^{16}O$ in minerals and rocks. In: P.H. Abelson (Editor), *Researches in Geochemistry.* Wiley, New York, N.Y., 29–62.

Fyfe, W.S., 1971. Some thoughts on granitic magmas. In: G. Newall and N. Rast (Editors), *Mechanism of Igneous Intrusion.* Gallery Press, Liverpool, pp. 201–216.

Fyfe, W.S., 1972. The granulite facies; partial melting and the Archaean crust. *Philos. Trans. R. Soc.* (in press).

Grout, F.F., 1945. Scale models of structures related to batholiths. *Am. J. Sci.*, 243A: 260–284.

Heier, K.S., 1972. Geochemistry of granulite facies rocks and problems of their origin. *Philos. Trans. R. Soc.* (in press).

Larsen, E.S., 1945. Time required for the crystallization of the great batholith of Southern and Lower California. *Am. J. Sci.*, 243A: 399–416.

MacGregor, A.M., 1951. Some milestones in the Precambrian of South Rhodesia. *Trans. Geol. Soc. S. Afr.*, 54 (27).

Ramberg, H., 1967. *Gravity, Deformation and the Earth's Crust.* Academic Press, London, 214 pp.

Runcorn, S.K., 1965. Changes in the convection pattern in the Earth's mantle and continental drift: evidence for a cold origin of the earth. *Philos. Trans. R. Soc.*, 258: 228–251.

Scarfe, C.M., 1972. *Viscosity and Related Properties of Basic Magmas.* Dissertation, Leeds University, 220 pp.

Segnit, R.E. and Kennedy, G.C., 1961. Reactions and melting relations in the system muscovite-quartz at high pressures. *Am. J. Sci.*, 259: 280–287.

Turi, B. and Taylor, H.P., 1971. An oxygen and hydrogen isotope study of a granodiorite pluton from the Southern California batholith. *Geochim. Cosmochim. Acta*, 35: 383–406.

Verhoogen, J., Turner, F.J., Weiss, L., Wahrhaftig, C. and Fyfe, W.S., 1970. *The Earth.* Holt, Rinehart and Winston, New York, N.Y., 748 pp.

Viljoen, M.J. and Viljoen, R.P., 1970. Archaean vulcanicity and continental evolution in the Barberton Region, Transvaal. In: T.N. Clifford and I.G. Gass (Editors), *African Magmatism and Tectonics.* Oliver and Boyd, Edinburgh, pp. 27–50.

Winkler, H.G.F., 1967. *Petrogenesis of Metamorphic Rocks.* Springer-Verlag, Berlin, 2nd. ed., 237 pp.

Wyllie, P.J., 1971. *The Dynamic Earth.* Wiley, New York, N.Y., 416 pp.

CHAPTER 5

DePaolo, D.J. (1981) Neodymium isotopes in the Colorado Front Range and crust–mantle evolution in the Proterozoic. *Nature*, **291**, 193–196.

This paper does not deal directly with granitic rocks, but its impact on studies of granite petrogenesis has, nevertheless, been tremendous. DePaolo and Wasserburg (1976a,b) had shown that, in a $^{143}Nd/^{144}Nd$ *vs.* time diagram of the evolution line of a given rock, the intersection with the evolution line for average chondrite, provides a good estimate of the time at which the rock material was first segregated from the mantle. This method assumed that the mantle is a reservoir of uniform chondritic composition (CHUR). The age thus calculated was formerly called "crust-formation age" or "crust-extraction age", and is currently known as "Nd chondritic model age", T_{CHUR}.

The CHUR model was originally built upon data from Archaean rocks, and it was quickly found to yield unreliable results for younger materials. This is because the Nd composition of post-Archaean mantle is severely depleted in *LREE* and has higher Sm/Nd than the chondrites. This was revealed by the fact that relatively young mantle-derived materials, such as mid-ocean basalts and island arc rocks, departed by about +7 to +12 εNd from CHUR, where $εNd_{(t)} = [(^{143}Nd/^{144}Nd_{rock}/^{143}Nd/^{144}Nd_{CHUR})_{(t)} - 1] \times 10^4$. It was therefore necessary to build a model for the evolution of Nd isotopes in the mantle (as an alternative to CHUR) capable of yielding realistic estimates of the "crust-formation age" of post-Archaean rocks. However, this task was seriously hampered by the scarcity of Nd isotope data for rocks with ages intermediate between Archaean and recent. The gap was partially filled by DePaolo in this paper about the Proterozoic rocks from the Colorado Front Range.

De Paolo found that six samples of basement gneisses, geographically remote from each other, fitted a Sm-Nd isochron at 1800 ± 90 Ma, and that their $εNd_{(1800Ma)}$ were remarkably uniform at around +3.7. Therefore, the author assumed that the gneisses were derived directly from the mantle 1800 My ago and used their $εNd_{(1800Ma)}$ plus the $εNd_{(0Ma)}$ value of recent island-arc rocks, to calculate a mantle magma source vector that approaches the Nd isotopic evolution of the mantle source of continental crust over the past 2 Gy. This vector was calculated from just two points, and for reasons not made clear in the paper, the vector was calculated as a quadratic curve: $εNd(T) = 0.25T^2 - 3T + 8.5$.

The Nd model ages calculated from this curve are called depleted mantle model ages, T_{DM}, and have proved to yield reliable estimates of the crust-formation age for many rocks from Proterozoic terranes. One the of the mostly successful approaches has been to use the Nd compositions of crustal granites, because the T_{DM} of each pluton can be determined easily with a few analyses and may represent the average of a large volume of their protolith terrane. This has permitted the age structure of large areas of the crust to be mapped with minimum effort, as shown by Bennet and DePaolo (1987) and DePaolo (1988), and increased the interest in granites as markers of crustal processes. In any case, it is important to remember that Nd model ages can be interpreted as the times of crust-mantle segregation only if supported by other geological and geochronological information. Otherwise they will lead to incorrect interpretations of crustal history and the nature of granite magma sources (Arndt and Goldstein, 1987).

Nature Vol. 291   21 May 1981

ARTICLES

Neodymium isotopes in the Colorado Front Range and crust–mantle evolution in the Proterozoic

Donald J. DePaolo

Department of Earth and Space Sciences, University of California, Los Angeles, California 90024, USA

Initial $^{143}Nd/^{144}Nd$ determined for major rock types of the basement underlying the Rocky Mountains in Colorado indicate that this segment of continental crust was formed from a homogeneous and previously depleted source material in the upper mantle 1,800 Myr ago and contains no identifiable component of older crust. Subsequent magmatic events at 1,670, 1,400 and 1,000 Myr represent mainly intracrustal differentiation with small or no further additions to the crust from the mantle. These results imply that a major increase in the mass of the North American continent occurred over a narrow time interval 1,800 Myr ago. From the isotopic data an empirical model is constructed that may provide accurate crust-formation ages for rocks of all ages. Preliminary results suggest that existing estimates of crustal age distributions may require substantial revision.

For financial reasons, it has not been possible to reproduce this paper in full here. Readers are directed instead to the original paper in the journal.

CHAPTER 6

England, P.C. and Thompson, A.B. (1986) Some thermal and tectonic models for crustal melting in continental collision zones. *Geological Society Special Publication*, **19**, 83–94.

After the work of Fyfe (1973a,b), the latter included here as a landmark paper, Wyllie *et al.* (1976) produced an excellent summary of what was then known about where granitic magmas came from, and what their compositions should be, especially their H_2O contents. There remained the questions of how much granitic liquid could be produced by melting of the continental crust, in what tectonic settings this could occur and the degree to which extracrustal heat sources would be necessary for such metamorphism and melting. The choice of England and Thompson (1986) as a landmark is a simple and logical one, because it was these authors who made the first serious attempt to forge the vital link between the petrological and phase equilibrium constraints on granite magma genesis, on one hand, and crustal heat production, heat flow and tectonic regimes on the other.

England and Thompson focused on continental collision regimes, tracking the likely thermal consequences of crustal thickening and subsequent extensional thinning. They sought to determine the heat flux that would be necessary to initiate crustal melting, and so to answer the question of whether heat from the mantle needs to be involved in the formation of granitic magmas, in these tectonic settings. Part of their reasoning involved the idea that large amounts of granitic magma are generated in continental collision settings. Indeed, some leucogranitic magmas are formed this way, but by far the majority of granitic magmas are found in other settings. Nevertheless, the approaches presented in this paper remain valid, and they have formed the basis for numerous subsequent models.

England and Thompson developed thermal models for thickening of normal continental crust (i.e. crust with average heat production) by thrust faulting and by homogeneous strain (both rapid and protracted). They found that, if H_2O-rich fluid were in excess, the lower parts of these thickening orogens would partially melt, especially if the crust were dominated by quartzofeldspathic and metasedimentary rocks. Influenced particularly by the ideas of Chappell (1984), England and Thompson thought that S-type granitic magmas were likely to have been low-

temperature, H_2O-saturated melts of metasediments, so they concluded that large volumes of S-type magma might be produced in this way. The obvious difficulty with this idea lies in the requirement for large volumes of free aqueous fluid to be available. In restricted instances, this may indeed occur, at least initially, but it is unlikely to be generally the case (see e.g. Clemens and Watkins, 2001). H_2O released by dehydration of metasediments, etc., is likely to be strongly channelled and chemically unavailable for melting reactions.

England and Thompson also paid attention to the case of fluid-absent partial melting through the breakdown of micas and amphiboles at higher temperatures than the wet solidus. They recognized that higher than average heat flux, in a tectonically young area would be necessary to initiate fluid-absent partial melting of even thick crust, and then only in the deepest parts of the orogen. We now appreciate that this severely limits the production of granitic magma in collisional settings, as the majority of granitic magmas were clearly initially H_2O-undersaturated. Nevertheless England and Thompson noted that metasediment-derived magmas (S-type granites) would greatly predominate in collisional settings. This is because the higher temperatures required to partially melt the sources of I-type magmas would not commonly be realized. Even in the fluid-present cases, they predicted that melt volumes from metabasaltic sources would be low, due to the intrinsic infertility of such rock types. On the role of mantle heat, advected by the emplacement of basaltic magma, England and Thompson emphasize caution, but they clearly did not think that this extra heat was necessary. Later work by Thompson showed that this was a somewhat mistaken view (e.g. Thompson and Connolly, 1995).

As the catalyst to much further research into the formation of granitic magmas, for their novel thermal modelling approach, and for the fact that their evolving geotherms are still used in the teaching of metamorphism and collision tectonics, we believe that England and Thompson should be recognized as a significant landmark in the evolution of scientific thought in these areas.

Some thermal and tectonic models for crustal melting in continental collision zones

P. C. England & A. Thompson

SUMMARY: Calculated geotherms and the pressure–temperature–time (*PTt*) paths followed by rocks during continental thickening episodes are interpreted with respect to the volumes of crustal melt that may be formed during orogenesis in the absence of heat transfer by mantle-derived melts. Particular attention is paid to a tectonic history that may characterize wider orogenic belts, such as are represented most obviously at present by Tibet. This comprises a period of crustal thickening, followed by an interval during which the crust is thinned by extensional strain, rather than by erosion. The amount of crustal melt produced depends strongly on the amount of water (free, and in hydrated minerals) contained in the lower crust. However, we may expect several (1–5) km^3 of crustal melt per km^2 of orogen if a crust of around average continental surface heat flux (60–70 mW m^{-2}) is thickened by a factor of two. For the lower surface heat flux, partial melting of a sedimentary source would produce predominantly S-type granites and, with slightly higher geotherms, doubling of crustal thickness can lead to partial melting of amphibolites to give I-type granitic activity and calc-alkaline volcanism.

The purpose of this paper is to discuss the thermal and tectonic regimes that are responsible for the generation of high-grade metamorphism and igneous activity in continental collision zones. It is commonly remarked (e.g. Wyllie 1977) that the *PT* conditions for high-grade regional metamorphism and crustal melting lie several hundred degrees to the high side of 'normal' continental geotherms. However, England & Richardson (1977) pointed out the pitfalls involved in relating *PT* conditions recorded in transient thermal regimes to steady-state geotherms; in this paper we discuss the transient geotherms that may be produced in a variety of tectonic settings and investigate the kinds of crustal melting that may result from them. Although quantitative data on heat transfer processes above subducted slabs are scarce, it is generally believed that the supply of volatiles and heat from a subduction zone is ample to produce crustal and mantle melting above such a system. In this paper we try to address the more interesting problem of placing *lower* bounds on the heat flux required to produce crustal melts in some tectonic situations; in particular we investigate the degree to which abnormal heat transfer from the mantle needs to be invoked to explain crustal anatexis.

Geological models

The thermal development of orogenic belts has usually been treated in terms of a conceptual model based on the supposed behaviour of Alpine-type belts: the thickening phase of orogenesis is assumed to be a thrusting event that happens so rapidly that negligible thermal relaxation takes place, while the erosional phase occurs over a time interval that is long enough to permit significant thermal relaxation and metamorphism (e.g. Oxburgh & Turcotte 1974; Bickle *et al.* 1975; Richardson & Powell 1976; England & Richardson 1977; England 1978; England & Thompson 1984).

This appears to be a reasonable approximation for narrow belts where the collision is of relatively short duration (<10 Ma) and where the exhumation of rocks is accomplished primarily by erosion. However, there are many mountain belts, most noticeably the Cordilleran belts of N and S America and the India-Asia collision zone, in which the deformation is taking place on a larger scale and has continued over several tens of millions of years. Locally such deformation may occur as several discrete episodes of folding or thrusting, but the quantitative treatment of such episodes requires a detailed knowledge of their timing, which is generally lacking. We shall investigate the simpler case of homogeneous thickening over a chosen time interval.

The second way in which the larger collision zones differ in behaviour from the narrow Alpine-type zones is that significant crustal thinning may occur without appreciable erosion. The active faulting of the elevated plateaux of Tibet (e.g. Molnar & Tapponnier 1975) and the Andes (Dalmayrac & Molnar 1981) shows that they are extending under their own weight despite their being close to convergent boundaries (see also England & McKenzie 1982, 1983). In these cases the faulting began comparatively recently and has

From COWARD, M. P. & RIES, A. C. (eds), 1986, *Collision Tectonics*, Geological Society Special Publication No. 19, pp. 83–94.

83

84 *P. C. England & A. Thompson*

accomplished little total extension as yet. We would expect, though, that when the compressive stresses generated by convergence are released, the extension of these areas under their own weight will be more rapid.

The duration of extension in the Basin and Range province of N America is considerably shorter than the usual estimate of ~100 Ma for the reduction of a mountain range by erosion; in addition the geometry of crustal thinning differs radically from that of the erosional process (see Fig. 1). The Basin and Range has undergone at least 35% extension since the Oligocene (Proffett 1971) and perhaps as much as 60–100% extension in its central portion (Wernicke *et al.* 1982). As the present crustal thickness there is between 25 and 35 km (Smith 1978) the region must orig-

inally have resembled closely in elevation and crustal thickness those plateaux which at the present day are just beginning to extend.

For these reasons we consider two basic types of continental thickening processes (Fig. 1). The first involves crustal thickening by a single thrust and the second involves homogeneous thickening of the entire lithosphere. These result in different thermal conditions immediately following thickening; in the first case the conductive gradient at the base of the crust is unaltered, whilst in the second case it is diminished in proportion to the amount of thickening. We should, therefore, expect different rates of temperature increase in the crust for these cases. In line with the studies mentioned above, we assume that the thrusting episode takes place instantaneously (Fig. 1A)

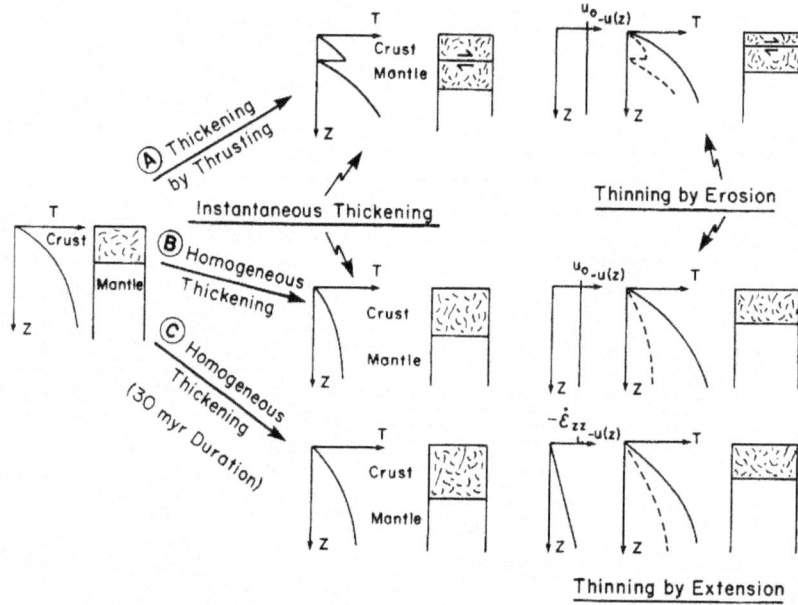

FIG. 1. Sketches of the geometries of crustal thickening and thinning processes discussed in this paper. The pre-thickening conditions in each case are the same (Equation 6) and are shown on the left. (A–C) These show the immediately post-thickening states (next two columns); the velocities, temperatures and crustal thicknesses at subsequent times are shown in the righthand three columns. (A) Thickening is accomplished instantaneously by a single thrust that emplaces 35 km of continental crust over another 35 km of crust and its subjacent lithosphere without perturbing their temperatures. This burden is then removed by erosion over the next 120 Ma, at a constant rate u_o. Results in Fig. 2 (a, b). (B) As (A), except thickening is accomplished instantaneously by homogeneous strain. Results in Fig. 2 (c, d). (C) Thickening is by homogeneous strain over a 30 Ma time interval and is followed by crustal thinning which occurs by homogeneous extensional strain at a constant rate. This takes place either at a constant rate over 60 Ma (results in Fig. 3c, d) or at a constant rate over 30 Ma, after a 30 Ma pause (Figs 3a,b and 4). Note that in (A) and (B), the upward velocity of the rocks is constant during the erosional phase, whereas in (C), both the extensional and the compressional histories require velocities that are depth-dependent (Equation 3).

and consider two other cases where homogeneous thickening takes place either instantaneously or over an interval of 30 Ma (Fig. 1B, C).

We also consider two modes of crustal thinning, each of which returns the crust to its original thickness. The first is erosion at a constant rate over 120 Ma for each of the cases where crustal thickening was instantaneous (Fig. 1A, B). In the second mode, crustal thinning takes place purely by extension and we allow this to take place over intervals of 30 or 60 Ma (Fig. 1C).

For the cases in which changes in crustal thickness are accomplished by distributed strain, rather than by thrusting (thickening) or erosion (thinning), the change in crustal thickness is assumed to take place by homogeneous strain throughout the lithosphere (see Fig. 1B, C), so that, for example, a doubling of crustal thickness is accompanied by a doubling in thickness of the entire lithosphere. Homogeneous strain is commonly assumed for extensional deformation during sedimentary basin formation (e.g. McKenzie 1978; Christie & Sclater 1980). It is clear that the brittle portion of the lithosphere accommodates such strain by failing in a discontinuous manner (faulting) but the assumption of homogeneous strain is made because any such faulting may only be treated in an *ad hoc* fashion in this kind of study. Because the seismogenic layer in active terranes usually extends less than about 20 km from the surface (Chen & Molnar 1983) it is assumed that a more continuous form of deformation operates below this level. As is discussed later (see Results), it is simple to extend qualitatively our results based on this assumption to take account of the rapid unroofing that would result from the kind of large-scale inhomogeneous extension suggested by Wernicke (1981).

Mathematical model

Heat transfer

We consider the transport of heat only by conduction and by the movement of solid rock with respect to our coordinate frame. We thus explicitly exclude the advection of heat by the melts that may be generated and deflection towards the solidi of the *PTt* paths owing to the latent heat of fusion. We are concerned primarily to determine the conditions under which large volumes of the lower crust can be heated above their solidi, rather than with the details of subsequent *PTt* paths. The discus-

sions (see Results and Discussion) are made with this uncertainty in mind. The coordinate frame is fixed with the z-direction vertical and the $x-y$ plane on the land surface, so the advection of heat by rock movement will accompany erosion or deformation of the rock. We assume that horizontal thermal gradients are negligible compared with vertical ones and thus we need to solve the one-dimensional advection-diffusion equation:

$$\frac{\partial T}{\partial t} = -u(z)\frac{\partial T}{\partial z} + \kappa\frac{\partial^2 T}{\partial z^2} + \frac{A(z)}{\rho c}, \qquad (1)$$

(e.g. Bird *et al.* 1960; p. 315) where T is temperature and z is the depth coordinate (positive downwards). κ is thermal diffusivity, ρ is density, c is specific heat, all assumed constant; A is the rate of internal heat generation and u is the velocity of the medium with respect to the land surface. When erosion operates $u(z)$ is a constant, u_0, equal to the negative of the erosion rate at the land surface. In this paper we concentrate on the influence of compressional and extensional strain on the thermal regimes of orogenic belts; we treat these as homogeneous strains in the vertical direction and define a vertical strain rate, taken to be independent of depth

$$\dot{\varepsilon}_{zz} = \frac{du}{dz} \qquad (2)$$

We fix the vertical velocity at the land surface to be zero, so that

$$u(z) = \dot{\varepsilon}_{zz}z, \qquad (3)$$

and $u(z)$ is positive for a thickening strain rate (see Fig. 1).

We may reduce the number of variables we need consider by making Equation 1 non-dimensional using: $z' = z/l$; $t' = \kappa t/l^2$; $T' = T/T_0$; $A' = Al^2/\kappa\rho c T_0$ where l and T_0 are a length and a temperature to be chosen shortly.

This gives:

$$\frac{\partial T'}{\partial t'} = -u'(z)\frac{\partial T'}{\partial z'} + \frac{\partial^2 T'}{\partial z'^2} + A'(z'), \qquad (4)$$

where $u'(z)$, the non-dimensional velocity, is either:

$$u'(z') = \frac{u_0 l}{\kappa}, \qquad (5a)$$

for the case of movement of rock by erosion, or

$$u'(z') = \frac{\dot{\varepsilon}_{zz}l^2}{\kappa}z', \qquad (5b)$$

86   *P. C. England & A. Thompson*

in the case of rock motion by compression or extension (Equation 3).

Geometry

The natural length-scale for this problem is the thickness of the lithosphere and we choose *l* to be 125 km, assumed to be the *undisturbed* lithosphere thickness. We need also to define an initial condition for this problem and we assume for this that in all cases the continental lithosphere lies on a steady-state geotherm immediately before thickening occurs:

$$T(z) = A_0 z \frac{(D - z/2)}{K} + \frac{Q \cdot z}{K} \qquad z < D$$

$$= \frac{A_0 D^2}{2K} + \frac{Q \cdot z}{K} \qquad D \leq z \leq l. \qquad (6)$$

This is the steady-state regime for a medium with heat production at the rate A_0 in the depth interval $0 \leq z \leq D$ and with a constant heat flux, $Q \cdot$, below the depth D. K is the thermal conductivity of the medium. This suggests a natural choice for T_0:

$$T_0 = Q \cdot l / K. \qquad (7)$$

Dealing with a lithosphere that changes rapidly in thickness involves a problem with the lower thermal boundary condition. Clearly neither a constant temperature nor a constant flux at a fixed depth (the usual choices in this kind of study) is an appropriate condition. Recognizing that the thermal lithosphere is defined as the region in which heat is predominantly transferred by conduction, and is underlain by the asthenosphere in which heat is predominantly transferred by convection, we have chosen to apply a constant flux at a horizon fixed to the rocks at the original base of the conducting lithosphere. This level migrates as the lithosphere thickens or thins.

Equation 4 is solved using standard finite difference techniques on a mesh whose spacing is 1 km; the interval for time stepping using an implicit (Crank–Nicolson) scheme is 0.5 Ma.

Parameter values

Once the geometry of crustal thickening has been specified, the principal variables that govern the thermal history of the belt are the supply of heat to the lithosphere, from radiogenic heat sources within the crust and from the upper mantle, the degree of thickening that occurs and the timescale over which the orogenic episode take place.

The average continental surface heat flux (Sclater *et al.* 1980) is about 60 mW m^{-2} and we choose this observation to fix the thermal conditions of the continental lithosphere. The parameters chosen for the pre-thickening geotherms (Equation 6) are listed in Table 1 and the geotherm appears as the curves

TABLE 1. *Parameters used in the calculations of Figs 2–4*

Conductivity (*K*)	2.5 W m^{-1} K^{-1}
Heat production depth scale (*D*)	35 km
Crustal heat production (*A*$_o$)	0.86 µW m^{-3}
	(Figs 2, 3, 4b,c,d)
	1.29 µW m^{-3}
	(Fig. 4a)
Basal heat flux (*Q*·)	30 mW m^{-2}
	(Figs 2, 3, 4a,c,d)
	45 mW m^{-2}
	(Fig. 4b)
$\frac{\mu_0 \ell}{K}$ (Equation 5a) Péclet number for erosion	1.15
$\frac{\dot{\varepsilon}_{zz} \ell}{K}$ (Equation 5b) Péclet number for thickening	11.4
	(Figs 3 & 4)
Péclet number for extension	11.4
	(Figs 3a,b & 4)
	5.7
	(Fig. 3c,d)

A, D and Q· refer to the actual distribution of heat sources in the calculations; these parameters also define the continental geotherm before thickening (Equation 6), except in Fig. 4(c,d) where the pre-thickening geotherm is hotter than the steady-state geotherm that would be supported by the heat source distribution (see text).

marked '0' in Fig. 2. As the real geometry corresponding to our assumption of homogeneous thickening (Fig. 1B, C) is more likely to be characterized by imbricate thrusting than by genuine pure shear, we have chosen a large value for the scale length (*D*) of radioactive heat production distribution so that in each case of Fig. 1, the heat production is distributed throughout the thickened crust rather than being concentrated in a surface layer. The steady-state surface flux has a contribution of 30 mW m^{-2} from each of the mantle and the crust.

In the illustrations used in this paper, a constant conductivity of 2.5 W m^{-1} K^{-1} is assumed throughout the lithosphere; a more reasonable value for the conductivity of the mantle portion of the lithosphere might be around 3.5 W m^{-1} K^{-1}. The introduction of this contrast into the system would not appreciably affect the crustal temperatures calcu-

lated here, but the discrepancy should be borne in mind when comparing sub-crustal temperatures in these models with those that would exist in a real situation.

The durations of the crustal thickening and thinning episodes are discussed above (Geological models). In each case we assume a maximum thickening of 100%; when this occurs by homogeneous compression in 30 Ma (Fig. 1C) it corresponds to a strain rate of 7.3×10^{-16} s^{-1}. The original crustal thickness is taken to be 35 km, so this is the length scale involved in the doubling of crustal thickness by thrusting (Fig. 1A). At the end of the thinning stage the crust is returned to its original thickness. This is accomplished either by the erosion of 35 km of material in 120 Ma (at 0.29 mm yr^{-1}—see Figs 1A, B and 2) or by an extensional strain rate of 3.7×10^{-16} s^{-1} over 60 Ma (Figs 1C and 3c, d) or by a 30 Ma period of no strain followed by 30 Ma of extensional

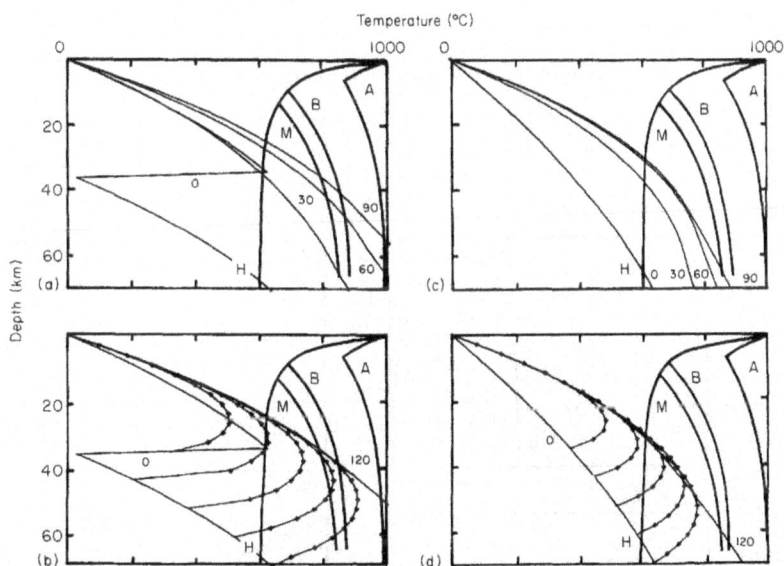

FIG. 2. (a) and (b): Geotherms and pressure–temperature–time (*PTt*) paths for the thickening/thinning geometry illustrated in Fig. 1(A). (a) shows the geotherms immediately after thrusting (marked 0) and at 30, 60 and 90 Ma later. In this, and all subsequent figures, only the upper 70 km are illustrated. The thick lines marked H, M, B, A are, respectively, the approximate curves for the melting of crustal rocks in the presence of excess water (H) and for the onset of melting in the fluid-absent system accompanying the breakdown of muscovite and biotite in metapelites (M), biotite and hornblende in granodiorites and tonalites (B) and hornblende in amphibolites (A). (b) shows the *PTt* paths (marked by crosses at 10 Ma intervals) for rocks that were at the top of the crust, at its base (35 km) and at three-quarters, one-half and one-quarter of this depth, before the start of the thickening event. Geotherms for 0 and 120 Ma are shown. Erosion of 35 km occurs in this 120 Ma. (c) and (d) are as (a) and (b) except that thickening is by homogeneous strain (Fig. 1B). Pressure–temperature–time paths are shown for rocks that, immediately after thickening, were buried at 35, 43.75, 52.5, 61.25 and 70 km.

P. C. England & A. Thompson

strain at $7.3 \times 10^{-16} s^{-1}$ (Figs 1C, 3a, b and 4).

It should be emphasized that considerable latitude is possible in the choice of these parameters and the results which follow ought to be regarded as illustrative of the processes discussed and not as the only possible paths for regional metamorphism under these circumstances—see England & Thompson (1984) for discussion of different parameter combinations.

Results

Figures 2 and 3 show the thermal development of the two systems discussed above, in terms of the geotherms plotted at various times after the beginning of the orogenic episodes and of the pressure–temperature–time (*PTt*) paths followed by rocks by those originally on the land surface and that were originally at a depth of 35 km (the original crustal thickness) and at one-quarter, one-half and three-quarters of that depth.

Also shown on these figures are curves for the beginning of melting under water-saturated conditions for crustal rocks (curve H after Wyllie 1977, Fig. 15). As noted by Yoder & Tilley (1962, p. 463), Wyllie (1977) and

Burnham (1979) among others, the majority of crustal rocks (including metasediments, granites and basalts) undergo water-saturated melting at very similar temperatures (600–700°C) at pressures corresponding to about 35 km burial. Such melting conditions are clearly only applicable to geological situations if a free fluid is normally present at these depths to promote water-saturated melting.

More generally applicable are the dehydration-melting curves appropriate to fluid-absent melting reactions involving muscovite, biotite and amphibole. Curve M is taken from Wyllie (1977, Fig. 9a) and from Thompson (1982, Fig. 7) and indicates the beginning of dehydration melting of muscovite and biotite in metapelites and peraluminous granites. Curve B is taken from Wyllie (1977, Fig. 9b) and indicates the beginning of fluid-absent melting of biotite and hornblende in granodiorite tonalite. Curve A is taken from Wyllie (1977, Fig. 9c) and from Burnham (1979, Figs 3.3 and 3.4) and indicates the beginning of dehydration melting of hornblende in amphibolites.

It is apparent from Figs 2 and 3 that the style of tectonics involved in a crustal-thickening/thinning episode exerts a major influence on

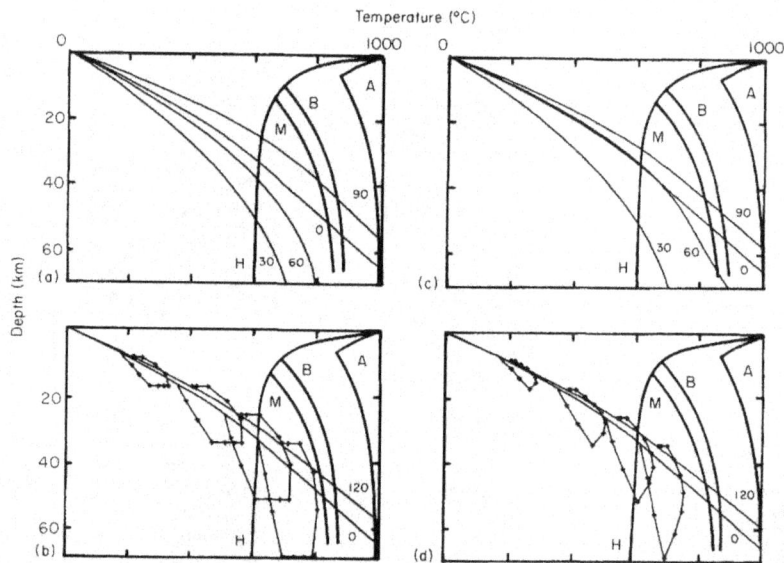

FIG. 3. As Fig. 2 (c and d) except the homogeneous strain occurs over 30 Ma (Fig. 1C). Geotherms are plotted at 0, 30, 60 and 90 Ma in (a) and (c), and at 0 and 120 Ma in (b) and (d). The crust is doubled in thickness in the interval 0–30 Ma and brought back to its original thickness in the interval 60–90 Ma (a and b) and in the interval 30–90 Ma (c and d). Pressure–temperature–time paths are shown for rocks that were buried at 8.75, 17.5, 26.25 and 35 km *before* thickening occurred.

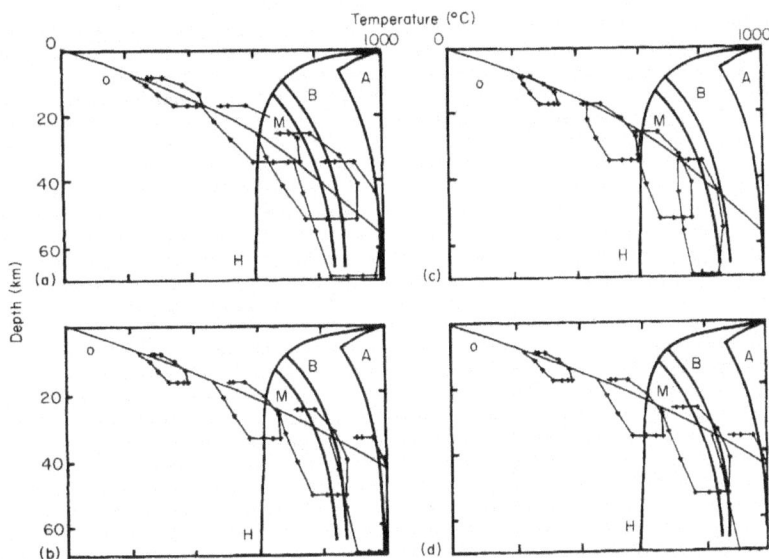

FIG. 4. As Fig. 3(b), except that the initial geotherm is now that corresponding to a surface heat flux of 75 mW m^{-2}. The four figures correspond to different assumptions about A_o and Q_* during the metamorphism (see text and Table 1).

the geotherms and on the *PTt* paths followed by rocks involved in the orogeny. Although the initial conditions and the total crustal thickening in each figure are the same, the mode and timing of the thickening and thinning processes differ importantly. The highest geotherms are experienced in the systems where the subcrustal heat flux after thickening is high (Fig. 2a, b) or there is a sizeable interval between thickening and the onset of crustal thinning (Fig. 3a, b).

Water-saturated crustal melting

The common features of these paths are that the lower half to one-third of the thickened crust passes through the conditions for water-saturated melting of crustal rocks. This is not surprising, as the base of the crust in the undisturbed state is close to the water-saturated solidus (e.g. curve '0', Fig. 2a, b), but the subsequent thermal relaxation puts the bottom 15–20 km of the thickened crust above the solidus (Figs 2b, d and 3b, d). Even if the original geotherms were considerably colder than those illustrated, Figs. 2 and 3 show that the temperature rises of several hundred degrees that are experienced by all these deeply buried rocks would be sufficient to

cause water-saturated minimum melting in the lowest portions of the thickened crust. This may be seen most easily by following the pressure–temperature–time (*PTt*) paths of Figs 2(b, d) and 3(b, d). The rate of temperature increase indicated by these paths in the order of 5–10°C per million years, indicates the time lapse that would be involved before the onset of water-saturated melting if the original geotherm lay considerably below this solidus. For example, if the base of the crust were originally 100°C below the solidus in Fig. 2b, it would take about 10 Ma for melting to start there. In each of the cases illustrated, water-saturated melting would occur almost immediately on crustal thickening at the base of the crust, and would occur progressively later at shallower levels within the crust.

Were water abundantly available as a free fluid within the lower crust then extensive melting would be the result of each continental collision. Because the temperature difference between the water-saturated solidus and liquidus for many sialic rocks is only about 50°C (see e.g. Harris *et al.* 1970, Fig. 8b) most metasediments, granites and other sialic rocks (comprising perhaps up to 50% of the volume of the lower crust) would be fully molten S-type granite (Chappell & White 1974) at 700°C and depths of 35 km. The water-saturated sialic melts could ascend and result in S-type granite plutonism until they freeze at the water-saturated solidus at quite shallow depths.

For basalt or amphibolite with excess water at 35 km depth the water-saturated solidus is at about 630°C and the corresponding liquidus is close to 1100°C (Yoder & Tilley 1962, Fig. 33; Harris *et al.* 1970, Fig. 8a). The experiments of Helz (1976, Fig. 1) indicate that 30–60% water-saturated 'granitic' melt (I-type: Chappell & White 1974) results from the melting of basalt at 5 kbar P_{H_2O} and 1000°C. Thus at 700°C and 35 km, water-saturated melting of basalt would generate less than 10% melt. The volumes of these melts would be much less than those produced from sialic rocks under the same conditions.

It is often argued, on the basis of porosity estimates on high-grade rocks, that large volumes (>1% porosity) of free fluid are unlikely in the lower crust. However, the tectonics envisaged here involve the burial by up to 35 km of rocks which were originally at much shallower levels and their metamorphism to temperatures above 600°C. The consequent dehydration reactions are capable of releasing 2–3 wt% H_2O (see e.g. Walther & Orville 1982) which would, if it were all retained in the rock, represent 6–9% porosity at 600°C and 5 kbar. It is unlikely such a high pore volume would be maintained in a metamorphic rock, but it is reasonable to expect a higher pore volume to be supported during volatile production than is supported once production has ceased; thus the low porosity of high-grade rocks that reach the surface may not be an adequate measure of their porosity while they were dehydrating. We can make estimates of the largest amount of water-saturated minimum melt to be expected by noting that the solubility of water in water-saturated crustal melts is between 12 and 15 wt% at 10 kbar (~70 mol%; Burnham 1979, Fig. 3.1) so that for every 1 wt% of water present in pore volume, we should expect to produce 7–8 wt% water-saturated minimum melt from sialic rocks. In Figs 2 and 3, the bottom 20–30 km of the thickened crust passes above the water-saturated solidus, so this calculation implies that about 1.5–2.5 km³ of sialic crustal melt would be produced per km² for every 1 wt% of water retained in the rock.

In contrast, as argued above, we should expect only small quantities of melt from the water-saturated melting of any amphibolites which may be in the lower crust (Burnham 1978, p. 95).

Fluid-absent crustal melting (dehydration melting)

For the geothermal regime chosen for Figs 2 and 3 (undisturbed surface heat flux 60 mW m⁻²), only the paths illustrated in Fig. 2(b) intersect the fluid-absent muscovite and biotite dehydration-melting curve (M) allowing the generation of S-type granites from metapelites; only the deepest of these paths intersect the fluid-absent biotite and hornblende dehydration melting curve (B), allowing the regeneration of I-type granites from granodiorites and tonalites.

In this system (Fig. 2a, b) the subcrustal contribution to heat flux while still entirely conductive, is considerably higher than the other cases. The crustal thickening event has taken place without diminishing the subcrustal heat flux (Figs 1A and 2a, curve '0'). Figure 4 illustrates the *PTt* paths for a system like that of Figs 1C and 3(a, b), in which the crust is homogeneously thickened by a factor of two in 30 Ma and, after a 30 Ma interval, is homogeneously thinned by the same amount over the next 30 Ma. The difference between this figure and Fig. 3(a, b) lies in the fact that the geotherms immediately *before* thickening (curves '0' in Fig. 4) correspond to a surface heat flux of 75 mW m⁻², the average for continental crust of tectonic age <250 Ma (Sclater *et al.* 1980). In each case the geotherm before thickening is obtained by evaluating Equation 6 with D = 35 km, K = 2.5 W m⁻¹ K⁻¹, Q. 30 mW m⁻² (Fig. 4a, c) or 45 mW m⁻² (Fig. 4b, d) and A_0 1.29 μW m⁻³ (Fig. 4a, c) or 0.86 μW m⁻³ (Fig. 4b, d). Note, that in Fig. 4(c, d), although the pre-thickening geotherm is evaluated as just described, this is presumed to reflect a heat-flow regime perturbed by tectonic activity and the supply of heat to the thickened pile is *the same* as in Fig. 3(a, b). These four models are not intended to be exact representations of any tectonic situation but instead investigate a number of regimes in which recent (<200 Ma) tectonism might result in a heat flow higher than the continental average; this is done in a fashion that gives easy comparison with the cases examined in Figs 2 and 3.

Higher-than-average surface heat flux in a tectonically young area might result from the fact that the crust is still thicker than average and/or contains a greater amount of internal heat generation (e.g. Vitorello & Pollack 1980, Figs 4 and 5). This is the case in Fig. 4(a), where the surface heat flux is enhanced by 15 mW m⁻² from radiogenic contributions in the crust over the equivalent situation in Fig. 3 (a, b).

Higher heat flux may also result from the erosion of a terrane (e.g. England & Richardson 1980), and in this case the near surface gradients are enhanced, while those at depth remain unaffected; this is represented in Fig. 4(c) where the pre-thickening geotherm also reflects a higher surface heat flux, but the internal heat supply is *the same* as in Fig. 3(a, b). Thus the initial geotherm is steeper than would be supported in steady state by the heat supply.

Other possibilities are that the principal thermal effect remaining from the recent tectonic activity is a geotherm with an enhanced mantle contribution to the heat flow, either supported by an equal heat flow during the metamorphism (Fig. 4b where the basal heat supply is kept at 45 mW m^{-2}) or with a 'normal' mantle heat flow (Fig. 4d where the basal heat supply is the same as in Fig. 3a, b, 30 mW m^{-2}).

The *PTt* paths in Fig. 4 illustrate several principles about the thermal evolution of thickened crust; these are discussed in detail by England & Thompson (1984) but we summarize here those features relevant to crustal melting.

First, the initial heat content of the crust does not, by itself, significantly influence the metamorphic grade attained. Comparison of Figs 3(b) and 4(c) shows that the extra heat content of the crust in Fig. 4(c) raises the peak metamorphic temperatures by little more than 50°C in the lower crust. In contrast, an increase in internal radiogenic heat production (compare Fig. 4a, c) raises peak temperatures by more than 150°C and the whole of the bottom half of the crust passes above the dehydration melting curves *M* and *B* in Fig. 4(a). Finally, if the whole of the lithosphere is thermally perturbed by the tectonic event (as in Fig. 4b, d) where the initial geotherm is steepened by an additional mantle contribution of 15 mW m^{-2}, the time scale for the orogenic event is short enough that the thermal regime in the crust is not sensitive to the heat supply at the *base* of the lithosphere. In Fig. 4(b), this is kept at 45 mW m^{-2} throughout, while in Fig. 4(d) it is kept at 30 mW m^{-2}, yet the *PTt* paths for the two regimes are almost identical.

Figure 4(a, b, d) illustrates regimes in which the thickening of continental lithosphere, with a higher than average surface heat flux (75 mW m^{-2}), results in the lower 20–30 km of the thickened crust passing into the field for dehydration melting of crustal rocks. In the cases illustrated here, because of the positive slope of the dehydration melting curves, the almost isothermal ascent of the deeper crustal levels during the extensional phase influences considerably the amount of melt we would expect. It should be noted that there is much less cooling during exhumation on these paths than occurs in comparable sections of paths for rocks exhumed by erosion (Fig. 2). In these regimes it would be possible to generate significantly more melt than envisaged here if the low-angle normal faulting described, for example, by Wernicke (1981) were to produce rapid, localized exhumation, with even steeper uplift paths than are illustrated in Fig. 4.

The most deeply buried portions of the crust, in Fig. 4(b, d), pass into *PT* regimes that could result in the dehydration melting of hornblende in amphibolites to give calc-alkaline magmas.

A secondary feature of the *PTt* paths for the setting illustrated in Figs 3 and 4 (extension after thickening) is that, although the original geotherm and the final steady-state geotherms are the same, immediately after extension ends the most deeply buried rocks are much hotter than the steady-state geotherm. This is particularly clear in Fig. 4, where the rocks follow closed *PTt* paths of which the last phase is *isobaric cooling*. The more deeply buried rocks in the case illustrated in Fig. 2(b) also experience a phase of isobaric cooling at the end of their *PTt* cycle, but in this case it occurs at much lower temperatures than the maximum the rocks experienced. In Figs 3 and 4, this phase of isobaric cooling starts very close to the maximum temperature the rocks experience and we expect it to be the most significant portion of the retrograde path to be recorded. Such a path might account for some mineralogical features in granulite-facies rocks, such as the 'corona' structures that are described from some deeply eroded terranes (e.g. Griffin & Heier 1973); however, because of the uncertainties in mineral kinetics, some corona structures could just as well reflect the nearly isothermal decompression segment of the *PTt* path prior to final isobaric cooling. Note that contrary to the suggestion of England & Richardson (1977), isobaric cooling cannot be regarded as diagnostic of magmatic augmentation of the thermal budget of an orogen and may equally well occur as the last stage in the metamorphic evolution of an extending region like the Basin and Range.

Discussion and conclusions

Crustal thickening can lead to widespread crustal melting without the involvement of material from the mantle; the amount of melt produced depends critically on the water available to the system. If the lower crust in such thickened belts contains pelitic lithologies with excess fluid, we expect water-saturated melting of these rocks to begin at the deepest levels shortly after crustal thickening starts and to migrate upwards as the metamorphism progresses. Under conditions of normal mantle heat flux we may expect approximately 2 km^3 of crustal melts per km^2 of surface from this process for every 1 wt% of water retained in the pore volume of the metamorphosing rocks (see Results). These melts would give rise to predominantly S-type granitic plutonism, with a negligible amount of I-type granitic activity unless the geotherms become hot enough in the lower crust to cross the vapour-absent melting curves. This occurs in only one case for the thermal regime with surface heat flux of 60 mW m^{-2} (Fig. 2b), and happens some 40 Ma after the thickening event.

However, if crust with a thermal regime corresponding to that regarded as average for Mesozoic–Cenozoic terranes (Sclater *et al.* 1980) becomes involved in an orogenic event, much of the lower crust may pass above the dehydration-melting curves for crustal rocks (Fig. 4). The volumes of melt expected in this case are hard to estimate, not least because the effects of melt migration are excluded from the simple calculations made here (see Geological models). It would not be unreasonable to expect 20% melt from the most deeply buried crustal rocks in Fig. 4 (Wyllie 1977; Burnham 1979) and an average of 5% melt for the lower half of the crust in this regime would give a further 1–2 km^3 per km^2 of orogenic belt.

It is to be expected that the dominant early product of dehydration melting in this tectonic setting will be S-type granite, produced by dehydration melting of metasediment in the thickened crust. Subordinate amounts of I-type granite are to be expected, produced by dehydration melting of older calc-alkaline magmatic rocks or from incongruent dehydration melting of biotite + plagioclase + quartz gneiss to yield granite with residual amphibole. This is succeeded by dehydration melting of amphibolite and in the cases shown here, this would occur some 20–40 Ma after the onset of dehydration melting in the most deeply buried sialic rocks.

The abundances of I-type granites will increase as the metamorphism progresses; in Fig. 4 the main dehydration melting of amphibolite occurs 20–60 Ma after the end of the thickening event.

There is little doubt that much of the calc-alkaline volcanism that is spatially related to Benioff Zones, has its origin within the subduction zone and/or the overlying mantle wedge. However, Fig. 4(b, d) shows that thickening of the crust with an initial geotherm corresponding to a surface heat flux of 75 mW m^{-2}, is capable of producing dehydration melting of amphibolite in continental crust with consequent production of calc-alkaline magmas. The degree of melt involved may be several tens of percent (Burnham 1979, p. 97) but as this applies only to the lowest few km of the pile, the absolute volumes may well be small. That some calc-alkaline batholiths are derived in substantial part from old crust, rather than being freshly fractionated from hydrous basalt, is evident from isotopic investigations (Allégre & Ben Othman 1980; Farmer & DePaolo 1983). If it can be demonstrated that calc-alkaline magmatic rocks in a continental collision zone were not produced in the subduction zone before collision, then they may reflect partial melting in the collision zone with an enhanced mantle heat flux. This is an important difference from requiring mantle involvement in the production of calc-alkaline magmas, because simple calculations (e.g. Burnham 1979) indicate that if anatexis occurs solely by the injection of mafic magma into the continental crust, each volume of crustal melt requires approximately the same volume of mafic magma to generate it.

In the conditions examined here, where mantle melt is not involved, the greatest volumes of melt are to be expected in a regime that permits a delay of a few tens of millions of years between crustal thickening and the onset of extension (Figs 3a, b and 4), in a regime in which the crust is thickened without a comparable increase in total lithosphere thickness (e.g. Fig. 2a, b) or in a regime in which recently active continental lithosphere is involved in a thickening event. The histories of such wide belts as the India–Asia collision or the Basin and Range of N America provide evidence that each of these possibilities is reasonable and thus that extension following such collision may permit melts of batholithic proportions to be generated. There was an interval of 20–30 Ma between the end of the Laramide and Sevier Orogenies, which produced much of the crustal thickening in what is now the Basin and Range, and the onset of extension there in Eocene and Oligocene times; the India—Asia collision has been continuing for 30—40 Ma and extension in Tibet has been happening only since the Late Miocene (Armijo *et al.* 1982).

The only model in this paper in which crustal thickening is not accompanied by a comparable thickening in the rest of the lithosphere is the thrust model of Fig. 2. However, Houseman *et al.* (1981) consider the thermal evolution of a thickened continental lithosphere overlying a convecting mantle and show that for likely upper-mantle conditions the thermal boundary layer beneath the continent is unstable and can drop off, to be replaced by hot material from the upper mantle, in a time that is short compared with the thermal conduction time of the lithosphere. It is likely, then, that the assumption underlying Figs 3 and 4, that the lithosphere increases in thickness in proportion to the crust, *underestimates* the subcrustal contribution to heat flux in continental collision zones.

Finally, there is ample evidence of repeated tectonic events occurring at continental edges; the Taconic and Acadian Orogenies are a well-known example, and Molnar & Tapponnier (1981) review the evidence for extensive Late Palaeozoic and Mesozoic tectonic activity in that region of Southern Asia which is currently deforming in response to the India—Asia collision.

The suggestion that crustal melting can arise from the formation and thermal relaxation of thick piles of continental rocks can be traced back at least as far as Bott (1954) and Tuttle & Bowen (1958), although the mechanism has been adumbrated earlier than this (e.g. Bowen 1928). Another school of thought (e.g. Brown

& Hennessey 1978) holds either that thermal gradients are too shallow or thermal relaxation is too slow for the production of crustal melts to occur in such environments and that mantle involvement is a necessary condition for crustal melting. One purpose of this paper is to counter the latter notion and to demonstrate the diversity of crustal magmatism that is possible during the thickening of continental crust.

We do not wish to imply that all igneous rocks produced in orogenic belts have purely crustal origins; what we have attempted here is to make quantitative estimates of the amounts of melt that would be generated in an environment where mantle magmatism does not contribute to the thermal budget, and to illustrate the *PT* paths that would be followed by rocks that generated such melts. In our current state of knowledge it would be unwise to make dogmatic statements about the volumes of crustal melt that can be produced without involvement of mantle-derived magma, but it appears from the discussion in this section that at least $2-5$ km^3 per km^2 of orogenic belt may be generated without the heat supply from mafic magmas. We wish particularly to emphasize the diversity of magma type that can result from continental crustal thickening and the importance of detailed isotopic studies of magmas in orogenic terranes in determining the degree of mantle involvement in the thermal budget of such terranes.

ACKNOWLEDGMENTS: We are grateful to Howard Day, Ben Harte, Roger Powell and Brian Wernicke for helpful comments. This work was supported in part by NSF grant EAR81-07659.

References

ALLÉGRE, C. J. & BEN OTHMAN, D. 1980. Nd-Sr isotopic relationship in granitoid rocks and continental crust development: a chemical approach to orogenesis. *Nature, Lond.* **286**, 335—42.

ARMIJO, R., TAPPONNIER, P., MERCIER, J. L. & TONGLIN, H. 1982. A field study of the Pleistocene Rifts in Tibet. *EOS, Trans. Am. Geophys. Union*, **63**, 1093.

BICKLE, M. J., HAWKESWORTH, C. J., ENGLAND, P. C. & ATHEY, D. 1975. A preliminary thermal model for regional metamorphism in the Eastern Alps. *Earth planet. Sci. Lett.* **26**, 13—28.

BIRD, R. B., STEWART, W. E. & LIGHTFOOT, E. N. 1960. *Transport Phenomena*, 780 pp. John Wiley, New York.

BOTT, M. H. P. 1954. Interpretation of the gravity field of the Eastern Alps. *Geol. Mag.* **41**, 377—83.

BOWEN, N. L. 1928. *The Evolution of the Igneous Rocks*, 332 pp. Princeton University Press, Princeton.

BROWN, G. C. & HENNESSEY, J. 1978. The initiation and thermal diversity of granite magmatism. *Phil. Trans. R. Soc.* **288**, 631—43.

BURNHAM, C. W. 1979. Magmas and hydrothermal fluids. *In*: BARNES, H. L. (ed.) *Geochemistry of Hydrothermal Ore Deposits*, 2nd edn, pp. 71—136. John Wiley, New York.

CHAPPELL, B. W. & WHITE, A. J. R. 1974. Two contrasting granite types. *Pacif. Geol.* **8**, 173—74.

CHEN, W. P. & MOLNAR, P. 1983. Focal depths of intracontinental and interplate earthquakes and

94   *P. C. England & A. Thompson*

their implications for the thermal and mechanical properties of the lithosphere. *J. geophys. Res.* **83**, 4183.

CHRISTIE, P. A. F. & SCLATER, J. G. 1980. An extensional origin for the Buchan and Witchground Graben in the North Sea. *Nature, Lond.* **283**, 279.

DALMAYRAC, B. & MOLNAR, P. 1981. Parallel thrust and normal faulting in Peru, and constraints on the state of stress. *Earth planet. Sci. Lett.* **55**, 473.

ENGLAND, P. C. 1978. Some thermal considerations of the Alpine metamorphism, past, present and future. *Tectonophysics*, **46**, 21.

—— & MCKENZIE, D. P. 1982. A thin viscous sheet model for continental deformation. *Geophys. J. R. astr. Soc.* **70**, 295–321.

—— & ——. 1983. Correction to: A thin viscous sheet model for continental deformation. *Geophys. J. R. astr. Soc.* **73**, 523.

—— & RICHARDSON, S. W. 1977. The influence of erosion upon the mineral facies of rocks from different metamorphic environments. *J. geol. Soc. London*, **134**, 201–13.

—— & ——. 1980. Erosion and the age dependence of continental heat flow. *Geophys. J. R. astr. Soc.* **62**, 421–37.

—— & THOMPSON, A. B. 1984. Pressure–temperature–time paths of regional metamorphism. I. Heat transfer during the evolution of regions of thickened continental crust. *J. Petrol.* **25**, 894–928.

FARMER, G. L. & DEPAOLO, D. J. 1983. Origin of Mesozoic and Tertiary granite in the western United States and implications for pre-Mesozoic crustal structure. I. Nd and Sr isotopic studies in the geocline of the Northern Great Basin. *J. geophys. Res.* **88**, 3379–401.

GRIFFIN, W. L. & HEIER, K. S. 1973. Petrological implications of some corona structures. *Lithos*, **6**, 315–35.

HARRIS, P. G., KENNEDY, W. Q. & SCARFE, C. M. 1970. Volcanism versus plutonism—the effect of chemical composition. *In*: NEWAU, G. & RAST, N. (eds) *Mechanism of Igneous Intrusion. Lrpool Manchr. geol. J. Spec. Issue.* **2**, Ch. 14.

HELZ, R. T. 1976. Phase relations of basalts in their melting ranges at $PH_2O = 5Kb$. II. Melt compositions. *J. Petrol.* **17**, 139–93.

HOUSEMAN, G. A., MCKENZIE, D. P. & MOLNAR, P. 1981. Convective instability of a thickened boundary layer and its relevance for thermal evolution of continental convergent belts. *J. geophys. Res.* **86**, 6115–32.

MCKENZIE, D. P. 1978. Some remarks on the development of sedimentary basins. *Earth planet. Sci. Lett.* **40**, 25–32.

MOLNAR, P. & TAPPONNIER, P. 1975. Cenozoic

tectonics of Asia: consequences and implications of a continental collision. *Science*, **189**, 419–26.

—— & ——. 1981. A possible dependence of the tectonic strength on the age of the crust in Asia. *Earth planet. Sci. Lett.* **52**, 107–14.

OXBURGH, E. R. & TURCOTTE, D. L. 1974. Thermal gradients and regional metamorphism in overthrust terrains with special reference to the Eastern Alps. *Schweiz. miner. petrog. Mitt.* **54**, 641–62.

PROFFETT, J. M. JR 1971. Late Cenozoic structure in the Yerington district, Nevada, and the origin of the Great Basin. *Abstr. with Programs Geol. Soc. Am.* **3**, 181.

RICHARDSON, S. W. & POWELL, R. 1976. Causes of Dalradian metamorphism. *Scott. J. Geol.* **12**, 237–68.

SCLATER, J. G. JAUPART, C. & GALSON, D. 1980. Heat flow through oceanic and continental crust, and heat loss of the earth. *Rev. Geophys. Space. Phys.* **81**, 269–312.

SMITH, R. B. 1978. Seismicity, crustal structure, and intraplate tectonics of the interior of the Western Cordillera. *In*: SMITH, R. B. & EATON, G. P. (eds) *Cenozoic Tectonics and Regional Geophysics of the Western Cordillera. Geol. Soc. Am. Mem.* **152**, 111–44.

THOMPSON, A. B. 1982. Dehydration melting of pelitic rocks and the generation of H_2O-undersaturated granitic liquids. *Am. J. Sci.* **282**, 1567–95.

TUTTLE, O. F. & BOWEN, N. L. 1958. Origin of granite in the light of experimental studies in the system $NaAlSi_3O_8–KAlSi_3O_8–SiO_2–H_2O$. *Geol. Soc. Am. Mem.* **74**.

VITORELLO, I. & POLLACK, H. N. 1980. On the secular variation of continental heat flow and the thermal evolution of continents. *J. geophys. Res.* **85**, 983–95.

WALTHER, J. V. & ORVILLE, P. M. 1982. Volatile production and transport in regional metamorphism. *Contrib. Miner. Petrol.* **79**, 252–7.

WERNICKE, B. P. 1981. Low angle normal faults in the Basin and Range Province: Nappe tectonics in an extending orogen. *Nature, Lond.* **291**, 645–8.

WERNICKE, B. P., SPENCER, J. E., BURCHFIEL, B. C. & GUTH, P. L., 1982. Magnitude of crustal extension in the southern Great Basin. *Geology*, **10**, 499–502.

WYLLIE, P. J. 1977. Crustal anatexis: an experimental review. *Tectonophysics*, **43**, 41–71.

YODER, H. S. & TILLEY, C. E. 1962. Origin of basalt magmas: an experimental study of natural and synthetic rock systems. *J. Petrol.* **3**, 342–52.

PHILIP C. ENGLAND, Department of Geological Sciences, Harvard University, Cambridge, MA 02138, U.S.A.
ALAN THOMPSON, Department für Erdwissenschaften, ETH, Zürich, Switzerland.

CHAPTER 7

Huppert, H.E. and Sparks, R.S.J. (1988) The generation of granitic magmas by intrusion of basalt into continental crust. *Journal of Petrology*, **29**, 599–624.

The thermal state of the stable continental crust is such that crustal geotherms never intersect the solidi of crustal rocks (Chapman and Furlong, 1992). Consequently, the generally accepted view that most granitic magmas were produced by partial melting of the continental crust poses the question about the source of heat for high-grade metamorphism and crustal partial melting. Indeed, this not a trivial question. The high latent heats of melting of silicates ($\approx 400,000$ J/kg) compared to their specific heats (≈ 850 J/kg/K) means that a rock heated to its solidus temperature still requires an enormous amount of additional heat for melting to proceed. In principle there are three mechanisms capable of heating sectors of the lower to middle continental crust to the solidus temperatures: advective heat-transport by mantle-derived mafic magmas, conductive heat-transport from the mantle (due to thermal relaxation after a tectonic perturbation), or self-generated radiogenic heating (in which materials with elevated heat production are buried deep in the crust).

This paper by Huppert and Sparks is devoted to modelling the first mechanism, and is one of the first serious attempts to understand quantitatively how granitic magmas are produced. The authors studied how a sill of basaltic magma could cause melting of its host roof rocks when emplaced into silicic crustal rocks that have a relatively low solidus temperature. They approached the problem using experiments with analogue materials — polyethylene glycol waxes and aqueous solutions, complemented by numerical modelling based on fluid mechanics and heat-transfer processes. These studies predict that the silicic roof of the mafic sill will melt quickly (with a time constant of ~10^2 to 10^3 y) after the emplacement. The melting process would produce a low-density melt layer with negligible mixing with the underlying mafic magma, and the felsic melt could then be transported to higher crustal levels.

The Huppert and Sparks model explains many facts observed for silicic magmatism, at least in some zones, and was enthusiastically embraced by many granite petrologists. Its tectonic implications were discussed by Warren and Ellis (1996), and its efficacy in producing granitic magmas was studied experimentally by McCarthy and Patiño Douce (1997) and numerically by Raia and Spera (1997). Such a mechanism has been proposed as the main driving engine for granite generation in the Ivrea Zone, NW Italy (Voshage *et al.*, 1990; Barboza and Bergantz, 1996; Henk *et al.*, 1997), the Amazonian Craton (Dallagnol *et al.*, 1994), the European Variscides (Williamson *et al.*, 1992; Buttner and Kruhl, 1997), the Northern China Craton (Zhao *et al.*, 1999), Central Asia (Jahn *et al.*, 2000), etc. The Ivrea zone, in particular, has been considered as a paradigm of granite generation by underplating (*ops. cit.*, see also Knesel and Davidson, 1999), but it is also where the model has generated major criticisms. Barboza and Bergantz (1996) concluded that "extensive, vigorous convection of partially molten rocks above mafic bodies is unlikely" a major conclusion of the Huppert and Sparks model. Furthermore, Barboza *et al.* (1999) and Barboza and Bergantz (2000) concluded that "magmatic accretion may not inexorably cause regional metamorphism and crustal anatexis".

Despite this criticism, the mechanism of granite generation by magma underplating as proposed by Huppert and Sparks is solidly founded. Not all model implications proposed by these authors are equally defendable twenty years on, but the essence of the model is, in our opinion, unquestionable. The intrusion of hot mafic magma into the continental crust is, by far, the most efficient mechanism for rapid heat transfer from the mantle to the crust, and is the only feasible explanation for rapidly generated granitic magmas (e.g. Bea *et al.*, 2007). Certainly, not all granites have been generated in this way but it is probably one of the major mechanisms for granite generation within the continental crust. More recent work, inspired by the Huppert and Sparks treatment (e.g. Petford and Gallagher, 2001), has shown that the Huppert and Sparks model is sound but just requires the addition of periodic intrusion of mafic magma to make it thermally efficient. Thus, the Huppert and Sparks paper represents a definite landmark for our understanding of the source of heat for crustal melting.

The Generation of Granitic Magmas by Intrusion of Basalt into Continental Crust

by HERBERT E. HUPPERT[1] AND R. STEPHEN J. SPARKS[2]

[1]*Department of Applied Mathematics and Theoretical Physics, Silver Street, Cambridge CB3 9EW*

[2]*Department of Earth Sciences, Downing Street, Cambridge CB2 3EQ*

(*Received 20 July 1987; revised typescript accepted 14 January 1988*)

ABSTRACT

When basalt magmas are emplaced into continental crust, melting and generation of silicic magma can be expected. The fluid dynamical and heat transfer processes at the roof of a basaltic sill in which the wall rock melts are investigated theoretically and also experimentally using waxes and aqueous solutions. At the roof, the low density melt forms a stable melt layer with negligible mixing with the underlying hot liquid. A quantitative theory for the roof melting case has been developed. When applied to basalt sills in hot crust, the theory predicts that basalt sills of thicknesses from 10 to 1500 m require only 1 to 270 y to solidify and would form voluminous overlying layers of convecting silicic magma. For example, for a 500 m sill with a crustal melting temperature of 850 °C, the thickness of the silicic magma layer generated ranges from 300 to 1000 m for country rock temperatures from 500 to 850 °C. The temperatures of the crustal melt layers at the time that the basalt solidifies are high (900–950 °C) so that the process can produce magmas representing large degrees of partial fusion of the crust. Melting occurs in the solid roof and the adjacent thermal boundary layer, while at the same time there is crystallization in the convecting interior. Thus the magmas formed can be highly porphyritic. Our calculations also indicate that such magmas can contain significant proportions of restite crystals. Much of the refractory components of the crust are dissolved and then re-precipitated to form genuine igneous phenocrysts. Normally zoned plagioclase feldspar phenocrysts with discrete calcic cores are commonly observed in many granitoids and silicic volcanic rocks. Such patterns would be expected in crustal melting, where *simultaneous* crystallization is an inevitable consequence of the fluid dynamics.

The time-scales for melting and crystallization in basalt-induced crustal melting (10^2–10^3 y) are very short compared to the lifetimes of large silicic magma systems ($>10^6$ y) or to the time-scale for thermal relaxation of the continental crust ($>10^7$ y). Several of the features of silicic igneous systems can be explained without requiring large, high-level, long-lived magma chambers. Cycles of mafic to increasingly large volumes of silicic magma with time are commonly observed in many systems. These can be interpreted as progressive heating of the crust until the source region is partially molten and basalt can no longer penetrate. Every input of basalt triggers rapid formation of silicic magma in the source region. This magma will freeze again in time-scales of order 10^2–10^3 y unless it ascends to higher levels. Crystallization can occur in the source region during melting, and eruption of porphyritic magmas does not require a shallow magma chamber, although such chambers may develop as magma is intruded into high levels in the crust. For typical compositions of upper crustal rocks, the model predicts that dacitic volcanic rocks and granodiorite/tonalite plutons would be the dominant rock types and that these would ascend from the source region and form magmas ranging from those with high temperature and low crystal content to those with high crystal content and a significant proportion of restite.

INTRODUCTION

One of the central questions in igneous petrology concerns the generation of silicic magmas. There is now convincing evidence that most of the large plutonic complexes of granite in the continental crust are the result of crustal anatexis (Pitcher, 1987). There is also

[*Journal of Petrology*, Vol. 29, Part 3, pp 599–624, 1988]

HERBERT E. HUPPERT AND R. STEPHEN J. SPARKS

widespread evidence that basaltic magma from the mantle is often intimately associated with the generation of silicic magmas (Hildreth, 1981). This association of mafic and silicic magmas can occur in orogenic belts above subduction zones, in continental hot-spots, and in regions of crustal extension. In plutonic complexes, mafic and intermediate igneous activity are recorded in contemporaneous dyke swarms, small satellite intrusions, and in mafic enclaves within the granites (Vernon, 1983; Pitcher, 1986, 1987). In silicic volcanic centres, evidence of basaltic magmatism is found in satellite lava fields and cinder cones, early lava shields and stratovolcano complexes prior to the main silicic volcanism (Lipman, 1984), and as mafic inclusions and bands within the silicic volcanic rocks (Smith, 1979; Bacon, 1986). Petrological and geochemical features of many silicic igneous rocks are also convincingly explained by admixture of a mantle-derived (mafic) component with a crustal melt. Regions of high temperature and low pressure metamorphism are commonly associated with granite plutonism and a plausible explanation of this association is that basalt is intruded into the crust, causing melting and high heat flow. Indeed basalt underplating of the crust is a currently popular idea to explain both large scale crustal melting and the strongly layered character of the lower crust. While there may be some silicic magmas that are generated by processes without the aid of basaltic input, such as tectonic thickening of radioactive crust (England & Thompson, 1984; Pitcher, 1987), this paper takes the position that in many cases the additional thermal energy of basalt is essential.

The continental crust is strongly layered in terms of its composition, density, and mechanical behaviour. The upper crust is cold and brittle whereas the lower crust is hotter, has a higher density, deforms in a ductile manner, and is commonly characterized by prominent horizontal layering. Basalt magma can be emplaced into the continental crust as dykes and sills and, in some cases where the rate of magma input is high, these intrusions can coalesce to form larger magma chambers.

Dyke emplacement does not seem an efficient way of generating large volumes of silicic magma, because dykes are usually small in width and much of the potential heat for melting will not be utilized if the mafic magma erupts. Sills provide a more promising situation in which *extensive* crustal melting can occur. Horizontal intrusions concentrate their heat at a particular level in the crust and do not dissipate their heat over a large depth range. Sills are intrinsically more efficient than dykes in this respect. Dykes may play an important role in heating up the crust to initiate melting. However, once a region of the crust has become hot, ductile, and partially molten, conditions for dyke propagation become less favourable. A layer or region of partially molten crust provides an effective density barrier and we suggest that basalt magma reaching such a level will spread out as horizontal intrusions. An additional factor which promotes sill formation in the lower parts of the crust is its strongly layered character providing a structural environment in which horizontal intrusions are favoured. For these reasons this paper is concerned principally with the heat transfer and fluid dynamics of sills intruded into hot continental crust.

We consider the cooling and crystallization of basaltic sills emplaced into the continental crust. In particular, we emphasize the situation where the roof of the sill is composed of rock which has a fusion temperature that is lower than the magma temperature and the roof rock consequently melts. This is likely to be the normal situation where basalt intrudes into the typical rock types of mature and ancient middle and upper crust which are already at high temperature. However, the concepts developed in this paper are also likely to be applicable to conditions in immature continental crust such as in island arcs, to more refractory lower crust and to lower crust formed by slightly older or even contemporaneous episodes of basalt underplating. In each of these latter cases, lithologies which have relatively low fusion temperatures can form by differentiation processes and can be remelted by further intrusion of basalt. Thus the model is not confined to the origin of granites, but should be relevant to the origins of intermediate rocks such as tonalites and evolved alkaline rocks such as syenite.

We present experimental studies on the melting of the roof of a sill. We develop a quantitative model of the melting process at the roof, which describes the rates at which a new layer of roof melt forms and the rates at which the underlying liquid layer solidifies. We discuss possible mechanisms by which the melts can be mixed together and also their implications for magma genesis within the continental crust. A companion paper (Huppert & Sparks, 1988a) describes the melting of the roof of a chamber from a detailed fluid mechanical point of view.

Throughout this paper the magma will be considered to be Newtonian. Although magma in reality can be non-Newtonian, especially when it is rich in crystals (McBirney & Murase, 1984) its nonlinear rheological properties and the consequences of its non-Newtonian rheology are poorly understood. Two effects may be evident: there may exist a yield strength, so that for a sufficiently low applied stress the magma will not move; and the nonlinear viscosity may alter the heat flux transferred by a convecting magma. Because of the relatively large values of the Rayleigh number that result in most of our calculations, we anticipate that the yield strength will be exceeded by quite a margin. The alterations in the heat flux are at the moment difficult to anticipate and we suggest that the reader views our quantitative results as an indication of the calculated quantity rather than as a precise value. It may be possible to examine non-Newtonian effects with greater insight in the future, but a Newtonian description illuminates many of the fundamental effects and is a necessary first step in order to form the basis for any comparison.

EXPERIMENTAL STUDIES

The geological problem is to consider how a sill of basaltic magma will melt its roof when emplaced in crustal silicic rocks with low fusion temperatures. We have investigated some of the essential physical features of this problem using polyethylene glycol (PEG) waxes to simulate the wallrocks and aqueous solutions to simulate the magma.

The experiments consisted of placing a wax roof, with a thickness of 15 cm, at the top of a perspex container $20 \times 20 \times 40$ cm high. A hot aqueous solution of $NaNO_3$ was introduced at the base of the container until it was completely filled. The melting temperature of the PEG 1000 wax used is 37–40 °C and the initial temperature of the hot solution was around 70 °C. The density of the molten wax decreases fairly linearly between $1 \cdot 11$ $g\,cm^{-3}$ at 40 °C to $1 \cdot 09$ $g\,cm^{-3}$ at 70 °C. The density of the hot fluid could be varied by altering its concentration. In the present study the density of the aqueous solution exceeded the density of the molten wax. In the experiments the melt formed a separate layer between the roof and the hot fluid. The layer was initially quiescent and transferred heat by conduction, though subsequently convection also occurred (Fig. 1). Qualitatively, the experiment displays simple physical principles which describe the melting of a roof. If the melt is light it forms a discrete layer separated from underlying fluid by a sharp interface. There is negligible mixing of fluids between the two layers.

Experiments and some of the qualitative observations described here have been obtained by Campbell & Turner (1987) using aqueous solutions and their solid equivalents. Their experiments also involved saturated solutions which crystallized as they melted the solid boundaries. The addition of crystallization to some of their experiments did not modify the basic features described here.

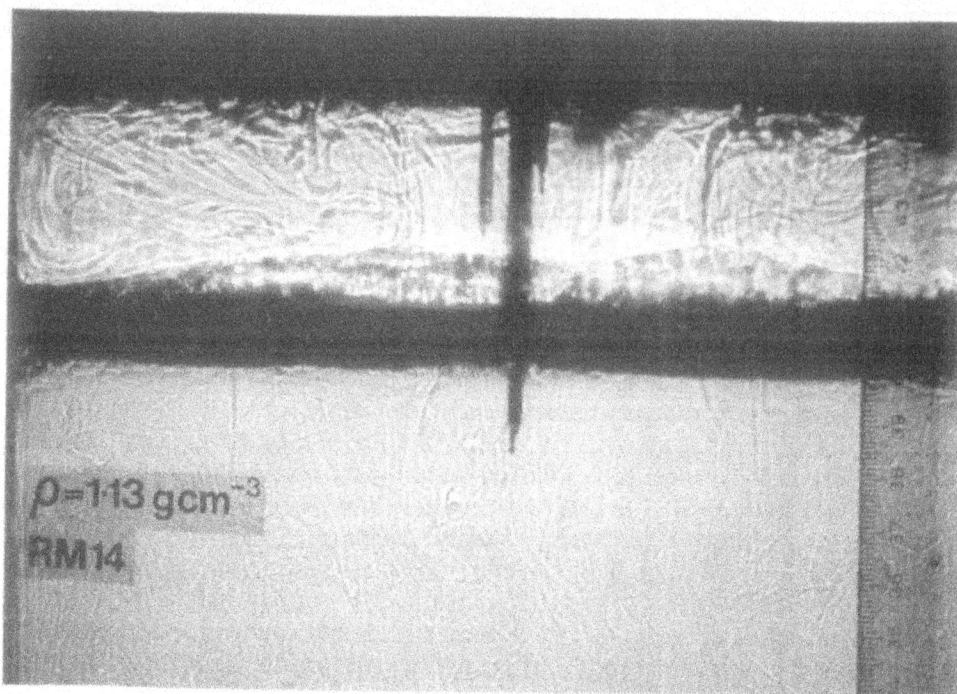

FIG. 1. The photograph shows a layer of molten PEG wax beneath a melting roof. The molten wax is separated by a sharp interface from the underlying hot, NaNO$_3$ solution. The interface appears as a dark band on the photograph due to refraction of light, but is in fact knife sharp

Melting at a roof can also occur when the melt generated is heavier than the underlying solution. In this case the melt sinks immediately into the solution and mixes with it. Experimental and theoretical studies of this case are given in Huppert & Sparks (1988a), but are not considered further here as the situation of crustal rocks generating melt which is denser than basalt is not likely to be common. Melting can also occur at the floor of a sill. Experimental studies of floor melting are presented by Huppert & Sparks (1988b). In this case a theoretical treatment has yet to be developed, because it is complicated by the opposing compositional and thermal effects on density at a floor where light melt is generated. Although there are important geological applications, the discussion of this case must await further research.

Finally, the wax in the experiments has a single melting temperature and thus behaves like a eutectic composition rock. In natural situations the additional complexities of a melting interval will have to be considered. In addition basalt will initially chill against country rock. However, the thermal energy in the convecting basaltic layer will quite quickly melt the chill and we hence omit further discussion of it from this paper. A very brief discussion, which presents the time scale to melt the chill, for a similar situation appears in Huppert (1986) and a more extensive description is currently in preparation (Huppert, 1988).

THEORY OF A MELTING ROOF

General considerations

A theoretical description is outlined for melting of a roof when a stable layer of low density, silicic magma is generated. A detailed theoretical analysis of the general case of roof

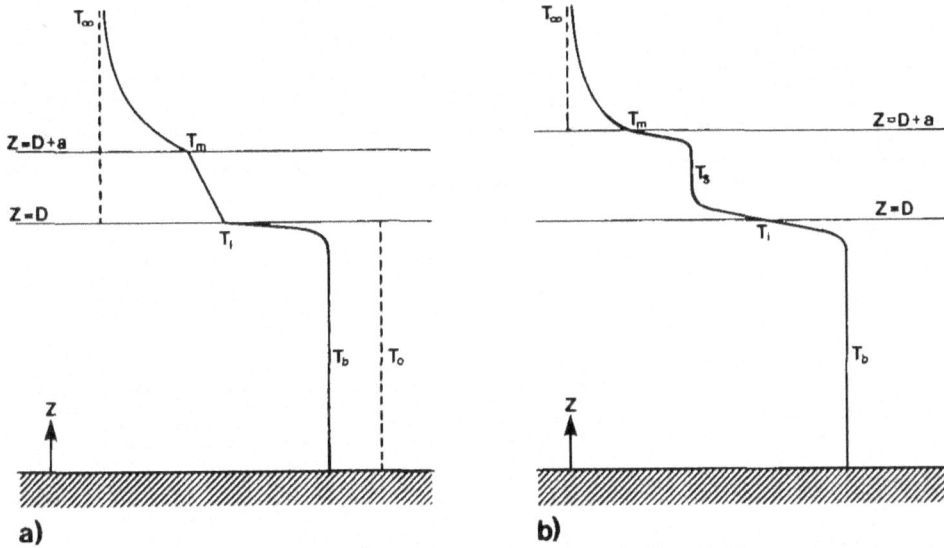

FIG. 2. Boundary conditions of a melting roof region showing temperature variations with height (a) Initial stage in which the melt layer is stable and heat is transferred by conduction from the underlying hot convecting fluid to the melting roof. (b) Later state in which the melt layer convects and has a uniform internal temperature.

melting is presented elsewhere (Huppert & Sparks, 1988a). Only the essential results of the analysis, modified for the geological application, will be given here.

We consider a layer of hot basaltic magma with an initial thickness D, and temperature $T_b(0)$, emplaced beneath a solid roof and we assume that the solid roof above it has a very large height and has an initial uniform temperature T_0. The roof melts at a temperature T_m which is attained at the contact with the underlying basaltic layer after a short time. Thereafter a stable layer of silicic magma forms and grows with time (Fig. 2). Initially the silicic layer is too thin to convect and heat is transferred by conduction across the layer (Fig. 2a). The temperature at the interface between the silicic magma and the basaltic layer is $T_i(t)$. It is assumed that the interface temperature T_i is too high for a chill to form and this will be justified a posteriori. Eventually the silicic magma layer increases in thickness to the stage where the critical thermal Rayleigh number is exceeded and convection begins. From then on the silicic magma layer is considered to become thermally well-mixed within its interior (Fig. 2b), and can be characterized by a single temperature $T_s(t)$. The thickness of the melt layer will be denoted by $a(t)$. The basaltic layer, which cools and crystallizes as it loses heat, eventually becomes so crystal-rich that it effectively becomes a solid and convection stops. However, the overlying silicic layer continues to convect and to melt its roof. Finally, the silicic layer itself becomes so crystal-rich and viscous that convection also ceases.

The requirements of a theoretical model are to predict the temperature of the interface between the magma layers, $T_i(t)$, the temperatures of the two magma layers, $T_b(t)$ and $T_s(t)$, and the thickness of the melt layer $a(t)$, as functions of time. It is also pertinent to calculate the time at which convection begins in the silicic magma layer, the time convection ceases in the basaltic layer and the time that convection ceases in the silicic magma layer.

With the assumption of active convection in the basaltic magma, the heat flux to the silicic magma above is given by

$$F_b = \rho_b c_b J_b (T_b - T_i)^{4/3}, \qquad (1)$$

where ρ_b is the density and c_b the specific heat of the basalt and J_b is defined by

$$J_b = 0 \cdot 1 (\alpha_b g \kappa_b^2 / v_b)^{1/3}, \tag{2}$$

where α_b is the coefficient of thermal expansion of the basalt, g the acceleration due to gravity, κ_b the thermal diffusivity of the basalt, and v_b the kinematic viscosity of the basalt. Since the silicic magma layer thickness, a, is always significantly less than the length scale of thermal diffusion $(\kappa_s t)^{1/2}$, where κ_s is the thermal diffusivity of the silicic magma, we can use the linear conductive profile to describe the temperature in the silicic magma before it begins to convect. The conductive flux through the silicic magma is then given by

$$F_s = k_s (T_i - T_m)/a, \tag{3}$$

where k_s is the thermal conductivity of the silicic magma. The conservation of heat in the basaltic layer of thickness D, across the interface between the two magmas and at the solid–melt interface at the roof, then becomes:

$$\rho_b c_b D \frac{dT_b}{dt} = -F_b + \rho_b L_b D x_b'(T_b) \frac{dT_b}{dt}, \tag{4}$$

where $x_b'(T)$ is the derivative of the crystal content, x_b, with respect to temperature, and L_b is the latent heat of crystallization of the basalt;

$$F_b = F_s; \tag{5}$$

and

$$\frac{da}{dt} = k_s H_s^{-1}(T_i - T_m)/a, \tag{6}$$

where

$$H_s = \rho_s [c_s (T_m - T_0) + L_s] \tag{7}$$

and ρ_s is the density of the solid, c_s its specific heat, and L_s the heat of fusion. H_s is the heat needed to raise unit volume of solid crust from its far-field temperature to its melting temperature and also to melt it. The initial conditions

$$T_b = T_b(0), \quad T_i = T_m, \quad \text{and} \quad a = 0 \quad (t = 0) \tag{8a,b,c}$$

complete the specification of the problem.

Substituting (3)–(5) into (6), and using (8a,b) in taking the first integral, we obtain

$$a = H_s^{-1}\{\rho_b c_b D[T_b(0) - T_b] + \rho_b L_b D x_b(T_b)\}. \tag{9}$$

Substituting (1), (3), and (9) into (5), we obtain the following implicit relationship between T_b and T_i

$$T_i = T_m + \rho_b^2 c_b^2 D J_b H_s^{-1} k_s^{-1} [T_b(0) - T_b + L_b c_b^{-1} x_b(T_b)](T_b - T_i)^{4/3}. \tag{10}$$

Finally (1) and (4) can be written

$$\frac{dT_b}{dt} = -(J_b/D)(T_b - T_i)^{4/3}/[1 - L_b c_b^{-1} x_b'(T)]. \tag{11}$$

Equations (8)–(11) describe the heat transfer and melting process while the silicic magma layer is stable to convection and transfers heat by conduction. The solution and derivation of the equations are discussed in more detail in Huppert & Sparks (1988a). They cease to be a valid description when the Rayleigh number of the melt layer exceeds a critical value of order 10^3, where the Rayleigh number is defined as

$$Ra_s = \alpha_s g(T_i - T_m)a^3/\kappa_s \nu_s. \tag{12}$$

Convection in the melt layer is then initiated and both (3) and (6) need to be altered. It is possible, however, that R_s, which is small both initially and after a long time, never exceeds the critical value.

Generally, however, convection in the silicic magma layer will set in and this is demonstrated below. Initially the convection will be weak, but as the layer thickness increases it will build up to such a strength that a turbulent layer of uniform temperature forms for which the transfer of heat can be described by the four-thirds relation. We now analyse this situation.

Consider the situation sketched in Fig. 2b. The temperature of the lower, hot basaltic layer is denoted by $T_b(t)$ and that of the silicic magma layer of thickness, $a(t)$, by $T_s(t)$. The heat flux from the basaltic layer is given by equation (1) while that from the silicic melt to the roof is given by

$$F_s = \rho_s c_s J_s (T_s - T_m)^{4/3}. \tag{13}$$

The heat flux into the silicic layer has to be equal to the heat flux out of the basaltic layer and thus F_b can also be defined as

$$F_b = \rho_s c_s J_s (T_i - T_s)^{4/3}.$$

Thus

$$\rho_b c_b J_b (T_b - T_i)^{4/3} = \rho_s c_s J_s (T_i - T_s)^{4/3}, \tag{14}$$

from which we deduce that

$$T_i = (T_b + yT_s)/(1 + y), \tag{15}$$

where

$$y = (\rho_s c_s J_s / \rho_b c_b J_b)^{3/4}. \tag{16}$$

Conservation of heat in the basaltic and silicic melt layers requires that

$$\frac{dT_b}{dt} = -(J_b/D)(T_b - T_i)^{4/3}/[1 - L_b c_b^{-1} x_b'(T_b)] \tag{17}$$

$$\frac{dT_s}{dt} = (J_s/a)\left\{[(T_i - T_s)^{4/3} - (T_s - T_m)^{4/3}] - \frac{da}{dt}[T_s - T_m - L_s c_s^{-1} x_s(T_s)]\right\} \Big/$$
$$\times [1 - L_s c_s^{-1} x_s'(T_s)], \tag{18}$$

where $x_s(T_s)$ is the crystal content of the silicic magma layer and $x_s'(T_s)$ is its derivative with respect to temperature. Conservation of heat at the silicic solid/melt interface becomes

$$L_s \frac{da}{dt} = c_s J_s (T_s - T_m)^{4/3} + c_s \kappa_s \frac{\partial \theta}{\partial z}(a, t), \tag{19}$$

where $\theta(z, t)$ is the (conductive) temperature in the solid country rock with respect to a fixed, vertically upwards z-axis. This temperature profile is governed by the diffusion equation

$$\frac{\partial \theta}{\partial t} = \kappa_s \frac{\partial^2 \theta}{\partial z^2} \tag{20}$$

$$\theta(z, 0) = T_0 \tag{21}$$

$$\theta(a, t) = T_{\mathrm{m}} \tag{22}$$

$$\theta \to T_0 \quad (z \to \infty). \tag{23}$$

If the roof melts at a steady rate, the solution of (19)–(23) can be expressed as (Huppert, 1986)

$$\frac{da}{dt} = F_s/H_s. \tag{24}$$

We determined numerical solutions of (15)–(18) in addition to either (19)–(23) or merely (24). The differences were negligible.

The final phase occurs when the basaltic magma layer cools to a temperature at which there are so many crystals present that the viscosity becomes very large and convection in the basalt ceases. However, the silicic magma will continue to convect, and will generate further melt at the roof. The silicic magma will continue to cool and crystallize until it also becomes so crystal-rich that convection ceases altogether. The amount of heat lost by the convecting silicic magma melting its roof is very large compared with the heat gained from the underlying, solidified basalt layer by conduction. Consequently, as a first approximation, the heat flux from the basalt can be neglected. The equations for temperature and thickness variations are then obtained by neglecting (17) and the term $(T_i - T_s)^{4/3}$ in (18). At this stage the conduction of heat into the roof becomes increasingly important as the convection weakens and the conductive terms dominate eventually.

A previous study of the melting of the roof of a magma chamber was presented by Irvine (1970) who modelled the process in terms of thermal conduction. He demonstrated that substantial volumes of silicic magma could be generated by this mechanism. The major difference in the new treatment is the incorporation of convection which dramatically increases the rate of melting and influences the thermal and fluid dynamical history of the silicic melt layer.

Modifications and assumptions for geological application

For application of the roof melting theory to crustal melting by basalt, some additional relationships are required, as well as discussion of under what circumstances the theory can apply in nature.

Melting temperature of crustal rocks

When crustal rocks progressively melt they can be regarded as partially molten solids with mechanical strength when the fraction of melt is low (Shaw, 1980). At this stage the crystalline phases form an interconnected framework. However, beyond some critical melt fraction the connectedness of the crystalline framework is destroyed and the rock is said to have been converted into magma. Experimental and theoretical studies (van der Molen & Paterson, 1979; Shaw, 1980; March, 1981) indicate that there is typically a very large variation in viscosity by up to ten orders of magnitude for small changes of melt fraction and temperature in the vicinity of the critical melt fraction. Various estimates from laboratory and theoretical studies indicate that between 30 and 50% melt is a reasonable range for the critical melt fraction (Marsh, 1981; Wickham, 1987). We assume that the temperature at which the critical melt fraction is attained is a sensible value to take as the effective fusion temperature of the crust roof rock in applying the theory.

Temperatures at which crustal rocks reach the critical melt fraction will vary according to the composition of the rock and the availability of volatiles, principally water. Upper crustal rocks contain an intrinsic water content, in the form of hydrous minerals such as mica and hornblende, which vary from a fraction of a percent to 2% depending on the rock type. The breakdown of these minerals can be the cause of substantial melting over a narrow temperature interval (Wyllie, 1984). On the whole, most crustal rocks have a low intrinsic water content (typically less than 1%) so their melting is close to the anhydrous case. On the other hand water may be added to the region that is melting. Degassing of basaltic magma could add substantial amounts of CO_2 and H_2O to the overlying silicic magmas. Circulating hydrous metamorphic fluids could also be added to increase the water content during melting (Wickham, 1987).

Since nature may provide situations varying from melting of completely dry lithologies to melting of water-saturated rock, we have chosen four different cases to cover the spectrum of possibilities. For most of the calculations (the standard case) we assume a granodiorite composition, since this is thought to be the average of the upper crust (Taylor & McLennan, 1985). Experimental studies (Wyllie, 1984) at pressures appropriate to the mid-crust suggest that a fusion temperature of 850 °C is a reasonable estimate for a water content of 2%. For the second 'wet' case we assume a water content of 6% which is an appropriate value for water-saturated granite which would be completely molten at 850 °C and 5 kb pressure. The main influence of water is to decrease viscosity substantially. For the third 'dry' case, we assume no intrinsic water and set the effective fusion temperature at 950 °C. These calculations would be suitable for considering the melting of relatively anhydrous rock or more refractory lithologies. In the dry case the viscosities are somewhat higher due to the absence of water, but this is somewhat offset by the high temperatures of the magmas. Finally, the fourth case is for a granite (*sensu stricto*) which displays eutectic melting behaviour at 850 °C.

Relationships between temperature, crystal content and viscosity

The detailed phase relationships of basalt magmas and crustal melts are complicated and depend on composition, total pressure and volatile partial pressures. The crystallinity is generally a nonlinear function of temperature between solidus and liquidus. A parameterization of one magma or rock type under specified conditions would be difficult to apply to another rock or magma or to different conditions. At this stage we consider that the broad features of crustal melting are best examined with the simplest possible parameterization. In actual fact there is very little information available in the petrological literature on how crystal content varies with temperature in silicic systems. We have chosen the following empirical formulae as representative, based on experimental information such as that presented by Wyllie (1984).

$$x_b = 7200\,T^{-1} - 6 \qquad (1091 < T < 1200) \qquad (25)$$

and
$$x_s = 0.65(1000 - T)/150 \qquad (850 < T < 1000) \qquad (26a)$$

$$x_s = 0.65(1100 - T)/150 \qquad (950 < T < 1100), \qquad (26b)$$

where x_b and x_s are the fractional crystal contents of the basalt and silicic magmas and T is the temperature in degrees centigrade. Equation (26a) represents the standard and wet cases and (26b) represents the completely dry case.

In most of the calculations presented in this paper we assume that the crystals formed in both the basaltic and silicic magmas are small enough to remain suspended in the turbulently convecting magma (Sparks *et al.*, 1984). They thus contribute to an increase in the viscosity of the magmas as the temperature decreases. The empirical relationships that describe the variation of viscosity with temperature and crystal content are based on the

HERBERT E. HUPPERT AND R STEPHEN J SPARKS

work of Shaw (1963, 1972, 1980) and Marsh (1981), and are

$$v_b = 10^3 (1 - 1.67 x_b)^{-2.5} \, \text{cm}^2 \, \text{s}^{-1} \quad (1091 < T < 1200) \tag{27}$$

and

$$v_s = \gamma_0 e^{1.85 \times 10^4/(T+273)} (1 - 1.53 \, x_s)^{-2.5} \, \text{cm}^2 \, \text{s}^{-1} \quad (850 < T < 1000), \tag{28}$$

where v_b and v_s are the kinematic viscosities of the basalt and silicic magma, γ_0 is a pre-exponential constant which has a value of 62 for the completely dry case, 0.62 for the case with 2% H_2O, and 0.62×10^{-3} for the wet case with 6% H_2O.

If the country rock is composed of granite (*sensu stricto*) with eutectic behaviour, then crystallization occurs at a single temperature. This case has been examined by modifying equation (28) so that no crystals are present above 850 °C

$$v_s = 0.62 e^{1.85 \times 10^4/(T+273)} \, \text{cm}^2 \, \text{s}^{-1} \quad (T > 850). \tag{29}$$

Table 1 lists values of the various physical parameters used in the calculations that follow.

TABLE 1

Value of physical parameters used in calculations

α_s, α_b	$5 \times 10^{-5} \, \text{K}^{-1}$
g	$981 \, \text{cm} \, \text{s}^{-2}$
κ_b, κ_s	$8 \times 10^{-3} \, \text{cm}^2 \, \text{s}^{-1}$
v_b	$10^3 \, \text{cm}^2 \, \text{s}^{-1}$
ρ_s	$2.3 \, \text{gm} \, \text{cm}^{-3}$
ρ_b	$2.7 \, \text{gm} \, \text{cm}^{-3}$
c_s, c_b	$0.32 \, \text{cal} \, \text{gm}^{-1} \, \text{K}^{-1}$
L_b	$100 \, \text{cal} \, \text{gm}^{-1}$
L_s	$70 \, \text{cal} \, \text{gm}^{-1}$

APPLICATION TO CRUSTAL MELTING

The standard case

We first consider a typical calculation to illustrate the general features which are displayed by all our results. We consider a 500-m sill of basalt at a temperature of 1200 °C emplaced into crust with a fusion temperature of 850 °C and an initial temperature of 500 °C. Initially the interface at the roof will be brought to an intermediate temperature of somewhat over 850 °C. Any chilled basaltic crust, initially formed against the roof, will melt before the overlying roof rocks melt and this dense crust will be mixed back into the convecting layer of basalt. The interface temperature will thus increase to near the basalt temperature. A layer of magma will develop between the interface at 1200 °C and the 850 °C isotherm. Initially this magma layer will be stable and conduct heat, but when the Rayleigh number exceeds about 2000 convection will begin. With this critical Rayleigh number we calculate that the magma layer will begin to convect after 7 days when its thickness will be only 1.2 m. The Rayleigh number exceeds 10^6, and the model becomes strictly valid, after 1 y when the melt layer thickness is approximately 7 m.

There are at least four reasons for believing that our calculation somewhat under-estimates the time-scale and thickness of the layer when convection is initiated. First, there is a strong viscosity gradient across the silicic magma layer reflecting the substantial change in temperature and crystal content across the layer. The thickness of the thermal boundary layer needed to initiate convection may well be somewhat greater if there is a strong viscosity gradient (Richter *et al.*, 1983; Jaupart & Parsons, 1985). Second, it is probable that the crystal-rich parts of the melting region have non-Newtonian properties. In particular a yield strength would inhibit convective instabilities in the thermal boundary layer. Available data on yield strengths in crystal-rich magmas (McBirney & Murase, 1984) do not suggest that yield strength will have any important influence on the criteria developed here. Third, many crustal rocks are likely to have a strongly anisotropic layered structure (as in many migmatites) and the occurrence of continuous layers of refractory lithologies in the zone of melting could inhibit melting for periods. Fourth, the formation of a chilled crust of basalt may temporarily inhibit heat transfer until it breaks away. Although we consider that such a crust is inherently unstable, its behaviour may be complicated and would be impossible to model in detail. However, the important point to emphasize is that the time-scale is very short compared to the overall evolution of the system and that only a thin crustal melt layer is required for convection to begin.

When both the basalt and silicic magma layers are convecting three temperatures characterize the system. Figure 3 shows the variation of the interior temperatures of the convecting magma layers and the interface temperature as functions of time. The silicic magma layer decreases in temperature continuously with time, because melt generated at the roof is being incorporated continuously. The evolution of the silicic magma must thus involve simultaneous melting and crystallization which at first sight appears a contradiction. However, melting is confined to the thermal boundary layer adjacent to the solid roof. Any parcel of fluid in this thermal boundary layer must experience a temporal increase in temperature causing melting to occur. Once a particular region of the thermal boundary layer becomes unstable it will be mixed into the underlying convecting magma where it will attain the mean internal temperature. However, the interior decreases in temperature with

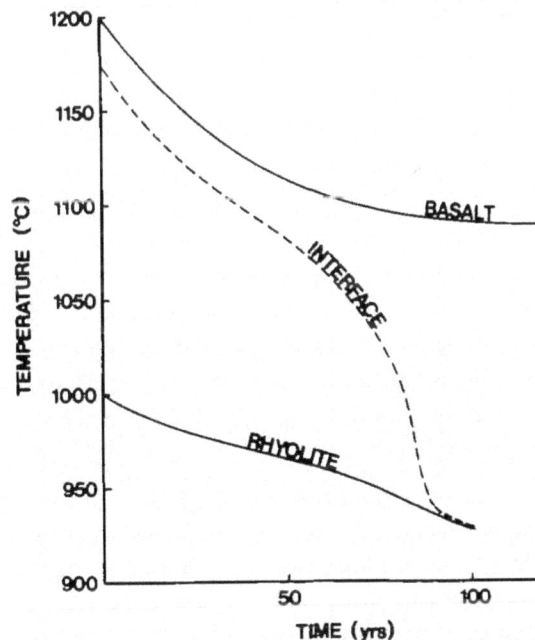

FIG. 3. Variation of the temperatures of the basalt and silicic magma layer with time for a 500 m sill emplaced into crust at 500 °C. The interface temperature between the two magma layers is also indicated.

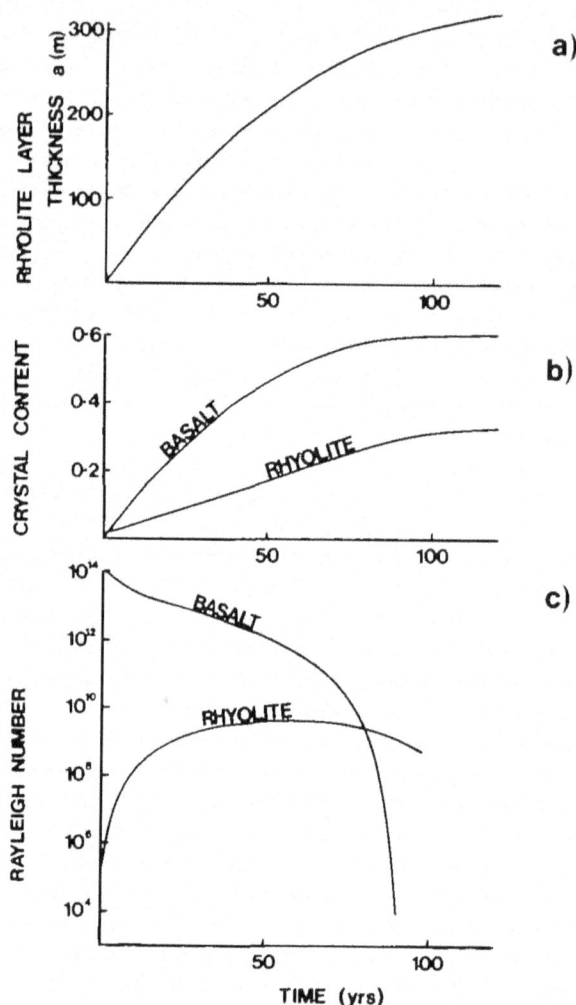

FiG. 4. Variations of (a) the silicic magma layer thickness, (b) the crystal contents of the magmas and (c) the Rayleigh number of the magma layers with time for a 500 m thick sill emplaced into crust at 500°C.

time and is thus a region of cooling and crystallization (Figure 4). Possible petrological consequences of this convecting structure are discussed below.

The temperature at the interface between the magma layers is initially much closer to the temperature of the basalt than of the silicic magma. This result reflects the difference in viscosity between the magmas. The higher viscosity magma requires a considerably larger temperature difference across the layer than the lower viscosity magma to sustain the same heat flux. Consequently the interface temperature is much closer to the lower viscosity, basaltic magma than to the silicic magma. After approximately 45 y, however, the interface temperature falls below the temperature at which the basaltic magma contains an excess of 60% crystals. After this time the interface temperature rapidly decreases towards the temperature of the silicic magma layer. This occurs because as the basalt layer becomes highly crystalline it becomes more viscous than the silicic magma layer. After approximately 90 y the basalt magma becomes so crystal rich and viscous that convection in it ceases. The cross-over in viscosity of the magma layers is also reflected in the variation of Rayleigh

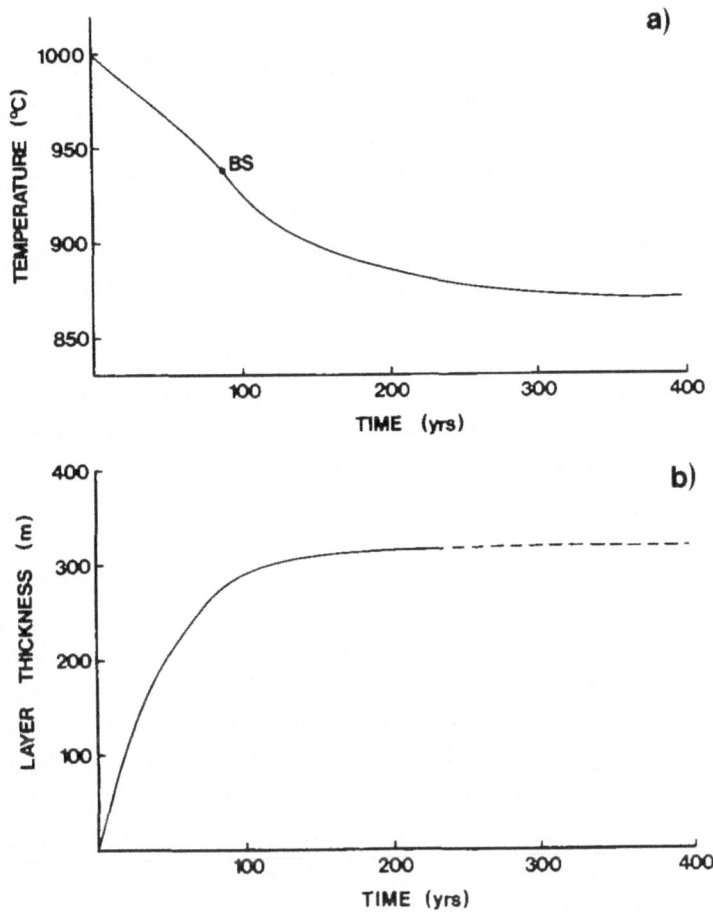

FIG 5 Variations of (a) temperature, and (b) thickness of the silicic magma layer with time for a 500 m thick sill emplaced into crust at 500 °C. The point BS marks when the basalt ceases to convect, because it is too crystal rich

number with time (Fig. 4c). The Rayleigh number of the basalt layer decreases with time as crystallization proceeds and viscosity increases. The Rayleigh number of the silicic magma layer initially increases as the layer increases in thickness, reaching a maximum after approximately 65 y.

Once convection has ceased in the basaltic layer, after 90 y, the silicic magma still has a temperature of 934 °C and a Rayleigh number of $1\cdot1 \times 10^9$, with a large temperature difference between the melting roof and the interior. The silicic magma layer will continue to melt its roof, to increase in thickness, to decrease in temperature and to increase in crystal content. Figure 5 shows some of these variations with time. Eventually the silicic magma layer will become so crystal-rich and viscous that the Rayleigh number will fall beneath the critical value and this layer will once again be a partially molten solid rather than magma. The time for the Rayleigh number to fall beneath 10^3 is about 1000 y as the crystallinity approaches 65%. However, this estimate is quite sensitive to slight changes in the empirical constants used in expressions defining relationships between crystallinity, temperature and viscosity (equations (26)–(28)); and these empirical constants are only known approximately. Furthermore, some extra heat will be conducted away from the underlying basalt layer and this has not been considered. A better indication of the time-scales involved is to choose some lower crystallinity where variations with time are still quite significant. Most silicic magmas erupt with crystal contents below 50%. The time taken to come to 50% crystallinity is 215 y, showing that most of the crystallization and cooling in the silicic magma takes place on a very short time-scale. The thickness of the silicic layer at this stage is 315 m. Little further growth will occur after this time because most of the heat convected out of the silicic layer is subsequently used to heat the roof above the melting front (see below).

Further cooling and crystallization of the non-convecting, partially molten silicic layer will occur by conduction over a much longer time period. We estimate that this layer would take at least 10^4 y to cool down to the ambient temperature of 500 °C. Thus the emplacement of the basalt sill creates a long-lived region of hot, partially molten crust. If another sill of basalt is emplaced within 10^4 y then a larger layer of silicic magma will be generated, because the crust is hotter and already partially molten.

Finally we consider the amount of heat conducted into the roof ahead of the melting boundary. The distance, δ, over which there exists a conductive temperature profile can be defined as

$$\delta = \frac{1}{(T_m - T_0)} \int\limits_{a(t)}^{\infty} (T_b - T_0)\,dz. \tag{30}$$

This distance is plotted as a function of time in Fig. 6 for the case of a 500 m sill. Much beyond this distance the temperature is ambient at 500 °C. The proportion of the thermal energy of the convecting silicic layer which is transferred into the overlying region by conduction is indicated. While the layer is still convecting, the conductive layer is thin (a few metres) and the amount of heat lost to this region represents a small proportion of the total heat budget. However, once the basalt solidifies, an increasing proportion of the heat convected out of the silicic layer is conducted away by the country rock rather than melting it. Concurrently the thickness of the conductive profile increases and the roof melting eventually ceases. As indicated in Fig. 5 after about 250 y the silicic layer has ceased to grow.

Variations in the main-parameters

This section explores how variations in the initial conditions influence the evolution of the magma layers.

Basalt layer thickness

The effect of changing the thickness of the basalt layer can be illustrated by calculating the time taken to reach 60% crystallinity in the basalt and by calculating the thickness and temperature of the silicic magma layer at that time. The results are listed in Table 2 which shows that the time-scale of solidification (τ_1) increases with the basalt layer thickness, which reflects the increasing heat content. However, even for a 1·5 km sill the time is still only 272 y, confirming that the melting of the crust has a strong cooling effect on basalt magma.

The calculated temperatures of the silicic magma at τ_1 for different basalt layer thicknesses are virtually identical at approximately 935 °C showing that the silicic magma layer will in all situations be heated up to a high temperature and that the process is closer to total fusion of the crust than to partial melting. The time-scale (τ_2) for reaching 50% crystallinity in the silicic magma layer is also listed in Table 2, and shows that the time-scale for most of the crystallization in the silicic magma layer is also short. The implication is that when silicic magmas are surrounded by rocks of similar composition and melting relations they

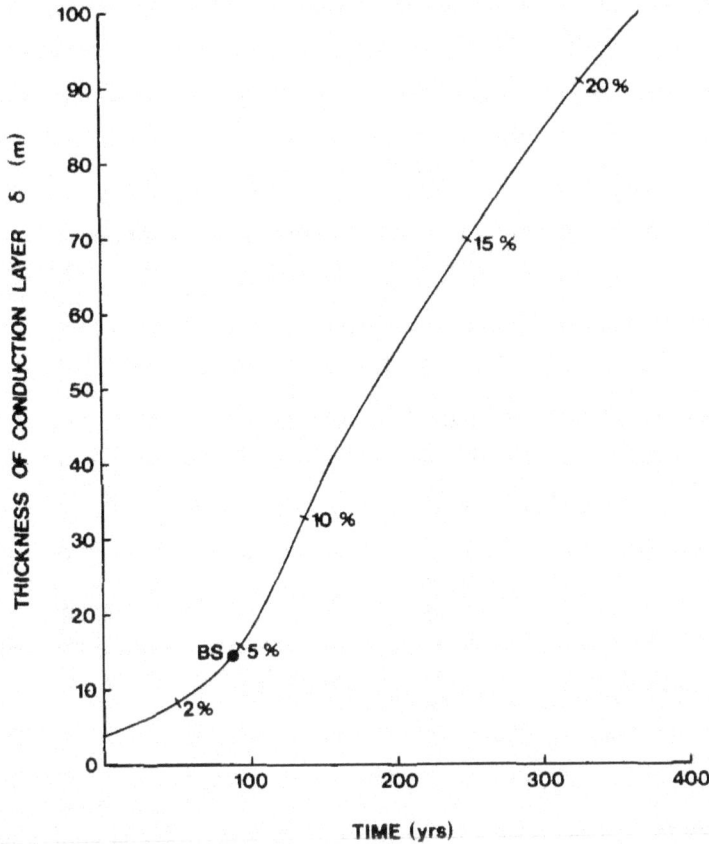

FIG. 6. Variation of the thickness, δ, of the conduction layer in overlying roof rocks as a function of time. The percentage values indicate the amount of heat that is conducted into the roof as a fraction of that in the silicic magma layer.

TABLE 2

The table lists the time (τ_1) at which the basalt layer becomes 60% crystalline and ceases convection for different basalt layer thicknesses. The silicic magma layer thickness and temperature at τ_1 are also listed. The time τ_2 is that at which the silicic magma layer becomes 50% crystalline. The country rock temperature is 500°C.

Basalt layer thickness (m)	$\tau_1(y)$	$a(\tau_1)$m	$T(\tau_1)$°C	$\tau_2(y)$
10	1·6	5·2	943	3·0
50	8·6	27	937	17·5
150	26·5	84	935	58
500	89·7	289	934	215
1500	271·8	878	933	650

cannot exist as magmas for long periods of time. They cool themselves off rapidly by melting their boundaries.

If the sills are too small, the model is not appropriate, because the Rayleigh number of the silicic layer does not reach the critical value. The smallest sill for which calculations are presented is 10 m and we estimate that sills smaller than 3 or 4 m in thickness cannot generate a convecting silicic magma.

Country rock temperature

The influence of country rock temperature on the evolution is illustrated in Table 3 for a 500 m sill. As this temperature increases the amount of heat required to melt a given mass of crust decreases, and so the silicic magma layer thickness which evolves increases. It can reach 1·2 km at 50% crystallinity if the crust is already partially molten at a temperature of 850°C. However, the time-scale and variations of temperature and crystallinity of the silicic magma layer are negligibly affected by changing the country rock temperature.

Table 3 suggests some of the effects which will arise if there are multiple injections of basalt into the crust. If repeated injections are made on a time-scale that is short compared to the overall conductive cooling time-scale ($> 10^4$ y) then each successive injection will form a greater amount of crustal melt. For example if the crust warms up by earlier basaltic intrusions from 500 °C to being in a partially molten state at 850°C the amount of silicic magma generated increases by a factor of four for a given volume of injected basalt.

We emphasize that the model discussed here is unlikely to be appropriate if the country rock is much colder than 500°C, because the contact temperature can never reach the effective fusion temperature of the country rock. Thus we would not expect this model to be appropriate in cold crust where a chilled margin in the basalt sill will effectively insulate the country rock from reaching high temperatures. The model is basically applicable to deeper levels of the crust that are already hot and to upper levels of the crust where the rocks have already been pre-heated by earlier intrusions.

Lithology and water content

The rocks of the continental crust vary considerably in their melting and crystallization behaviour and in their viscosity. Fusion temperature and the water content are likely to be the most important variables, because these parameters control the boundary conditions at the roof and the viscosity of the silicic magma layer. Various possible cases are considered and some typical results are listed in Table 4. A wet magma with a water content of 6% and effective melting temperature of 850°C has a considerably lower viscosity (equation (28)). The wet water-saturated magma consequently convects more readily and melts the roof and cools more rapidly. Thus it takes a shorter time to reach the condition where the basalt

TABLE 3

The table lists the time (τ_1) at which the basalt layer becomes 60% crystalline and ceases convection for different country rock temperatures in the case of a 500 m basalt sill. The silicic magma layer thickness, $a(\tau_1)$ and its temperature, $T(\tau_1)$ are listed. The time τ_2 is that at which the silicic magma layer becomes 50% crystalline

Country rock temperature °C	$\tau_1(y)$	$a(\tau_1)$	$T(\tau_1)$	$\tau_2(y)$
500	89·4	294·8	934	230
750	90·2	498·0	939	385
850	90·6	688·0	943	600
850*	91·0	986·0	941	795

*850 assumes that the crust is already 35% partially molten with a reduced heat of fusion ($L_s = 42$ cal gm^{-1})

TABLE 4

The table lists the time (τ_1) at which the basalt layer becomes 60% crystalline and ceases convection. The table compares the results for the dry case, the eutectic case, the standard case, and finally the case where the silicic magma layer is always water-saturated at 6% H_2O. The silicic magma layer thickness and temperature at τ_1 are also listed. The time τ_2 is that at which the silicic magma layer becomes 50% crystalline. Results for different country rock temperature are presented

Country rock temperature °C	$\tau_1(y)$	$a(\tau_1)m$	$T(\tau_1)°C$	$\tau_2(y)$
500 (dry)	658	197	1010	805
500 (standard)	89·7	289	934	215
500 (eutectic)	76·0	276	910	195
500 (wet)	25 1	337	894	32
750 (wet)	25·4	629	903	45
850 (wet)	25·6	917	908	65
* 850	25·7	1365	909	85

*850 as described in Table 3.

solidifies (Table 4). At this time the temperature of the magma is lower, which can again be attributed to its lower viscosity. The time for the magma to reach 50% crystallinity is likewise much shorter than the dry case. We conclude that if the magmas produced by melting have a lower viscosity than the standard case then the melting process takes place more rapidly, but cooler magmas are generated at any given stage.

On the other hand, for melting an anhydrous lithology with an effective fusion temperature of 950°C (the completely dry case) the viscosity is considerably greater (equation (28)). The time to generate the melt layer and solidify the basalt is increased and very high temperature magmas are generated.

Although the results for the extreme cases of completely dry and water-saturated depart from the standard case, they represent only relatively small changes in the time-scales involved or in the temperature and crystallinity histories of the magmas. We conclude that irrespective of the lithology of the crust, the water content of the magmas, the thickness of the basalt sill, and temperature of the country rock, the time-scale for solidification of basaltic sills in the crust and melting and crystallization of the adjacent crust is only tens or a few hundred years.

If the crustal rocks were granite (*sensu stricto*) with a minimum melt composition and a single 'eutectic' melting temperature, the time-scales or amount of melt generated would not be significantly affected. This case has been considered using equation (29) and the results are given in Table 4 as the eutectic case. At temperatures of 850°C or greater the rhyolitic magma would be entirely molten or even superheated. In comparison to the standard case, the time-scale τ_1 for melting the magma, the temperature at time τ_1 and the magma layer thickness are all slightly less. This is expected as more heat is required to melt the country rock totally and the crystal-free magma has lower viscosity at a given temperature. Thus a granodiorite composition wallrock would tend to form porphyritic dacites whereas truly minimum melt granitic wall rocks would tend to produce crystal-poor or crystal-free rhyolitic magmas.

Finally, we consider the case where the wall rocks are entirely refractory and do not melt. If it is assumed that the basalt convects freely and does not solidify against the roof, the time to reach 60% crystallinity is 700 y and would be longer if a roof chill formed. These results show that the cooling rate of basalt sills typically increases by about an order of magnitude when its roof melts. This is discussed in greater detail in Huppert & Sparks (1988a), particularly in the Appendix.

Other complicating factors

The roof melting model does not incorporate all the factors and complexities of nature. We can identify a number of effects which will modify the evolution of the system, but at the moment we only make qualitative statements on their influence.

The crustal rocks could be strongly heterogeneous with interbedded layers of refractory rock with high melting temperatures as is commonly the case in gneisses and regional metamorphic rocks. If the boundary between the melting roof and silicic magma layer reaches a thick refractory layer further melting could be inhibited or even stopped. In migmatites mafic refractory layers typically disrupt into blocks which may resist melting and become enclaves within the silicic magma layer. Some of these dense enclaves could be sufficiently large and dense to fall through the interface into the basalt layer where they may be partially or wholly resorbed or undergo further chemical reactions with a basalt. Such a process would also complicate the thermal and compositional evolution of both magma layers, as well as having interesting geochemical and petrological consequences.

During the evolution of the basalt layer the interface temperature can reach a value where a crystal-rich crust may develop between the layers through which heat must be conducted. Although such a layer would slow heat transfer down the two magma layers could still convect vigorously and the rates would still be controlled by convection.

Simultaneous crystallization and melting

The physical structure and dynamical behaviour of the melting roof region is illustrated in Fig. 7. The effects expected in this region represent a conceptual change in thinking about the relationship between melting and crystallization in magmas. An axiom of many petrological models is that melting occurs in one place (the source) and crystallization occurs at a later time usually in another place (a magma chamber). This is not, however, what happens in crustal melting as described here, where melting and crystallization of the crust occur simultaneously in the source region. The roof region consists of three zones: a region of partial melting, a thermal boundary layer and an internal hot convecting region. The behaviour predicted in each of these regions is now discussed in turn.

In the partially molten roof the temperature gradient decreases from the effective fusion temperature to the ambient country rock temperature over a distance which is determined by the rate of melting of the roof and the conduction of heat. The situation is profoundly different to a magma chamber where crystallization occurs at the margins. In this latter case, the heat loss from the magma chamber and the thermal gradient in the country rock are entirely determined by conduction in the country rock. In a melting roof the heat loss from the magma body is determined by the rate of melting at the roof and therefore by the boundary condition for melting. There is only a very weak dependence due to conductive heat loss through the country rock. The heat leaked into the roof zone is typically only a few per cent of the total heat transferred from the basalt to the silicic magma. The thermal gradient in the partially molten roof is in fact determined by the melting conditions at the roof.

Beneath the roof there is a thermal boundary layer in which further melting must occur. The thermal boundary is a dynamical region where there is a very steep temperature

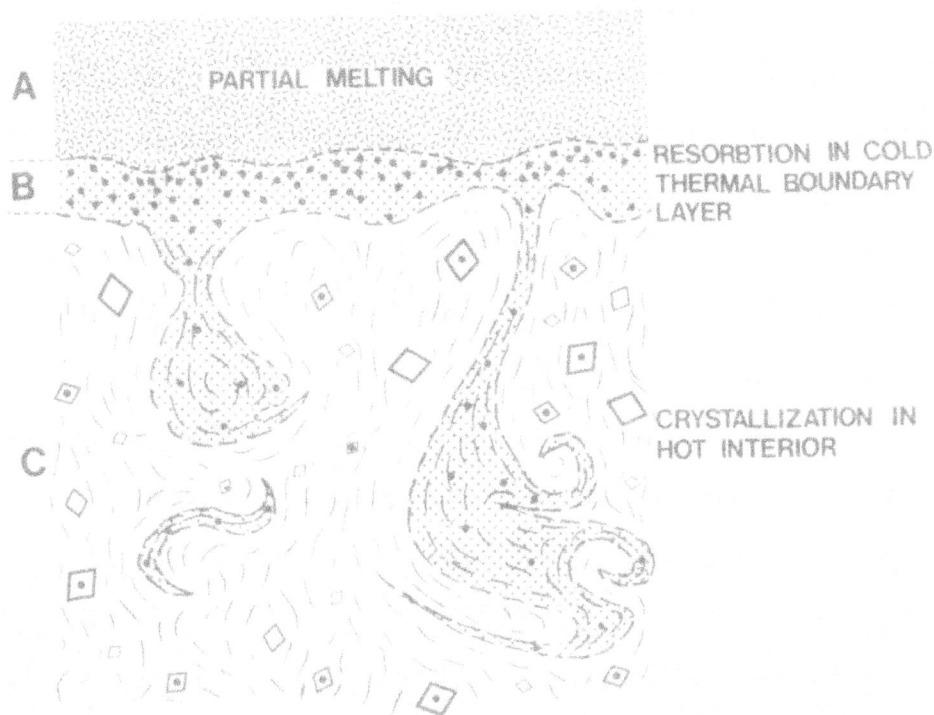

FIG. 7. Schematic diagram illustrating thermal and fluid dynamical structure of melting roof. Partial melting occurs in a roof-zone (A). The thermal boundary (B) is a region of heating and strong thermal gradients in which crystals (restite) are resorbed. The thermal boundary randomly detaches to form plumes which move downwards and mix into the hot interior. The convecting interior (C) is a region of small temperature variations where crystallization occurs due to mixing in of cold magma from above.

gradient from the melting temperature at 850 °C to the interior temperature. Where the thermal boundary layer becomes unstable locally a plume of magma descends into the underlying hot convecting magma. The thermal boundary and the detaching plumes must also be regions of melting since any fluid parcel must be heated from an initial value of 850 °C to the internal magma temperature. Restitic crystals will be resorbed in this region.

The interior of the convecting magma layer is the hottest region, but paradoxically is being cooled. Therefore, crystallization must also be occurring in the interior. An important consideration in this process is the ratio of crystals which form by crystallization in the hot interior to relict crystals derived from the roof. The former are igneous phenocrysts and the latter are restite crystals. At any given temperature between the liquidus and the effective solidification temperature (65% crystallinity) crystals from the roof rocks will not be totally melted, but will be mixed in as a restite component. The amount of restite in the crustal melt layer can be calculated from our numerical results (Table 5).

The percentage of restite decreases with time as crystallization in the magma becomes more important than melting at the roof. For crystal contents between 20 and 50% the proportion of restite ranges from 0·45 to 0·3. These calculations only provide a broad indication of the proportions since at any stage the magma may ascend and crystallize further at shallow depth in which case the restite component will be smaller. As an example consider a magma which at 10% crystallinity contains approximately 5·5% restite and 4·5% phenocrysts. If this magma ascended and crystallized to 50% in a magma chamber

TABLE 5

The table lists calculations of the total percentage of crystals, the percentage of restite and temperature for various times for the silicic melt layer. The calculations are for a 500 m basalt sill and country rock temperature of 500 °C. However, calculations for different sill thicknesses and different country rock temperatures lead to very similar proportions of restite at a given total crystal content. When the silicic magma approaches 50% crystallinity approximately one third of the crystals are restite, defined as solid crystals from the roof that have been mixed into the melt layer without melting.

Time (y)	Total crystals (%)	Restite (%)	Temperature (°C)
30	11·1	6·0	974
60	18·6	9·5	957
90	29·7	12·2	931
150	45·4	15·3	895
200	50·1	16·6	884

there would be 44·5% phenocrysts and still 5·5% restite. Also the melting relationships of some lithologies may be strongly nonlinear in contrast to the simple linear parameters used here, so that the proportions of restite may well vary even more widely than indicated in these calculations. The curves probably give an indication of the maximum amount of restite. The calculations support the view of Chappell (1984) and Wyborn & Chappell (1986) that restite can be a significant component of silicic magmas. However, the proportion of restite in the form of unmelted crystals is significantly less in this model than in that of Wyborn & Chappell (1986), who appear to regard the bulk of the crystals as restite. Because the silicic layer goes through a high temperature evolution a significant fraction of the refractory components of the crust will be melted and then re-precipitated as the layer cools and crystallizes.

A specific application of these ideas and calculations can be made to the interpretation of plagioclase feldspars in silicic magmas. Plagioclase is a particularly useful mineral as it strongly resists re-equilibration and preserves details of magmatic histories in zoning patterns. If the country rock consists of plagioclases with a range of anorthite contents then these will be partially dissolved but will also provide a nucleus for new plagioclase growth in the convecting and cooling interior. We would expect to observe a wide range of plagioclase cores with resorbed outline mantled by normally zoned rims. The cores could be both more calcic or more sodic than the rims if there is a spectrum of rock types in the melting country rock. Temperature fluctuations in the convecting interior could be sufficient to cause oscillatory zoning (Fig. 7). This interpretation is essentially that given by Wyborn & Chappell (1986) and is consistent with observations of dacitic volcanic rocks in the Andes (Francis et al., 1988). Hybridism of crustal magmas with basalt is a closely related phenomenon and can also produce mixed feldspar populations and these zoning patterns. Magma mixing and restite/phenocryst mixing are *both* effects which can be anticipated in the source region of granites.

GEOLOGICAL IMPLICATIONS

Before developing a general discussion of the implications of the model, we emphasize that the model is for a single intrusive event. In reality, a silicic magma system probably develops by large numbers of individual intrusive events over a prolonged period. However, we believe that the phenomena described for single events captures the essence of the physics for a simple case. More complex models which consider either the progressive intrusion of magma or the cumulative effects of large numbers of intrusive events are clearly worth developing in the future.

Current models of silicic magmatism

There are a number of diverse concepts in the literature on processes involved in the formation of silicic magma systems and the origin of granitoid plutons. Perspectives on current and past ideas are provided by Lipman (1984) and Pitcher (1987). One possible view is that silicic magmatism involves partial melting of crustal source rocks, and ascent of these melts to higher crustal levels where long-lived magma chambers are formed. In the conceptual model envisaged by Smith (1979) and Hildreth (1981), crystallization and differentiation occur in pluton-sized shallow magma chambers and compositional zoning is a common consequence. In these magma chambers it is argued that a zoned silicic cap develops beneath a dominant mafic volume, and the involvement of mafic magma in the development of such systems is regarded as crucial (Hildreth, 1981).

Not all silicic volcanic systems appear to be strongly zoned and some are characterised by eruption of large volumes of rather homogeneous and usually crystal-rich magma. These include the Fish Canyon Tuff (Stormer & Whitney, 1985; Whitney & Stormer, 1985), some volcanic units in the Lachlan Fold Belt of southeast Australia (Wyborn & Chappell, 1986) and the Cerro Galan ignimbrites, in northwest Argentina (Francis et al., 1988) Whitney & Stormer (1985) have proposed the controversial idea that crystallization in the Fish Canyon Tuff occurred in a deep chamber at high pressure before ascent to a high level chamber. Objections to this idea are partly based on questioning the validity of the geobarometric calculations (Grunder & Boden, 1987), but also appear to be based on the opinion that crystallization predominantly occurs in shallow chambers.

Yet another view of silicic magmatism is provided by Chappell (1984) and Wyborn & Chappell (1986) based on studies of S and I type granitoid plutons and associated volcanic rocks in the Lachlan fold belt of southeast Australia. According to them, silicic magmas represent mixtures of restite crystals and partial melt which rise from the source region and thus can be close to the crustal source composition. In some cases mixtures of restite and melt ascend to form homogeneous plutons and crystal-rich volcanic units. In other cases partial melt separates to form crystal-poor magmas which ascend and crystallize along the margins of the chamber to form zoned plutons and zoned volcanic units. Vernon (1983) has criticized the restite hypothesis in its simplest form pointing out that there is involvement of basaltic magma in at least I-type plutons. The mafic enclaves in many of these plutons are more convincingly interpreted as originating by mixing of basaltic magma than of restite.

The crystallization and cooling of shallow silicic magma chambers have been studied by theoretical calculations and experiments. Heat loss is limited by conduction through the margins of the chamber and cooling times are in the range 10^4 to over 10^5 y for pluton-sized bodies (Spera, 1979). Crystallization is thus assumed to be a slow process. Brandeis & Jaupart (1986) have argued that in silicic magmas crystallization occurs predominantly at the margins of the chamber because of high magma viscosity. They calculated that the characteristic time-scales for crystal nucleation and growth are short compared to the characteristic time-scales for detachment of the thermal boundary layer. Most of the boundary layer solidifies in the margin. This argument would suggest that internal convection is weak and that little crystallization can occur in the interior. Crystallization at the margin can also generate low density compositional boundary layer flows and this is currently the most popular hypothesis for causing compositional zoning (McBirney *et al.*, 1985).

There are thus a diversity of ideas on how silicic magmas are generated and differentiate. Some of the concepts appear to be in conflict with one another. We suggest that none of these concepts have a monopoly on the truth and that there is merit in them all. However, the results reported introduce new concepts on the time-scales and mechanisms of melting crust and crystallization in silicic magmas. We present an alternative model based on our crustal melting model and which incorporates several of the concepts mentioned in this short review.

General evolution

In normal continental crust with a geothermal gradient of $20\,°C\,km^{-1}$ rocks in the crust will have temperatures well below $450\,°C$ (Sclater *et al.*, 1980). When an episode of basaltic magmatism is initiated, the crust should often be cold enough to allow some of the basalt to penetrate to the surface. As our calculations indicate, rather little melt will be generated while the crustal rocks are cold and the chilling effect along the margins of dykes and sills will inhibit rapid heat transfer and melting.

However, if the emplacement of basaltic intrusions adds heat to the crust faster than conduction along the geotherm can extract heat, the interior of the crust will heat up and the geotherm will steepen. The heating effect will be accentuated if most intrusions are concentrated at favourable depths governed by density, the mechanical properties of the crust and the stress regime (Ryan, 1987). The focus of intrusion, for example, might be either the Moho or in the middle crust at the transition from dense lower crust to lower density upper crust. As the rocks at the focus of intrusion approach their melting temperature the probability of substantial melting associated with emplacement of a basalt intrusion increases. As argued above, if the country rock temperature exceeds $500\,°C$ the contact temperature for a $1200\,°C$ sill will be of order $850\,°C$ and partial melting will often be sufficient to form magma. As convection breaks up any chilled crust then the interface temperature approaches within a few tens of degrees of the basalt and there is a rapid transfer of heat by convection between basalt and crust.

Once extensive partial melting of the crust has been initiated, we suggest that it becomes difficult or impossible for basalt to penetrate through to the surface. Essentially, basic dykes cannot propagate through a layer of ductile, partially molten crust. When this stage is reached, the heating of the crustal interior must accelerate and the formation of horizontal intrusions beneath the region of partial melting is favoured. While some basalt can reach the surface only a part of the heat is used. However, once the crust becomes too hot and ductile all the basalt will be trapped and so the heat available for melting will approach a maximum. Each individual input of basalt increases the heat content of the crust. Again, if the crust becomes hotter with time, the volume of crustal melting associated with each successive basaltic input must also increase with time (Table 3).

Lipman (1984) has emphasized a common sequence of events in large silicic caldera complexes, which are initiated with basic to intermediate lava shields and then evolve to more silicic explosive eruptions. Such systems often develop to a final stage of a small number of very large volume silicic ignimbrite eruptions and the formation of calderas. Mafic to felsic sequences are also common in plutonic systems (Presnall & Bateman, 1983). Such sequences can be interpreted in terms of progressive heating of the crust. The shield stage represents cold crust which allows basaltic differentiates and/or hybrid andesites (between crust and basalt) to reach the surface (Fig. 8a). As the heat content of the crust

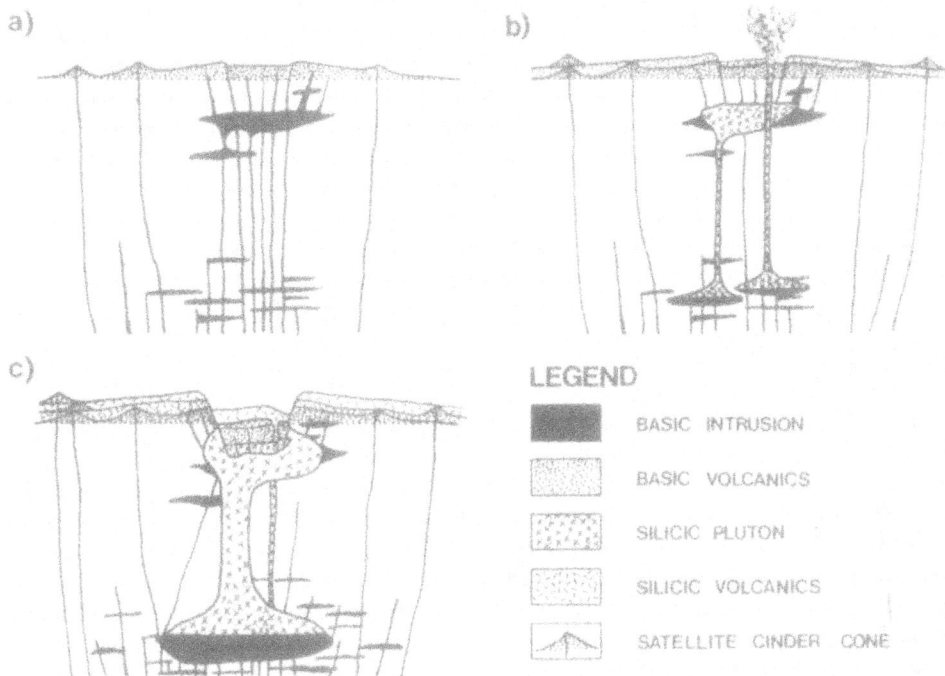

FIG. 8. Evolutionary scheme for a silicic magma system formed by emplacement of basalt into the crust. (a) Early stage where crust is cold and most basalt reaches surface to form volcanic shield or cinder cone field. (b) Crust reaches high temperature so that melting is initiated. Basalt is now mostly trapped within the crust and the silicic magma produced by episodes of intrusion either erupts directly at the surface or forms shallow intrusions. (c) When a large region of the crust is close to melting, large magma bodies can be generated which ascend to the surface causing major ignimbrite eruptions, caldera collapse, and large plutonic units. Basalt can still reach the surface in peripheral regions.

builds up beneath the volcanic system and efficiency of trapping basalt in the hot crust increases, the system evolves predominantly to erupting crustal silicic magmas (Fig. 8b). The volumes of silicic magma should increase with time reaching the climax of large volume ignimbrite eruptions and high level pluton emplacement (Fig. 8c).

There are a number of important differences between this melting model and previous ideas. First there is no necessity to involve shallow long-lived magma chambers at all, although there is no reason to suggest that they cannot form as a consequence of magma generation. Crustal melting and crystallization occur on a very short time-scale (10^2–10^3 y) compared to the lifetime of a large silicic magmatic system (10^6 y). Thus there may be no permanent magma chamber and much of the time the root zone of a silicic system may be partially molten hot crust. Every time a batch of basalt arrives in this root zone a melting event is triggered and this magma could rise straight to the surface or be emplaced at shallow levels to either differentiate, to generate more evolved magma, or to freeze as plutons. (Fig. 8b).

Another important difference is that extensive and rapid crystallization can occur in the source region. A fundamental feature of the model is that melting occurs in the thermal boundary layer and simultaneously crystallization must occur in the interior of the melt layer *as it forms*. The idea that melting occurs in the source, but crystallization must occur after magma ascends to a shallow level in a cooling environment is invalidated. Objections to deep crystallization are removed. The model predicts that large volumes of crystal-rich magma could be generated in the source region in very short time periods. Brandeis & Jaupart (1986) have identified important theoretical difficulties with forming crystals in the interior of solidifying magma chambers. We suggest that phenocryst-rich silicic magmas can form in the source region during melting. High level zoned plutons and volcanic units can form by marginal crystallization of crystal-poor magmas that ascend from the source region. Again this view is close to that advanced by Wyborn & Chappell (1986).

Our calculations suggest that crustal melt layers will initially be at high temperatures. For the standard case the crustal melt is between 900 and 950 °C when the basalt solidifies. For completely dry lithologies even higher temperature silicic magmas could form (Table 4). Thus the model can produce magmas which represent large degrees of partial fusion of the crust. The silicic magmas could thus represent the source composition as argued by Wyborn & Chappell (1986). Previously many petrologists have objected to the restite hypothesis because many crystals in silicic rocks look like phenocrysts grown from a melt and not like metamorphic relicts. We suggest that the results in this paper rationalize this apparent problem. Because the crust is largely melted in the initial stages even refractory components are dissolved. Thus the 'restite' component is melted and then reprecipitated. This effect can be observed in Table 5 where the proportion of restite decreases with crystallization. The views of Wyborn & Chappell are thus given a sound physical basis. Many crystals are of genuinely igneous origin, but, from a geochemist's viewpoint, represent refractory components of the crust.

Other implications

The model envisaged in this paper has some other possible consequences for the evolution of silicic-mafic systems. We do not have the space to explore all these in detail, but offer the following as ideas which are worthy of further work.

(1) The paper has emphasized melting of granodioritic material because this is essentially the average composition of the upper crust. If minimum-melt granitic compositions are melted, then superheated, crystal-poor rhyolitic magmas will have formed at temperatures of 900–950 °C when the basalt solidifies (Table 4). The tendency for rhyolites to be crystal-poor or erupted as obsidians finds a simple explanation. Such minimum melt eutectic compositions are either entirely liquid or entirely solid.

(2) The case for development of compositional zoning in shallow-level magma chambers by fractional crystallization coupled with other effects is overwhelming (Hildreth, 1981). Indeed there is nothing in our models which precludes shallow magma chambers or extensive crystallization within them. Crystal-poor silicic magmas can rise to shallow levels before they have time to crystallize at depth (Fig. 8b). However, the view that all zoned ignimbrites are formed in this way must be questioned. If a heterogeneous crust is being melted it is plausible that zoned systems can be generated in the source region during melting. The components of the crust with the lowest melting temperatures produce melts with lowest density and there is a strong tendency for light fluid to rise to the top, to be the last compositions to solidify and the first to melt on further influx of basalt. While experiments or theory for melting heterogeneous material have not yet been carried out, zoning is a probable outcome.

(3) The model suggests that the cooling history of silicic magmas can depend critically on whether or not melting of the surrounding wallrocks occurs. In the hot source region silicic magmas can cool quite rapidly if surrounded by materials of the same composition and melting relations. For example the silicic magmas effectively resolidify in periods of only a few hundred years by continuing to melt their surroundings. However, in the cold upper crust where heat is extracted slowly by conduction through the boundaries and where

crystallization rather than melting occurs, the cooling rate may be much slower, allowing opportunities for extensive differentiation.

(4) When roof melting is dominant in heterogeneous crustal rocks, refractory dense lithologies may resist melting in the silicic magma layer. An important process of chemical transfer may be the foundering of the refractory blocks through the silicic magma into the underlying basalt. The geochemical consequences of removal of some refractory components from the silicic magma and selective contamination by relatively refractory components of the crust into basaltic magmas should also be assessed.

(5) Another important effect may relate to the role of water in the basalt magma and in the crust. Crustal lithologies have quite low intrinsic water contents and so melting may commonly be under dry conditions generating quite hot magmas. However, some basalt magmas, particularly those associated with subduction zones, contain significant water which is further concentrated during differentiation. Since water has such a large effect on density, it is likely that the silicic differentiates of basalt intrusions can be lower in density than the dry silicic crustal melts. The underlying basalt intrusions could generate water-rich differentiated magmas which could mix with denser overlying crustal melts. Some of these volatiles could also be transferred to overlying crustal melts.

(6) The concepts developed here are not confined to the case of melting old pre-existing crustal rocks. If basalt is repeatedly intruded into the same region during an episode of underplating, later intrusions are very likely to encounter the more silicic or felsic differentiated products of earlier intrusive events. Repeated remelting of such lithologies could generate felsic magmas such as trachytes, phonolites, and alkali rhyolites from basalt magmas of alkaline affinity as well as silicic magmas of tholeiitic or calcalkaline affinity.

ACKNOWLEDGEMENTS

R.S.J.S. acknowledges the helpful discussions with colleagues while staying at the California Institute of Technology, notably, M. Rutter, E. Stolper, S. M. Wickham, and P. J. Wyllie. We are grateful to Mark Hallworth and Joyce Wheeler, who provided a great deal of support in their respective help with the experimental work and in carrying out the numerical calculations. We acknowledge constructive reviews of the paper by C. Jaupart, C. J. N. Wilson and S. M. Wickham. Sandra Last typed the manuscript. Our work is supported by grants from the BP Venture Research Unit and from NERC.

REFERENCES

Bacon, C. R., 1986. Magmatic inclusions in silicic and intermediate volcanic rocks. *J. geophys. Res.* 91, 6091–112
Brandeis, G., & Jaupart, C., 1986. On the interaction between convection and crystallization in cooling magma chambers. *Earth planet Sci. Lett.* 82, 345–62.
Campbell, I. H., & Turner, J. S., 1987. A laboratory investigation of assimilation at the top of a basaltic magma chamber. *J. Geol.* 95, 155–73.
Chappell, B. W., 1984. Source rocks of I- and S-type granites in the Lachlan Fold Belt, southeastern Australia. *Phil. Trans. R. Soc. Lond.* A310, 693–707.
England, P. C., & Thompson, A. B., 1984. Pressure-temperature-time paths of regional metamorphism. I. Heat transfer during evolution of regions of thickened continental crust. *J. Petrology*, 25, 894–928.
Francis, P. W., Sparks, R. S. J., Hawkesworth, C. J., Thorpe, R. S., Pyle, D. M., & Tait, S. R., 1988. Petrogenesis of dacitic magmas of the Cerro Galan Caldera, N.W. Argentina. *Geol. Mag.* (subjudice).
Grunder, A. L., & Boden, D. R., 1987. Comment on 'Magmatic conditions of the Fish Canyon Tuff, Central San Juan Volcanic Field, Colorado'. *J. Petrology*, 28, 737–46.
Hildreth, W., 1981. Gradients in silicic magma chambers: implications for lithospheric magmatism. *J. geophys. Res.* 86, 10153–92.
Huppert, H. E., 1986. The intrusion of fluid mechanics into geology. *J. Fluid Mech.* 173, 557–94.
——— 1988. The response to the initiation of a hot, turbulent flow over a cold, solid surface. *J. Fluid. Mech.* (subjudice)
——— Sparks, R. S. J., 1988a. Melting the roof of a chamber containing a hot, turbulently convecting fluid, *ibid.* 188, 107–31.

—— — - 1988b. The fluid dynamics of crustal melting by injection of basaltic sills. *Phil. Trans. R Soc. Edinburgh. Earth Sci.* (in press).

Irvine, T. N, 1970 Heat transfer during solidification of layered intrusions. I. Sheets and sills. *Can J. Earth Sci.* **7**, 1031–61

Jaupart, C., & Parsons, B, 1985. Convective instabilities in a variable viscosity fluid cooled from above. *Phys. Earth planet Interiors* **39**, 14–32.

Lipman, P. W., 1984. The roots of ash-flow calderas in Western North America. windows into the tops of granitic batholiths *J. geophys. Res.* **89**, 8801–41.

Marsh, B. D., 1981. On the crystallinity, probability of occurrence and rheology of lava and magma *Contr. Miner. Petrol.* **78**, 85–98

McBirney, A. R., Baker, B. H., & Nilson, R. H., 1985. Liquid fractionation. Part 1: Basic principles and experimental simulations *J Volcanol. geotherm. Res.* **24**, 1–24

McBirney, A. R., & Murase, T., 1984 Rheological properties of magmas *Ann. Rev. Earth planet Sci.* **12**, 337–58.

Molen, van der, I, & Paterson, M. S., 1979. Experimental deformation of partially-melted granite. *Contr Miner Petrol.* **70**, 299–318.

Pitcher, W. S., 1986. A multiple and composite batholith In: Pitcher, W. S., Atherton, M P., Cobbing, E J., & Beckinsale, R D. (eds) *Magmatism at a Plate Edge* Blackie Halsted Press, 93–101

—— —— 1987 Granites and yet more granites forty years on. *Geol. Rundschau* **76**, 51–79

Presnall, D. C., & Bateman, P C., 1973. Fusion relations in the system $NaAlSi_3O_8$–$CaAl_2Si_2$–$KAlSi_3O_8$–SiO_2–H_2O and generation of granite magmas in the Sierra Nevada Batholith *Geol. Soc. Am. Bull* **84**, 3181–202.

Richter, F. M., Nataf, H.-C., & Daly, S. F., 1983. Heat transfer and horizontally averaged temperature of convection with large viscosity variations. *J. Fluid Mech.* **129**, 173–92.

Ryan, M. P., 1987. Neutral buoyancy and the mechanical evolution of magmatic systems. In: Mysen, B O. (ed) *Magmatic Processes: Physicochemical Principles* Special Publication No. 1 of The Geochemical Society, 259–88.

Scalter, J. G, Jaupart, C., & Galson, D., 1980 The heat flow through oceanic and continental crust and the heat loss of the Earth. *Rev. Geophys Space Phys.* **18**, 269–311.

Shaw, H. R., 1963. Obsidian-H_2O viscosities at 1000 and 2000 bars in the temperature range 700 to 900 °C. *J. geophys. Res.* **68**, 6337–44

—— 1972. Viscosities of magmatic liquids: an empirical method of prediction *Am J Sci* **272**, 870–93.

—— 1980 The fracture mechanisms of magma transport from the mantle to the surface In: Hargreaves, R. B. (ed) *Physics of Magmatic Processes.* Princeton University Press, 201–64.

Smith, R. L, 1979 Ash-flow magmatism. *Geol. Soc Am. Special Paper* 180, 5–28.

Sparks, R. S. J., Huppert, H. E., & Turner, J. S, 1984 The fluid dynamics of evolving magma chambers. *Phil Trans. R. Soc. Lond.* **A310**, 511–34.

Spera, F., 1979 Thermal evolution of plutons: a parameterized approach. *Science* **297**, 299–301.

Stormer, J C, & Whitney, J. A., 1985. Two feldspar and iron-titanium oxide equilibria in silicic magmas and the depth of origin of large volume ash-flow tuffs. *Am Miner.* **70**, 52–64.

Taylor, S. R., & McLennan, S. M., 1985. *The Continental Crust, Its Composition and Evolution.* Oxford: Blackwell Scientific Publications, 312 pp.

Vernon, R. H., 1983. Restite, xenoliths and microgranitoid enclaves in granites. *J. Proc. R. Soc. NSW* **116**, 77–103.

Whitney, J. A., & Stormer, J. C, 1985. Mineralogy, petrology and magmatic conditions from the Fish Canyon Tuff, Central San Juan mountain field, Colorado. *J. Petrology,* **26**, 726–62.

Wickham, S. M., 1987. The segregation and emplacement of granitic magmas. *J. Geol. Soc. Lond.* **144**, 281–98

Wyborn, D., & Chappell, B. W, 1986. The petrogenetic significance of chemically related plutonic and volcanic rock units *Geol. Mag* **123**, 619–28.

Wyllie, P. J., 1984. Constraints imposed by experimental petrology on possible and impossible magma sources and products *Phil Trans R Soc. Lond.* **A310**, 439–56

CHAPTER 8

Campbell, I.H. and Taylor, S.R. (1983) No water, no granites – no oceans, no continents. *Geophysical Research Letters*, **10**, 1061–1064.

In the 1980s, planetary geology really took off as a discipline. The various manned space missions, telescopic observations, satellite fly-bys, radar imaging, etc. produced an incredibly rich database of images and spectral data, generating numerous important and fascinating problems that planetary scientists could study. Many bizarre planetary features, totally unlike any known on Earth, were discovered, studied, modelled and speculated upon. However, in comparing Earth to the other terrestrial planets of our solar system, one of the hot topics was the search for features that could be equated with Earth's continental masses.

Now, if we define a continent strictly, and not simply as a large plateau-like structure elevated above the general level of the planet's surface, we find a remarkable absence of such things on Venus and Mars, the most Earth-like of the rocky inner planets. To be sure there are elevated, plateau-like regions on both these. However, Earth's continents are granitic in composition. Our planet has plate tectonic processes. Venus and Mars may have had something similar going on at some stage too, but their elevated regions, likened by some to our continents, are not granitic and do not seem to have been created by plate-tectonic processes.

It was Campbell and Taylor (1983) who looked at this question and put the simple rules together for us. Although they wrote about other planets, what they had to say is very relevant to our understanding of how continental nuclei were created in Earth's Early Archaean and how they grew quickly to their more-or-less steady-state condition. The title of their paper essentially spells it out: "No water, no granites – no oceans, no continents".

Water is the key. If a planet has liquid water on its surface, in large enough quantity, there will be oceans. If the planet is geologically active, with structures analogous to mid-ocean ridges then hydrothermal alteration and sea-floor weathering will create hydrated and altered ocean-floor basaltic rocks. Low-pressure heating and partial melting of these will produce sodic felsic melts that solidify to primitive plagiogranites. These masses will aggregate to form the continental nuclei and, because of their low density, they will not easily be recycled into the mantle, for example by subduction-like processes. These granitic masses will continue to grow and emerge from the oceans to form protocontinents. These, and other more mafic rocks will weather and erode, be deposited as sediments and potentially recycled again and again, producing granitic crust that evolves toward the modern more potassic compositions. This chain of events apparently never got properly underway on Venus or Mars because the presence of liquid water on both these bodies was ephemeral. Thus there are no real continents on these planets, no plate tectonics and no granites worthy of the name.

Perhaps what I (J.D. Clemens) have written above is biased spin. Why not look at what Campbell and Taylor actually had to say and decide for yourself whether this paper is important enough to occupy a place in this volume?

GEOPHYSICAL RESEARCH LETTERS, VOL. 10, NO. 11, PAGES 1061-1064, NOVEMBER 1983

NO WATER, NO GRANITES - NO OCEANS, NO CONTINENTS

I.H. Campbell* and S.R. Taylor

Research School of Earth Sciences, Australian National University,
Canberra, ACT 2600, Australia

Abstract. Water is essential for the formation of granite and granite, in turn, is essential
for the formation of continents. Earth, the only
inner planet with abundant water, is the only
planet with granite and continents. The Moon and
the other inner planets have little or no water
and no granites or continents.

CHAPTER 9

Hutton, D.H.W. (1988) Granite emplacement mechanisms and tectonic controls: inferences from deformation studies. *Transactions of the Royal Society of Edinburgh: Earth Sciences*, **79**, 245–255.

This work is not only a research paper but also constitutes an excellent text dealing with the emplacement of granitic magmas. It is quite suitable for undergraduate students because of its clarity and very didactical approach in applying the methods and logic of structural geology to the problem of granite emplacement. The paper opens with a review of the methods and a discussion of the approach of Cloos to the classification of the structures present in igneous rock units (Cloos, 1936; Balk, 1937). Hutton then recommends that this approach be abandoned and replaced with a new and less genetically based terminology. In the second part of the paper, Hutton reviews the literature on the tectonic influences on granite emplacement, using a series of well studied cases as type examples. The final and most important part of the paper is a synthesis in which Hutton proposes that the diverse mechanisms of granite emplacement represent the interplay between the magma buoyancy forces, the rate of tectonic cavity opening and the existence of conduits provided by pre-existing structures. A major conclusion is that, in the presence of regional tectonics, the emplacement of granitic magma does not pose any sort of "space problem". This idea was later taken up and explored thoroughly by the author (Hutton *et al.*, 1990; Hutton, 1996, 1997), and also forms a cornerstone of our current understanding of how granite magmas migrate from their sources to their final places of crystallization (e.g. Clemens and Mawer, 1992; Petford *et al.*, 1994; Clemens, 1998; Cruden, 1998; Solar *et al.*, 1998; Brown and Solar, 1999; Vigneresse, 1999; Weinberg, 1999; Moyen *et al.*, 2003; Brown, 2004, etc.). The landmark status of this paper rests on its clear reasoning, the fact that it draws on existing knowledge and then uses this to synthesize a new way of looking at the problem. Whether subsequent workers realized it or not, it is sure that some of their thinking about granitic magma migration and emplacement was conditioned by having been exposed to Hutton (1988).

Transactions of the Royal Society of Edinburgh: Earth Sciences, **79**, 245–255, 1988

This paper was presented at a Symposium on the Origin of Granites organised jointly by The Royal Society of Edinburgh and The Royal Society of London and held in Edinburgh, 14–16 September 1987.

Granite emplacement mechanisms and tectonic controls: inferences from deformation studies

Donald H. W. Hutton

ABSTRACT: This paper is a structural and tectonic approach to the emplacement and deformation of granitoids. The main methods available in structural geology are briefly reviewed and this emphasises that (a) a wealth of data, particularly strain and shear sense, which pertain to these problems, can be determined in and around plutons; (b) given the nature, unlike many other crustal rock types, of granites to crystallise from isotropic through weakly anisotropic crystal suspension fluids, that deformation which has occurred in these states may not be well preserved; and (c) it is entirely possible, using this methodology, to separate deformation resulting from externally originating tectonic stresses from that which is associated with internal magma-related stresses. It is also recommended that the genetically-based Cloosian classification of granite fabrics and structures into "primary" (magmatic flow/magmatic flow current) and "secondary", be abandoned and that a more observationally-based approach which classifies granite deformation fabrics and structures according to their time of occurrence relative to the crystallisation state of the congealing magma, be adopted (i.e. pre-full crystallisation deformation and crystal plastic strain deformation).

Examples of recent, structurally based, studies of emplacement mechanisms of plutons within tectonic settings are described and these show that, in general, space for magma can be created by the combination of tectonically-created cavities and internal magma-related buoyancy. This occurs in both transcurrent and extensional tectonic settings and there is no reason to doubt that it can happen in compressive-contractional regimes. It is concluded that transient and permanent space creation, such as may be exploited by available magmas, is a typical feature of the tectonically stressed and deforming lithosphere and this, in combination with the natural buoyancy and ascending tendency of magmas, can generate the varied emplacement mechanisms of granites.

KEY WORDS: magma rheology, strain, tectonics.

Emplacement mechanism is traditionally classified into two basic types: "forceful", including doming, diapirism and ballooning; and "passive" or "permitted", involving stoping, cauldron subsidence and sheeting. The relationship between these two types is obscure. Most geologists would accept that magma probably rises through the lithosphere because of buoyancy (the density differential between the magma and the integrated wall rock column (Roberts 1970)) and this therefore suggests a natural "forceful" aspect to intrusion. On the other hand at very high levels in the crust, the close proximity of the earth's surface as a free surface seems a logical reason for cauldron and related caldera behaviour. Yet at deeper levels, probably remote from the free surface, passive mechanisms seem as common as the more easily understood forceful mechanisms. While Read (1957) and Buddington (1959) have suggested that crustal level and elapsed time in a batholith emplacement sequence are the controlling factors in this bimodal emplacement behaviour, there are situations, such as in Donegal, where both mechanisms occur virtually simultaneously in time and space. The probable reason for this, and, in my view, the reason in general for variety in emplacement mechanism, is that it is a function of the interaction between natural

magma buoyancy forces and ambient tectonic forces. A qualitative model for this is developed in the final section of this paper. The appreciation of tectonic controls comes most easily from the study of the deformation aspects of granites and their surroundings. Therefore, before proceeding to describe examples of tectonic control on emplacement, the basic methodology which can be applied to the deformation and structure of igneous rocks is briefly reviewed.

1. Review of methodology

In general the techniques that are used in the study of granite deformation and structure come from current methods in structural geology. In most cases the techniques can be applied without modification, but in other situations the unique material behaviour of igneous rocks compared to other crustal rocks needs to be taken into account.

1.1. Deformation fabrics and strain
The deformation fabrics or foliations of granitoids as either crystal alignment or crystal shape and alignment structures occur as planar, linear or planar and linear and can be

qualitatively related to the shape of the strain ellipsoid (K value) as S, L, or LS fabrics in the Flinn nomenclature (Flinn 1965; Pitcher & Berger 1972; Fig. 1a, b, c, d.). This is a measure of the time-averaged relative magnitudes of the strain rates in three orthogonal directions and, when mapped over an entire pluton together with the varying orientations of the planar and linear components, is a basic starting point in any study.

Both K value and strain intensity are determinable, at least semi-quantitatively, using a variety of methods, the most important of which comes from the shape and orientation of deformed objects (such as xenoliths and deformed quartz aggregates) and also the spatial relationship of pre-deformational objects (especially phenocryst phases). There are many methods applicable to the deformation of xenoliths (see Ramsay & Huber 1983 for a review). The simplest techniques involve measurement of the long and short axes of two-dimensional sections through xenoliths on two or three orthogonal planes which correspond to the principal strain planes (XY, YZ, XZ; Fig. 1a, b, c). Measurement of thirty or more shape ratios can yield a mean ratio which is a measure of the strain ratio (X/Y, Y/Z, X/Z) on that particular plane. The orientation of individual long axes may be combined with the shape ratio of a more precise measure of strain ratio (the R_f/ϕ technique). Strains from any two principal strain planes can be combined to determine the K value. The main limiting factor with this method in granites is that the natural erosion surfaces available for measurement of xenoliths may not be precisely parallel to the principal strain planes (Hutton 1982b). For sensible and useful results to be obtained, therefore, care must be taken to measure xenolith axial lengths in natural surfaces which lie in or very close to these principal strain planes. Non-spherical initial shapes of xenoliths can be accounted for in many of the techniques and viscosity contrast between xenolith and host can be approximately determined (Hutton 1982a). In the present author's experience deformed mafic xenoliths (i.e. "clots", "autoliths" or "cognate" xenoliths, as they are variously called), which occur commonly in many granite types, produce the most consistent strain results: country rock xenoliths in general show much more variability (Hutton 1982b).

A recently-developed technique which is ideally suited to deformed granites is the "nearest neighbour" method of Fry (Hanna & Fry 1979; Fry 1979a,b). This can be quickly and easily applied on outcrops and cut sections to crystal separations and can provide a much wider converage of strain data for any pluton than that provided by the often less regularly occurring xenoliths. This method can also be

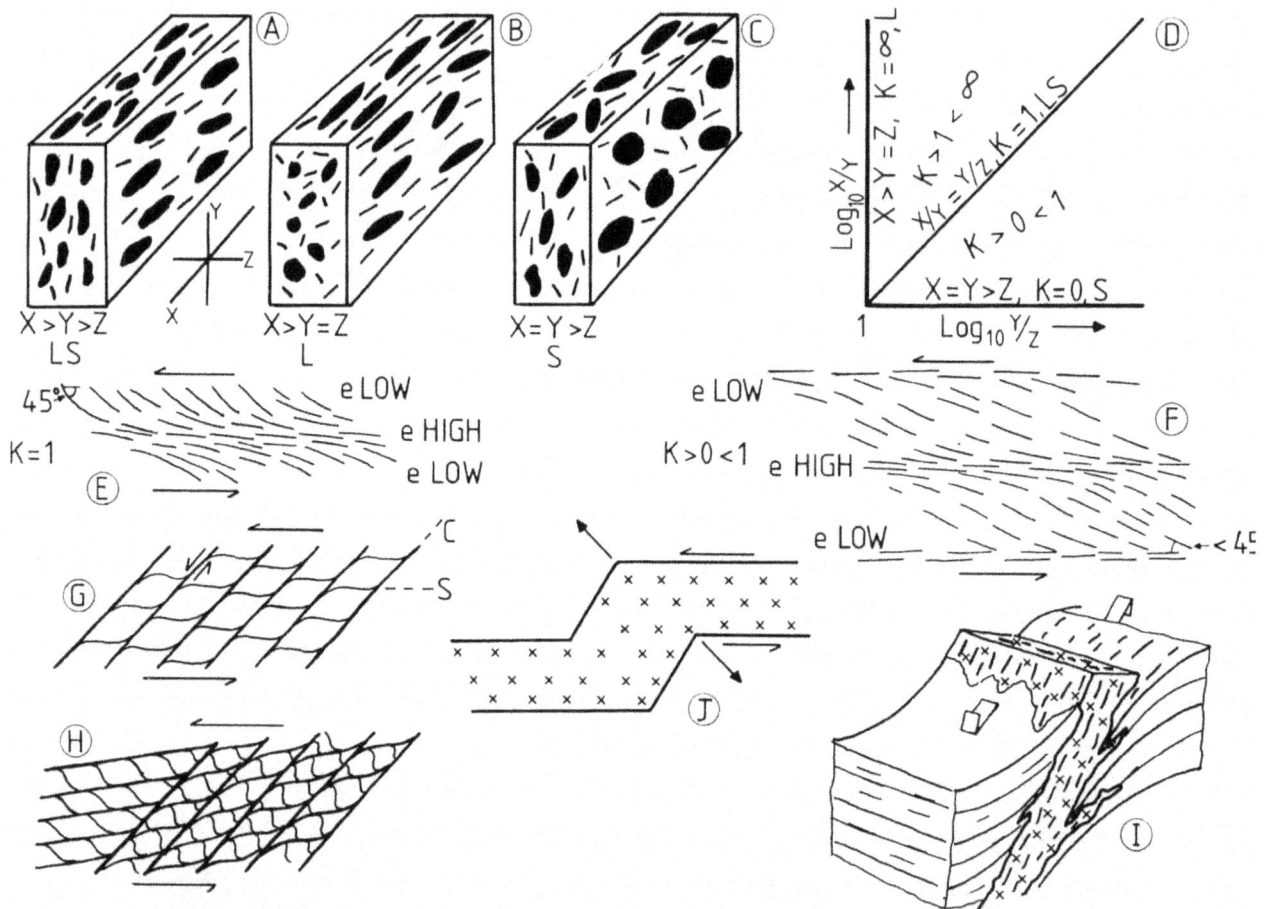

Figure 1 Some aspects of methodology referred to in text: A–C Basic strain types showing qualitative variations in shape change (deformed xenoliths) and orientation (xenolith long axes and phenocryst alignments) on orthogonal XY, YZ and XZ principal strain planes for (A) plane strain ($K = 1$); (B) prolate, constrictional strain ($1 < K < \infty$); (C) pure flattening strain ($K = 0$); (D) typical method of graphically representing strain data (Long-Flinn Diagram); (E) simple shear zone foliation trajectories; (F) transpressional shear zone foliation trajectories; (G) S-C fabrics, S in the granite context represents the main penetrative foliation; (H) higher shear strain situation in which the original C planes have back-rotated towards S and been cut by a new generation of C-planes; (I) granite vein synchronously intruding into, and being deformed in an extensional shear; (J) asymmetric dilational dyke giving shear sense during dilation.

used to determine shear sense (see below) (Saltzer & Hodges 1988).

Strain can be determined at many localities in a pluton and maps of K value and strain intensity can be generated and combined with fabric maps. Although, naturally, the strain measurement literature emphasises the need for precise strain determination (and indeed this is highly necessary when calculating inflation volumes in the Holder/Ramsay balloon model (Holder 1979)) often the most useful ideas on overall deformation and emplacement mechanism come from the general nature of the strain patterns and strain gradients rather than from the actual strain values themselves.

1.2. Shear sense

This can be determined on a variety of scales. At the largest scale gross fabric and stratigraphic contact re-orientations, when combined with knowledge of the tectonic transport azimuth (from mineral lineations and mean xenolith X-axis orientations) and characteristic strain patterns, can be correlated with regional to pluton scale simple shear zones (Ramsay & Graham 1970; Fig. 1e) or transpressive shear zones (Sanderson & Marchini 1984; Soper & Hutton 1984; Fig. 1f).

On a much smaller scale the recognition and appreciation of asymmetric S-C fabrics (Berthe et al. 1979; Fig. 1) has recently revolutionised shear sense determination in plastically deformed rocks (see Choukroune et al. 1987; and White et al. 1986 for reviews). In this terminology S refers to "schistosité" (in the granite context the main grain shape and alignment foliation) and C refers to "cisaillement" (the small-scale cross-cutting crenulation cleavage planes which effect an asymmetric shear and extension of the main S fabric; Fig. 1g, h). These types of fabrics are often best seen in plastically deformed granites and, where shear strains are moderately high, the asymmetry of the oblique C planes (whose shear sense is considered to be synthetic with the shear sense on the main S fabric) can be easily read. At lower shear strains deformation may give rise to higher angle structures of similar morphology (extensional crenulation cleavage). These may often occur in conjugate sets both slightly asymmetric to the main fabric. In this situation the antithetic set is often much less common than the synthetic set (Platt & Vissers 1980; Hutton 1977, 1982a). These types of study are often combined with observations of the asymmetry of rotation and deformation of individual crystals, aggregates, augen, boudins, etc. and also by the asymmetry of C-axis orientations in quartz and other minerals (Passchier & Simpson 1986; Simpson & Schmidt 1983; Lister & Hobbs 1980; Bouchez et al. 1983). Although both high and low temperature shear sense criteria are identifiable, very few of these asymmetric deformation indicators occur, in my experience, in granites which underwent deformation before full crystallisation. Exceptions to this are the development of outcrop scale discrete extensional shears and asymmetric boudin necks whose geometry mimics that of synthetic extensional crenulation cleavage (Fig. 1i). Such structures often occur in pre-full crystallisation deformation conditions in, and around, and crosscutting country rock xenoliths and internal granitic sheets, veins and bands: in effect associated with any planar mechanical heterogeneities. The structures are often best developed in granite contact zones and inner aureoles where granite and pegmatite veins intrude synchronously along the extensional shear planes, much in the manner of some neosomes in migmatites (Hollister & Crawford 1986). Also

in this category comes any shear sense during the dilation associated with sheeted granite bodies. This can be determined from consistent asymmetric offsets within and between the component sheets (Fig. 1i).

The orientations of extended and folded cross cutting vein systems in deformed granitoids can be used to determine K value (Talbot 1970) and shear sense (Hutton 1982a), although these sometimes require careful interpretation.

Shear sense determinations in deformed granites and their synchronously deformed aureoles are in many ways central to recent interpretations of emplacement mechanism. They record, in general, two important features: (1) the relative movement between granites and their wall rocks; (2) movements which are imposed on granites and wall rocks alike. These types of data have allowed interpretation of the emplacement mechanisms of a number of individual plutons and batholiths, e.g. the oblique diapirism of the Criffel granite, Scotland (Courrioux 1987), the transcurrent shear zone cavity opening models for Donegal (Hutton 1982a) and Mortagne (Guineberteau et al. 1987); the crack opening and rotation model for the plutons of the Central Extramadura batholith of Spain (Castro 1986).

1.3. Deformation in aureoles and wall rocks

Granite emplacement studies should not stop at the granite contacts. In syn-kinematic or ballooning/diapiric plutons the wall rocks, especially if metasediments, record in different ways to the pluton itself much of the strain and kinematics associated with emplacement. Moreover, they often record the critical time relationships between intrusion, deformation and metamorphism and in the case of external tectonic control on emplacement their study is probably the only way of separating the nature of the external tectonic forces from the granite buoyancy forces (Hutton 1977, 1982a; Castro 1986). The problem with this type of study is that emplacement related mechanisms and structures may be superimposed on penetrative regional polyphase deformation events which entirely pre-date and are unrelated to the emplacement episode (Hutton 1977). This requires careful, painstaking structural work to establish chronology, structural geometries, kinematics and strain superposition histories (Sanderson 1976) before these earlier events can be properly isolated from the granite emplacement deformation. In apparently "passive" or "permitted" intrusions the orientation and movement sense of syn-magmatic fracture systems at wall rock-pluton boundaries (or elsewhere) can be analysed to determine the orientation of the syn-magmatic stress field (Pitcher 1952; Berger 1971; Castro 1984).

1.4. Time of deformation relative to crystallisation state in deformed granitoids

As will be obvious from much of the philosophy of the previous sections this author regards the vast majority of fabrics in granitoids to be the result of combinations of: change in shape, change in orientation, change in length; in other words to be due to deformation or strain that in itself is the result of applied stresses. Unlike the Cloosian paradigm (Balk 1937) this can happen at any time during or after crystallisation. The relationship between time of deformation and crystallisation state can be broadly determined from the textures and fabrics themselves and two basic types (which are often in effect end-members of a series) can be distinguished.

1.4.1. Pre-full crystallisation fabrics. As the term implies this is deformation that occurs before all of the phases have

248   DONALD H. W. HUTTON

crystallised and we can envisage a particle suspension of crystallised, relatively rigid, early phases (often feldspar and mafic phenocrysts) being rotated and aligned in an uncrystallised matrix. If deformation then ceases and the remaining melt crystallises, a fabric results which is composed of aligned euhedral phenocrysts, internally undeformed, surrounded by essentially unaligned internally undeformed late matrix crystals (typically much of the quartz; Fig. 2a).

1.4.2. Crystal plastic strain fabrics. In this situation we can envisage deformation affecting a granite after all the phases have crystallised but while there is still enough heat in the system to allow crystal alignment to occur by ductile plastic deformation mechanisms. The crystals are aligned but show evidence of internal plastic deformation in both early phenocrysts and late crystallising phases. Quartz, rather than forming the equant knots of the pre-full crystallisation situation, is often lenticular or ovoid (Fig. 2b). This type of fabric is very similar to the schistosities or foliations of metamorphically deformed granular sediments and the term "metamorphic fabric" is a useful synonym with the understanding that the heat source is the pluton itself rather than an external source.

A. Pre-full crystallisation fabric

B. Crystal plastic strain fabric

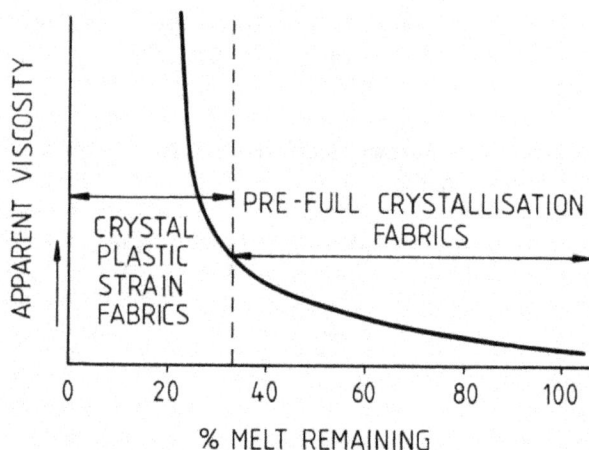

Figure 2 Time of deformation relative to crystallisation state in granitoids: (A) pre-full crystallisation fabric; (B) crystal plastic strain fabric–rectangular crystals are feldspar phenocrysts, long thin black crystals are hornblende/pyroxene/micas, groundmass is quartz etc.; (C) fabric types in relation to Arzi type diagram.

A useful way to visualise the relationship between time of deformation and crystallisation state, and indeed to visualise deformation in granites as a whole, is with the Arzi "critical rheological melt-percentage" curve (Arzi 1978; Van der Molen & Patterson 1979). In general this curve relates to the belief that as a magma crystallises and the volume of solid material builds up (remaining melt diminishes), the apparent viscosity of the system does not increase linearly, but at a critical remaining melt percentage jumps by many orders of magnitude. At this point and beyond the crystal-laden system 'locks up' and stress is transmitted through the system across crystal contacts creating plastic strain fabrics. Before the critical melt percentage is reached, deformation creates the pre-full crystallisation fabrics in the particle suspension (Fig. 2c). At very low crystal contents deformation may also occur. This, because of the very small number of aligned crystals and the generally random alignments of later crystallising more numerous crystals, may be extremely difficult to recognise: the rock may appear "undeformed". Likely examples of this are the Arran granite in Scotland (Woodcock & Underhill 1987) which is a diapir/balloon (it generates a well-developed rim syncline in the country rocks, but it has no internal fabric (R. England, pers. comm); in Saudi Arabia Davies (1982) has described a similar diapir (Ash Sha'b pluton) with, again, no internal fabric. A corollary of these observations is that geologists should be very wary of describing a pluton as "passively emplaced" because it has no internal fabric: granites, unlike many other rocks, can lose their "memory" of deformation rather easily.

1.4.3. Syn kinematic strain patterns. The Arzi curve may also be used to visualise the nature of syn kinematic intrusion. In a pluton which is emplaced synchronously with deformation in the country rocks, whether it be as an expanding balloon or passively within a regional shear zone, much of its deformation will take place in the pre-full crystallisation state. It will not start to record crystal plastic deformation strains (for example in the shape of deforming country rock xenoliths) until after "lock-up" occurs. In the wall rocks, on the other hand, crystal plastic deformation will have occurred throughout the period of deformation. Therefore, when one compares crystal plastic strain magnitudes between wall rocks and granite, a distinctive drop in plastic strain magnitude should occur for a syn kinematically emplaced pluton. This is the case with the Main Donegal granite (Hutton 1982a), the Ardara pluton, Donegal (Sanderson & Meneilly 1981) and is a feature of the rapakivi granites in South Greenland (Hutton unpublished).

1.4.4. Comment on Cloosian ideas. Some may seek to see parallels between the pre-full crystallisation/crystal plastic strain fabrics used here and the primary/secondary fabrics of Cloos (Balk 1937). Although the textural criteria used by Cloos to distinguish primary fabrics are the same as those for pre-full crystallisation fabrics the term 'flow' or 'magmatic flow' is deliberately avoided here in this context. Cloos used the term 'flow' in an entirely genetic way to imply the operation of 'magmatic currents' moving against wall rocks often by direct analogy with water flowing in streams and conduits, with much of the emphasis on (in essence) simple shear mechanisms. (See also Marre (1986) for a recent reiteration of this philosophy.) Further, the term 'deformation' was apparently reserved for later imposed (cf. tectonic) fabrics. While Cloos, in fairness, supported the view that primary fabrics were due to internal processes because they were concordant with granite contacts and otherwise spatially closely related to the plutons, this, as a basic assumption, may be incorrect (Berger & Pitcher 1970). The main difficulty with the Cloosian approach is that if you allow the possibility that a pluton may be affected by tectonically originating forces either singly or in combination with internal buoyancy forces before full crystallisation, the primary/secondary methodology will be unable to yield such a result. The present author's suggestion is, therefore, that the terms "primary" and "secondary" be abandoned and that a less genetic system, such as is used here, be employed. (See also Hibbard 1987.) This current methodology seeks to describe the deformation patterns in and around a pluton, attempts to allot the various increments in the deformation history to the changing rheology of the magma and finally, using various obvious precepts, models this deformation in terms of tectonic externally originating strains, internal buoyancy related strains, or combinations of both.

2. Tectonic influences on granite emplacement

Although deformation of plutons adjacent to major faults and shear zones has long been recognised and although it has often been suggested that major structures control the siting of plutons and batholiths (e.g. Pitcher & Russell 1977; Leake 1978), rarely, until recently has any attempt been made to establish a kinematic link between major structures and emplacement mechanisms. The following section describes a number of specific examples of this.

2.1. Emplacement associated with transcurrent shear zones

2.1.1. The Donegal granites. In this well-known Caledonian batholith (Pitcher & Berger 1972; Fig. 3) the plutons, which are radiometrically virtually synchronous (c. 400 Ma) and emplaced at the same level in the crust, exhibit a number of emplacement mechanisms which vary apparently randomly in space and elapsed time sequence in the batholith emplacement history. A probable reason for this is a major regional sinistral shear zone which lies along the southern side of the batholith and contains within it the penetratively deformed Main Donegal granite (Hutton 1982a).

The synchroneity of the Main granite emplacement and deformation with the country rock deformation in the shear zone is indicated by (a) the synkinematic growth of the thermal prophyroblasts in the aureole; (b) the continuity of the penetrative granite crystal plastic strain fabric with the external shear zone fabric; (c) the overlap in timing of small scale intrusions and deformation features in the granite contact zones; (d) a strain drop in crystal plastic strain magnitude between wall rocks and pluton. Pre-full crystallisation deformation is probably represented in the pluton by strongly overprinted but still recognisable early phenocryst fabrics (Hutton 1982a) and by many of the early banding features (Berger 1971).

Plastic strain magnitudes and sinistral displacements decrease with distance SW in the external shear zone and a similar strain gradient is seen (from deformed xenoliths) in the pluton itself (Fig. 3a). Such a displacement gradient along a shear zone will cause the zone to bend and distort

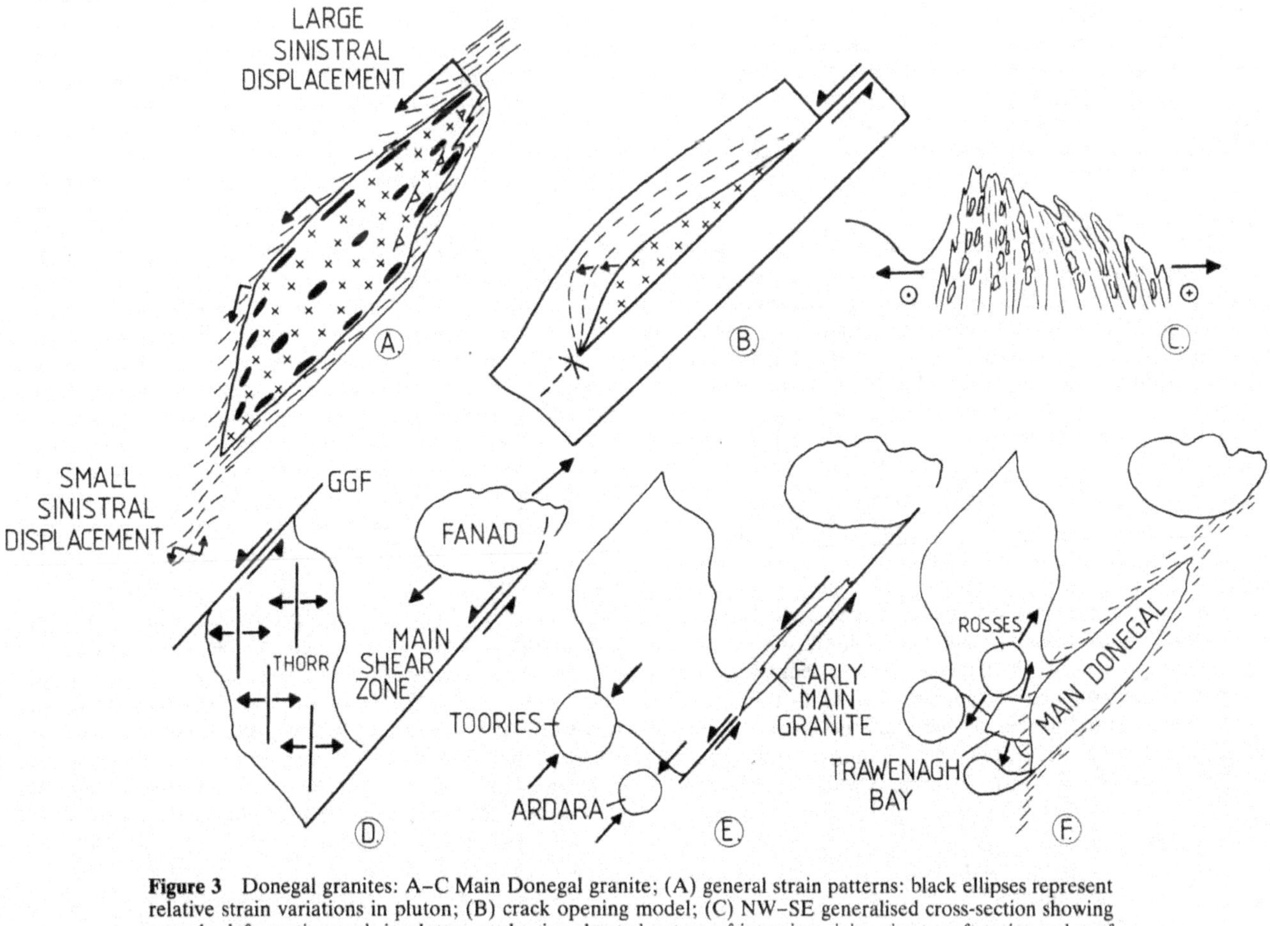

Figure 3 Donegal granites: A–C Main Donegal granite; (A) general strain patterns: black ellipses represent relative strain variations in pluton; (B) crack opening model; (C) NW–SE generalised cross-section showing aureole deformation and, in pluton, coalescing sheeted nature of intrusion giving rise to raft trains and roof pendants. D–F evolution of Donegal batholith; see text for details.

during shear movement (Coward 1976). In the Donegal situation the bend is on one side of the shear zone only and between the bend and the southern straight margin the Main Donegal pluton is located. This suggests that at an early stage, after initial formation, the shear zone split axially and the northern wall only moved away, accommodating the displacement gradient and providing a potential hole or cavity for the emplacement of the pluton (Fig. 3b).

On a more local scale emplacement was achieved by a coalescing series of longitudinal granite sheets emplaced within the transtentional environment. This sheeting process created the roof septa near the granite margins and, deeper within the pluton, the remarkable raft trains with their "ghost stratigraphy" (Fig. 3c). The body force created by this accession of magma is represented by a modification of the external transcurrent strains in the innermost granite aureole and by a weak residual coaxial flattening strain component in the granite itself. The conclusion is that the pluton was largely accommodated in the zone and that magma buoyancy forces have only played a very small part in creating space.

The reason for the shear zone displacement gradient can only be guessed at but it seems reasonable to suggest that the very earliest Main granite magma intruded into one end of the already established planar shear zone and locally increased the slip and displacement rates in that area. The instability, once created, drew new magma into the opening cavity, deforming it in the process. This continued until the pluton crystallised and the large-scale viscosity contrast within the shear zone disappeared.

The stress conditions varying in time and space around the developing shear zone can explain the varying emplacement histories of the other virtually synchronous plutons in the batholith (Fig. 3d, e, f): Thorr, in an extensional zone between the main shear zone and the (offshore) Great Glen fault, Fanad in a dilational shear zone tip area (White & Hutton 1985), Ardara and Toories in a compressional tip area, Rosses in outer arc extensional stresses around the shear zone bend, and Trawenagh Bay as Main Granite magma emplaced outside the shear zone, in outer arc tension similar to Rosses.

2.1.2. Other examples. General examples of syn kinematic intrusion into major tectonic shear zones are described by Brun and Pons (1981) from the Sierra Morena and Courrioux (1982) from the Puentedeume granite in Galicia, both in Spain, and from Devonian shear zones in Newfoundland (Hanmer 1981). In the Sierra Morena plutons Bruns and Pons (1981) have detailed typical foliation patterns produced by ballooning into synchronous transcurrent and thrust type shear zones. In both types there is geometric continuity in foliation trends between plutons and wall rocks: in the former the plutons develop characteristic helicoidal patterns; in the latter sub concordant but eccentric patterns. In both situations characteristic foliation triple points are developed because of the interference of regional and ballooning strain fields.

Examples of more detailed kinematic emplacement mechanisms are provided by Guineberteau et al. (1987) for the Mortagne pluton in Vendee. This lozenge-shaped pluton was emplaced syn kinematically into a sinistral shear zone associated with the southern Armorican Shear Zone. The orientation and deflection of early pre-full crystallisation fabrics together with later overprints by crystal plastic strain S–C structures is consistent with a model of the emplacement of the pluton into a pull apart void formed by an E–W jog (Sibson 1986) in the NW–SE-trending sinistral shear zone.

In the Central Extramadura batholith Castro (1986) has modelled the emplacement of the Hercynian plutons in lensoid extension cracks orientated at 45° to the shearing direction in a major dextral transcurrent shear zone. Late emplaced plutons preserve this geometry together with internal balloon-related pre-full crystallisation fabrics, whereas the earlier plutons have been dextrally rotated from this original angle and deformed in a crystal plastic strain regime. Similarly Natal'in et al. (1986) mention a set of plutons in the Stanovoy range, Soviet Far East, which occur as echelons oblique and related to movement in the southern Yakutsk dextral fault system.

In Ireland the Caledonian Leinster granite appears to lie in a major contemporaneous sinistral shear zone (Cooper & Bruck 1983; Murphy 1987). The units of the batholith lie en echelon with long axes oblique but at less than 45° to the shear zone trend and a model can be suggested similar to that proposed by Castro (1986) for the Central Extramadura but involving ballooning into a transtensional shear zone. Also in Ireland, the Caledonian Ox Mountains granite was emplaced synchronously into a major sinistral shear zone which is the side wall of the probable Irish extension of the Highland Boundary Fault (Hutton & Dewey 1986). Mechanisms of emplacement can be demonstrated on a local scale in this sheeted pluton by asymmetric sinistral oblique transform offsets between and within the sheets (K. McCaffrey, pers. comm.) indicating that emplacement occurred in a sinistral dilational environment.

Finally Davies (1982) has described different emplacement mechanisms in contemporaneous Pan African plutons associated with a major shear zone in Saudi Arabia. In a manner reminiscent of the Donegal situation, passively emplaced cauldrons occur in the walls of the shear zone and balloons/diapirs occur within the shear zone itself. Detailed strain studies show that the shear zone was associated with a volume increase, some of which can be attributed to the diapiric emplacement of granite.

2.2. Emplacement in extensional shear zones

2.2.1. Strontian. The emplacement of this pluton will be described in detail elsewhere (Hutton 1988; unpublished) and in this contribution it is intended only to summarise the main findings and emphasise the applicability of the emplacement processes that this pluton illustrates.

The pluton (Fig. 4a) contains three units, which in order of emplacement are an outer tonalite, an inner granodiorite and an innermost biotite granite (Sabine 1963; Munro 1965, 1973). Emplacement occurred in two distinct phases with the intrusion of the tonalite and granodiorite in the first phase and the biotite granite in the second. Internal structures related to the first phase are modified in the vicinity of the biotite granite and are best preserved in the main northern area of tonalite and granodiorite. The foliation in this area, which is a pre-full crystallisation fabric, is concordant with the outer contact and dips inwards all around the marginal granite zone, although it is nearly vertical in the W. The foliation dips are flat in the middle and so the general pattern is that of an asymmetric synform plunging southwards (Fig. 4b). Strain determinations from numerous deformed mafic xenoliths indicate that deformation was highest around the contact, and X directions within the body are orientated approximately N–S (Fig. 4b). Shear sense indicators (particularly extensional shears in the wall rocks which are synchronously intruded by minor granite veins) and similar structures within the granites themselves show consistent top towards the S movement. Although this pluton has been treated as a diapir/balloon (Hall 1987) the X direction and shear sense are incompatible with such a model and, moreover, the marginal zone of concordant deformation in the wall rocks is too narrow (<100 m) and not highly enough deformed to have allowed significant *in situ* expansion of the pluton. The pluton lies in a major southward-plunging synform in the country rocks and therefore another model would be that it was originally emplaced as a flat sheet and was then folded with the country rocks during regional Caledonian orogenesis in Scotland. This model is contradicted by the shear sense data, which are incompatible with such folding, and also by the fact that although the pluton is generally concordant with the regional fold it does in fact truncate some large stratigraphic units in the wall rocks (Fig. 4b). The conclusion is, therefore, that the pluton was intruded as a synformal shape, although it was probably sited in a pre-existing country rock fold; therefore one must seek other mechanisms for the emplacement.

In N–S section, parallel to the X strain direction, the foliation has a listric profile. This, combined with the shear sense, suggests deformation by horizontal extension and the relative movement between an overlying hanging wall country rock block and the magma/country rock

footwall in a geometric scheme analogous to the ramp/flat listric shapes of extensional fault systems (Gibbs 1984; Fig. 4c, d). Strain determinations in the granite, which relate to pre-full crystallisation deformation, show that the inner flat foliation zone has relatively high K values ($K \approx 1$) and regular N–S-oriented X directions. However, around this inner zone, in the more highly inclined foliations, K values are much lower ($0 < K \ll 1$) and X directions diverge from N–S (Fig. 4b). These data are consistent with emplacement into, and simultaneous deformation of magma in, a listric extensional fault where space is largely created by the relative movement of hanging wall and footwall. The deformation in such a situation will tend to reflect the relative shape of the hanging wall, its movement relative to the underlying magma and also the tendency of the magma to fill the spaces as they become available. Space between the hanging wall and footwall ramp will open and change shape more rapidly than the space between the hanging wall and footwall flat (Fig. 4d). This more tightly constrained movement between near parallel sided walls in the lower "flat space" will generate flat fabrics with more regular parallel orientated and larger maximum strain rate directions (hence K close to 1, and X direction parallel). In the upper "ramp space" the foliations will be more steeply

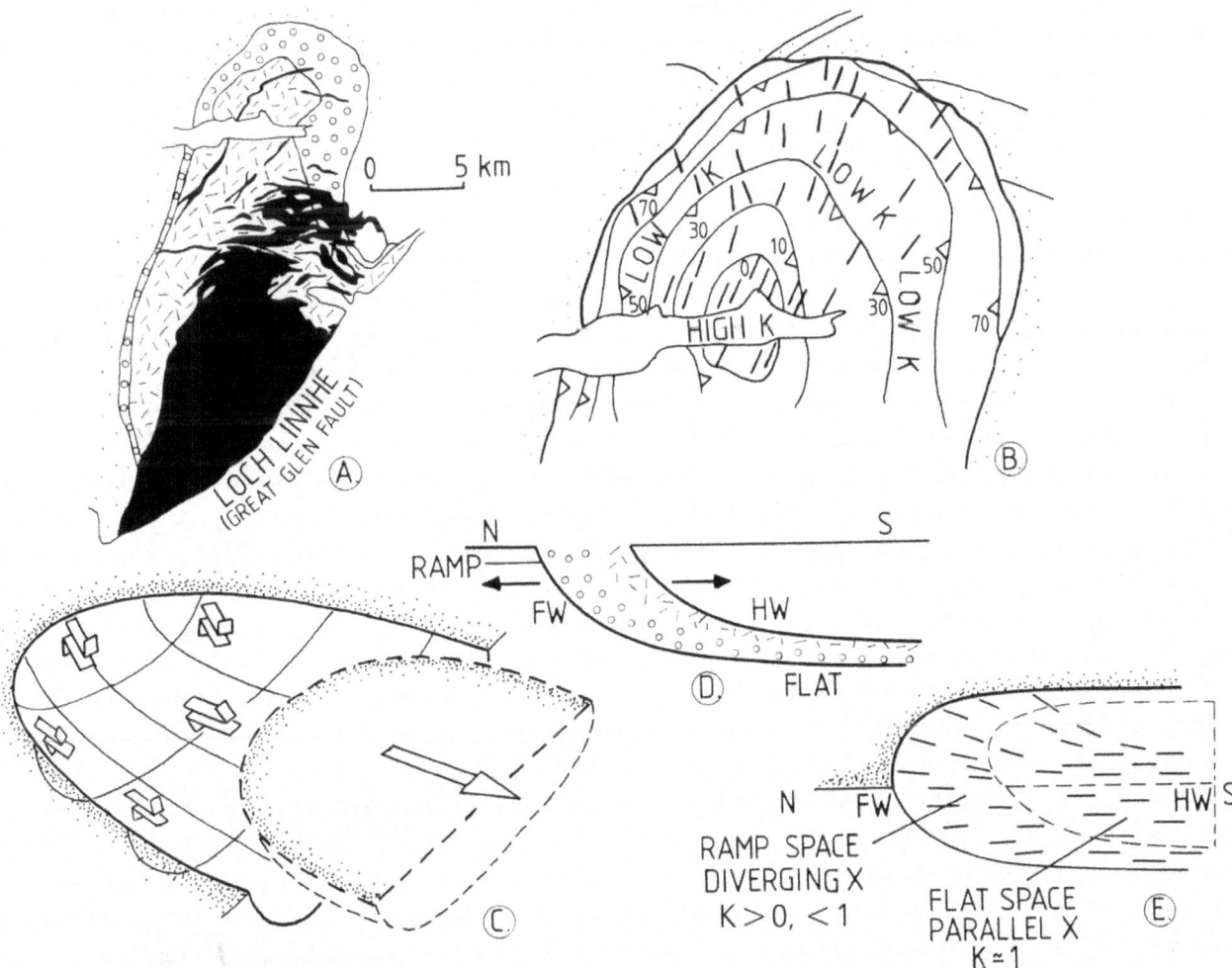

Figure 4 Strontian granite: (A) general geology: country rock (dotted ornament); tonalite (circles); granodiorite (dashes); biotite granite (black); (B) northern part of pluton showing truncation of country rock stratigraphy at contacts, generalised foliation strikes and dips in pluton, orientation of xenolith long axes and mineral alignments, relative variations in K values; (C) block diagram showing nature of footwall to intrusion with shear sense observations indicating the horizontal southward movement of an overlying hanging wall block of country rock; (D) generalised N–S section—HW = hanging wall, FW = footwall; (E) cut away block diagram showing X direction variation between "flat space" and "ramp space".

252   DONALD H. W. HUTTON

inclined, and a type of dispersive flow deformation will
occur with divergent and weakening maximum strain rate
giving way to more nearly equal strain rates in all directions
in the foliation planes (hence $K > 0 \ll 1$ and non-parallel,
divergent X directions) (Fig. 4c).

Although the data are consistent with such a model the
main weakness is that the hanging wall country rocks are not
available for inspection at Strontian (they have presumably
been eroded). However, more recent work in Greenland
(Brown, Dempster and Hutton, unpublished) shows ample
evidence of Strontian-type processes in the emplacement of
rapakivi granite sheets with the displaced hanging walls
intact.

The biotite granite at Strontian (Hutton 1988) was
emplaced later than the tonalite and granodiorite units, but
its shape broadly follows that of the earlier intrusion. In the
area W of the long western contact of the biotite granite the
tonalite and granodiorite pre-full crystallisation fabrics are
strongly modified by crystal plastic strain deformation
related to a NNE-trending dextral shear zone. Xenolith
strain studies suggest that this shear zone extended
eastwards into the site of the biotite granite before this body
was emplaced, although some movements continued
afterwards to produce a weak pre-full crystallisation fabric
in the marginal part of the later granite. Dextral shearing
and associated strains diminish northwards and are absent
around the northern contact of the biotite granite which
here is an arcuate zone of steeply-inclined sheets indicating
N–S horizontal extension (Fig. 4a). The combination of data
is strongly suggestive of an extensional termination to a
dextral shear zone with biotite granite emplaced and then
marginally deformed in the space created (Fig. 5a).
Following the geometry of the earlier tonalite and
granodiorite, the steep lateral shear zone and steep curved
northern contact would merge at depth into a lower flat
contact. The general model for Strontian, therefore, is that
of a two phase, near horizontal removal, by extensional
stresses, of a segment of country rock out of the core of a
large pre-existing plunging synform. Magma exploited the
space that was created and was deformed in the process. On
a larger scale the siting of the pluton may have been
controlled by magma feeding up a dextral releasing bend
cavity in the adjacent Great Glen fault before being
emplaced sideways into the adjacent synform cavity (Fig.
5b).

3. General conclusions and synthesis

The preceding examples illustrate two general points:
(1) space can be made available within tectonic structures
for the emplacement of magmas, and (2) the sign and
magnitude of tectonically-originating stresses may otherwise
determine pluton emplacement mechanism.

These are oversimplifications and it is more likely that
natural magma buoyancy forces *combine* with the above to
create space. The physical reality of this general relationship
has been known about for some time. We are most familiar
with the concept in the theoretical and model studies of the
emplacement of dykes and sills (Anderson 1936; Robson &
Barr 1963; Murrell 1970; Roberts 1970; Pollard & Johnson
1973; Dixon & Simpson 1987), where intrusion shape and
orientation are understood in terms of the interaction
between magma buoyancy stress, lithostatic stress, tectonic
deviatoric stress and the material parameters for both
isotropic and anisotropic crust. If we shift this general and
oversimplified philosophy to a more realistic complexly

Figure 5 (A) Summary diagram of bitotite granite emplacement
mechanism showing diminution of dextral shear northwards into
arcuate extensional zone of sheeting; (B) general siting model for
the Strontian complex (tonalite, granodiorite and biotite granite):
dextral shear on the Great Glen Fault creates an extensional
component at a releasing bend; this pulls a flat segment out of the
core of a pre-existing Caledonian synform; magma comes up the
releasing bend hole and emplaces sideways into the synform cavity.

anisotropic, broken and flawed continental crust, then I
envisage simply that the mechanisms become more
complicated and the emplacement geometries more
irregular and more varied, where tectonics tend to create
dilation either on a regional scale or locally within otherwise
compressional areas. The basis for expecting transient and

permanent dilation and cavities to exist in tectonic regimes comes from any review of the last few decades of structural geology and tectonic research. It now seems clear that deformation in the continental crust, because of such things as anisotrophy/viscosity contrast, contemporaneous fractures, and pre-existing structures including lineamental fractures, proceeds in a far from homogenous way. Shear systems in shortening, transcurrent and extensional regimes

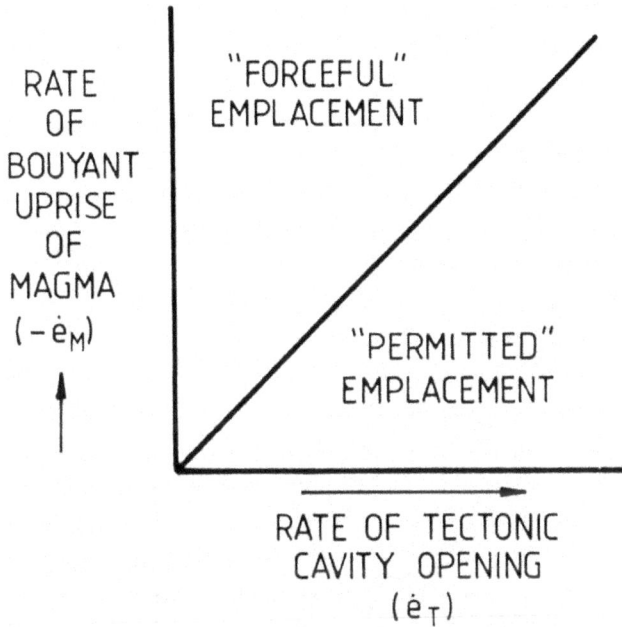

Figure 6 Simplified view of the relationship between magma related buoyancy forces and strain rates and tectonic 'cavity creating' rates leading to a spectrum of emplacement types with two basic end members.

and on a *complete* array of scale may operate in a complex linked network of failure planes and zones where displacement is transferred and accommodated. Within this, gaps and areas of dilation open transiently or permanently. These processes are held to be generally common in the anisotropic seismic upper crust and many current studies in the exposed parts of old middle and lower crust suggest that similar processes, associated with aseismic creep and ductile shear, are no less common there. While more igneous orientated studies suggest that once magma becomes involved in these systems it can affect the development of the structures themselves, it seems clear that many of the basic structural patterns, known from structural and tectonic studies, can still be identified as such (e.g. Mortagne, Central Extramadura, Strontian). As a general statement, therefore: if magma is available, then in the presence of regional tectonics, there need be no *a priori* "space problem".

Broadening this discussion to the question of the general variety of emplacement behaviour (Fig. 5), if there is a physical problem about granite diapirs ascending to sufficiently high levels in the crust to be observable (Anderson 1981; Marsh 1982; Miller *et al.* 1988) then it seems reasonable to follow Bateman (1984) in saying that fracture "conduits" may allow buoyancy-driven ascent to higher levels. Piecemeal accession of small magma batches may lead to ponding/cavity infill at high levels in the crust where the buoyancy head *then* expresses itself in the *in situ* ballooning or even late unidirectional and limited diapiric behaviour of plutons. The observation of such conduits may be difficult, yet there are some cases such as the Puentedeume granite (Courrioux 1982) where a linear pluton emplaced in a shear zone subsequently balloons *in situ*.

If buoyancy allows the more fluid magmas to rise, via

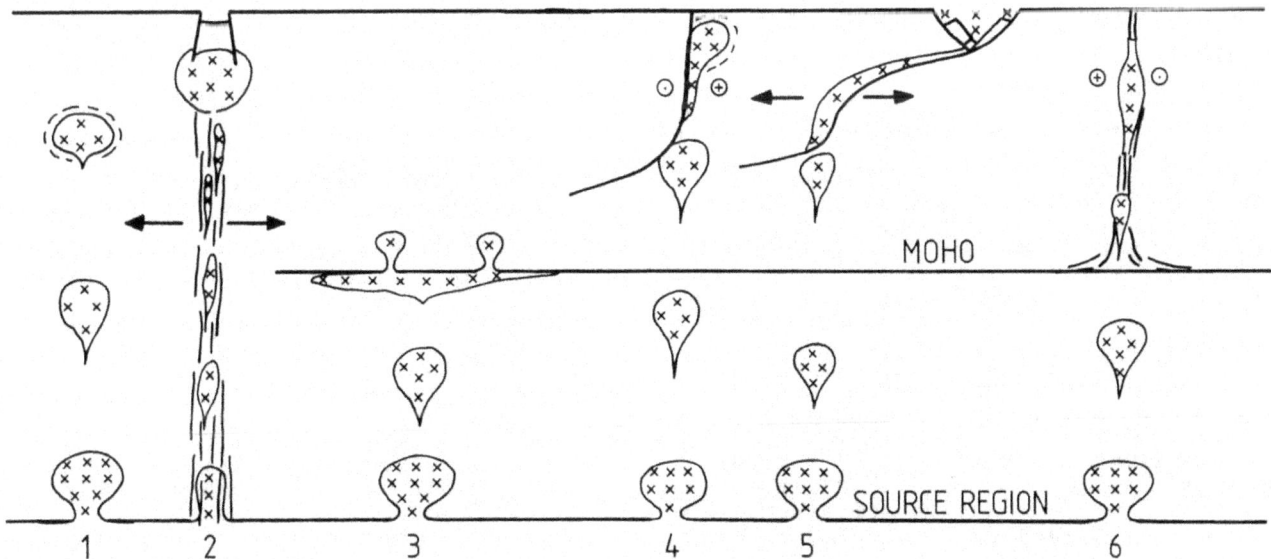

Figure 7 Six generalised modes of ascent and emplacement of granitoids; all begin with initial diapiric detachment and uprise of melts: (1) continued diapric uprise, in the absence of tectonics, leading to final arrest due to density equilibration followed by late ballooning; (2) uprise into major vertical tectonic extensional fault system, magmas rise to high levels, with uppermost crustal ponding and cauldron/caldera behaviour; (3) diapiric uprise arrested by viscosity/strength changes at Moho—this leads to lateral spreading with possible late spawning of crustal plutons; (4) diapiric rise into middle crust, intercepting an intra crustal strike slip fault zone leading to elongate plutons with late ballooning; (5) diapiric uprise intercepting intra crustal listric extensional fault/shear zone leading to listric granite sheets and possible generation of asymmetric cauldrons and calderas; (6) uprising melts intercept trans crustal vertical transcurrent fault/shear zone—jogs, pull apart and large tension gash features, etc., create space for magma ponding; note that in all these scenarios the "source region" is arbitrarily located in the lithospheric mantle.

fracture and cavity pathways, close to the earth's surface, then the releasing effect of that free surface allows the formation of high level cauldrons and related surface calderas, as has long been realised (Anderson 1936). The tectonic control is present here: the "anorogenic" plutons are typical of continental extensional regimes.

Deeper in the crust, away from the direct influence of the free surface, emplacement mechanisms can be generally accounted for by the interaction of natural magma buoyancy forces and tectonic "forces" (Fig. 6). If in a tectonically-created cavity/volume of dilation the extensional strain rate $(\dot{e}_T)$ for the volume exceeds that of the compressional strain rate $(-\dot{e}_M)$ produced by the buoyancy of the emplacing magma volume, then net extension will occur and "passive" or "permitted" emplacement will take place. On the other hand, if the tectonic volumetric extension rate is less than the magmatic volumetric compression rate, then "forceful" emplacement will occur. This model predicts, apart from the celebrated end members, the occurrence of many plutons which exhibit, in varying degrees, both "forceful" and "passive" behaviour: I believe this is what we see. Finally Figure 7 is an attempt to show six basic ascent/emplacement schemes for magmas.

4. Acknowledgements

I would like to take this opportunity to acknowledge the interest, help and discussion that I have benefited from over the last few years with colleagues and workers in the granite field, particularly: W. S. Pitcher, P. Bateman, D. Brew, R. Bateman, A. Castro, G. Courrioux, L. T. Silver, B. Chappell, E. Stephens, P. Brown, K. McCaffrey, R. England, A. Halliday, E. An-Zen and M. Atherton. I would also like to thank J. Preston and I. Meighan who first got me interested in granites. The work on Strontian was supported by NERC grant GR3/4633.

5. References

Anderson, E. M. 1936. The dynamics of the formation of conesheets, ring dykes, and cauldron subsidence. PROC R SOC EDINBURGH 56, 128–163.

Anderson, D. L. 1981. Rise of deep diapirs. GEOLOGY 9, 7–9.

Arzi, A. A. 1978. Critical phenomena in the rheology of partially melted rocks. TECTONOPHYSICS 44, 173–184.

Balk, R. 1937. Structural behaviour of igneous rocks. MEM GEOL SOC AM 5, 177 pp.

Bateman, R. 1984. On the role of diapirism in the segregation, ascent and final emplacement of granitoids. TECTONO-PHYSICS 110, 211–231.

Berger, A. R. 1971. The origin of banding in the Main Donegal Granite, N.W. Ireland. GEOL J 7, 347–358.

Berger, A. R. & Pitcher, W. S. 1970. Structures in granitic rocks. A commentary and a critique on granite tectonics. PROC GEOL ASSOC LONDON 81, 441–461.

Berthe, D., Choukroune, P. & Jegouzo, P. 1979. Orthogeneiss, mylonites and non coaxial deformation of granites: the example of the South Armorican shear zone. J STRUCT GEOL 1, 31–42.

Bouchez, J. L., Lister, G. S. & Nicolas, A. 1983. Fabric asymmetry and shear sense in movement zones. GEOL RUNDSCH 72, 401–419.

Brun, J. P. & Pons, J. 1981. Strain patterns of pluton emplacement in a crust undergoing noncoaxial deformation. J STRUCT GEOL 3, 219–229.

Buddington, A. F. 1959. Granite emplacement with special reference to North America. BULL GEOL SOC AM 70, 671–747.

Castro, A. 1984. Emplacement fractures in granite plutons (Central Extramadura batholith, Spain). GEOL RUNDSCH 73, 869–880.

Castro, A. 1986. Structural pattern and ascent model in the Central Extramadura batholith, Hercynian belt, Spain. J STRUCT GEOL 8, 633–645.

Choukroune, P., Gapais, D. & Merle, O. 1987. Shear criteria and structural symmetry. J STRUCT GEOL 9, 525–530.

Cooper, M. A. & Bruck, P. M. 1983. Tectonic relationships of the Leinster Granite, Ireland. GEOL J 18, 351–360.

Courrioux, G. 1982. Exemple de mise en place d'un leucogranite pendant le fonctionnement d'une zone de cissaillement: le granite hercynien de Puentedeume (Galice, Espagne). BULL SOC GEOL FR 25, 301–307.

Courrioux, G. 1987. Oblique diapirism: the Criffel granodiorite/granite zoned pluton (southwest Scotland). J STRUCT GEOL 9, 313–330.

Coward, M. P. 1976. Strain within ductile shear zones. TECTONOPHYSICS 28, 89–124.

Davies, F. B. 1982. Pan-African granite intrusion in response to tectonic volume changes in a ductile shear zone from northern Saudi Arabia. J GEOL 90, 467–483.

Dixon, J. M. & Simpson, D. G. 1987. Centrifuge modelling of laccolith intrusion. J STRUCT GEOL 9, 87–103.

Flinn, D. 1965. On the symmetry principle and the deformation ellipsoid. GEOL MAG 102, 36–45.

Fry, N. 1979a. Density distribution techniques and strained length methods for determination of finite strains. J STRUCT GEOL 1, 221–229.

Fry, N. 1979b. Random point distributions and strain measurements in rocks. TECTONOPHYSICS 60, 89–105.

Gibbs, A. D. 1984. Structural evolution of extensional basin margins. J GEOL SOC LONDON 141, 608–620.

Guineberteau, B., Bouchez, J-L, & Vigneresse, J-L. 1987. The Mortagne granite pluton (France) emplaced by pull-apart along a shear zone: structural and gravimetric arguments, regional implication. BULL GEOL SOC AM (in press).

Hall, A. 1987. Igneous Petrology. London: Longman.

Hanmer, S. 1981. Tectonic significance of the northeastern Gander Zone, Newfoundland: an Acadian ductile shear zone. CAN J EARTH SCI 18, 120–135.

Hanna, S. S. & Fry, N. 1979. A comparison of methods of strain determination in rocks from southwest Dyfed (Pembrokeshire) and adjacent areas. J STRUCT GEOL 1, 155–162.

Hibbard, M. J. 1987. Deformation of incompletely crystallised magma systems: granitic gneisses and their tectonic implications. J GEOL 95, 543–561.

Holder, M. T. 1979. An emplacement mechanism for post-tectonic granites and its implication for their geochemical features. In Atherton, M. P. & Tarney, O. O. (eds) Origin of granite batholiths: geochemical evidence, 116–128. Nantwich: Shiva.

Hollister, L. & Crawford, M. L. 1986. Melt-enhanced deformation: a major tectonic process. GEOL 14, 558–561.

Hutton, D. H. W. 1977. A structural cross-section from the aureole of the Main Donegal granite GEOL J 12, 99–112.

Hutton, D. H. W. 1982a. A tectonic model for the emplacement of the Main Donegal granite, NW Ireland. J GEOL SOC LOND 139, 615–631.

Hutton, D. H. W. 1982b. A method for the determination of the initial shapes of deformed xenoliths in granitoids. TECTONOPHYSICS 85, T45–50.

Hutton, D. H. W. 1988. Igneous emplacement in a shear zone termination: the biotite granite at Strontian, Scotland. BULL GEOL SOC AM (in press).

Hutton, D. H. W. & Dewey, J. F. 1986. Paleozoic terrane accretion in the western Irish Caledonides. TECTONICS 5, 1115–1124.

Leake, B. E. 1978. Granite emplacement: the granites of Ireland and their origin. In Bowes, D. R. & Leake, B. E. (eds) Crustal evolution in northwestern Britain and adjacent regions. GEOL J SPEC PUBL 10, 221–248.

Lister, G. S. & Hobbs, B. E. 1980. The simulation of fabric development during plastic deformation and its application to quartzite: the influence of deformation history. J STRUCT GEOL 2, 355–370.

Marre, J. 1986. The structural analysis of granitic rocks. Oxford: North Oxford Academic.

Marsh, B. D. 1982. On the mechanics of igneous diapirism, stoping and zone melting. AM J SCI 282, 808–855.

Miller, C. F., Watson, E. B. & Harrison, T. M. 1988. Perspectives on the source, segregation and transport of granitoid magmas. TRANS R SOC EDINBURGH EARTH SCI 79, 135–156.

Munro, M. 1965. Some structural features of the Caledonian granitic complex at Strontian, Argyllshire. SCOTT J GEOL 1, 152–175.

Munro, M. 1973. Structures in the south-eastern portion of the Strontian granitic complex, Argyllshire. SCOTT J GEOL **9**, 99–108.

Murphy, F. C. 1987. Late Caledonian granitoids and timing of deformation in the Iapetus suture zone of eastern Ireland. GEOL MAG **124**, 135–142.

Murrell, S. A. F. 1970. Global tectonics, rock mechanics, and the mechanism of volcanic intrusions. *In* Newall, G., & Rast, N. (eds) *Mechanisms of igneous intrusion.* GEOL J SPEC PUBL **2**, 231–244.

Natal'in, B. A., Parfenov, L. M., Vrubleusky, A. A., Karsakov, L. P., & Yushmanov, V. V. 1986. Main fault systems of the Soviet Far East. *In* Reading, H. G., Watterson, J., & White, S. H. (eds) *Major crustal lineaments and their influence on the geological history of the continental lithosphere,* 267–275. London: Royal Society of London.

Passchier, C. W. & Simpson, C. 1986. Porphyroclast systems as kinematic indicators. J STRUCT GEOL **8**, 831–843.

Pitcher, W. S. 1952. The Rosses granitic ring complex, County Donegal, Eire. PROC GEOL ASSOC LONDON **64**, 153–183.

Pitcher, W. S. & Berger, A. R. 1972. *The geology of Donegal: a study of granite emplacement and unroofing.* London: Wiley Interscience.

Pitcher, W. S. & Bussell, M. A. 1977. Structural control of batholith emplacement in Peru: a review. J GEOL SOC LOND **133**, 249–256.

Platt, J. P. & Vissers, R. L. M. 1980. Extensional structures in anisotropic rocks. J STRUCT GEOL **2**, 397–410.

Pollard, D. D. & Johnston, A. M. 1973. Mechanics of growth of some laccolithic intrusions in the Henry mountains, Utah-II. Bending and failure of overburden layers and sill formation. TECTONOPHYSICS **18**, 311–345.

Ramsay, J. G. & Graham, R. H. 1970. Strain variation in shear belts. CAN J EARTH SCI **7**, 786–813.

Ramsay, J. G. & Huber, M. I. 1983. *The techniques of modern structural geology, Volume 1: strain analysis.* London: Academic Press.

Read, H. H. 1957. *The Granite Controversy.* London: Thomas Murby & Co.

Roberts, J. L. 1970. The intrusion of magma into brittle rocks. *In* Newall, G. & Rast, N. (eds) *Mechanisms of igneous intrusion.* GEOL J SPEC PUBL **2**, 287–338.

Robson, G. R. & Barr, K. G. 1963. The effect of stress on faulting and minor intrusions in the vicinity of a magma body. I.U.G.G. **13**, 315–330.

Sabine, P. A. 1963. The Strontian granite complex, Argyllshire. BULL GEOL SURV GB **20**, 6–42.

Saltzer, S. D. & Hodges, K. V. 1988. The Middle Mountain shear zone, southern Idaho: Kinematic analysis of an early Tertiary high temperature detachment. BULL GEOL SOC AM **100**, 96–103.

Sanderson, D. J. 1976. The superposition of compaction and plane strain. TECTONOPHYSICS **30**, 35–54.

Sanderson, D. J. & Marchini, D. 1984. Transpression. J STRUCT GEOL **6**, 449–458.

Sanderson, D. J. & Meneilly, A. W. 1981. Strain modified uniform distributions: andalusites from a granite aureole. J STRUCT GEOL **3**, 109–116.

Shaw, H. R. 1980. The fracture mechanisms of magma transport from the mantle to the surface. *In* Hargreaves, R. B. (ed.) *Physics of magmatic processes,* 201–264. Princeton, N. J: Princeton Press.

Sibson, R. H. 1986. Earthquakes and lineament infrastructure. *In* Reading, H. G., Watterson, J. & White, S. H. (eds) *Major crustal lineaments and their influence on the geological history of the continental lithosphere,* 63–79. London: Royal Society of London.

Simpson, C. & Schmid, S. M. 1983. An evaluation of criteria to deduce the sense of movement in sheared rocks. BULL GEOL SOC AM **94**, 1291–1288.

Soper, N. J. & Hutton, D. H. W. 1984. Late Caledonian sinistral displacements in Britain: implications for a three plate collision model. TECTONICS **3**, 781–794.

Talbot, C. J. 1970. The minimum strain ellipsoid using deformed quartz veins. TECTONOPHYSICS **9**, 47–76.

Van der Molen, I. & Peterson, M. S. 1979. Experimental deformation of partly-melted granite. CONTRIB MINERAL PETROL **70**, 299–318.

White, N. & Hutton, D. H. W. 1985. The structure of the Dalradian rocks in west Fanad, County Donegal. IR J EARTH SCI **7**, 79–92.

White, S. H., Brefan, P. G. & Rutter, E. H. 1986. Fault zone reactivation: kinematics and mechanisms. *In* Reading, H. G., Watterson, J. & White, S. H. *Major crustal lineaments and their influence on the geological history of the continental lithosphere,* 81–97. London: Royal Society of London.

Woodcock, N. H. & Underhill, J. R. 1987. Emplacement related fault patterns around the Northern Granite; Arran, Scotland. BULL GEOL SOC AM **98**, 515–527.

DONALD H. W. HUTTON, Department of Geological Sciences, Durham University, Durham DH1 3LE, U.K.

MS received 23 December 1987. Accepted for publication 10 May 1988.

CHAPTER 10

Clemens, J.D. and Mawer, C.K. (1992) Granitic magma transport by fracture propagation. *Tectonophysics*, **204**, 339−360.

One of the late 20th century's major shifts in thinking about granitic magmatism was the virtual abandonment of the concept of diapirism as a major process in the ascent and emplacement of granitic magmas. This concept, apparently so attractive to the geological mind, was born in the 1920s, when analogies were made between the outcrop shapes of salt domes and granitic plutons. The assumption was also that granitic plutons are vertically extensive bodies and that the shape of the pluton, in plan, had something to do with the ascent mechanism. The idea was that granitic magmas rose to their emplacement levels as huge buoyant blobs that softened and deformed the surrounding crust to make way for themselves. This gave rise to heated debates about 'the room problem' − where did the displaced crust go? In turn, this fuelled transformist theories because *in situ* replacement and granitization offered a ready answer to this 'problem'.

From a range of studies − structural, stratigraphic and geophysical − we now understand that granitic plutons are most commonly sheet-like or tabular in form, with rather lower volumes than was once thought (e.g. McCaffrey and Petford, 1997; Haederle and Atherton, 2002). Writing from our present perspective, the evidence seems overwhelming that most of these plutons were constructed from hundreds to thousands of incremental magma additions or pulses. Diapirs are thermally and mechanically inefficient (or even ineffectual) at accomplishing this and so do not form part of the current mainstream of granitological opinion. Given the diapir's former premier position in the magmatist view of how granitic magmas ascend and are emplaced, how did this relatively rapid revolution occur? To understand this we have to be precise about what the modern view is, and we will attempt to express that now. Granitic magmas are best modelled as having ascended, due to buoyancy, by means of initially self-propagating hydrofractures (dykes) that rapidly feed growing plutons by delivering magma batches in solitary waves.

The idea that fracture systems could be important in emplacement of granitic magmas has been around for a long time, well over a century. The evidence is apparent from field studies (e.g. Pitcher and Berger, 1972; Haederle and Atherton, 2002). However, we are not talking about emplacement here. Emplacement is the group of mechanisms by which space is created for a growing pluton. Rather, we are talking about ascent − how the magma

traverses sometimes tens of kilometres before it reaches the site of emplacement. The mathematical formalism relating to the physics of liquid movement, under gravity, in cracks was certainly developed by the 1970s (e.g. Cook and Gordon, 1964; Weertman, 1971, 1977; Pollard, 1977). The late 1980s and earliest 1990s saw more specific treatments of the problem (e.g. Wilson and Head, 1981; Sleep, 1988; Lister and Kerr, 1991). However, nobody applied any of this to granitic magmas because diapirism held absolute sway in thinking about granitic magma ascent. Basaltic magmas clearly exploited fracture systems but granitic magmas were different, they were too cool and viscous to move in that way. The catalyst that changed all this was the publication of Clemens and Mawer (1992).

Previous work on fracture ascent of magmas gave Clemens and Mawer the tools they needed to take a fresh look at granitic magma ascent. They realized first that the field evidence for granitic diapirism was flimsy, at best. Previous work had also taught them that granitic magmas were commonly much higher-temperature entities than had been commonly supposed. Thus, although these magmas were H_2O-undersaturated, their viscosities were nothing like as high as was commonly assumed. They calculated that granitic magma batches could ascend through the crust in less than a month and, given sufficient magma supply at the source, fractures only a few metres wide could supply enough magma to build a batholith in a few hundreds to a few thousands of years. These geologically insignificant time spans were a surprise to many. The space problem evaporated because the volumes were lower than previously supposed and because space only needed to be created at the emplacement site, and this could be accomplished by many different mechanisms. Thus, what was newly introduced by Clemens and Mawer was the concept that long-distance granitic magma transport could be rapid and controlled by fractures.

Clemens and Mawer thought that brittle crack formation was probably critical in the whole process, from melt segregation to long-range magma transport and final emplacement. It is now clear that ductile fracture, during deformation, is a more important process in and near the magma source region (e.g. Brown, 2007; Hobbs and Ord, 2009). Clemens and Mawer also stated that the magma would be delivered to a growing pluton in pulses. This conclusion stemmed from the physics of liquid-filled cracks

(Weertman, 1971), which, when scaled to granitic magma ascent in the crust, dictate that a granite-filled fracture could not remain open from source to emplacement site (if the distance is more than a few km), and that magma pulses will travel up these pathways as solitary waves. Thus, the source must be drained incrementally and the pluton fed incrementally (see also Glazner et al., 2004).

Clemens and Mawer (1992) was "in preparation" from 1984, and it is to their discredit that they failed to bring it forth sooner. Their work was followed up by Petford et al. (1993), and there were elements of disagreement, such as whether the magma-transporting fractures were pre-existing or were generated by the buoyancy forces, and whether pulses of magma were inherently going to exist. Nevertheless the two groups joined forces in the battle against the old diapir paradigm.

Today we talk and write about granitic magma ascent in dykes and pulsed magma delivery as if we had always known these things must occur. Physical and geochemical evidence mounts for their occurrence. However, these ideas start somewhere and, for granites, they did so with Clemens and Mawer. The history of this is nicely covered in the book *Mind over Magma* (Young, 2003 pp. 605–607). Perhaps the nub of the residual arguments lies in the questions of how fast a pluton can grow (related to deformation rates) and whether the assembled magma chamber (pluton) ever undergoes chamber-wide magmatic processes, such as differentiation. These are connected concepts. People also differ on whether differentiation (at emplacement level) is of any great significance for granites. Whatever the answers to these questions, and since it represents a watershed or catalyst for paradigm shift, Clemens and Mawer (1992) deserves its place in this volume.

Tectonophysics, 204 (1992) 339–360
Elsevier Science Publishers B.V., Amsterdam

339

Granitic magma transport by fracture propagation

J.D. Clemens [a] and C.K. Mawer [b,1]

[a] *Department of Geology, The University, Manchester, M13 9PL, UK*
[b] *Department of Geology, University of New Mexico, Albuquerque, NM 87131, USA*

(Received September 20, 1991; revised version accepted January 21, 1992)

ABSTRACT

Clemens, J.D. and Mawer, C.K., 1992. Granitic magma transport by fracture propagation. *Tectonophysics*, 204: 339–360.

Granitic magmas commonly ascend tens of kilometres from their source terranes to upper crustal emplacement levels, or to the Earth's surface. Apart from its obvious bearing on the interpretation of the geology and geochemistry of felsic igneous rocks, the magma ascent mechanism critically affects any modelling of the metamorphic evolution of the upper lithosphere as well as its rheology. We propose that, in general, granitic magmas ascend via propagating fractures, as dykes, in extremely short time periods. Long-distance diapiric transport of granitoid magmas, through crustal sections, is not viable on thermal or mechanical grounds, and there is an apparent total absence of field evidence for diapiric rise of such magmas, even in supposed "type" localities. Neither the shapes nor internal or external characteristics of granitic plutons necessarily reveal anything about the transport of their precursor magmas; these are purely arrival phenomena dictated by local structure, kinematics, and stress states. Based on existing numerical treatments of the problem, we show that granitic magmas are apparently sufficiently inviscid to travel through fractures, to high crustal levels, without suffering thermal death by freezing. For example, a 2000-km^3 granitic batholith can be inflated by a 1 km × 3 m × 20 km-deep dyke system in less than 900 years. The model proposed has numerous, far-reaching, petrological and rheological consequences, some of which are outlined.

Introduction

Granitoid magmatism is fundamentally a crustal phenomenon. When mantle-derived magmas invade fertile crust, and raise its temperature above about 650°C, crustal melting is the natural consequence. Melting may also occur in thickened continental crust, without extraneous heat input (e.g., Patiño Douce et al., 1990). The visible products of such partial fusion are migmatites, granitoids, silicic volcanic rocks and exhumed restitic granulites. There is abundant evidence that granitoid magmas (tonalitic to granitic, *sensu stricto*) can be generated by relatively small degrees of partial fusion, and that they have commonly risen many kilometres from their sources

to emplacement levels. As pointed out by Miller et al. (1988, p. 136), "... the absence of a physical model to explain this... is... a missing link in understanding granitoid magmatism". We believe that this "missing link" is fundamental also to a much wider variety of Earth processes.

The similar ranges of temperatures, water contents and initial crystallinities of both volcanic and plutonic granitoid magmas (e.g., Clemens and Wall, 1981; Clemens, 1988), provides evidence that emplacement is probably entirely controlled by local structural factors plus properties developed late in the crystallization of the magmas (Clemens and Vielzeuf, 1987). If this is true, emplacement style may provide little evidence, if any, of the mechanisms of magma segregation or ascent.

A popular choice for an ascent mechanism is diapirism. In this model, a sizeable plutonic mass of granitoid magma ascends by virtue of a density contrast with its wall rocks. In order to rise, the

---

[1] Present address: M.I.M. Exploration, G.P.O. Box 1042, Brisbane, Qld. 4001, Australia

Correspondence to: J.D. Clemens, Department of Geology, The University, Manchester, M13 9PL, U.K..

diapir must heat and soften those wall rocks, expending thermal energy in the process. There are several reasons why diapirism has been so popularly invoked. Observational factors include the facts that many granitoid plutons are roughly circular in plan, and that some are interpreted to have inflated, like balloons, to produce marginal wall-rock deformation reminiscent of that expected in diapirism (Holder, 1978; Schmeling et al., 1988; Ramsay, 1989; Cruden, 1990; though see Bateman, 1985; Paterson, 1988; etc.). Leake (1978), however, sounded a note of caution in pointing out that emplacement geometry is not necessarily a reliable indication of ascent mechanism. Until recently, a major factor in the acceptance of diapirism has been the lack of a demonstrably viable alternative, derived from either experiments or modelling. However, perhaps the greatest impediments to progress in our understanding of granitic magma transport have been the historically ubiquitous cartoons of granitoid diapirs ascending through the crust as inverted tear-drops or prolonged carrots.

Diapirism versus the evidence

Diapiric ascent of a granitoid magma body must produce characteristic deformation patterns in the surrounding country rocks (Schmeling et al., 1988; Cruden, 1990). Late in its crystallization, the magma itself must also undergo non-recoverable deformation (mesoscopic—Marsh, 1984; Bateman, 1985; and microscopic—Paterson et al., 1989). Preservation and recognition of such internal fabrics would depend on the crystallization history of the pluton and the presence of suitable strain markers. However, refractory phases that resist low-temperature recrystallization and annealing (such as amphiboles and plagioclase) ought to preserve evidence of high-temperature intracrystalline deformation. Outside the pluton there should be rim synclines, intense local flattening parallel to the pluton margin and a cylindrical shear zone about the pluton (with pronounced vertical extension lineation and ubiquitous pluton-up kinematic indicators) extending deep into the underlying crust. As noted previously (e.g., Bateman, 1985), such structures are

exceedingly uncommonly, if ever, encountered in exposed sections of crust that must have been traversed by batholithic volumes of granitic magmas. We, along with many others, have made detailed studies of mid-crustal terranes of all ages across the world, with both petrological and structural eyes, and have yet to see any indication of such a structure. Furthermore, a diapir ascending via a softening deformation zone will lose 3–10 times as much heat as would a static magma body the same size and shape (Miller et al., 1988). The "contact" metamorphic effects would be correspondingly more intense.

Schwerdtner (1990) used detailed structural analyses to test the existing hypothesis that gneissic granitoid rocks in Ontario were emplaced as diapirs. All of the studied complexes failed tests for solid-state diapirism and tensile bending tests for magmatic diapirism. We believe that similar close examination of most putative diapirs will probably yield similar negative results.

The most sophisticated and detailed modelling of granitoid diapirism to date (Mahon et al., 1988) has shown that such bodies suffer thermal death and lock up solid in the middle crust. This seems to be true no matter how the magma body volume, temperature and starting depth are varied, no matter what the density contrast between the diapir and the enveloping crust may be, and irrespective of plausible variations in geothermal gradient. Thus, the common occurrence of large, shallow silicic magma chambers and batholithic granitoids emplaced at $P \leqslant 100$ MPa (e.g., SE Australia—Phillips et al., 1981; Wyborn et al., 1981; Noyes et al., 1983; Clemens and Wall, 1984) suggests that diapirism is not the mechanism by which such magmas travel most of the distance through the crust (commonly $\geqslant 20$ km; e.g., Miller et al., 1988). An alternative explanation, adopted by Miller et al. (1989), is that the continental crust must be between 10 and 100 times less viscous than indicated by numerous well-constrained experimental measurements (e.g., Kirby, 1985). This is highly unlikely.

Granitoid diapirs must be fluid and undergo continual, vigorous, internal convection. This is required in order that heat be effectively transferred to the margins of the plutons, to heat and soften the wall rocks and so allow the diapir to proceed upward. Modelling (e.g., Cruden, 1988, 1990) has shown that forces operating within such a magma blob will cause convective overturn several times during ascent. Indeed, as advocated by Wickham (1987), convective overturn of the partially molten source region is the preferred mechanism by which a diapir is spawned. Convective overturn will result in a classic "fold-and-stretch" chaos pattern (Mandelbrot, 1982; Ottino et al., 1988), and should blur or, when more vigorous, could even obliterate source-inherited geochemical and isotopic heterogeneities. However, as Miller et al. (1988) note, granitoids very commonly show much greater scatter in various isotopic ratios than could be explained by analytical uncertainties. It is generally agreed that such variability *is* source-inherited. It might be argued that an heterogeneous pluton could be built by the agglomeration of numerous smaller diapirs. One serious problem with this solution is that the small diapirs would solidify too rapidly to reach the emplacement level. Thus, the preservation of source-inherited heterogeneity is further evidence against the universality of diapirism.

The wider importance of actualistic models for granitoid magma ascent can begin to be appreciated by considering the effects of false assumptions regarding melt proportions and ascent mechanisms on thermal and tectonic modelling of high-grade metamorphic terranes.

Alternative mechanisms

Marsh (1984) and Miller et al. (1988) have reviewed the various mechanisms proposed to account for granitoid magma ascent. These authors conclude that domal uplift and ring-dyke intrusion are only important at shallow crustal levels, that block choking disqualifies stoping as a mechanism for significant ascent, and that zone melting is ruled out by both energy considerations and observational evidence. Fracture transport is dismissed as simply improbable, though compelling reasons for this conclusion are not presented.

Some batholiths have basal feeder dykes (see, e.g., LeFort, 1981; John, 1988). Glassy rhyolitic to

dacitic magmas have been erupted or emplaced as high-level dykes, carrying phenocryst assemblages produced at pressures of up to 1300 MPa (see, e.g., epidote dacite—Evans and Vance, 1987; peraluminous rhyolite—Pichavant et al., 1988a,b). Brun et al. (1990) studied the small Flamanville Granite in France. Based on the petrological, structural and geophysical evidence, they concluded (p. 271) that emplacement of high-level granitic plutons into upper crustal rocks "... most likely results from lateral expansion of magma injected through brittle crust than from ballooning of a diapiric body". On page 282 they comment that the magma for this very small pluton "... must have travelled through about 15 km or more of brittle crust before it was emplaced within the shallow sedimentary cover". These authors believe that the differential stress at the head of a cold diapiric granite can never be large enough to deform the brittle crust, and that a hot diapir this size would not contain sufficient heat to soften a conduit through the wall rocks. Their final proposition is that this magma must have risen along a "narrow channel", i.e. a fracture system. We suggest that a mechanism of rapid magma transport through self-propagating fractures deserves more serious consideration, for the crustal-scale transport of granitoid magmas, than it has previously received.

Magma fracture as a process: melting, fracturing and melt segregation

The melting of fertile source rocks and melt segregation

Most large granitoids and silicic volcanic accumulations have crystallized from hot, "dry" magmas and were emplaced during late-orogenic extension (e.g., Hutton et al., 1990) or in so-called "anorogenic" settings (e.g. Anderson, 1983; but see Mawer et al., 1989). This may be true, also, of "arc-related" plutons such as in the Sierra Nevada or the Andean chain (Atherton, 1990). Major proportions of the magmas were formed by fluid-absent partial fusion of fertile metamorphic rocks, in the lower crust (Fyfe, 1973; White and Chappell, 1977; Clemens and Vielzeuf, 1987;

Rutter and Wyllie, 1988; Le Breton and Thompson, 1988; Clemens, 1990). The responsible reactions involve decomposition of biotite and hornblende, and experiments on synthetic and natural rock systems, as well as model calculations, show that these fluid-absent melting reactions will commonly produce 20–50 vol% of water-undersaturated melt. They occur at temperatures between ~ 850 and 950°C and, except at mantle pressures, have vertical or positive dP/dT slopes, implying positive ΔV of reaction. For rocks with densities of 2600–2800 kg m^{-3}, and containing 10–40 wt.% biotite, we calculate that the volume change accompanying melting would range from +2.4 to 18 vol.%. This modelling assumes biotite breakdown in reactions with the following approximate stoichiometries:

Bt + Sill + 5Qtz = Grt + melt

and

Bt + 6Qtz = 3Opx + melt

where the melt is considered to be Or + 3Qtz + xH$_2$O, in both cases. The partial molar volumes of H$_2$O and metal oxide components in the melts were taken from the work Burnham and Davis (1974) and Lange and Carmichael (1987).

Jurewicz and Watson (1985) showed that wetting angles for granitoid melts, even in static quartzo-feldspathic aggregates, are 45–60°. Mawer (1992) used dynamic melting experiments on rock analogue materials to show that, in deforming aggregates, wetting angles are unstable and commonly approach 0°, with corresponding high connectivity of melt. Consequently, even for very small melt fractions, the melt will form an interconnected network of grain-edge channels, perhaps in a zone of limited extent.

Miller et al. (1988) dealt with the contiguity of partially molten aggregates. They define contiguity, in the present context, as the fraction of the surface area of solid grains, in a partially molten system, that is shared with other solid grains. In the static case, for a well-annealed fabric, the continuous, self-supporting skeleton of solid grains apparently breaks down when the contiguity falls to values below 0.15–0.2. Figure 1 is modified from figure 3 of Miller et al. (1988), showing contiguity as a function of melt fraction

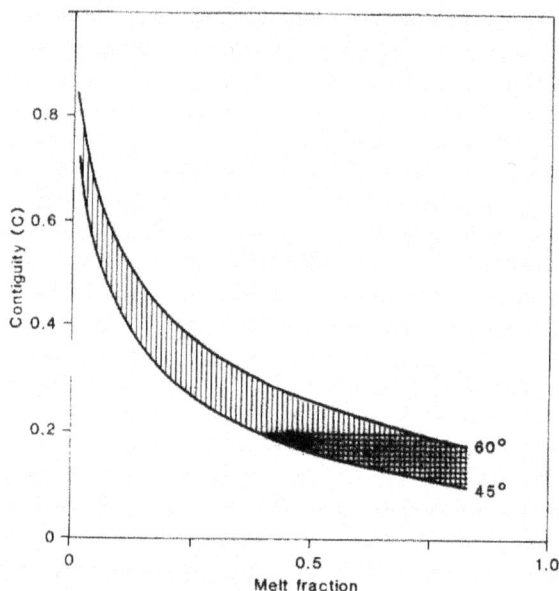

Fig. 1. Plot of contiguity (C) as a function of melt fraction for wetting angles (θ) considered characteristic of crustal fusion. Modified from fig. 3 of Miller et al. (1988). The lower limits of contiguity (~ 0.2) for retention of the solid skeleton of the partially molten mass are also shown, along with the common range of melt fractions ($\geqslant$ 0.4) formed in fluid-absent partial fusion of common crustal rock types. The small black triangular area is therefore the expected region in which fluid-absent partially molten crustal rocks will begin to lose their solid frameworks.

and wetting angle, for the static case. This graph implies that, unless the wetting angle is atypically low and the melt fraction less than 40%, fluid-absent crustal fusion should preserve the integrity of the solid skeletons of the rocks. Following the relationship suggested by Miller et al. (1988), the strengths of the partially molten rocks will be roughly 0.2–0.4 those of the unmelted equivalents; the partially melted source rocks will behave as solids with significantly reduced strengths. However, they will be able to sustain only compressive and shear stresses, with negligible tensile strength (e.g., Hibbard and Watters, 1985). If the rocks at the source level are actively deforming, however, even these values are overestimates; the rocks will behave as fluidized aggregates with effectively no long-term strength in tension, compression or shear.

Stevenson (1989) showed that in a partially molten rock mass undergoing deformation, the melt will migrate along the direction parallel to the axis of least compressive stress and accumulate in veins. Furthermore, the melt will flow into regions where it is already concentrated, so that the more fertile regions of a magma source will act as melt sinks.

For typical source rock compositions, the rates of melt production during fluid-absent crustal fusion are highly non-linear. Experiments (Rutter and Wyllie, 1988; Vielzeuf and Holloway, 1988; Rushmer, 1991) have shown that the melting rates, as a function of increasing temperature, are generally discontinuous and usually show remarkable bursts of melt production over narrow temperature intervals (generally $\leqslant 10°C$ wide). Thus, we believe that the rate-limiting step in this process will be thermal diffusion from the heat source rather than the rate of the melting reaction. Equations governing this process (Carslaw and Jaeger, 1980) imply that partially molten zones hundreds or thousands of metres in thickness could be generated in tens to hundreds of years. This conclusion assumes heat flow properties and boundary conditions similar to those envisaged by Huppert and Sparks (1988), and accounts for the probable latent heats of partial fusion of typical fertile source rocks (e.g., Vielzeuf et al., 1990). Large volumes of basaltic magma (heat source) are assumed to be emplaced essentially instantaneously into the crust, and to convect during the ensuing stages of the process. Effective tensile stresses (from positive ΔV of melting) would be developed as rapidly as thermal diffusion will allow. High volumetric strain rates ($\geqslant 10^{-7}$) and maintenance of the solid framework of the rock would result in the development of high pore fluid pressures in melt pockets, lowering of effective normal stresses, and thus brittle failure (e.g., Etheridge et al., 1983). Syn-melting deformation can only increase this tendency towards fracturing (e.g., Mawer, 1992). Stress concentrations at the fracture tips will be more than adequate to overcome the tensile strength of any crustal rock type.

From the above considerations, we consider it highly likely that fracturing will play a major role in producing a porosity in the source rocks which will greatly assist in the initial segregation of partial melt. In addition, it is conceivable that melt pooling could also be (at least partially) a chemically-induced phenomenon related to nucleation and growth of the refractory products of the partial melting reactions plus differences in equilibration volumes for melt and restite components (Powell and Downes, 1990). Whatever the initial segregation mechanism, it would seem essential that the melt collects, at least briefly, in volumes greater than those generated at the sites of initial melting. Progressively larger veins and dykelets (e.g., Sleep, 1988) forming a progressively larger proportion of the rock mass would constitute such a reservoir.

Once the melt has effectively segregated from the restite, the solidus for the melt fraction (and not of the original rock) is relevant to considerations of mineral and melt stability. The melt solidus will commonly be at least 100°C below that of the original source rock. Thus, a large range of potential cooling paths exists, for which the magma may ascend significant distances without solidifying.

The tensile strengths of rocks

Rocks are notoriously weak in tension (e.g., Donath, 1961; Gretener, 1969; Etheridge, 1983). Even for intact, unmelted rock, tensile failure stresses are probably only on the order of 10 MPa (Etheridge, 1983). For rocks with pre-existing mechanical anisotropies (such as foliations), tensile strengths are even lower (Gretener, 1969). Moreover, the mode of tensile failure in rocks, even at elevated temperatures, is commonly by fracturing (witness the near-ubiquitous presence of dykes, even in granulite facies terrains, and the common occurrence of dyking in incompletely solidified granitoid magmas). The presence of a rising fluid or melt phase enhances this tendency, both by corrosive effects at the crack tips and by wedging-apart of the fracture walls due to melt or fluid volume increases.

Stress orientations in the crust

Deep-crustal stress orientations are not well-constrained. It is assumed that one principal stress is vertical (and hence the remaining two are horizontal) throughout at least the upper 15 km of the crust (Zoback and Zoback, 1980), in both compressional and tensional environments. It seems reasonable to assume that a similar stress pattern obtains at the somewhat deeper levels pertinent to generation of mobile granitoid magmas. If this is the case, then fractures in both tensional and compressional environments should commonly be oriented vertically, essentially parallel to the maximum principal compressive stress orientation. Stress patterns in the partially-molten source zone may be more complex. If the melt fraction in the zone is large, roughly lithostatic stress conditions might develop there. This is, however, unlikely in the case where the zone is actively deforming. Due to the complexities of thermal structure in granitoid source regions, the source zone is unlikely to be entirely horizontal. If the zone (or parts of it) is not perpendicular to one principal stress (i.e. not horizontal), shear stresses must operate across the inclined section(s) and, as the strength of the zone is less than the surrounding unmelted rocks, a melt-lubricated zone of shearing will develop (Hollister and Crawford, 1986; Mawer, 1989, 1992). If the overall stress environment is tensional, then vertical fractures should be able to tap the melt zone, and allow magma to escape upwards. If compressional, then the melt may be unable to ascend as easily, since fracturing may be suppressed by the compressive character of all three principal stresses. In any case, for fluid-absent melting of fertile granitoid source regions, with positive ΔV of melting reaction, effective stress magnitudes will be different from the bulk stress magnitudes. Such reactions will increase effective tensile stresses and decrease effective compressive stresses, as shown by the principle of effective stress (Hubert and Rubey, 1959); high-pressure impels normal stress to decrease, and can lead to the development of local tensional stress, even within an overall compressive environment.

There are further thermal considerations in this process. As shown by Huppert and Sparks (1988), the intrusion of large, hot, basaltic "sills" into the lower crust can be an efficient means of raising the temperature above the solidi for many common rock-types. The geologically sudden emplacement of hot mantle-derived magma (at perhaps 1200°C) into cooler crust (~ 500°C) must initially produce some extreme thermal effects. Stresses of the order of 10 to 100 MPa would be developed in rocks across which there was a thermal gradient of only 100°C (Gerla, 1988). As a consequence, tensile fractures would again be promoted. Though subsequently closed, these pre-existing fractures could provide sites for the nucleation of other fractures later in the thermal evolution of the terrane.

All of the above suggests that, during fluid-absent metamorphism and partial fusion of crustal rocks, there would commonly be networks of subvertical tensile fractures penetrating the partially-molten rocks and their envelopes. Such features are not to be sought in mid-crustal migmatite terranes since, in contrast to the sources of high-level granitoids, these are usually the products of relatively low-temperature, fluid-present melting (e.g., Clemens, 1990).

Miller et al. (1988) and Ribe (1987) reviewed melt segregation mechanisms. Much attention has been focussed on porous flow and matrix compaction (e.g., McKenzie, 1984, 1987). These processes may be important, but most of the analysis to date has assumed entirely plastic behaviour in isotropic media, and that the driving force for compaction, and hence melt migration, is gravity. The nets of fractures developed in the partially molten rocks would rapidly become melt-filled veins. The driving force for vein filling would be the pressure difference between the opening vein and the surrounding rock. Since the melt outside the veins will form a continuous, three-dimensional network, and melt-filled porosity will be relatively high (up to 50%), porous flow into the veins will be rather efficient. Assuming a pressure gradient of 10 MPa m^{-1} (Etheridge et al., 1984), a grain-size of 1–2.5 mm and McKenzie's (1984) assumed relation between porosity (0.2–0.5) and permeability, the Fletcher and Hoffman (1974) equations predict that melts with viscosities of 10^4–10^6 Pa s would only take between 17 h and 127 a to flow 1 m into the opened fractures. Thus, providing that the solid skeleton could compact by plastic flow, the whole process of vein segregation ought to be highly efficient. If the melt generation zones are indeed actively deforming, as argued above, then the deforming solid skeleton would be "squeezing" melt into the

"sucking" veins. Another factor in the efficiency of vein segregation is fracture spacing. The spacing must be less than, or of the same order as, the characteristic compaction length (about 2.6 m for

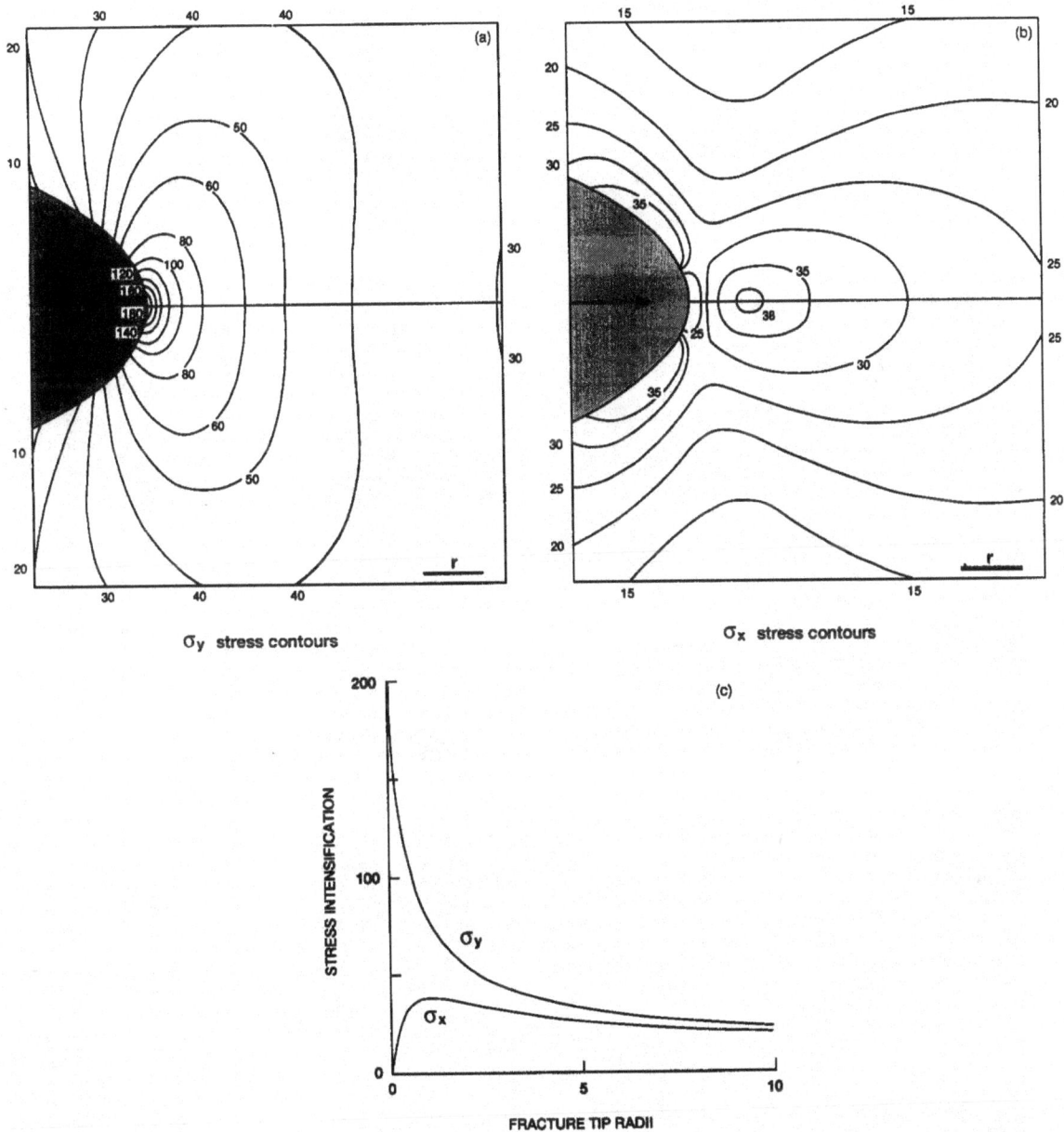

σ$_y$ stress contours

σ$_x$ stress contours

(c)

Fig. 2. Tensile isostress contours at the tip region of a propagating magma-filled fracture (modified after after Cook and Gordon, 1964, figs 3, 4, and 7). Numbers on contours represent stress concentration factors. (a) Contours of σ$_y$, the tensile stress component operating perpendicular to the travelling fracture; maximum stress intensification can be over 200 times the far-field stress, and occurs at the fracture tip. (b) Contours of σ$_x$, the tensile stress component operating parallel to the travelling fracture; maximum stress intensification can be over 40 times the far-field stress, and occurs within the unfractured rock in advance of the fracture tip. (c) Graph of variation in σ$_x$ and σ$_y$ in a plane parallel to the travelling fracture, in material immediately ahead of the fracture tip.

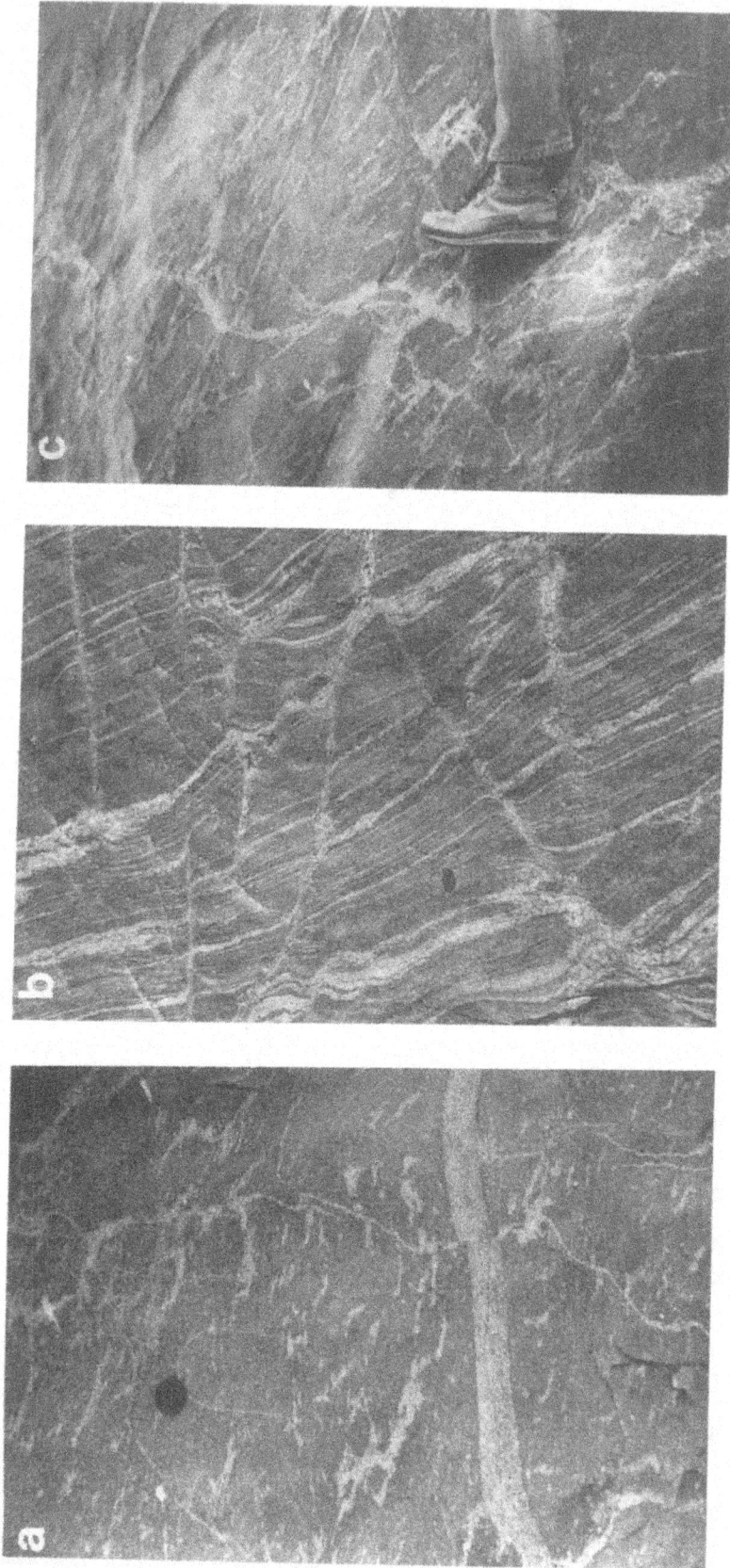

PLATE 1

Photographs of outcrops of Archaean Sand River Gneiss, in the bed of the Sand River upstream from the Messina–Tshipise road, central zone, Limpopo mobile belt, South Africa. The rocks are tonalitic gneisses in which the fluid-absent reaction $Bt + Pl + Qtz = Opx + Kfs + M$ has occurred but not gone to completion. Felsic veins and dykes abound and are up to 4 m in width. These bodies contain quartz, plagioclase, microcline and minor biotite, and appear to represent minor segregations of locally-derived melt. The larger felsic bodies contain angular enclaves of the host rock and seem to have been emplaced in a brittle regime (Fripp, 1983).
(a) Schistosity-parallel melt pods and intersecting veinlets. The broader felsic body is a small dyke whose emplacement predated the metamorphic event responsible for the formation of the pods and veinlets.
(b) Schistosity-parallel melt segregations and cross-cutting veins and shear zones filled with melt products.
(c) Network of melt-filled veins feeding larger veins and stringers.

a system with 30% melt, in which the solid aggregate and melt viscosities are 10^{16} Pa s and 10^5 Pa s respectively).

Melt tapping by veins and dykes is expected to be an efficient process (Sleep, 1988). Sleep's model is partly based on the field observations and earlier models of Nicolas (1986) and Nicolas and Jackson (1982). It is envisaged that melt flows small distances (here $\leqslant 3$ m), by porous flow, into mesoscopic veins. These may intersect to form larger veins and/or be tapped by dykes that ascend through the overlying crust. Plate 1 shows a field example of the early stages of the processes. The importance of fracture coalescence in magma transport has been stressed by Takada (1989). Other discontinuities that could play roles as melt sinks and transport paths include shear zones, shear bands and lithological contacts. However, most important would be the interactions between fracture formation, porous flow and matrix compaction.

Stresses at fracture tips

Stress intensification at fracture tips arises from the fact that the radius of tip curvature is small compared to the fracture length. This leads to local intensification of far-field applied stresses, by a factor which depends on the tip radius (see, for example, Lawn and Wilshaw, 1975). This intensification is commonly large enough, even in the case of regional compressive stress fields, to exceed the tensile strength of any crustal rock. The rock will split, and the fracture will become self-propagating. The magma-filled fracture *must* assume a sheet-like form, and thus become a dyke (Emerman and Marrett, 1990).

By this process, several distinct stress concentrations form at the tips of travelling fractures (Fig. 2a). The best-known involves a tensile stress oriented perpendicular to the fracture. Here, intensification can be very large, up to about 200 times the far-field stress. A second tensile stress concentration occurs at some distance ahead of the propagating crack tip, in the intact material. This is oriented parallel to the fracture, and produces an intensification of about 40 times the far-field stress (Fig. 2b). There are also two conjugate shear stress concentrations, emanating

from the fracture tip, oriented at $\sim 45°$ to the fracture. These have been modeled numerically by Cook and Gordon (1964); see also Pollard (1977). It is worthwhile noting that each of the tensile stress concentrations is more than large enough, individually, to fracture intact rock (Etheridge, 1983), while the shear stress concentrations are almost certainly enough to overcome the shear strength of any pre-existing mechanical discontinuity, such as a joint or a fault, or an anisotropy, such as bedding or foliation. Thus, the propagation of a magma-filled fracture will usually be catastrophic.

Weertman (1971) noted that, when magma is frozen in at the top of a crack that has propagated to the emplacement site of a large intrusion, tensile stresses will be increased in the region below. This will make it more likely that a new crack will nucleate below an old one. Hence, once a system of cracks connects a pluton to its underlying magma source, new pulses of magma will probably follow the already established plumbing, even if it becomes temporarily blocked by solidified magma.

Magma ascent

Once formed, dykes filled with granitoid magma will be self-propagating. This is due to two main effects. The density contrast between magma and rock will produce a buoyancy drive. Even magma slightly denser than its wall rocks can ascend if tensile stress increases upwards, as in regional doming of the crust in extensional settings (Takada, 1989, 1990). Ascending hydrous magma will expand, in response to decompression. This is mostly the result of the increase in the partial molar volume of "water" dissolved in the melt. As pressure decreases, the magma will tend to wedge apart the dyke walls, further increasing the tensile stress concentrations at the dyke tip (see also Knapp and Knight, 1977). Furthermore, as Lister and Kerr (1991) point out, in the tip of an extending, magma-filled crack, there is a small region across which there is an essentially infinite pressure drop and, as a consequence, volatiles ought to exsolve from this small portion of the magma. Any such exsolution of volatiles will be accompanied by an even more impressive, positive ΔV, further aiding fracture propagation. As the volatiles will be mostly hydrous, there may be accompanying chemical effects ("hydrolytic weakening") that decrease the fracture toughness of the wall rocks, further promoting easy fracturing (e.g., Atkinson, 1984).

Ramberg (1967) points out that, in order to be correctly scaled in his experiments (where putty is used to model the solid crust), the viscosity of a model "magma" would have to be 10^{-14} times that of a real magma. As he notes, this leads to impossibly low viscosities for any laboratory liquids used to model granitoid magmas. The least viscous model magma he could use was water which, at 20°C, would be equivalent to an extremely viscous real magma (10^{13} Pa s). On pages 131–132 (Ramberg, 1967) he comments that "Owing to the high fluidity of the magma, in contrast to the rigidity of the crystalline crust, the melt can rise with considerable speed..., even through a narrow and irregular path. The subsidence of the crust... and the rise of the magma through narrow channels are energetically efficient processes.".

Can dyking then efficiently transport relatively viscous felsic magmas the required distances? In the following example we have used the equations of Turcotte (1987) and Spence and Turcotte (1990) for buoyancy-driven elastic fracture propagation, and neglect the magma expansion effect which would further increase propagation rates. We assume that a dyke 1 km long (in plan) has grown, and that a continuous fracture network will eventually connect this dyke to the magma source. The magma has a viscosity of 10^5 Pa s and a density of 2400 kg m^{-3}, and melt flow within the dyke will be laminar. The wall rocks have an average density of 2750 kg m^{-3} and Poisson's ratio is taken to be 0.28. For wall rocks with an average shear modulus of 3.1×10^{10} Pa s, the calculated stress intensity factors (Turcotte, 1987), for dykes with tail widths of $\geqslant 10$ cm, exceed critical values by factors of at least 4. Dyke propagation is therefore catastrophic. In Figure 3 we show a plot of the time taken for a dyke to propagate 20 km upward and the calculated time needed for this single dyke, once open,

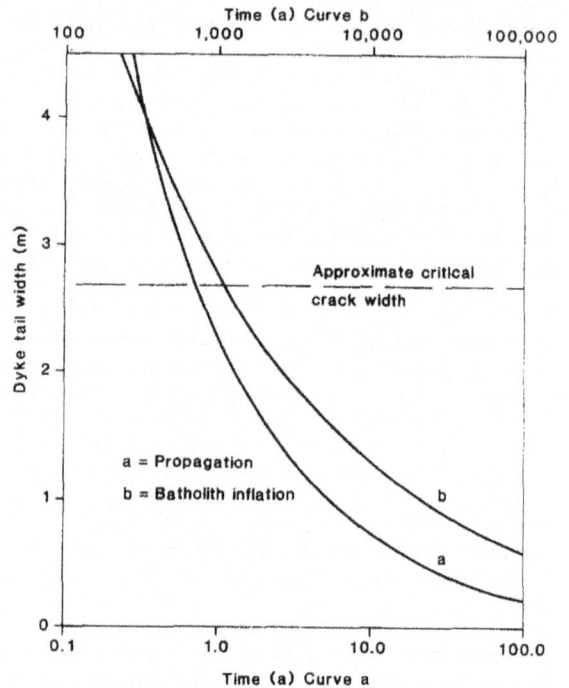

Fig. 3. Graph showing the relationships between dyke tail width (at infinite distance behind the propagating tip) and the time taken for dykes to propagate 20 km upward through the crust (curve a), or the time required to inflate a 2000-km³ batholith at the upper end of a 20-km-long fracture system, once open (curve b). Calculations are based on the equations of Spence and Turcotte (1990). See text for details.

to feed the growth of a 2000-km³ batholith. The smaller the dyke, the longer the time. However, the times needed are impressively brief. A single dyke 1 km long and 3 m wide could propagate 20 km in about 8 months and inflate the batholith in under 900 years. This mechanism would not, however, preclude slow batholith formation by intermittent, small magma additions over millions of years. Such sluggish batholith assembly is probably rare but has been documented in the Himalayas (Deniel et al., 1987). Indeed, this mode of batholith formation would seem to require dyke propagation as the mode of magma transport (see below).

A critical requirement for successful transport of magma through a fracture system is that its rate of ascent be fast enough to prevent thermal conduction (to the wall rocks) from causing solidification of the magma. Wilson and Head (1981) provide an expression for calculating the critical crack width below which buoyant magma ascent could theoretically not occur. We assume a magma with a viscosity of 10^5 Pa s, negligible yield strength, a density of 2400 kg m^{-3} and a thermal diffusivity of 6.6×10^{-7} m^2 s^{-1}, rising buoyantly through a 20-km-long fracture system, by laminar flow, through rocks with a density of 2750 kg m^{-3}. If the magma would become effectively solid in cooling from 900 to 790°C, the calculated critical crack width is 2.7 m. Changing the solidification interval has little effect on the calculation but increasing the magma viscosity by an order of magnitude would result in a critical width of only 4.9 m. Thus, fractures of the order of 3 m or so wide would be adequate to transport granitic magmas. These dimensions are in accord with the observation of Corry (1988) that most felsic dykes are 1–4 m wide, while horizontal lengths are up to 1 km. Note that *preserved* dyke thicknesses are almost certainly less, and perhaps far less, than the widths of their active magma-channelling fractures (e.g., the drained dykelets in the breccia at the base of the pluton described by Bédard and Sawyer, 1991). Due to magma cooling and potential solidification, a dyke cannot propagate further than some critical distance. Lister and Kerr (1991) present an approximate equation which conservatively estimates this "thermal entry length". Using their equation 61, with a magma flow rate of 10^{-2} m s^{-1}, a dyke width of 3 m and a wall-rock thermal diffusivity of 10^{-6} m^2 s^{-1}, the predicted critical length is around 23 km. Thus, such a granitic dyke would seem to be well able to traverse the crust without suffering thermal lock-up. The above calculations are probably on the pessimistic side, since they take no account of decreased thermal diffusivity of the wall rocks as they heat up, or the probable non-linear physical behaviour of the granitoid magmas. Delaney and Pollard (1981) note that a propagating dyke may become unstable if the regional stress field does not remain constant in orientation throughout the vertical propagation interval. A single dyke can split into a series of *en échelon* segments (Delaney and Pollard, 1981, pp. 37–38), with the overall trend of the array aligned parallel to the trend of the parent dyke. Takada (1990) noted similar behaviour in experimental studies of dyke propagation, and related the phenomenon to the achievement of some critical volume in the growing crack (see below). These effects seem to occur close to the termination of a dyke and, therefore, should not effect our arguments regarding pluton growth. Note, however, that this behaviour may explain the occurrence of roughly periodically spaced silicic volcanic centres aligned on linear trends (N.H. Sleep, written commun., 1991).

Since batholiths might be fed by a number of dykes, it appears that fracture propagation is a remarkably efficient means of rapidly transporting large volumes of granitic magma through the crust and building batholiths or volcanic piles. However, it is worth examining the relative efficacy of dyke propagation, as compared with diapirism, with respect to the cooling rate of magma flowing through such a dyke. Marsh (1984) concluded that a given volume of magma flowing through a dyke must move about 10^4 times as fast as an equivalent volume of the same magma moving as a sphere ($\sim$ diapir), in order to reach a given emplacement depth, from the same starting depth, at the same temperature. Mahon et al. (1988) showed that granitoid diapirs probably ascend no faster than about 10^{-8} m s^{-1}. Using the Spence and Turcotte (1990) equations for ascent velocity, with magma and wall-rock properties as in the previous example, this criterion is satisfied for all dykes with widths greater than a few centimetres. For a 3-m-wide dyke, ascent velocity would be about 10^{-3} m s^{-1}, around 10^5 times faster than the swiftest diapir modelled by Mahon et al. (1988).

Weertman (1971) modelled propagating liquid-filled fractures as dislocations and gave an equation for calculating the maximum length of fracture that can remain instantaneously open. Again, assuming the magma has a density contrast of 350 kg m^{-3} with its wall rocks, that Poisson's ratio is 0.28 and that the shear modulus of the wall rocks is 3.1×10^{10} Pa s, we calculate that a fracture 3 m $\times$ 1 km in plan could have a maximum vertical extent of 5.8 km. This would reduce to 4.0 km for a 1-m-wide fracture. For a 20-km-deep fracture system only about 5 km could, therefore, be open as a single magma-filled fracture at any one time.

This implies that granitoid magmas would be delivered to emplacement levels in a series of pulses rather than as a constant flow. These predicted pulsations could be partially responsible for the observed preservation of source-inherited isotopic and chemical heterogeneities in granitoid plutons. Separate pulses could be derived from different parts of the partially-molten source region, and the lack of efficient convective stirring at any level could result in the partial preservation of heterogeneities. For the 3-m-wide fracture the number of pulses needed to inflate a 2000-km^3 batholith would be about 3.33×10^4 or about 40 pulses per year, for the minimum inflation period of about 800 a.

From his experiments, Takada (1990) concluded that the directions of fracture growth and propagation depend on the density difference ($\Delta\rho$) between the matrix and the "fluid" in the fracture. If $\Delta\rho$ is small and positive, the fracture will grow upward and expand. If, as is more likely in magma generation, $\Delta\rho$ is large and positive, propagation will occur as an *isolated*, magma-filled fracture. The more vertically extensive this fracture, the longer and wider it will be, and the greater the radius of curvature of the fracture tip. When the fracture volume increases to some critical value, the fracture will break up and develop several daughter segments *en échelon*, but only the central parent fracture will continue upward. Takada derived an expression relating the width–height ratio of a fracture to its height. Applying this to wall rocks with a Young's modulus of 6.9×10^9 kg m^{-3}, the maximum width of fractures (1–4 km deep) will be of the order of a few metres. These calculations seem to suggest that the postulated dimensions of fractures for the transport of granitic magma are reasonable, and they also agree with field observations on the dimensions of granitoid feeder dykes (Corry, 1988).

Structural and metamorphic evidence

As noted above, expected signs of granitoid diapir ascent are essentially lacking in exposed crustal sections. What then should we expect to see if the granitoid magmas moved by fracture propagation?

If magma segregation occurs by fracture processes then the mesoscopic fabrics (e.g., layering, schistosity, graded bedding, etc.) in the source rocks would not necessarily be greatly disturbed beyond the extent usual in high-grade tectonites. As pointed out by Sawyer (1992), as long as melt can readily escape, large amounts of partial melting could take place without the permeability threshold value ever being exceeded (see also Dell'Angelo and Tullis, 1988; Mawer, 1989). The most efficient and simple way that melt and magma may be drained from their source volume is through an anastomosing mesh of interconnected grain-edge fractures, in an actively deforming zone (e.g., Mawer, 1989, 1992). These would feed into local dilatant sites, where an hydraulically critical magma volume would collect, tensile failure of the overlying rocks would take place, and a series of essentially vertical veins would develop. Veins may be efficiently drained of their melt by flow into larger veins and dykes, along shear zones or into lithological contacts. The drained veins would simply close down and any subsequent grain growth, recrystallization or deformation would further mask their presence. It may be difficult, but probably not impossible, to find physical evidence for the operation of this segregation mechanism among the restitic solids.

The predicted high contiguity of crust partially fused in fluid-absent reactions could mean that convective overturn, homogenization and wholesale source mobilization (e.g., Wickham, 1987) should not generally occur. Field evidence from restitic granulites that were formed in the lower crust (e.g., in the Ivrea zone, Italy; J.D. Clemens and C.K. Mawer, unpubl. data) supports this prediction. In some of these restitic rocks, premetamorphic structures, such as transposed meso-scale bedding, have survived melting and melt extraction, relatively unscathed.

Narrow feeder dykes, drained of their melt, would also neck down. Small amounts of chilled magma from the dyke walls, and pods of residual melt, may be left as legacies of the process. However, faulting, folding, etc. could dismember these meagre remains, as well as larger feeder dykes (at higher levels). In any case, a metre-scale granitoid dyke might easily appear geologically trivial in comparison to a neighbouring batholith, even though it could have completely fed that body in a geologically short period of time. In passing, we note the ubiquity of deformed granitoid dykes and dyke remnants in exposed mid- to lower-crustal sections.

Thermal effects

The contact metamorphic effects surrounding feeder dykes for granitoid plutons can be roughly

predicted using data in Carslaw and Jaeger (1980). A feeder dyke will have magma flowing through it, at essentially constant temperature, for as long as it takes to inflate the pluton perched on top of it. Thus, the thermal aureole surrounding a feeder will be broader than one surrounding a similar dyke which was not a feeder, but in which a single batch of static magma cooled and crystallized. Figure 4 shows the predicted maximum extents of thermal effects surrounding a 3-m-wide dyke which transported granitoid magma at 900°C for a period of 800 a. If the thermal effects would become readily apparent in the wall rocks at around 500°C, this means that one should expect

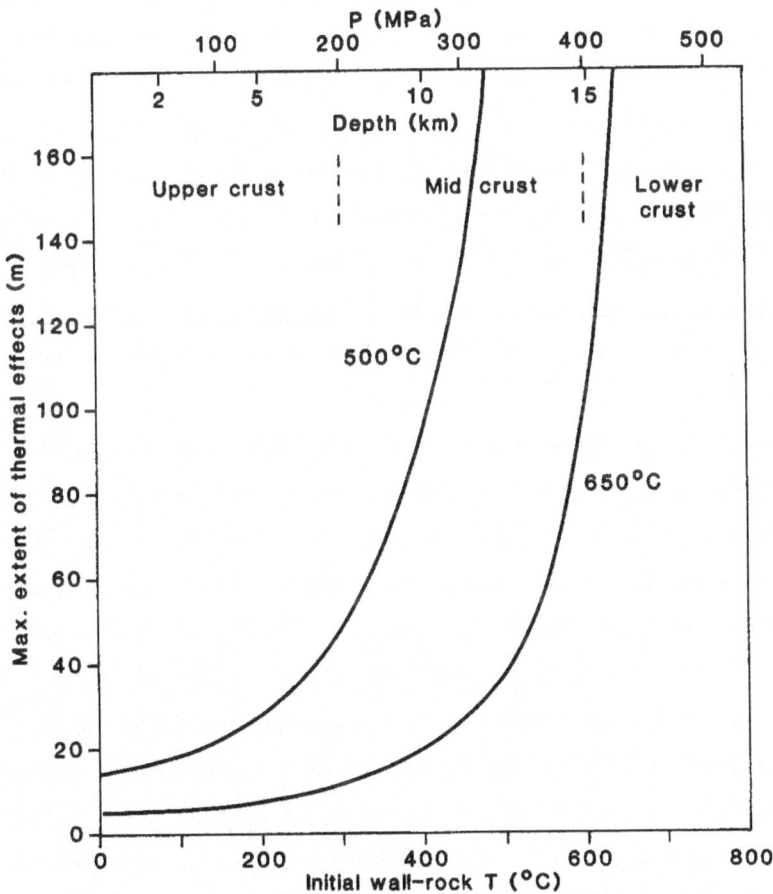

Fig. 4. Extents of thermal effects surrounding a 3-m-wide granitic dyke acting as a feeder for magma at 900°C over a period of 800 a. Distance into the country rock is plotted against the initial temperature of the wall rocks. Two isotherms are shown, for 500°C (assumed minimum T for detection of thermal metamorphism) and 650°C (assumed minimum T needed for partial anatexis). The wall rocks have a constant thermal diffusivity of 1.18×10^{-6} m^2 s^{-1} and no account has been taken of heat absorbed in metamorphic recrystallization, mineral reactions or partial melting, or liberated in partial crystallization of the magma (data taken from fig. 41 of Carslaw and Jaeger, 1980). The approximate pressures on the upper axis were calculated assuming an enhanced average geothermal gradient of 1.5°C MPa^{-1}. Corresponding depths assume an average crustal density of 2750 kg m^{-3}, while division into crustal zones is arbitrary.

to find thermal aureoles tens of metres wide, in upper crustal rocks, around such feeder dykes. Furthermore, if the wall rocks were of suitable composition and if there were a source of H_2O-rich fluid available, one ought to see an inner zone, several metres wide, in which the wall rocks were partially melted. This partial melting ought to be most apparent at mid-crustal levels. The widths of the effects in Figure 4 are probably significantly overestimated (perhaps by a factor of 2) since no account has been taken of the heat energy absorbed in recrystallization, metamorphic reactions, or partial fusion. Nevertheless, the aureole width would be very much greater than expected for crystallization of a static magma in an ephemeral dyke. In the latter case, the 500°C isotherm would be considerably less than 1 m from the contact.

An example of these phenomena is the ring dyke associated with the Glencoe cauldron subsidence in Scotland (Bailey, 1960). Garnham (1988) recognized this structure as the feeder for a felsic volcanic system and documented the occurrence of high-grade contact metamorphism and partial melting of the country rocks surrounding the dyke.

Another probable example of the predicted kind of aureole is discussed by Didier et al. (1987). These authors describe a disrupted dioritic dyke, chilled against an older enclosing granite and cropping out as *en échelon* masses 10–100 m across, over a zone nearly 600 m long. This probably represents the exposed roots and feeder of a much larger pluton removed by erosion. The diorite is surrounded by a broad zone in which the granite has been partially remelted. From the descriptions and map, it is difficult to gauge the true extent of the anatectic zone, but it appears to be at least 10–15 m wide, far too large for a static magma body of this size.

The formation of plutons: blunting of fracture-tips, cessation of upward propagation, and magma ponding

In order to lower stress intensification at a travelling fracture tip and halt its propagation, the radius (or effective radius) of tip curvature

must be increased. Several possibilities for this include:

(1) intersection of a highly ductile zone, such as a marble, limestone, or shale, or a water-saturated horizon, or perhaps a partially-molten zone at depth—this should stop the fracture by converting and dissipating the fracture propagation energy directly to anelastic strain (skarns around granitoid plutons could develop because of this);

(2) intersection of a very brittle, isotropic zone —this would cause a large process zone to develop around the fracture tip (Lawn and Wilshaw, 1969), thus robbing the parent fracture of its propagation energy;

(3) a "stress barrier" (Gretener, 1969)—it has been noted that the magnitudes of near-surface horizontal compressive stresses may overcome the vertical body force stresses within the travelling magma (Engelder and Sbar, 1984); however, field evidence shows that such stress barriers, if they exist at all, are uncommon (Corry, 1988). The travelling magma may, in some rare cases, create its own stress barrier at a rheological change (N.H. Sleep, written commun., 1991), but we believe that any such rheological effects would be negated by the continually increasing overpressure in the rising magma pulse, and the shear velocity of the propagating fracture;

(4) attainment of the neutrally- or negatively-buoyant elevation (Gilbert, 1877; Corry, 1988; Lister and Kerr, 1991)—this occurs when the travelling, magma-filled fracture reaches a crustal level such that the magma density is the same or lower than the weighted mean density of the country rocks. While commonly invoked as the major mechanism whereby rising magma is trapped in the crust, this does not, on the face of it, seem to explain either (a) why high-level mafic plutons develop at all, or (b) why any igneous extrusive rocks occur. However, Lister and Kerr (1991) suggest that elastic stress build-up may occur at the level of neutral buoyancy (LNB), and could cause the magma to overshoot this level, allowing volcanism to occur. This potential overshoot is one of the reasons why we are sceptical over the importance of the LNB, and believe that the level and style of emplacement of felsic mag-

mas are more related to effects of magma differentiation and local properties of the crust than to the ascent process (see later);

(5) intersection with a freely-slipping fracture (Weertman, 1980)—this seems mechanically valid, though Corry (1988, p.26) believes that stress reorientation caused by sill formation should then allow further vertical magma transport. His argument seems to be spurious—such a process has been modelled by Pollard (1977), and a mechanically homologous mechanism, albeit on a larger scale, has been proven to occur where a small granite intrusion has developed in a dilatant jog within an extensional ductile shear zone (Hutton et al., 1990). We would argue that the intrusion studied by Hutton et al. (1990) has been fed by a dyke or dykes ascending from a partially molten zone at depth, and intersecting an active extensional shear zone;

(6) the Cook-Gordon mechanism (Cook and Gordon, 1964)—in front of a travelling tensile fracture there are two tensile stress concentrations, one operating perpendicular and a second oriented parallel to the fracture (Fig. 2). There are also symmetrical shear stress concentrations oriented at $\pm45°$ to the fracture. While the stress intensification for the second tensile stress concentration is not as large as the better-known tensile stress intensification perpendicular to the fracture, it will still exceed the tensile strength of most intact rocks, and will certainly exceed the tensile strength of any roughly horizontal mechanical discontinuity (bedding planes, foliation surfaces, compositional layers) in the strata. This is the mechanism whereby glass-fibre-reinforced plastics arrest propagating fractures (Gordon, 1984). The operation of this mechanism is shown schematically in Figure 5.

The Cook-Gordon mechanism leads to a model which has, as its fundamental element, the development of magma ponds along roughly horizontal discontinuities. Re-use of the feeder dykes (Weertman, 1971; see above), by subsequent magma pulses, would lead to the construction of a lens-like plutonic edifice, composed of numerous, individual, sill-like bodies. If the feeder dykes were draining the same partially molten source rocks, then petrological differences between indi-

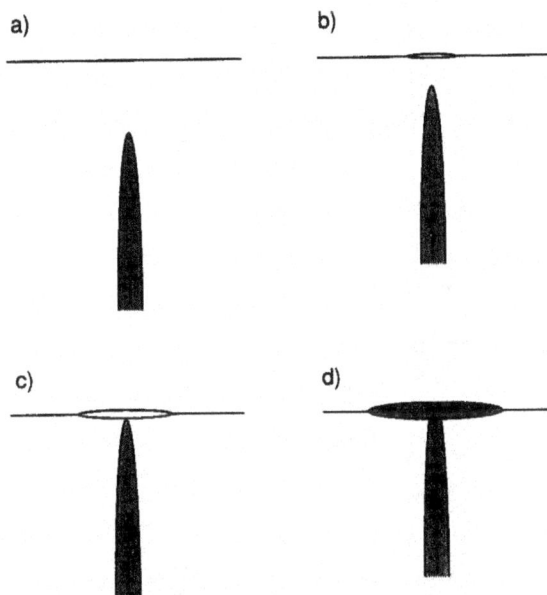

Fig. 5. Proposed mechanism for halting the further ascent of granitoid magma-filled fractures, leading to the development of lens-like or tabular plutons (after Cook and Gordon, 1964; see also Pollard, 1977); only one pulse of intrusion shown. See text for discussion.

vidual magma pulses might be rather subtle.

This model predicts that granitoid plutons should commonly be laccolithic or flattened/tabular in shape, perhaps with some boundary complications (Fig. 6), with horizontal dimensions far in excess of vertical thicknesses. This appears to be common among well-exposed examples (e.g., Corry, 1988; John, 1988; Hutton et al., 1990; Schwerdtner, 1990; Ryan, 1991; among others) and geophysical studies also support this conclusion (J.-L. Vigneresse, oral commun., 1988). Probable exceptions to the laccolithic or flattened/tabular shape would be those plutons developed in dilatant jogs and extensional sites in zones of strike-slip shearing (e.g., Castro, 1986; Mawer et al., 1989; Gleizes et al., 1991).

Figure 6 shows plausible pluton shapes and attendant syn-intrusion wall-rock structures (magma-accommodating shear zones/faults) developed according to the above model.

Finally, we speculate that some relatively unevolved felsic extrusive rocks are erupted because their feeder dykes have passed through crust that lacks large-scale, roughly horizontal, mechanical discontinuities. This type of crust would include

6a

6b

6c

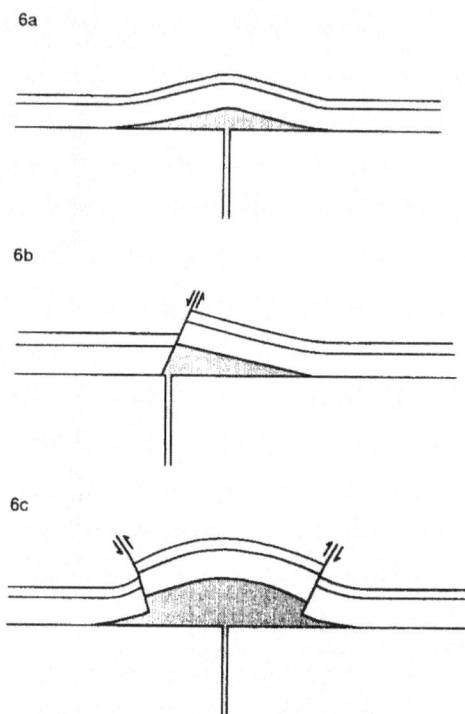

Fig. 6. Possible shapes for granitoid plutons, and syn-intrusion wall-rock structures. (a) simple "laccolithic" pluton; (b) asymmetric "laccolithic" pluton with accommodating shear zone/fault; (c) "cookie-cutter" pluton with accommodating shear zones/faults (these would be curved in three dimensions).

crystalline basement, strongly folded strata, high-grade metamorphic rocks, essentially unlayered rocks, or rocks with tightly welded layer boundaries. However, most felsic volcanic magmas undergo some fractionation prior to eruption, and commonly produce eruption products whose compositional zoning is interpreted as having been produced in large, usually shallow, magma chambers. This fact suggests that the surface eruption of such magmas depends not only on their innate ability to rise directly to the surface, but on the occurrence of volatile pressure build-up and local structural peculiarities that allow failure of the roofs of the magma chambers. There is no requirement that high-level plutonic rocks have volcanic equivalents. However, large-volume, silicic, volcanic deposits probably always have substantial plutonic equivalents, unless their feeder chambers are more or less completely drained.

Conclusions and implications

Each block of crust and each magma is a unique system and some aspects of its genesis and evolution must, necessarily, be unique. However, we suggest that fracture formation and propagation are critical processes in the segregation, ascent and, probably, emplacement of most granitoid magmas, and are critical phenomena in the chemical, metamorphic and histories of lower crustal granulite terranes. Given the accuracy of this conclusion, there are numerous wider implications for geology, geochemistry, petrogenesis and crustal rheology, some of which we outline below.

(1) Melt segregation is far more efficient than indicated by simple gravity compaction models. It is driven by shearing in the partially molten source regions, creating a mesh of melt-filled grain-edge fractures. The postulated significant small-scale fracture porosity means that the melts need only flow short distances, into larger fractures, to form coarse, interconnected meshes of veins. These veins will be oriented vertically in most crustal stress situations.

(2) The concept of the rheologically critical melt proportion (e.g., Arzi, 1978), apart from other complicated controls on its value, only has relevance when the melt segregation mechanism is specified. Gravitational rise of an homogenized restite-melt system may not be a factor of any major significance in the process of granitoid magma movement.

(3) The necking down of drained veins and dykes (e.g., Bédard and Sawyer, 1991) may mean that few or no traces of melt segregation will be preserved in the restitic remnants of magma sources. This is especially apparent when consideration is given to the fact that the melting episode is extremely brief in comparison with both the associated deformation and the metamorphism, which must continue in the source terrane long after melt extraction. Since the solid frameworks of the source rocks may remain essentially intact, meso-scale rock fabrics may survive these processes.

(4) Rather small dykes can feed batholithic plutons in geologically reasonable times. This

means that the only signs of the flux of huge volumes of granitoid magma (through the middle and upper crust) may be relatively narrow dykes. Even this evidence might be obscured or obliterated by further deformation.

(5) A characteristic feature of the postulated feeder dykes is that they should be surrounded by thermal aureoles which are apparently incommensurably large compared with the present dyke volume. Careful textural examination of the rocks in the inner zones of such aureoles should, in many cases, reveal evidence of partial melting. The prediction of the existence of such features provides a geological test for the feeder dyke hypothesis; several probable examples are reported in the literature.

(6) The present shape of a pluton is due to magma arrival and minor local accommodation effects (e.g., ballooning, domal uplift of overlying strata, ring fracturing, block lifting, cauldron subsidence, stoping, etc.), or some combination of these, plus any superimposed deformation. Almost nothing geologically observable in and around most granitoid plutons will have relevance for interpreting the mechanism of magma migration. Pluton and wall-rock structure, emplacement depth and style, crystal content, mineralogy, etc., are mostly local characteristics, controlled by pressure, temperature, structures and stress fields at the emplacement sites. The occurrence of several very different emplacement styles among the plutons of the Donegal batholith in Ireland (Pitcher and Berger, 1972), all within a limited span of space and time, argues powerfully against either predestination or regional-scale structure as a major control.

(7) Granitoid plutonism is a rapid process. From the inception of a thermal anomaly in the lower crust to the final consequent emplacement of a high-level granitoid batholith, the elapsed time may be as little as 10^2–10^4 years (see also Huppert and Sparks, 1988). In contrast, the metamorphic episode associated with the magma production might last for 10^6–10^7 years. This rapidity (and consequent small degree of magma cooling) would allow the high-level emplacement of granitoid magmas with very low crystal contents. Batholiths and silicic volcanic magma chambers could thus have protracted crystallization histories at high levels in the crust. Crystallization and differentiation would occur in the final or holding magma chambers. Most chemical and mineralogical variations are therefore unlikely to be primarily source-inherited. Chemical interactions between the magma and conduit walls would be minimal for fracture ascent. This means that many granitoid magmas may carry essentially unmodified isotopic (and some geochemical) signatures reflecting source provenance.

(8) A granitic magma with a viscosity of 10^5 Pa s, showing a density contrast of 350 kg m^{-3} with its wall rocks and flowing upward through an open fracture 3 m wide, will have an ascent velocity of about 2.57×10^{-2} m s^{-1}. Assuming the presence of suspended spherical particles with densities between 2700 and 2900 kg m^{-3}, calculations show that the Reynolds number is $\ll 0.1$ and Stokes's law must hold. Using 2.57×10^{-2} m s^{-1} as the critical particle settling velocity, above which enclaves will not be transported, it is simple to show that critical enclave diameters are always very large (metres). Hence, essentially any entrained particle would reach emplacement level unless: (a) it was dissolved during ascent; (b) flow segregation or filtering effects removed most such particles; or (c) they were removed by gravity settling in a relatively quiescent magma chamber. Filtering is the natural, near-source consequence of our proposed segregation and ascent mechanism, and it seems likely that lack of entrainment in the first place is the reason for the observed paucity of restite lumps and mid-crustal xenoliths in high-level granitoids. Additionally, rapid magma ascent and limited cooling would result in constitutional superheating of the magma with respect to the stability limits of many potential restite/xenocryst phases. This is a consequence of the positive dP/dT slopes of many silicate mineral saturation boundaries in water-undersaturated granitoid melts (e.g., quartz, feldspars, amphiboles, some pyroxenes, biotite and garnet). However, rapid ascent might allow such solid materials to persist metastably during magma rise. Thus, although the magmas might contain some restitic solids, their proportion is likely to be very minor. This means that the process of restite

unmixing (e.g., Chappell et al., 1987) cannot be a major factor in the high-level chemical evolution of most granitoid magmas.

(9) Melt will be rapidly and efficiently expelled via veins and dykes throughout a partially molten rock volume, and will not generally pond as a layer (pluton) overlying a "sill" of basalt emplaced into fertile crust (c.f. Huppert and Sparks, 1988; Fountain et al., 1989). This will mean that there may be little or no direct chemical interaction between such a mafic heat source and a derived felsic crustal melt. Although associated in space and time, mafic and silicic magmas may be unable to mix unless they share a high-level composite magma chamber. At this stage the two magmas would generally mingle rather than mix thoroughly (e.g., Vernon, 1984; Sparks and Marshall, 1986; Frost and Mahood, 1987).

(10) Large plutons are probably only developed at or near their final emplacement levels. Pulsating magma delivery seems to be a necessity in our dyke model for granitoid magma ascent. Many separate magma batches or pulses may coalesce to form a batholith, but mixing processes are likely to be inefficient in the final magma chamber (e.g., Vernon, 1984). In consequence, different parts of a batholith may undergo quasi-independent chemical evolution. Some source-inherited magma heterogeneities may persist and be recognizable in the solidified rock products. This may explain the occurrence of the separate mingled magma batches in many I-type plutons (e.g., Vernon, 1984). Lowell and Bergantz (1987) have showed that felsic magmas have to accumulate into bodies several metres thick for convective stirring to occur.

(11) Felsic magma ascent via fracture networks will potentially leave large volumes of granulite-facies restites in the lower crust. This situation contrasts markedly with diapirism, in which wholesale mobilization and ascent of partially molten source regions is postulated. One of the testable features of the fracture ascent model is that, where exhumed by major tectonism, samples of granulite-facies lower crust should commonly contain rocks with the mineralogical and chemical characteristics of restites, as well as large bodies of essentially anhydrous metagabbroic

rocks (e.g., Clemens, 1990; Vielzeuf et al., 1990). This seems to be the case in at least the Ivrea Zone of Italy (Rutter and Brodie, 1991).

(12) The ascent mechanism advocated here implies minimal contact between the travelling magma and wall rocks, thus recent suggestions that water-undersaturated granitic magmas could absorb water from surrounding rocks and aid production of anhydrous granulites are most unlikely. There are also other arguments, based on thermal and chemical diffusion kinetics, that militate powerfully against this notion.

(13) Efficient melt evacuation via self-propagating fractures would minimize retrograde rehydration due to in situ crystallization and eventual expulsion of aqueous fluid from the melt fraction. Thus, the highest grade granulites should generally be the least retrogressed, except where an externally derived hydrous fluid has been reintroduced.

(14) The heat consumed in the chemical work of partial fusion, and that rapidly and efficiently advected away as hot granitoid magma, may be sufficient to deplete, or even exhaust, the metamorphic heat source and initiate the retrograde metamorphic history (see, for example, Vielzeuf et al. (1990) for a detailed discussion). This conclusion does not assume the existence of an extracrustal heat source, and should hold equally, or more so, in cases where only intracrustal radioactive heat sources are involved.

(15) If the Cook-Gordon mechanism does operate as proposed above, then an injected, roughly horizontal magma body could contribute to the development of a crustal decoupling horizon, or could lubricate an existing active structure. This contrasts with schemes wherein rising diapirs "pin" low-angle normal and thrust faults.

Acknowledgements

Many colleagues over the years have shared their knowledge, prejudices, derisive remarks and gratuitous abuse with us, and we thank them all. Specifically, in alphabetical order, we thank: Lawford Anderson, Cal Barnes, Adrian Brearley, Weecha Crawford, Sandy Cruden, Mike Etheridge, Jeff Grambling, Mark Harrison, John

Holloway, Dana Johnston, Bruce Marsh, Win Means, Robin Nicholson, Wally Pitcher, Dave Rubie, Tracy Rushmer, Ed Sawyer, Chris Schmidt, Fried Schwerdtner, Alan Thompson, Amy Thompson, Ron Vernon, Daniel Vielzeuf and Vic Wall. The manuscript was reviewed by Gary Ernst and Norm Sleep, and we thank them for their excellent work. We also appreciate the superb editorial comments by Mel Friedman. This study was partly funded by providence, the French CNRS, the Université Blaise Pascal at Clermont-Ferrand and N.E.R.C. (U.K.) research grant GR3/7669 (J.D.C.), the NSF (Grant EAR 88-16402 to C.K.M. and J.A. Grambling, University of New Mexico), Sandia National Laboratories at Albuquerque (SURP Grant 86-5629, Task 7, to C.K.M.), and the Caswell Silver Foundation of the University of New Mexico (travel monies to C.K.M.).

References

Anderson, J.L., 1983. Proterozoic anorogenic granite plutonism of North America. In: L.G. Medaris Jr., C.W. Byers, D.M. Mickelson and W.C. Shanks (Editors), Proterozoic Geology. Geol. Soc. Am. Mem., 161: 131–154.

Arzi, A.A., 1978. Critical phenomena in the rheology of partially melted rocks. Tectonophysics, 44: 173–184.

Atherton, M.P., 1990. The Coastal Batholith of Peru: the product of rapid recycling of "new" crust formed within rifted continental margin. Geol. J., 25: 337–349.

Atkinson, B.K., 1984. Subcritical crack growth in geological materials. J. Geophys. Res., B89: 4077–4144.

Bailey, E.B., 1960. The Geology of Ben Nevis and Glencoe (Sheet 53), 2nd edn., Mem. Geol. Surv. U.K.

Bateman, R., 1985. Progressive crystallization of a granitoid diapir and its relationship to stages of emplacement. J. Geol., 93: 645–662.

Bédard, L.P. and Sawyer, E.W., 1991. Replenishment and melt expulsion: field evidence from northern Abitibi Greenstone Belt. Geol. Assoc. Can./Mineral. Assoc. Can./Soc. Econ. Geol. Joint Annu. Meet., Program with Abstracts, 16: A9.

Brun, J.P., Gapais, D., Cogne, J.P., Ledru, P. and Vigneresse, J.L., 1990. The Flamanville Granite (northwest France): an unequivocal example of a syntectonically expanding pluton. Geol. J., 25: 271–286.

Burnham, C.W. and Davis, N.F., 1974. The role of H_2O in silicate melts: II. Thermodynamic and phase relations in the system $NaAlSi_3O_8 \cdot H_2O$ to 10 kilobars, 700° to 1100°C. Am. J. Sci., 274: 902–940.

Carslaw, H.S. and Jaeger, J.C., 1980. Conduction of Heat in

Solids. Reprint of 2nd edition, Oxford University Press, Oxford, 510 pp.

Castro, A., 1986. Structural pattern and ascent model in the Central Extremadura batholith, Hercynian belt, Spain. J. Struct. Geol., 8: 633–645.

Chappell, B., White, A.J.R. and Wyborn, D., 1987. The importance of residual source material (restite) in granite petrogenesis. J. Petrol., 28: 1111–1138.

Clemens, J.D., 1988. Volume and composition relationships between granites and their lower crustal source regions: an example from central Victoria, Australia. Aust. J. Earth Sc., 35: 445–449.

Clemens, J.D., 1990. The granulite–granite connexion. In: D. Vielzeuf and Ph. Vidal (Editors), Granulites and Crustal Evolution. Kluwer Academic Publishers, Dordrecht, pp. 25–36.

Clemens, J.D. and Vielzeuf, D., 1987. Constraints on melting and magma production in the crust. Earth Planet. Sci. Lett., 86: 287–306.

Clemens, J.D. and Wall, V., 1981. Crystallization and origin of some peraluminous (S-type) granitic magmas. Can. Mineral., 19: 111–131.

Clemens, J.D. and Wall, V., 1984. Origin and evolution of a peraluminous silicic ignimbrite suite: the Violet Town Volcanics. Contrib. Mineral. Petrol., 88: 354–371.

Cook, J. and Gordon, J.E., 1964. A mechanism for the control of crack propagation in all-brittle systems. Proc. R. Soc. London, Ser. A, 282: 508–520.

Corry, C.E., 1988. Laccoliths: Mechanisms of emplacement and growth. Geol. Soc. Am., Spec. Pap. 220: 110 pp.

Cruden, A.R., 1988. Deformation around a rising diapir modeled by creeping flow past a sphere. Tectonics, 7: 1091–1101.

Cruden, A.R., 1990. Flow and fabric development during the diapiric rise of magma. J. Geol., 98: 681–698.

Delaney, P.T. and Pollard, D.D., 1981. Deformation of Host Rocks and Flow of Magma During Growth of Minette Dikes and Breccia-bearing Intrusions near Ship Rock, New Mexico. U.S. Geol. Surv., Washington, D.C., 61 pp.

Dell'Angelo, L.N. and Tullis, J., 1988. Experimental deformation of partially melted granitic aggregates. J. Metamorph. Geol., 6: 495–515.

Deniel, C., Vidal, Ph., Fernandez, A., LeFort, P. and Peucat, J.-J., 1987. Isotopic study of the Manaslu leucogranite (Himalaya, Nepal): inferences on the age and source of Himalayan leucogranites. Contrib. Mineral. Petrol., 96: 78–92.

Didier, J., El Mouraouah, A. and Fernandez, A., 1987. Microtexture de refusion dans le granite migmatitique du Velay autour de la diorite du Peyron près de Burzet (Ardèche, Massif Central français). C.R. Acad. Sci. Paris, 304, II: 1227–1232.

Donath, F.A., 1961. Experimental study of shear failure in anisotropic rocks. Bull. Geol. Soc. Am., 72: 985–990.

Emerman, S.H. and Marrett, R., 1990. Why dikes? Geology, 18: 231–233.

Engelder, T. and Sbar, M.L., 1984. Near-surface in situ stress; introduction. J. Geophys. Res., 89: 9321–9322.

Etheridge, M.A., 1983. Differential stress magnitudes during regional deformation and metamorphism: upper bound imposed by tensile fracturing. Geology, 11: 231–234.

Etheridge, M.A., Wall, V.J. and Vernon, R.H., 1983. The role of the fluid phase during regional metamorphism and deformation. J. Metamorph. Geol., 1: 205–226.

Etheridge, M.A., Wall, V.J. and Vernon, R.H., 1984. High fluid pressures during regional metamorphism and deformation: implications for mass transport and deformation mechanisms. J. Geophys. Res., B89: 4344–4358.

Evans, B.W. and Vance, J.A., 1987. Epidote phenocrysts in dacitic dykes, Boulder County, Colorado. Contrib. Mineral. Petrol., 96: 178–185.

Fletcher, R.C. and Hoffman, A.W., 1974. Simple models of diffusion and combined diffusion-infiltration metasomatism. In: A.W. Hoffman (Editor), Geochemical Transport and Kinetics. Carnegie Institute of Washington, D.C., pp. 242–262.

Fountain, J.C., Hodge, D.S. and Shaw, R.P., 1989. Melt segregation in anatectic granites: a thermo-mechanical model. J. Volcanol. Geotherm. Res., 39: 279–296.

Fripp, R.E.A., 1983. The Precambrian geology of an area around the Sand River near Messina, central zone, Limpopo mobile belt. Geol. Soc. S. Afr., Spec. Publ., 8: 89–102.

Frost, T.P. and Mahood, G.A., 1987. Field, chemical, and physical constraints on mafic–felsic magma interaction in the Lamarck Granodiorite, Sierra Nevada, California. Bull. Geol. Soc. Am., 99: 272–291.

Fyfe, W.S., 1973. The granulite facies, partial melting and the Archean crust. Philos. Trans. R. Soc. London, Ser. A, 273: 457–461.

Garnham, J.A., 1988. Ring-faulting and Associated Intrusions, Glencoe, Scotland. PhD thesis, Imperial College, London, 296 pp. (unpubl.).

Gerla, P.J., 1988. Stress and fracture evolution in a cooling pluton: an example from the Diamond Joe stock, western Arizona, U.S.A. J. Volcanol. Geotherm. Res., 34: 267–282.

Gilbert, G.K., 1877. Geology of the Henry Mountains, Utah. U.S. Geogr. Geol. Surv. Rocky Mountain Region, 170 pp.

Gleizes, G., Leblanc, D. and Bouchez, J.-L., 1991. Le pluton granitique de Bassies (Pyrenees ariegeoises): zonation, structure et mise en place. C.R. Acad. Sci., 312 (Ser. II): 755–762.

Gordon, J.E., 1984. The New Science of Strong Materials, or, Why You Don't Fall Through the Floor (2nd Edition). Princeton University Press, Princeton, N.J., 287 pp.

Gretener, P.E., 1969. On the mechanics of the intrusion of sills. Can. J. Earth Sci., 6: 1415–1419.

Hibbard, M.J. and Watters, R.J., 1985. Fracturing and diking in incompletely crystallized granitic plutons. Lithos, 18: 1–12.

Holder, M.T., 1978. Granite emplacement models. J. Geol. Soc. London, 135: 459–460.

Hollister, L.S. and Crawford, M.L., 1986. Melt-enhanced deformation: a major tectonic process. Geology, 14: 558–561.

Hubert, M.K. and Rubey, W.W., 1959. Role of fluid pressure in the mechanics of overthrust faulting. Bull. Geol. Soc. Am., 70: 115–205.

Huppert, H.E. and Sparks, R.S.J, 1988. The generation of granitic magmas by intrusion of basalt into continental crust. J. Petrol., 29: 599–624.

Hutton, D.H.W., Dempster, T.J. and Brown, P.E., 1990. A new mechanism of granite emplacement: rapakivi intrusions in active extensional shear zones. Nature, 343: 451–454.

John, B.E., 1988. Structural reconstruction and zonation of a tilted mid-crustal magma chamber: the felsic Chemehuevi Mountains plutonic suite. Geology, 16: 613–617.

Jurewicz, S.R. and Watson, E.B., 1985. The distribution of partial melt in a granitic system: the application of liquid phase sintering theory. Geochim. Cosmochim. Acta, 49: 1109–1121.

Kirby, S.H., 1985. Rock mechanics observations pertinent to the rheology of the continental lithosphere and the localization of strain along shear zones. Tectonophysics, 119: 1–27.

Knapp, R.B. and Knight, J.E., 1977. Differential thermal expansion of pore fluids: fracture propagation and microearthquake production in hot pluton environments. J. Geophys. Res., B82: 2515–2522.

Lange, R.A. and Carmichael, I.S.E., 1987. Densities of Na_2O-K_2O-CaO-MgO-FeO-Fe_2O_3-Al_2O_3-TiO_2-SiO_2 liquids: new measurements and derived partial molar properties. Geochim. Cosmochim. Acta, 51: 2931–2946.

Lawn, B.R. and Wilshaw, T.R., 1975. Fracture of Brittle Solids. Cambridge University Press, Cambridge, 204 pp.

Leake, B.E., 1978. Granite emplacement: the granites of Ireland and their origin. In: D.R. Bowes and B.E. Leake (Editors), Crustal Evolution of Northwestern Britain and Adjacent Regions. Geol. J. Spec. Issue, 10: 221–248.

Le Breton, N. and Thompson, A.B., 1988. Fluid-absent (dehydration) melting of biotite in metapelites in the early stages of crustal anatexis. Contrib. Mineral. Petrol., 99: 226–237.

LeFort, P., 1981. Manaslu Leucogranite: a collision signature of the Himalaya. A model for its genesis and emplacement. J. Geophys. Res., B86: 10,545–10,568.

Lister, J.R. and Kerr, R.C., 1991. Fluid-mechanical models of crack propagation and their application to magma transport in dykes. J. Geophys. Res., B96: 10,049–10,077.

Lowell, R.P. and Bergantz, G., 1987. Melt stability and compaction in a partially molten silicate layer heated from below. In: D.E. Loper (Editor), Structure and Dynamics of Partially Solidified Systems. Martinus Nijhoff Publishers, Dordrecht, pp. 383–400.

Mahon, K.I., Harrison, T.M. and Drew, D.A., 1988. Ascent of a granitoid diapir in a temperature varying medium. J. Geophys. Res., B93: 1174–1188.

Mandelbrot, B.B., 1982. Fractal Geometry in Nature. W.H. Freeman, New York, N.Y., 468 pp.

Marsh, B.D., 1984. Mechanics and energetics of magma formation and ascension. In: Explosive Volcanism: Inception, Evolution, and Hazards. National Academy Press, Washington D.C., pp. 67–83.

Mawer, C.K., 1989. Melt formation and migration in experimental and natural situations. 28th Int. Geol. Congr., Abstr. Vol., 2: 390.

Mawer, C.K., 1992. Small-scale melt migration: experiments, natural examples, and the development and significance of stromatic migmatites. Phys. Earth Planet. Inter., in press.

Mawer, C.K., Grambling, J.A. and Vernon, R.H., 1989. Syntectonic nature of the 1.45 Ga Sandia batholith, New Mexico. Geol. Soc. Am. Annu. Meet., Abstr. Programs, 21: A308.

Miller, C.F., Watson, E.B. and Harrison, T.M., 1988. Perspectives on the source, segregation and transport of granitoid magmas. Trans. R. Soc. Edinburgh: Earth Sci., 79: 135–156.

McKenzie, D., 1984. The generation and compaction of partially molten rock. J. Petrol., 25: 713–765.

McKenzie, D.P., 1987. The compaction of igneous and sedimentary rocks. J. Geol. Soc. London, 144: 299–307.

Nicolas, A., 1986. A melt extraction model based on structural studies in mantle peridotites. J. Petrol., 27: 999–1022.

Nicolas, A. and Jackson, M.,1982. High temperature dikes in peridotites: origin by hydraulic fracturing. J. Petrol., 23: 568–582.

Noyes, H.J., Wones, D.R. and Frey, F.A., 1983. A tale of two plutons: petrographic and mineralogic constraints on the petrogenesis of the Red Lake and Eagle Peak plutons, central Sierra Nevada, California. J. Geol., 91: 353–379.

Ottino, J.M., Leong, C.W., Rising, H. and Swanson, P.D., 1988. Morphological structures produced by mixing in chaotic flows. Nature, 333: 419–425.

Paterson, S., 1988. Cannibal Creek granite: post-tectonic "ballooning" pluton or pre-tectonic piercement diapir? J. Geol., 96: 730–736.

Paterson, S.R., Vernon, R.H. and Tobisch, O.T., 1989. A review of criteria for the identification of magmatic and tectonic foliations in granitoids. J. Struct. Geol., 11: 349–363.

Patiño Douce, A.E., Humphreys, E.D. and Johnston, A.D., 1990. Anatexis and metamorphism in tectonically thickened continental crust exemplified by the Sevier hinterland, western North America. Earth Planet. Sci. Lett., 97: 290–315.

Phillips, G.N., Wall, V.J. and Clemens, J.D., 1981. Petrology of the Strathlogie batholith: a cordierite-bearing granite. Can. Mineral., 19: 47–64.

Pichavant, M., Kontak, D.J., Briqueu, L., Herrera, J.V. and Clark, A.H., 1988a. The Miocene–Pliocene Macusani Volcanics, SE Peru ii. Geochemistry and origin of a felsic peraluminous magma. Contrib. Mineral. Petrol., 100: 325–338.

Pichavant, M., Kontak, D.J., Herrera, J.V. and Clark, A.H.,

1988b. The Miocene–Pliocene Macusani Volcanics, SE Peru i. Mineralogy and magmatic evolution of a two-mica aluminosilicate-bearing ignimbrite suite. Contrib. Mineral. Petrol., 100: 300–324.

Pitcher, W.S. and Berger, A.R., 1972. The Geology of Donegal: A Study of Granite Emplacement and Unroofing. Wiley Interscience, New York, N.Y., 435 pp.

Pollard, D.D., 1977. Derivation and evaluation of a mechanical model for sheet intrusions. Tectonophysics, 19: 233–269.

Powell, R. and Downes, J., 1990. Garnet porphyroblast-bearing leucosomes in metapelites: mechanisms, phase diagrams, and an example from Broken Hill, Australia. In: J.R. Ashworth and M. Brown (Editors), High-Temperature Metamorphism and Crustal Anatexis. Unwin–Hyman, London, pp. 105–123.

Ramberg, H., 1967. Gravity, Deformation and the Earth's Crust. Academic Press, London, 214 pp.

Ramsay, J.G., 1989. Emplacement kinematics of a granite diapir: the Chindamora batholith, Zimbabwe. J. Struct. Geol., 11: 191–209.

Ribe, N.M., 1987. Theory of melt segregation—a review. J. Volcanol. Geotherm. Res., 33: 241–253.

Rushmer, T., 1991. Partial melting of two amphibolites—contrasting experimental results under fluid-absent conditions. Contrib. Mineral. Petrol., 107: 41–59.

Rutter, E.H. and Brodie, K.H., 1992. Rheology of the lower crust. In: D.M. Fountain, R.J. Arculus and R.W. Kay (Editors), Continental Lower Crust. Elsevier, Amsterdam, in press.

Rutter, M.J. and Wyllie, P.J., 1988. Melting of vapour-absent tonalite at 10 kbar to simulate dehydration-melting in the deep crust. Nature, 331: 159–160.

Ryan, B., 1991. Makhavinekh Lake pluton, Labrador, Canada: geological setting, subdivisions, mode of emplacement, and a comparison with Finnish rapakivi granites. Precambrian Res., 51: 193–225.

Sawyer, E.W., 1992. Structure and composition of migmatite leucosomes: implications for the melt-residuum separation process. Phys. Earth Planet. Inter., in press.

Schmeling, H., Cruden, A.R. and Marquart, G., 1988. Finite deformation in and around a fluid sphere moving through a viscous medium: implications for diapiric ascent. Tectonophysics, 149: 17–34.

Schwerdtner, W.M., 1990. Structural tests of diapir hypotheses in Archean crust of Ontario. Can. J. Earth Sci., 27: 387–402.

Sleep, N.H., 1988. Tapping of melt by veins and dikes. J. Geophys. Res., B93: 10,255–10,272.

Sparks, R.S.J. and Marshall, L.A., 1986. Thermal and mechanical constraints on mixing between mafic and silicic magmas. J. Volcanol. Geotherm. Res., 29: 99–124.

Spence, D.A. and Turcotte, D.L., 1990. Buoyancy-driven magma fracture: a mechanism for ascent through the lithosphere and the emplacement of diamonds. J. Geophys. Res., B95: 5133–5139.

Stevenson, D.J., 1989. Spontaneous small-scale melt segregation in partial melts undergoing deformation. Geophys. Res. Lett., 16: 1067–1070.

Takada, A., 1989. Magma transport and reservoir formation by a system of propagating cracks. Bull. Volcanol., 52: 118–126.

Takada, A., 1990. Experimental study on propagation of liquid-filled crack in gelatin: shape and velocity in hydrostatic stress condition. J. Geophys. Res., B95: 8471–8481.

Turcotte, D.L., 1987. Physics of magma segregation processes. In: B.O. Mysen (Editor), Magmatic Processes: Physicochemical Principles. Spec. Publ. 1, The Geochemical Society, University Park, Penn., pp. 69–74.

Vernon, R.H., 1984. Microgranitoid enclaves in granites—globules of hybrid magma quenched in a plutonic environment. Nature, 309: 438–439.

Vielzeuf, D., Clemens, J.D., Pin, C. and Moinet, E., 1990. Granites, granulites and crustal differentiation. In: D. Vielzeuf and Ph. Vidal (Editors), Granulites and Crustal Evolution. Kluwer Academic Publishers, Dordrecht, pp. 59–85.

Vielzeuf, D. and Holloway, J.R., 1988. Experimental determination of the fluid-absent melting relations in the pelitic system. Consequences for crustal differentiation. Contrib. Mineral. Petrol., 98: 257–276.

Weertman, J., 1971. Theory of water-filled crevasses in glaciers applied to vertical magma transport beneath ocean ridges. J. Geophys. Res., B76: 1171–1183.

Weertman, J., 1980. The stopping of a rising, liquid-filled crack in the earth's crust by a freely slipping horizontal joint. J. Geophys. Res., B85: 967–976.

White, A.J.R. and Chappell, B.W., 1977. Ultrametamorphism and granitoid genesis. Tectonophysics, 43: 7–22.

Wickham, S.M., 1987. The segregation and emplacement of granitic magmas. J. Geol. Soc. London, 144: 281–297.

Wilson, L. and Head, J.W. (III), 1981. Ascent and eruption of basaltic magma on the Earth and Moon. J. Geophys. Res., B86: 2971–3001.

Wyborn, D., Chappell, B.W. and Johnston, R.M., 1981. Three S-type volcanic suites from the Lachlan Fold Belt, southeast Australia. J. Geophys. Res., B86: 10,335–10,348.

Zoback, M.L. and Zoback, M., 1980. State of stress in the conterminous United States. J. Geophys. Res., B85: 6113–6156.

CHAPTER 11

Chappell, B.W. and White, A.J.R. (1974) Two contrasting granite types. *Pacific Geology*, **8**, 173−174.

Everyone working on the petrology and petrogenesis of granitic rocks is aware of the classification of granites into I- and S-types. However, fewer people perhaps recognize the real importance of Bruce Chappell and Alan White's two-page extended abstract, published in 1974, in *Pacific Geology*.

By the early 1970s it would be true to say that work on granitic rocks was in something of a backwater. The main emphasis in petrology was the study of mafic volcanic rocks, their place in plate-tectonic environments and their meaning for mantle compositions and processes. Very little attention was paid to the genesis of felsic magmas. Most workers, excited by collisions, plumes and volcanic arcs spent all their efforts on the mafic to intermediate rocks. Ideas on the genesis of granitic magmas had not really changed much since the 1950s and 1960s. They were regarded by some as differentiates of mafic magmas and by others as melts of the middle to deep crust. Most people viewed them a low-temperature highly hydrous magmas that were generated, more or less, as petrological after-thoughts - the artefacts of more significant petrogenetic and tectonic processes.

Chappell and White had seen a lot of granitic rocks on various continents, particularly in North America, but took their example from the belts of granitic batholiths that run along the eastern seaboard of Australia. Chappell's geochemical expertise led him to recognize that some granites are weakly peraluminous to metaluminous and others are strongly peraluminous. Furthermore, this division could be correlated (in a slightly complex way) with the major and accessory mineralogy of the rocks, the types of enclaves that they contained, the range of SiO_2 contents, the Fe^{3+}/Fe^{2+}, K_2O/Na_2O and their Sr isotopic characteristics as well. It seemed logical that these two obvious divisions of granitic rocks (and therefore of their magmas) should have different origins. This reasoning ran parallel to the thoughts of Jean Didier whose book (Didier, 1973) contains similar observations and an interpretation of mantle and crustal components to the magmas, which led to his C (crustal), M (mantle) and CM (mixed) classification.

Chappell and White were not convinced of any direct involvement of mantle-derived components in the vast majority of granitic magmas, so they preferred crustal melting (or anatexis) as the origin of the magmas. This led them to hypothesize that the metaluminous to slightly peraluminous granitic rocks with igneous-looking enclaves

and a broad range of compositions were formed by partial melting of igneous protoliths. These were their I-types, abundant in places like the Sierra Nevada and the Andes. In an analogous way they recognized that the most likely source of strongly peraluminous granitic magmas is partial melting of sedimentary materials that had once contained aluminous clays. This group was designated as S-type, rocks that they had observed forming a large component of the Kosciusko and Berridale batholiths in southeastern Australia. The Didier classification, with its implications of mantle magma, fractionation and mixing with crustal melt never really caught on outside western Europe, and particularly France. In contrast, the crust-focused classification of White and Chappell took off, spread rapidly across the globe, spawned numerous extensions and additions, sub-types and a legion of misinterpretations and misassignments of rocks.

Part of the confusion surrounding some early applications of the S-I classification of granites was due to the mistaken idea that peraluminosity alone could be used to diagnose S- or I-type kinship. Further problems were caused when many workers (including Chappell and White) began to see the S-I classification as somehow inextricably entangled with the idea of restite unmixing (Chappell *et al.*, 1987) as a mechanism for producing compositional diversity in granitic series. Serious challenges were mounted to the restite-unmixing hypothesis (e.g. Wall *et al.*, 1987; Clemens, 1989) and it never gained wide acceptance. Somehow the S-I classification also suffered as a consequence, even though it was and is a completely separate concept. Other problems with it included things like the discovery of intermediate types, granites with clear textural and isotopic evidence for direct involvement of mafic magma, and persistent misuse of the classification. All these problems, imagined and real, have led to the partial to complete abandonment of the S-I classification, even though it still has some validity and utility. The mass of information learned about granitic rocks and magmas since 1974, which has led to the virtual demise of the S-I classification gives us a hint of the true importance of Chappell and White's paper.

So much for classification! This is not at all why Chappell and White (1974) is included here. The real importance of this diminutive paper, published 35 years ago in an almost unknown and now defunct journal, is that it provided the critical spark that ignited a petrological renaissance. Suddenly there were new ideas to discuss and new reasons to investigate granites. Suddenly granites became valuable geochemical and isotopic windows into the commonly unexposed deep continental and arc crust, and probes into the processes that occurred there. This paper made many petrologists realize that, far from being mere residues of fractionation or mineralogically tedious products of crustal melting, granites could be used to discover important things about the composition and structure of the crust, and that they were worthy of much closer attention than they had been receiving. You are reading this book probably because Chappell and White (1974) made granites interesting again.

PACIFIC GEOLOGY 8, 173~174 (1974)
© Tokai University Press, Tokyo, Japan.

TWO CONTRASTING GRANITE TYPES

B.W. CHAPPELL* and A.J.R. WHITE**

* Australian National University, Canberra, Australia.
** La Trobe University, Bundoora, Melbourne, Australia.

Expanded abstract

Granites of the major batholiths in the Tasman Orogenic Zone of eastern Australia are of two contrasting types which are of widespread occurrence and which may be distinguished by chemical, mineralogical, field and other criteria. We interpret these granites as being derived by partial melting of two different types of source material—igneous and sedimentary. Differences in the derived granites are inherited from the source rocks so that we recognize an I-type and an S-type respectively.

Some of the distinctive chemical properties of the two types are shown in the following table:—

I-types	S-types
Relatively high sodium, Na_2O normally $> 3.2\%$ in felsic varieties, decreasing to $> 2.2\%$ in more mafic types	Relatively low sodium, Na_2O normally $< 3.2\%$ in rocks with approx. 5% K_2O, decreasing to $< 2.2\%$ in rocks with approx. 2% K_2O
Mol $Al_2O_3/(Na_2O + K_2O + CaO) < 1.1$	Mol $Al_2O_3/(Na_2O + K_2O + CaO) > 1.1$
C.I.P.W. normative diopside *or* $< 1\%$ normative corundum	$> 1\%$ C.I.P.W. normative corundum
Broad spectrum of compositions from felsic to mafic	Relatively restricted in composition to high SiO_2 types
Regular inter-element variations within plutons; linear or near-linear variation diagrams	Variation diagrams more irregular

These chemical properties result from the removal of sodium into sea water (or evaporites) during sedimentary fractionation, and calcium into carbonates, with subsequent relative enrichment of the main sedimentary pile in aluminium. S-type granites come from a source that has been subjected to this prior chemical fractionation.

Petrographic features reflect the differences in chemical composition. Hornblende is common in the more mafic I-types and is generally present in felsic varieties, whereas hornblende is absent, but muscovite is common, in the more felsic S-types; biotite may be very abundant, up to 35%, in more mafic S-types. Sphene is a common accessory in the I-type granites whereas monazite may be found in S-types. Alumino-silicates, garnet and cordierite may occur in S-type xenoliths or in the granites themselves. All of these features result from the high aluminium content relative to alkalis and calcium in S-type granites and the converse in I-types. Apatite inclusions are common in biotite and hornblende of I-type granites whereas it occurs in larger discrete crystals in S-types.

Detailed studies in the Berridale Batholith (COMPSTON, SHIRAHASE, CHAPPELL & WHITE *in preparation*) have shown that strontium is more radiogenic in S-type granites (initial $Sr^{87}/Sr^{86} > 0.708$) because their source rocks had been through an earlier sedimentary cycle. I-types have initial Sr^{87}/Sr^{86} ratios in the range 0.704–0.706. Isochrons of

Received on Feb. 18, 1974

B.W. CHAPPELL and A.J.R. WHITE

I-types give a regular linear set of points whereas those of S-types show a scatter of points within a broad envelope, reflecting variations in the initial Sr^{87}/Sr^{86} within a single pluton as a consequence of more heterogeneous source material.

Field relationships of the two types may be distinctive. More mafic I-types contain mafic hornblende-bearing xenoliths of igneous appearance whereas hornblende-bearing xenoliths are rare in the S-types but metasedimentary xenoliths may be common. When both types occur together in composite batholiths the S-types are usually early in the intrusive sequence and they often have a strong secondary foliation truncated by later I-type intrusions which are either massive or have a dominant primary foliation.

Economic minerals are also different in their association with the two types. Tin mineralization appears to be confined to highly silicic S-type granites whereas tungsten and porphyry-type copper and molybdenum deposits are associated with I-types.

CHAPTER 12

Ishihara, S. (1977) The magnetite-series and ilmenite-series granitic rocks. *Mining Geology*, **27**, 293–305.

This paper, published in a special issue of *Mining Geology*, does not represent the first appearance of the terms "magnetite-series" and "ilmenite-series" as applied to granitic rocks. That primacy belongs to an earlier work, by the same author (Ishihara, 1975). The magnetite series contain magnetite ± ilmenite and the ilmenite series only ilmenite. Note that these series do not map onto the Chappell and White (1972) S-I classification scheme, which is also not referred to in Ishihara's paper. All of Ishihara's Japanese rocks would be considered I-types. What the present paper does is to expound on the recognition of these two series of granitic rocks and then go into the details of how these two series may have formed, and finally to make some useful comments on the associations between various types of ore deposits and the two series.

In his opening sentence, Ishihara remarks that the accessory minerals of granitic rocks are commonly ignored, but that they may have importance for petrogenetic processes and for economic mineral potential. Fundamental to the idea behind the two series is that Fe^{3+}/Fe^{2+} is higher in the magnetite-series rocks, and that this, in some way, reflects f_{O_2} in the magma. This is a simple view that can be criticised in view of the sensitivity of redox equilibria to f_{H_2O}, P and T. Nevertheless, it is a view that has utility, as long as the other factors are also borne in mind. Burnham and Ohmoto (1980) is the other paper in this collection that deals with the influence of oxidation state on the metal- and sulphur-bearing capacity of felsic magmas. Curiously, although that paper was also published in a special issue of *Mining Geology*, it makes no reference to the slightly earlier work of Ishihara.

Ishihara seems to ascribe the occurrence of ilmenite in ilmenite-series granitic rocks to late-stage processes, possibly the interaction between the magmas and graphite-bearing wall rocks into which the magmas were emplaced. It is concluded here that magnetite-series magmas are of deep, possibly mantle origin. There are reasons to doubt both these ideas. For example, the ilmenite crystals are commonly found included near the cores of early-crystallized phases, and the mantle origin of felsic magma still looks unlikely on volumetric grounds, despite the protestations of Sisson *et al.* (2004). It seems more likely that the sources of the two series had different Fe^{3+}/Fe^{2+}. The ilmenite series contain pyrrhotite as the accessory sulphide, while the magnetite-series rocks have pyrite and/or chalcopyrite (see the discussion in Burnham and Ohmoto, 1980).

Perhaps the most important part of Ishihara's paper is the section dealing with the metallogenic potential of the two series. Ishihara notes that, world-wide, the magnetite-series rocks are associated with vein deposits of scheelite-molybdenite and gold, and porphyry copper-molybdenum deposits. Likewise, he notes the association between ilmenite-series rocks and cassiterite-wolframite vein deposits. He goes on to make the astute comment that the "general scarcity of porphyry copper deposits in Precambrian terranes may be explained by the general paucity of the magnetite-series granitoids in the older continental crust". Though the details may not have been exactly correct, according to present understanding, this paper represented an early recognition that the environments of magma formation and emplacement could have an important bearing on the genesis of metallic deposits associated with granitic rocks.

Mining Geology, **27**, 293~305, 1977

The Magnetite-series and Ilmenite-series
Granitic Rocks*

Shunso Ishihara**

Abstract: Opaque minerals of common granitic rocks were studied microscopically. The granitoids were divided into (i) a magnetite-bearing magnetite-series and (ii) a magnetite-free ilmenite-series. Each series has the following characteristic assemblages of accessary minerals:

Magnetite-series: Magnetite (0.1–2 vol.%), ilmenite, hematite, pyrite, sphene, epidote, high ferric/ferrous (and high Mg/Fe) biotite;

Ilmenite-series: Ilmenite (less than 0.1 vol.%), pyrrhotite, graphite, muscovite, low ferric/ferrous (and low Mg/Fe) biotite.

The mineral assemblages imply a higher oxygen fugacity in the magnetite-series granitoids than in the ilmenite-series granitoids during solidification of the granitic magmas. The boundary separating the two series is probably near the Ni–NiO buffer.

The magnetite-series granitoids are considered to have been generated in a deep level (upper mantle and lowest crust) and not to have interacted with C-bearing materials; whereas the ilmenite-series granitoids were generated in the middle to lower continental crust and mixed with C-bearing metamorphic and sedimentary rocks at various stages in their igneous history. The former carries porphyry copper-molybdenum deposits and the latter accompanies greisen-type tin-wolframite deposits. Lack of porphyry copper deposits in the Mesozoic orogeny belts in East Asia is related to a general paucity of the magnetite-series granitoids in this terrane.

Introduction

In studies of granitic rocks, opaque oxide minerals are often ignored, although they may have an important bearing on both petrogenesis and ore genesis. In recent years there have been excellent studies on small pluton where oxidation and reduction of granitic magma can be analyzed in detail (e.g., Czamanske & Mihalik, 1972). However, the distribution and occurrence of the opaque minerals in common granitic rocks seems not yet well established, although it is generally understood that both magnetite and ilmenite are present as accessary constituents in common granitic rocks.

Studies of Japanese granitic rocks during the past years have revealed that species and amounts of the opaque minerals change from one granitic belt to the other, and that the rocks may be divided into two series: one containing magnetite and ilmenite, and the other in which no magnetite but very small amounts of ilmenite are observed. They may be called "magnetite-series" and "ilmenite-series" (Ishihara, 1975), referring, respectively, to magnetite-bearing series and magnetite-free series granitic rocks.

Previous papers (Ishihara, 1971a; Tsusue & Ishihara, 1974) have reported bulk chemistry of granitic rocks and mode of occurrence and some chemistry of Fe–Ti oxide minerals, mostly from southwestern Japan. Here, granitoids from the remaining part of the Japanese Islands and some foreign countries are considered. This paper briefly summarizes classification, distribution and

* Received June 3, 1977; in revised form July 7, 1977. Paper presented at the 13th Pacific Science Congress, Vancouver, B.C., Canada, August 1975, in a symposium organized by the Circum Pacific Plutonism Project of the International Geological Correlation Program and chaired by P. C. Bateman.

** Geological Survey of Japan, Hisamoto 135, Takatsu, Kawasaki, Japan.

Key words: Magnetite-series granitoid, Ilmenite-series granitoid, Oxygen fugacity, Sphene, Graphite, Porphyry Cu–Mo deposits, Tin–wolframite deposits.

293

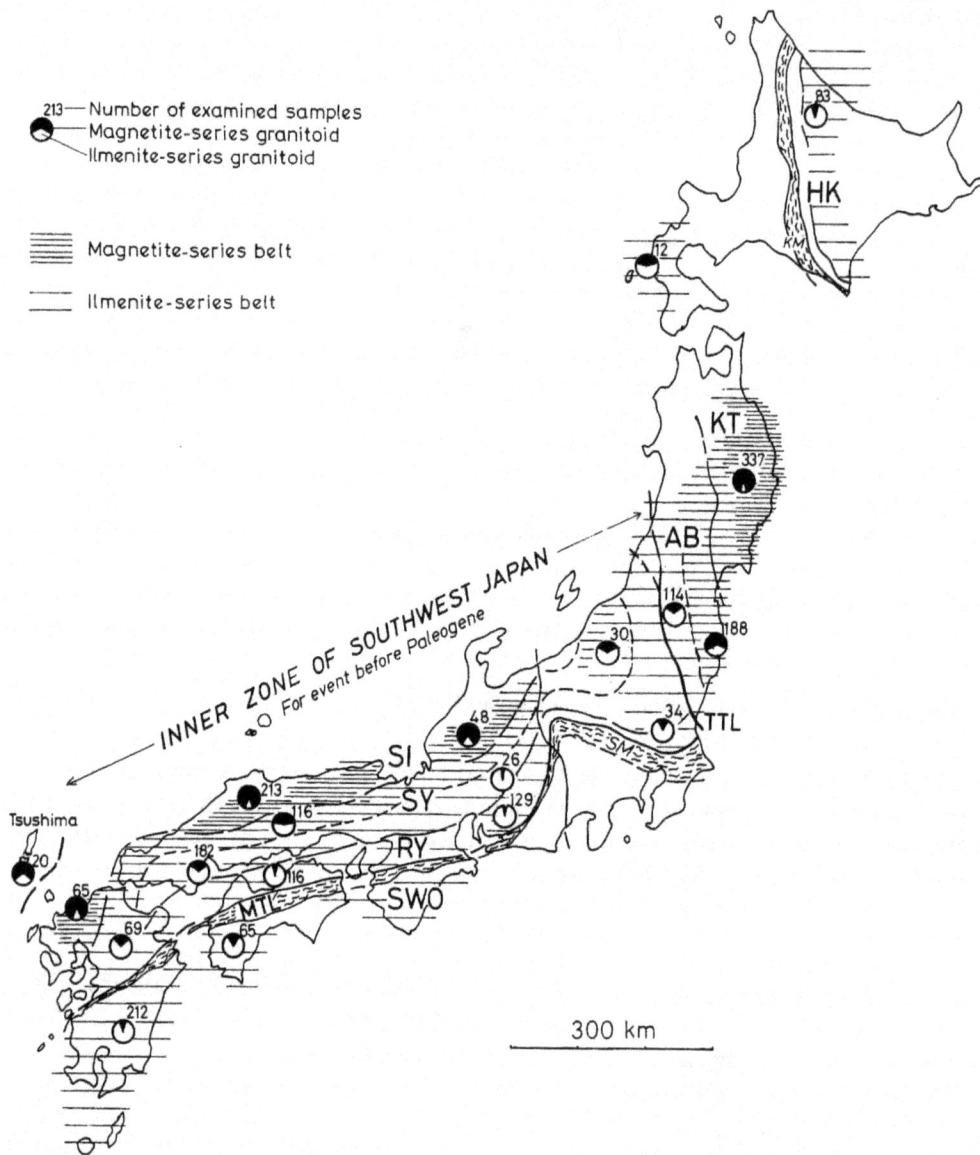

Fig. 1 Distribution of the magnetite-series and ilmenite-series granitoids in Japan. Ratios of the two series in one tectonic unit or one area are shown in circles. The Inner Zone of Southwest Japan is subdivided into northern Kyushu, Chugoku-Kinki, Chubu and Niigata-Kanto districts. Abbreviations: HK, Hidaka belt (Tertiary); KT, Kitakami belt (early Cretaceous); AB, Abukuma belt (Cretaceous and minor older rocks); RY, Ryoke belt (Cretaceous and minor older rocks); SY, Sanyo belt (Cretaceous-Paleogene); SL, Sanin belt (Cretaceous-Paleogene); SWO, Southwestern outer belt (Miocene); TTL, Tanakura tectonic line; MTL, Median tectonic line; SM, Sanbagawa metamorphic belt; KM, Kamuikotan metamorphic belt. Jurassic Funatsu granitoids and Miocene granitoids of the Green Tuff belt are not shown. Both consist of magnetite-series rocks. Tsushima is Miocene and probably an independent belt.

mineralogical characteristics of the two series of granitoids, and discusses some genetic implications for the formation of granitoids and proximal types of ore deposits. Detailed magnetic and mineralogical studies will be given in separate papers. Studied granitic rocks of both concordant and discordant types are divided into several belts as illustrated in Fig. 1.

Definition, Rock Types and Distribution

The magnetite-series and ilmenite-series granitoids were distinguished by the presence or absence, respectively, of magnetite in polished sections. As little as one grain of magnetite was sufficient, by definition, to classify a granitoid as belonging to the magnetite-series. Amounts of magnetite in the magnetite-series granitoids are not constant at a given silica content. If the magnetite content of a rock is low, one polished section may not be sufficient to determine whether or not the hand specimen belongs to the magnetite-series. However, most of magnetite-series granitoids studied had several grains of magnetite in one polished section. If one pluton or one geotectonic belt contains more magnetite-series rocks than ilmenite-series rocks, it is called magnetite-series pluton or magnetite-series belt.

Most of the Japanese granitoids are of the calc-alkaline suite of Cretaceous to Miocene age, and consist of hornblende-biotite granodiorite, biotite monzogranite and subordinate hornblende-biotite tonalite. Pyroxene may be present in mafic phases. The granitoids belong to the magnetite-series or ilmenite-series, as illustrated in Fig. 1. Monzogranite having roughly equal amounts of biotite and muscovite is most abundant in the Ryoke belt but is rare in the other belts. The biotite-muscovite and muscovite granites are free of magnetite as is also true of two-mica granodiorite in the Sierra Nevada of the United States (FD-20 of DODGE et al, 1969). But magnetite was reported in some two-mica granites of the Yenshanian cycle in southern China (WANG et al., 1975).

Alkalic and subalkalic plutonic rocks are very rare in the Japanese Islands. However, they do occur in limited extent in both the magnetite-series and ilmenite-series belts and are very magnetite-rich. The former examples are Zone IV plutons of KATADA (1974) of the Kitakami belt and monzonitic rocks at Hamanaka in easternmost Hokkaido (FUJIWARA, 1959). In the Kitakami belt the alkaline rocks have higher magnetite contents than the calc-alkaline rocks in the same belt.

Syenitic rocks of possible metasomatic origin are present sporadically in the ilmenite-series belt of southwestern Japan. Magnetite occurs as accessary constituents or as massive ores. The Ashizuri-misaki pluton is an exception to others of the southwestern outer belt in as much as it is magnetite-bearing. This pluton consists of monzogranite, syenogranite, quartz syenite and syenite. Residual iron sand deposits (HAYASHI et al., 1969) were formed by weathering of the pluton. Small lens-shaped magnetite deposits are known to occur in magnetite-free areas of the Sanyo belt, as exemplified at Yashiro, Okayama Prefecture (HENMI & NUMANO, 1966) and at Daniwa, Hiroshima Prefecture (SOEDA, 1964). Wherever they occur, syenitic rocks are present within calc-alkaline biotite monzogranite. Syenite is the host for the magnetite deposits.

Distribution of the magnetite-series and ilmenite-series granitoids on a regional scale in the Japanese Islands is illustrated in Fig. 1. Of the largest batholithic exposure of the Inner Zone of Southwest Japan (Cretaceous to Paleogene), granitoids of the Ryoke metamorphic belt and the southern part of the Sanyo belt are generally magnetite-free. Magnetite may or may not be present in those of the northern part of the Sanyo belt. The Sanin belt consists of the magnetite-series granitoids. The magnetite contents of the magnetite-series granitoids in these two belts gradually increase to north. The famed residual iron sand deposits in the Sanin belt, which had supplied raw materials for iron industry for most periods of Japan's history, were formed by weathering of these magnetite-rich granitoids.

In region to the east of the Tanakura tectonic line of Cretaceous granitoids terranes, the ilmenite-series granitoids predominate in the western part but the magnetite-series is dominant in the eastern part. Content of magnetite increases generally to the east. Granitoids of the Kitakami belt are mostly magnetite-bearing. The content is lowest in the Senmaya and Hitokabe plutons which are located in the westernmost part. A few residual iron sand deposits were mined in magnetite-rich magnetite-series plutons in the northern part of the Kitakami belt.

Magnetite is generally absent in Tertiary granitoids of the Hidaka belt. Among Neogene granitoids, almost all of those in the south-western outer belt are of the ilmenite-series, but all of the Green Tuff belt belong to the magnetite-series.

Magnetite-free granitoids do occur in magnetite-bearing belts. In typical magnetite-bearing belts, such as Kitakami and Sanin, 8 and 10 percent, respectively, of examined samples contain no magnetite. In the Kitakami belt, magnetite-free rocks tend to occur at the margin of individual plutons, especially along the eastern margin of the Goyosan pluton. In the Sanin belt, magnetite-free rocks occur in the Mochigase area in Tottori Prefecture and in the Awaradani pluton of Shirakawa area, Gifu Prefecture. In the Mochigase area, the magnetite-free rocks have older K–Ar mineral ages (about 10 m.y.) than the surrounding magnetite-bearing rocks (ISHIHARA & SHIBATA, unpublished data).

Mineralogical Characteristics

The most distinctive feature of the two series of granitoids is the difference in volume percentage of the total opaque minerals (Fig. 2). The magnetite-series granitoids have a much higher content of opaques than do the ilmenite-series granitoids. The magnetite-series granitoids contain 0.1 to 2 volume percent of opaque minerals, as determined by point counting thin sections. Most of this

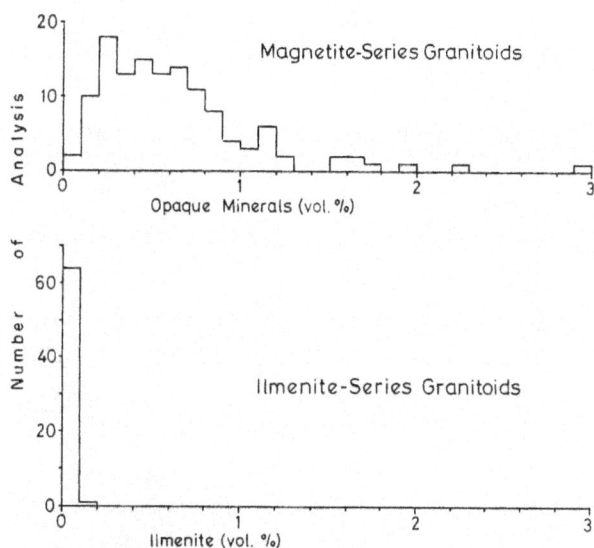

Fig. 2 Histograms of opaque mineral contents. The data source is the same as for Fig. 3.

content, more than 90 percent in general, is magnetite. Magnetite modes are as high as 3 percent in quartz diorite at Katsuraga-dani and 5 percent in quartz gabbro at Zakka, Shimane Prefecture, both of which are hosts for the Akome-type residual iron sand deposits (TSUSUE & ISHIHARA, 1975).

The ilmenite-series granitoids, on the other hand, contain less than 0.1 volume percent of stubby crystals of ilmenite. In another words, the ilmenite-series granitoids are practically free of opaque oxides. In some granitoids, those of the Hidaka belt and the southwestern outer belt for example, pyrrhotite and/or graphite may be more abundant than ilmenite. The areal extent of these rocks is very limited.

Magnetite contents of the magnetite-series rocks decrease with increase of potassium feldspar plus quartz or similar parameters indicative of magmatic differentiation (Fig. 3). The contents also depend upon location. Highest values were obtained in granitiods from the central Sanin belt where many residual iron sand deposits are distributed, and from the Tono-Kurihashi, Kesengawa, Miyako and Oura plutons in the Kitakami belt. Lowest values were found in those from the northern part of the Sanyo belt and the Abukuma belt.

Magnetite in the magnetite-series granitoids is generally euhedral and occurs within mafic silicates or closely associated with them, and in some instances with plagioclase. Blades of ilmenite and hematite may be seen in the magnetite. Ilmenite occurs as small euhedral to subhedral crystal in mafic silicates or sub-hedral to anhedral crystals coexisting with magnetite. Hematite-ilmenite intergrowths of various ratios occurs in small amounts in mafic rocks. Iron sulfide, if any, is generally pyrite. Chalcopyrite may be seen in small amount.

Martitization, hematite replacing magnetite, is common in the magnetite-series rocks, particular in the salic phase. Hematite replaces magnetite along grain margin and (111) cleavage planes. This is prominant in the Sanin belt where complete martitization is also seen. Some of these rocks have depleted values

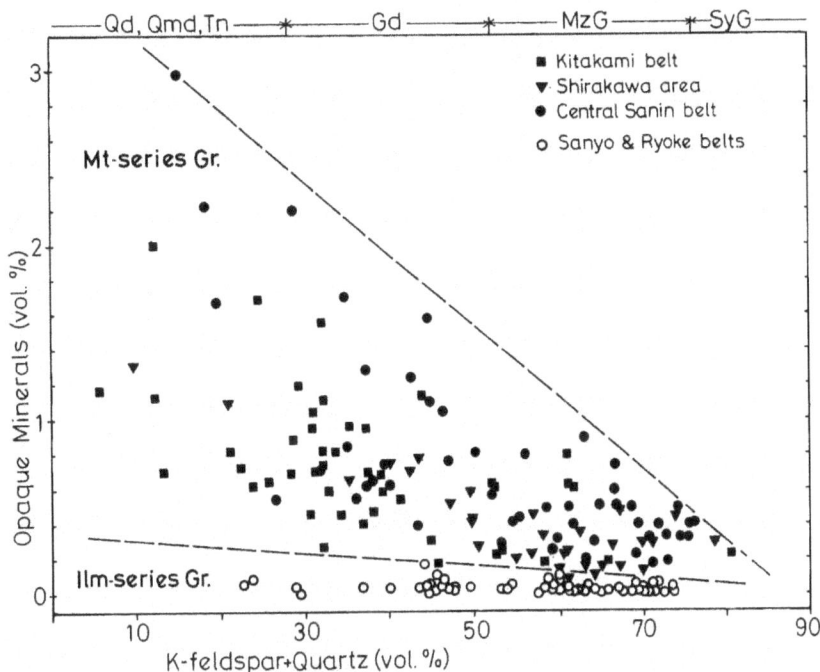

Fig. 3   Modal opaque minerals plotted against modal potassium feldspar plus quartz. All determined by point counter method described in the papers listed below. Ore-microscopic study on selected samples indicates that more than 90 percent of the opaque minerals of the magnetite-series granitoids consists of magnetite. Examined areas for the magnetite-series granitoids are the main plutons (except Hitokabe and Sanmaya plutons) of the Kitakami belt (n = 45, ISHIHARA & SUZUKI, 1974), Shirakawa granitoids (n = 30, ISHIHARA, 1971a), western Tottori Pref. (n= 5 HATTORI & SHIBATA, 1974), eastern Shimane Pref. (n = 38, ISHIHARA, 1971a), and northern Hiroshima Pref. (n = 11, ISHIHARA et al., 1969). Opaque minerals of the ilmenite-series granitoids were determined by an integration apparatus on polished sections. Examined areas are the southern part of the Sanyo belt (n = 34, ISHIHARA, 1971b) and the Ryoke belt of Chubu district (n = 26, ISHIHARA & TERASHIMA, 1977a). Abbreviations for rock names (nomenclature recommended by IUGS subcommision): Qd, quartz diorite; Qmd, quartz monzodiorite; Tn, tonalite; Gd, granodiorite; MzG, monzogranite; SyG, syenogranite.

of δ ^{18}O (ISHIHARA & MATSUHISA, 1977). In very small plutons in the Green Tuff belt (e.g., Ōe mine stock), magnetite is completely converted to hematite, and ilmenite is decomposed to mats of hematite and TiO_2 minerals. Strong martitization and break-down of ilmenite are observed commonly in small granitoid plutons in the western Cascade Range and the San Juan Mountains of the United States. These also exhibit ^{18}O depletion (TAYLOR, 1971, 1974). Magnetites in small stocks in porphyry copper areas of the southwestern United States and the Philippines are more or less martitized.

The ilmenite-series granitoids are completely free of magnetite under the ore-microscope with ordinary (100 ×) magnification. Small euhedral ilmenite occurs in mafic silicates. Secondary ilmenite fills

cleavages and interstices of mafic silicates. Examples are the Hidaka and Ryoke belts, southern part of the Sanyo belt, and the southwestern outer belt.

Sulfide minerals are generally pyrrhotite. The mineral has two types of occurrence, one in the main phase and the other in the most differentiated phase. The first type is common in tonalite that occurs closely associated with migmatite in the Hidaka belt and in small plutons in the southwestern outer belt where exnoliths of shale and sandstone from the intruded Shimanto Supergroup are dominant. The second type fills cavities of aplitic or pegmatitic clots and accompanies pyrite and other sulfides in some instances.

Graphite occurs in the ilmenite-series granitoids but is visible under the ore-microscope(100 ×) only in the migmatitic rocks

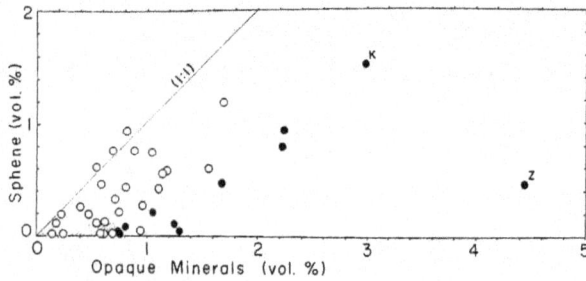

Fig. 4 Contents of sphene and opaque oxide minerals of some sphene-rich magnetite-series granitoids. Open circle, Kitakami belt including plutons of Miyako (n = 11), Yamada (n = 6), Tanohata (n = 4) and Oura (n = 6); solid circle, Kawai-type granitoids (quartz gabbro to monzogranite) from eastern Shimane Pref. Batholithic granodiorite and monzogranite from the same area is shaded. The data source is the same as for Fig. 3.

in the Hidaka belt and the xenoliths in the other magnetite-free belts. Graphite content of the migmatites is generally less than 0.2 wt. percent.

Mafic silicates have a certain correlation with these opaque minerals assemblages. Sphene is commonly present in the magnetite-series rocks. It occurs either as euhedral crystals which can be seen with the naked eye or as secondary aggregated after decomposition of mafic silicates and ilmenite. The amount of of sphene is generally correlated with that of magnetite (Fig. 4), but its occurrence is somewhat erratic. In the Kitakami belt, sphene is dominant in the Miyako (-Yamada), Tanohata and Oura (-Omoe) plutons but is rare in other major plutons. In the Sanin belt, sphene is most concentrated in the Kawai-type fine-grained rocks, at the margins of the coarse-grained batholithic plutons.

Epidote may be seen in the magnetite-series granitoids. It occurs generally as a secondary mineral after mafic silicates as, for example, in small plutons in the Taro zone of the Kitakami belt. Primary epidote is visible with the naked eye in the Ecstall pluton in the Coast Range batholith, Canada. This pluton consists of slightly magnetite-bearing granitoids. Since iron was consumed to form Fe–Ti oxides, biotite and hornblende are generally rich in magnesium in the magnetite-

series granitoids (see Fig. 5).

Genetic Considerations

Among several factors controlling crystallization of ferromagnesian minerals of granitoids (oxygen fugacity, temperature, bulk composition etc.), oxygen fugacity seems to be the most important variable in forming the magnetite-series and ilmenite-series granitoids. The bulk total iron contents of the two series are more or less similar at given silica contents but their ferric/ferrous ratios are distinctly different (ISHIHARA, 1971a; TSUSUE & ISHIHARA, 1974; ISHIHARA & TERASHIMA, 1977a). A permissible conclusion is that the magnetite-series granitoids were formed under conditions of higher oxygen fugacity than the ilmenite-series granitoids (ISHIHARA, 1971a),

Table 1 Average modal compositions of selected granodiorite and monzogranite batholiths (weight %)

	Granodiorite		Monzogranite	
	(1)	(2)	(3)	(4)
Plagioclase	51.2	49.8	33.6	33.9
Potassium feldspar	12.6	10.6	29.5	28.0
Quartz	23.2	25.0	34.5	34.6
Hornblende	4.5	4.2	none	none
Biotite	7.2	10.0	1.6	3.0
Magnetite or ilmenite	1.2	0.04	0.8	0.04
Sphene	0.2	none	nil	none
Others	nil	0.4[*1]	nil	0.5[*2]
Total Fe–Ti oxides/Fe–Mg silicates	0.10	0.00	0.50	0.01

Remnant bulk chemistry after subtracting Fe_2O_3 and FeO to form the Fe–Ti oxides

$Fe^{+3}/Fe^{+3} + Fe^{+2}$	0.28	0.16	0.48	0.19
Fe/Fe + Mg	0.31	0.39	0.47	0.44

(1) Magnetite-series, Daito granodiorite (n = 8, after ISHIHARA, 1971a). (2) Ilmenite-series, Sumikawa granodiorite (n = 4, after ISHIHARA & TERASHIMA, 1977a). (3) Magnetite-series, Yokota monzogranite (n = 7, after ISHIHARA, 1971a). (4) Ilmenite-series, Toki monzogranite (n = 5, after ISHIHARA & TERASHIMA, 1977a). [*1] Mainly clino-pyroxene, [*2] Mostly fluorite. $Fe^{+3}/Fe^{+3} + Fe^{+2}$ and Fe/Fe + Mg ratios indicate approximately those of hornblende + biotite or biotite

assuming that common granitoids were solidified in a closed environment (TSUSUE & ISHIHARA, 1974).

Table 1 gives modal composition of typical granodiorite and monzogranite batholiths. It is clear that the ratios of opaque oxides/ferromagnesian silicates is distinctly different between the two series of granitoids. Assumed $Fe^{+3}/Fe^{+3} + Fe^{+2}$ ratios of the silicates, which are obtained from the bulk ratios after subtracting amounts of Fe_2O_3 and FeO allocated to the modal oxides, are much higher in the magnetite-series granitoids than in the ilmenite-series ones, especially in those granitoids containing no other ferromagnesian silicates than biotite.

Biotite analyses from the magnetite-series granitoids coexisting with potassium feldspar and magnetite plot above the Ni–NiO buffer on WONES & EUGSTER's (1965) Fe^{+3} — Fe^{+2} — Mg diagram for biotite (DODGE et al., 1969; KANISAWA, 1972, 1974). Although the biotite-potassium feldspar-magnetite assemblage does not occur in the ilmenite-series granitoids, their biotites generally plot below (TSUBOI et al., 1938) or around (HONMA, 1974) the Ni–NiO buffer. It appears that a boundary between the two series of granitoids is near the Ni–NiO buffer (Fig. 5), which was also suggested by SHIMAZAKI (1976) and TSUSUE (1976).

Water is important in regulating oxygen fugacity in igneous processes (OSBORN, 1959, 1962; WONES & EUGSTER, 1965). CZAMANSKE and WONES (1973) pointed out that H_2O can act as an oxidizing medium through dissociation and loss of H_2 only after its separation from a silicate melt. Such a model can explain the magnetite-series granitoids of near surface intrusion in the Sanin, Kitakami and Green Tuff belts. Some external sources of oxidizing medium, such as free oxygen from air (MURAKAMI, 1969) and meteric ground water, may have accelarated the oxidation in the later stage of crystallization.

The above explanation meets difficulty with the small discordant plutons in the southwestern outer belt and the Hidaka belt. These are shallow level intrusions, some

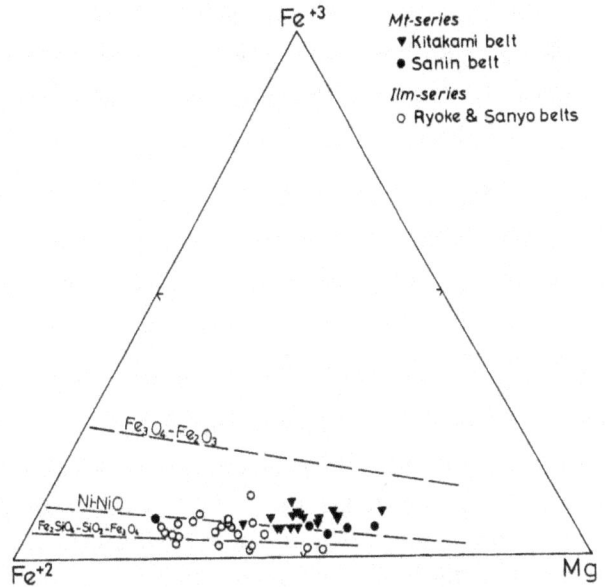

Fig. 5 Fe^{+3}–Fe^{+2}–Mg relation of Japanese biotites of calc-alkaline suite. Biotites from small stocks are excluded. Data source: KANISAWA (1972, 1974) for the Kitakami belt including Tono, Hitokabe, Kesengawa, Oura, Omoe and Taro; KANISAWA (1976) for the Sanin belt; TSUBOI et al. (1938) and HONMA (1974) for the Ryoke and Sanyo belts. Only chemical analyses whose magnetite bearing and free nature is confirmed from the listed localities are plotted. Broken lines represent compositions of buffered biotites in the ternary system $KFe_3^{+3}AlSi_3O_{12}H_{-1}$–$KFe_3^{+2}AlSi_3O_{10}(OH)_2$–$KMg_3AlSi_3O_{10}(OH)_2$ depicted by WONES & EUGSTER (1965).

of which are related to cauldron subsidence, yet they are free of magnetite. These granitoids contain evidence that their magmas were contaminated or had interacted with pelitic rocks which contain graphite and/or related carbonaceous matter. Moreover, magnetite-free rocks occur widely in the high T/P type metamorphic terranes. Thus, buffering of oxygen fugacity by graphite may be worth considering.

The role of graphite as a reducing agent in the earth's crust has been emphasized by metamorphic petrologists (e.g., MIYASHIRO, 1964). In common pelitic rocks, magnetite and graphite do not occur together and the assemblage graphite-pyrrhotite-ilmenite is common (KANEHIRA et al., 1964; MARIKO et al., 1975). FRENCH and EUGSTER (1965) stressed the importance of graphite-gas equilibrium on oxygen fugacity in both igneous and metamorphic processes, and further mentioned (p. 1537) that "Because of the low gram-molecular weight of graphite, even trace amounts of graphite will exert a very large buffering effect with respect to changes in the composition of the gas phase."

Foliated granitoids of the metamorphic terranes are generally free of magnetite and contain ilmenite only in very small amount. The assemblage graphite-pyrrhotite-ilmenite is seen in the migmatitic rocks. These are observed in the Ryoke and Hidaka belts. However, foliated granitoids of the Abukuma belt contain magnetite in some portions of the highest grade zone. Chemical analyses indicate that the metamorphic rocks of the Abukuma belt have much lower carbon contents than those of the Ryoke belt (Table 2). It is considered, therefore, that the magnetite-bearing and magnetite-free granitoids are a consequence of different graphite contents in the intruded metamorphic rocks.

Many pelitic xenoliths are found in massive, discordant-type granitoids in the southwestern outer belt. Graphite is commonly seen in the

Table 2 Average carbon contents of pelitic and semi-pelitic metamorphic rocks from the Ryoke and Abukuma belts

Area	n	C(wt %)	Reference
Ryoke belt			
Northern Kiso area	9	0.96	KATADA et al. (1964)
Takato area	9	0.79	MIYASHIRO (1964)
Tsukuba area	8	0.78	UNO (1961)
Abukuma belt			
Takanuki series	7	0.24	MIYASHIRO (1964)
	6	0.24	This study
Gozaisho series	15	0.12	MIYASHIRO (1964)
	13	0.10	This study

Note: Averages for mafic metamorphic rocks of the Abukuma belt, which are the majority in the belt, are 0.06%C (n = 3) for the Takanuki series and 0.08%C (n = 14) for the Gozaisho series. Samples analyzed in this study are those listed in ISHIHARA et al. (1973). Analyst for this study was T. NISHIMURA, Geological Survey of Japan

xenoliths and a graphite-bearing xenolithic block at Manguro in the Minami-osumi pluton was once mined for both graphite and associated sulfide minerals. Seven analyses of shales from the intruded Shimanto Supergroup contain an average of 0.85% C. Thus it is obvious that the granitic magma interacted with graphite of the intruded sedimentary rocks.

Interaction during magma emplacement is considered to have happened at several stages for the different types of contained xenoliths and even at the beginning of the magmatic history for the following reasons. In the southwestern outer belt, some granitoids intrude Sanbagawa metamorphic rocks which are predominantly mafic meta-igneous rocks, yet the granitoids are free of magnetite. In the Hidaka belt, foliated tonalites in the axial zone are associated with migmatite and pelitic metamorphic rocks. Other granitoids occur as discordant stocks and pelitic xenoliths are not common, yet these plutons are composed of magnetite-free granitoids.

Distribution of the magnetite-bearing and magnetite-free suites is observed on a regional scale, and magnetite-free rocks are generally salic in composition and are rich in F, Rb, Li, Sn and Be (ISHIHARA & TERASHIMA, 1977a,b). Thus it is speculated that most of the ilmenite-series granitoids were generated in salic continental materials and interacted with near-surface rocks before their solidification. This speculation is supported by oxygen isotope study (ISHIHARA & MATSUHISA, 1977).

Of the Japanese granitoids, the Miocene granitoids of the Green Tuff belt in northeastern Japan and the Cretaceous granitoids of the Kitakami belt were derived from a deep source region, probably from the upper mantle, as evidenced by their tectonic setting and initial strontium ratios (SHIBATA & ISHIHARA, 1976). These granitoids contain magnetite. It is concluded that such granitic magmas originated in deep regions where no carbonaceous matter is available form the magnetite-series granitoids. Sporadic distribution of magnetite free-rocks especially at the margins of magnetite-series plutons may be a result of minor interaction of the original magma with surrounding pelitic rocks.

Relation to Metallogenic Provinces

Tin, tungsten, molybdenum and porphyry copper deposits are known to occur spatially close to granitic rocks. These deposits can be correlated with the magnetite-series and ilmenite-series. In the Japanese metallogenic provinces (ISHIHARA & SASAKI, 1973), molybdenite and scheelite-gold deposits are distributed in the Sanin and Kitakami belts where related granitoids are of the magnetite-series. Tin and greisen-type wolframite deposits are located in the southwestern outer belt and

the southern part of the Sanyo belt where related granitoids are of the ilmenite-series.

Similar results from other regions around the world are shown in Fig. 6. In the regions of tin and greisen-type wolframite deposits, such as Erzgebirge, northern Portugal, Tasmania, Malay Peninsula, Seward Peninsula and Mt. McKinley area in Alaska, and Round Mountain area in Nevada, the granitoids consist of biotite granite or biotite-muscovite granite and contain very small amount of ilmenite; hence they belong to the ilmenite-series granitoids. Some contain magnetite but the content is as low as a few grains in one hand specimen. "Normal" magnetite-series granitoid occurs in a small tin granite stock of the Serpentine Hot

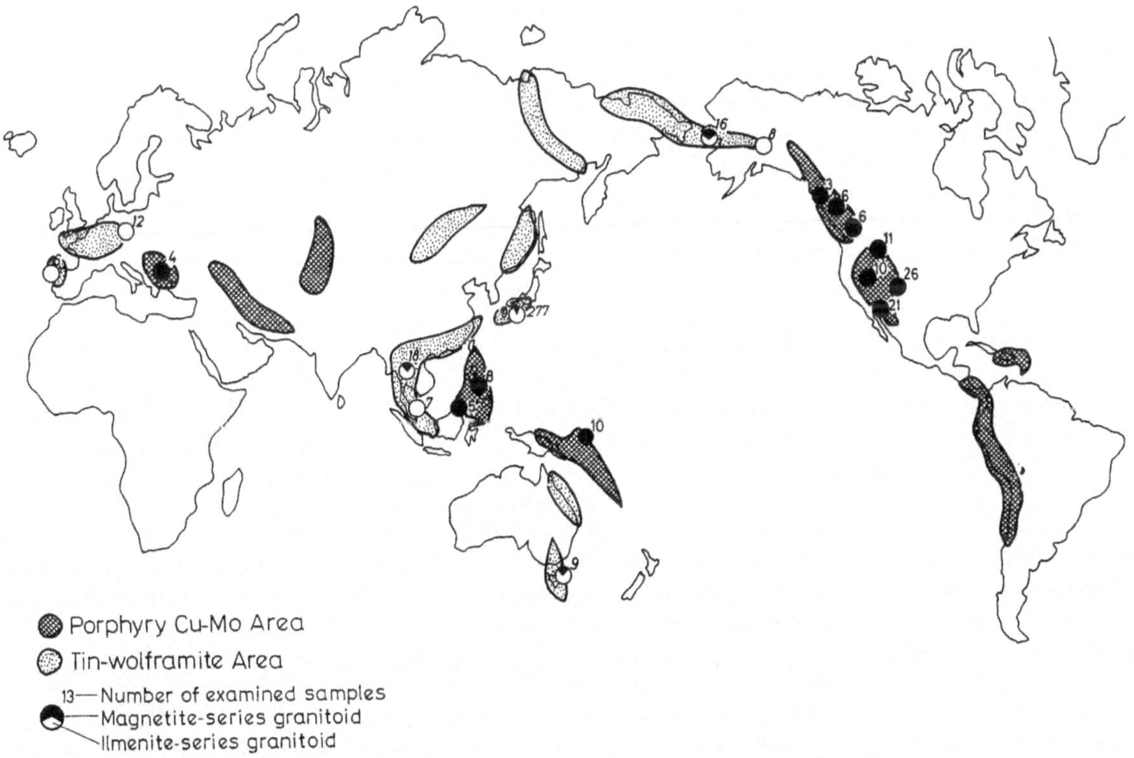

⊚ Porphyry Cu-Mo Area
◉ Tin-wolframite Area
13—Number of examined samples
 —Magnetite-series granitoid
 —Ilmenite-series granitoid

Fig. 6   Distribution of the magnetite-series and ilmenite-series granitoids in the major porphyry copper-molybdenum provinces and greisen-type tin-wolframite provinces. Examined samples include both small stocks related to the mineralization (designated as stock) and general rocks in the proper mining areas (designated as regional). *Cu–Mo areas*: Arizona (n = 21): Cornelia pluton (n = 8, stock), Patagonia and Santa Rita Mts. (n = 8, stock), Esperanza-Sierrita and Silver Bell (n = 5, stock); Front Range (n = 26): Climax (n = 3, stock), Questa (n = 23, stock); Nevada and Utah (n = 10): Bingham (n = 5, stock), Robinson (n = 2, stock), Yerington (n = 3, stock); Butte (n = 11, regional), Highland Valley (n = 6, regional), Endako (n = 6, regional), Alice Arm (n = 3, stock); Philippine (n = 8): Atlas (n = 2, stock), Sto. Thomas and Sto. Nino (n = 4, stock), Sipalay (n = 2, stock); Mamut (n = 5, regional); Panguna (n = 10, stock); Medet, Bulgaria (n = 4, stock). *Sn–W areas*: Mt. McKinley (n = 8, stock), Seward Peninsula (n = 16, stock), Southwestern outer belt (n = 277, regional and stock), Northern Thailand (n = 18, regional), Malaysia (n = 7, regional), Tasmania (n = 9, regional), Erzgebirge (n = 12, regional), Northern Portugal (n = 6, regional).

Springs area at Seward Peninsula in Alaska but its distribution is limited to the Zone 2 phase of porphyritic biotite granite.

Scheelite-gold deposits occurring in non-calcareous host rocks which are not accompanied by greisen-type alteration are rather rare, but are known in the southern Kitakami belt and Sierra Nevada (e.g., Atolia district, LEMMON & TWETO, 1962). Since the major part of the Sierra Nevada batholith seems to contain magnetite (DODGE et al., 1969), this mineralization is characteristic of the magnetite-series granitoids.

In porphyry copper deposits, magnetite, hematite, pyrite (instead of pyrrhotite) and anhydrite occur within the orebodies. These minerals suggest that the related intrusive rocks should be magnetite-series granitoids and, indeed, all examined specimens from various localities listed in Fig. 6 fall within the category of the magnetite-series rocks. A possible exception may be the Copper Canyon deposit (THEODORE & BLAKE, 1975) where pyrrhotite occurs in an altered stock. Porphyry-type molybdenite deposits are also related to the magnetite-series granitoids.

In the Mesozoic and Cenozoic orogenic belts in the Circum-Pacific region, porphyry-type copper and molybdenum deposits are abundant in North and South America, whereas tin-tungsten deposits of the greisen type are dominant in East Asia and Alaska. This regional pattern may imply that magnetite-series granitoids are more abundant on the American side than on the Asian side. Samples were examined from various widespread units ranging in age from Triassic to late Cretaceous from the Malaysia-Thailand region. Their ratios of magnetie-series/ilmenite-series rocks is very similar to that of the Japanese tin-wolframite region. Mesozoic granitoids of southeastern China (Nanking Univ., 1974) have similarity to those of the Malay Peninsula and the southern part of the Sanyo belt. Thus it seems likely that the magnetite-series/ilmenite-series ratio is below 1 in the Asian side and is above 1 in the American side of the Mesozoic to early Cenozoic orogeny belts. This difference

appears to be a prominent feature in the Circum-Pacific region and implies that a different tectonic history developed on either side of the Pacific Ocean, although both sides are located above the consuming plate margins.

The ratio of the magnetite-series/ilmenite-series granitoids may also be related to evolution of lithosphere and hydrosphere in the earth's history. Low Fe_2O_3/FeO ratios of Precambrian granitoids in the Canadian shield (FAHRIG & EADE, 1968) suggest that most of the Archean rocks are composed generally of the ilmenite-series. If this assumption is valid, a general scacity of porphyry copper deposits in Precambrian terranes may be explained by the general paucity of the magnetite-series granitoids in the older continental crust.

Concluding Remarks

Granitoid batholiths and stocks can be divided into (i) magnetite-bearing magnetite-series and (ii) magnetite-free ilmenite-series, and are considered to form as a result of exposure to different oxygen fugacity during the life of the granitic magmas. The magnetite-series granitic magma may have been generated at great depths where no carbonaceous material exists, whereas the ilmenite-series magma may have originated at a shallow level where small amounts of crustal carbon are present in the host country rocks. Both magmas were modified slightly in terms of oxygen fugacity at the shallowest level at which they solidified, yet the original characteristics are retained even through post-magmatic processes. Thus the two series of granitoids have an important bearing on both petrogenesis and metallogenesis. Recognition of these two fundamental types of granitoids is very useful in exploration for mineral deposits of granitic affinity.

Identification of the two series is easy in the field with the aid of a hand magnet and by observing heavy minerals concentrated on weathered surfaces. Typical rocks can be recognized by the mineralogical characteristics aforementioned. The most difficult cases of distinguishing the two types may be encountered in granitic massifs where biotite granites of low magnetite content are monotonously exposed (e.g., Hiroshima granite). In this case, quantitative analysis is necessary in the laboratory and one polished section per outcrop may not be enough. Magnetic susceptibility measurement (KANAYA & ISHIHARA, 1973; ISHIHARA & KANAYA, in preparation) and ferric/ferrous ratios of bulk analysis of granitoids are useful adjuncts to proper identification.

Acknowledgement: The writer is indebted greatly to the following people who provided some samples from foreign countries: Dr. G. R. BALCE, Bureau of Mines, Manila; Prof. J. M. GUILBERT, University of Arizona; Drs. W. E. HALL, T. L. HUDSON, B. L. REED, D. R. SHAWE and T. G. THEODORE, U. S. Geological Survey; Prof. W. C. KELLY, University of Michigan;Dr. C. NISHIWAKI, International Mineral Resources Development; Dr. J. R. RICHARDS, Australian National University; Dr. S. SUENSILPONG, Department of Mineral Resources, Bangkok; Prof. H. P. TAYLOR, Jr., California Institute of Technology; among my colleagues, T. NOZAWA and K. SHIBATA. Thanks are also due to Dr. G. W. WALKER, U. S. Geological Survey at Menlo Park and Prof. C. MEYER, University of California at Berkeley for generous permission to use their facilities to study some of the North American samples during a short visit in summer 1975. The visit was supported by a research grant from the Science and Technology Agency of Japan. Critical review by Prof. S. D. SCOTT, University of Toronto, has improved English of the original manuscript.

References

CZAMANSKE, G. K. and MIHALIK, P. (1972): Oxidation during magmatic differentiation, Finnmarka complex, Oslo area, Norway: Part 1, The opaque oxides. Jour. Petrol., **13**, 493 ~ 509.

CZAMANSKE, G. K. and WONES, D. R. (1973): Oxidation during magmatic differentiation, Finnmarka complex, Oslo area, Norway: Part 2, The mafic silicates. Jour. Petrol., **14**, 349 ~ 380.

DODGE, F.C.W., SMITH, V. C. and MAYS, R. E. (1969): Biotites from granitic rocks of the central Sierra Nevada batholith, California. Jour. Petrol., **10**, 250 ~ 271.

FAHRIG, W. F. and EADE, K. E. (1968): The chemical evolution of the Canadian Shield. Canadian Jour. Earth Sci., **5**, 1247 ~ 1252.

FRENCH, B. M. and EUGSTER, H. P. (1965): Experimental control of oxygen fugacities by graphite-gas equilibriums. Jour. Geophy. Res., **70**, 1529 ~ 1539.

FUJIWARA, T. (1959): On the igneous activities and the ore deposit at Hamanaka area, Hokkaido. Jour. Japan. Assoc. Min. Petr. Econ. Geol., **43**, 208 ~ 214.

HATTORI, H. and SHIBATA, K. (1974): Concordant K–Ar and Rb–Sr ages of the Tottori granite, western Japan. Bull. Geol. Surv. Japan, **25**, 157 ~ 173.

HAYASHI, S., ISHIHARA, S. and SAKAMAKI, Y. (1969): Uranium in the decomposed granitic rocks at the cape Ashizuri, Kochi Prefecture, with special reference to the green uranothorite. Geol. Surv. Japan, Pept. 232, 93 ~ 103.

HENMI, K. and NUMANO, T. (1966): Feldspar dikes in the Yamate-Kasaoka area, Okayama Prefecture. Rept. Earth Sci. Res. Okayama Univ., 1, 111 ~ 119.

HONMA, H. (1974): Chemical features of biotites from metamorphic and granitic rocks of the Yanai district in the Ryoke belt, Japan. Jour. Japan. Assoc. Min. Pet. Econ. Geol., **69**, 390 ~ 402.

ISHIHARA, S. (1971a): Major molybdenum deposits and related granitic rocks in Japan. Geol. Surv. Japan, Rept. 239, 1 ~ 178.

ISHIHARA, S. (1971b): Modal and chemical composition of the granitic rocks related to the major molybdenum and tungsten deposits in the Inner Zone of Southwest Japan. Jour. Geol. Soc. Japan, **77**, 441 ~ 452.

ISHIHARA, S. (1975): Acid magmatism and mineralization—Oxidation status of granitic magma and its relation to mineralization—. Marine Sci. Monthly, **7**, 756 ~ 759.

ISHIHARA, S. and SASAKI, A. (1973): Metallogenic map of Japan: Plutonism and mineralization (1), molybdenum, tungsten and tin. 1/2,000,000 Map Series, 15–1, Geol. Surv. Japan.

ISHIHARA, S. and SUZUKI, Y. (1974): Modal compositions of Cretaceous granitic rocks in the Kitakami Mountains. Geol. Surv. Japan, Rept. 251, 23 ~ 42.

ISHIHARA, S. and TERASHIMA, S. (1977a): Chemical variation of the Cretaceous granitoids across southwestern Japan. —Shirakawa-Toki-Okazaki

304   S. Ishihara   Mining Geology:

transection—. Jour. Geol. Soc. Japan, **83**, 1 ~ 18.

Ishihara, S. and Terashima, S. (1977b): Chlorine and fluorine contents of granitoids as indicators for base metal and tin mineralizations. Mining Geol., **27**, 191 ~ 199.

Ishihara, S. and Matsuhisa, Y. (1977): The magnetite-series/ilmenite-series granitoids and their $^{18}O/^{16}O$ ratios (abs). Abstract Issue, 84th Annual Mtg, Geol. Soc. Japan, 82.

Ishihara, S., Komura, K. and Murakami, T. (1969): Source rocks of Miocene bedded-type uraniferous deposits in northern Miyoshi district and genesis of uranium anomalies at Myoga, Shobara city, Hiroshima Pref., Japan. Bull. Geol. Surv. Japan, **20**, 161 ~ 172.

Ishihara, S., Hattori, H., Sakamaki, Y., Kanaya, H., Sato, T., Mochizuki, T. and Terashima, S. (1973): Lateral chemical variation of the granitic and metamorphic rocks across the central Abukuma highland—With emphasis on the contents of uranium, thorium and potassium. Bull. Geol. Surv. Japan, **24**, 269 ~ 284.

Kanaya, H. and Ishihara, S. (1973): Regional variation of magnetic susceptibility of the granitic rocks in Japan. Jour. Japan. Assoc. Min. Pter. Econ. Geol., **68**, 211 ~ 224.

Kanehira, K., Banno, S. and Nishida, K. (1964): Sulfide and oxide minerals in some metamorphic terranes in Japan. Jour. Geol. Geogr., **35**, 175 ~ 191.

Kanisawa, S. (1972): Coexisting biotites and hornblendes from some granitic rocks in southern Kitakami Mountains, Japan. Jour. Japan. Assoc. Min. Pet. Econ. Geol., **67**, 332 ~ 344.

Kanisawa S. (1974): Granitic rocks closely associated with the lower Cretaceous volcanic rocks in the Kitakami Mountains, Northeast Japan. Jour. Geol. Soc. Japan, **80**, 355 ~ 367.

Kanisawa, S. (1976): Chemistry of biotites and hornblendes of some granitic rocks in the San'in Zone, Southwest Japan. Jour. Geol. Soc. Japan, **82**, 543 ~ 548.

Katada, M. (1974): Granitic rocks in the southern Kitakami Mountains and zonal arrangement of granitic rocks in the entire Kitakami Mountains. Geol. Surv. Japan, Rept. 251, 121 ~ 133.

Katada, M., Isomi, H., Omori, E. and Yamada, T. (1964): Chemical composition of Paleozoic rocks from northern Kiso district and of Toyoma clay-slates in Kitakami mountainland: Supplement. Carbon and carbon dioxide. Jour. Japan. Assoc. Min. Pet. Econ. Geol., **52**, 217 ~ 221.

Lemmon, D. M. and Tweto, O. L. (1962): Tungsten in the United States. U.S. Geol. Surv. Miner. Invest. Res. Map MR-25, 1 ~ 25.

Mariko, T., Tanaka, K. and Itaya, T. (1975): Oxide and sulphide minerals in pelitic and psammltic schists from the Nagatoro district, Saitama prefecture, Japan. Jour. Japan. Assoc. Min. Petr. Econ. Geol., **70**, 413 ~ 423.

Miyashiro, A. (1964): Oxidation and reduction in the Earth's crust with special reference to the role of graphite. Geochim. Cosmochim. Acta, **28**, 717 ~ 729.

Murakami, N. (1969): Two contrastive trends of evolution of biotite in granitic rocks. Jour. Japan. Assoc. Min. Pet. Econ. Geol., **62**, 223 ~ 248.

Nanking Univ. (1974): Granitic rocks of different geological periods of southeastern China and their genetic relations to certain metallic mineral deposits. Sci. Sinica, **17**, 55 ~ 72.

Osborn, E. F. (1959): Role of oxygen pressure in the crystallization and differentiation of basaltic magma. Amer. Jour. Sci., **257**, 609 ~ 647.

Osborn, E. F. (1962): Reaction series for subalkaline igneous rocks based on different oxygen pressure conditions. Amer. Miner., **47**, 211 ~ 226.

Shibata, K. and Ishihara, S. (1976): Regional variation of the initial $^{87}Sr/^{86}Sr$ ratio of the Japanese plutonic rocks (abs). Abstract Issue, 83rd Annual Mtg., Geol. Soc. Japan, 307.

Shimazaki, H. (1976): Granitic magmas and ore deposits (2) Oxidation state of magmas and ore deposits. Mining Geol. Spec. Issue, 7, 25 ~ 35.

Soeda, A. (1964): Ore deposits of Hiroshima Prefecture. Explanetary text for 1/200,000 scale map, Hiroshima Pref., 125 ~ 154.

Taylor, H. P. Jr. (1971): Oxygen isotope evidence for large-scale interaction between meteoric ground waters and Tertiary granodiorite intrusions, western Cascade Range, Oregon. Jour. Geophy. Res., **76**, 7855 ~ 7874.

Taylor, H. P. Jr. (1974): Oxygen and hydrogen isotope evidence for large-scale circulation and interaction between ground waters and igneous intrusions, with particular reference to the San Juan volcanic field, Colorado. Carnegie Inst. Washington Pub. 634, 299 ~ 342.

Theodore, T. G. and Blake, D. W. (1975): Geology and geochemistry of the Copper Canyon porphyry copper deposits and surrounding area, Lander county, Nevada. U.S. Geol. Surv. Prof. Paper 798-B, 1 ~ 86.

Tsuboi, S., Sugi, K. and Iwao, S. (1938): On rock-forming biotites from Japan. Jour. Geol. Soc. Japan, **45**, 453 ~ 455.

TSUSUE, A. (1976): Granitic magmas and ore deposits (1) Especially those of Southwest Japan. Mining Geol. Spec. Issue, 7, 15 ~ 24.

TSUSUE, A. and ISHIHARA, S. (1974): The iron-titanium oxides in the granitic rocks of Southwest Japan. Mining Geol., **24**, 13 ~ 30.

TSUSUE, A. and ISHIHARA, S. (1975): "Residual" iron-sand deposits of Southwest Japan. Econ. Geol., **70**, 706 ~ 716.

UNO, T. (1961): Metamorphic rocks of the Tukuba district, Ibaraki Prefecture. Jour. Geol. Soc. Japan, **67**, 228 ~ 236.

WANG, L., ZHANG, Y. and LIU, S. (1975): Multiple emplacements and some geochemical characteristics of the Zhuguangshan granitic batholith, southern China. Geochimica, 3, 189 ~ 201.

WONES, D. R. and EUGSTER, H. P. (1965): Stability of biotite: Experiment, theory, and application. Amer. Jour. Sci., **50**, 1228 ~ 1272.

磁鉄鉱系花崗岩類とチタン鉄鉱系花崗岩類

石 原 舜 三

要　旨

表題の2組の花崗岩類について，主として鏡下観察結果から構成鉱物の特徴が記載され，分類の基準・両者の分布・成因・鉱化作用との関連性などがのべられた．2組の花崗岩類は一般の鏡下観察（100×）で磁鉄鉱が認められるか否かの点で分類され，磁鉄鉱系花崗岩類は0.1－2容量％の磁鉄鉱とごく少量のチタン鉄鉱を有し，チタン鉄鉱系花崗岩類は0.1容量％以下のチタン鉄鉱を伴うにすぎない．すなわち，両者は苦鉄質珪酸塩鉱物とFe-Ti酸化鉱物の量比において著しく異なり，チタン鉄鉱系花崗岩類はFe-Ti酸化鉱物に欠ける系列とみなしてよい．

このFe-Ti酸化鉱物に欠ける事実から，チタン鉄鉱系花崗岩類が磁鉄鉱系花崗岩類より低い酸素フュガシティの条件下で生成されたものと推論された．このように考えると，2組の花崗岩類にそれぞれ特徴的に認められる他の苦鉄鉱物や硫化鉱物の組合せが説明し易い．両者の酸素フュガシティを定量的に推定する共通の鉱物組合せは得られていないが，黒雲母のFe^{+3}/Fe^{+3}＋Fe^{+2}比から両者の境界はほぼNi-NiOバッファー付近と考えられた．

花崗岩類の生成時の酸素フュガシティを規制する要因としては花崗岩質マグマの発生から固結に至る過程における炭質物によるバッファーが重視され，H$_2$Oの解離とH$_2$の逸散は磁鉄鉱系花崗岩類の一部について考慮された．磁鉄鉱系花崗岩類は炭質物が存在しない深所起源であり，チタン鉄鉱系花崗岩類は炭質物を伴う大陸地殻起源であろうと考えられた．

花崗岩類中かその近傍に産出する鉱床においては花崗岩類にみられる性質が継続して認められ，たとえばポーフィリーカッパー鉱床では磁鉄鉱系花崗岩類と共通の鉱物組合せが産出する．2組の花崗岩類の性質はマグマ期末期から後マグマ期の一部に及んでおり，花崗岩類に密接な鉱床探査では両者を識別することが重要である．スズ－鉄マンガン重石鉱床がチタン鉄鉱系花崗岩類と密接な経験則から，環太平洋地域の西側では磁鉄鉱系花崗岩類に乏しいことが予想され，このことが沿海州―中国大陸南東部―マレー半島に至る中生代花崗岩類にポーフィリーカッパー鉱床が発見されない一因と考えられた．

CHAPTER 13

Watson, E.B. and Harrison, T.M. (1983) Zircon saturation revisited: temperature and compositional effects in a variety of crustal magma types. *Earth and Planetary Science Letters*, **64**, 295−304.

Zircon is an accessory mineral which has many applications in Earth Sciences because of a fortunate combination of properties:

1. It is ubiquitous; most rocks, including all granitoids, contain a minute modal fraction of zircon (<0.001)
2. Its crystal lattice permits the coordination of tetravalent ions with moderate octahedral crystal radius such as U^{4+} (1.03 Å) and Th^{4+} (1.08 Å) but does not easily accept larger bivalent ions such as Pb^{2+} (1.33 Å).
3. It has sluggish intracrystalline diffusion kinetics. Accordingly, zircon is capable of keeping its initial fine-scale distribution of trace elements and isotopes through a protracted thermal history.
4. It has high mechanical and chemical resilience.
5. It is the main repository of Zr and Hf.
6. Its solubility in common crustal magma types was accurately determined experimentally by E.B. Watson and M. Harrison in their 1983 paper, introduced here.

Watson and Harrison established that the solubility of zircon in non-peralkaline felsic to intermediate melts depends mainly on two parameters, temperature and melt composition, according to the expression:

$$\ln D_{zircon/melt} = (-3.8 - (0.85(M - 1)) + 12900/T \qquad (1)$$

where T is the absolute temperature, and M is a compositional parameter calculated as (Na+K+2Ca)/ (Al*Si/(Si+Ti+Al+Fe+Mg+Ca+Na+K)), where the element concentrations are expressed as atomic proportions (normalized to a total of 1.0).

The formulation of M is somewhat unusual and, therefore, commonly misunderstood. M depends mostly on SiO_2 and the aluminium saturation index (ASI = mol $Al_2O_3/(CaO+Na_2O+K_2O)$) so that, for non-peralkaline melts with SiO_2 > 60 wt.%, it can be quickly estimated as $M = 1.304 - 0.02*SiO_2 + 1.591/ASI$ (regression coefficient = 0.9935). Assuming that a stoichiometric zircon with an average hafnium concentration (Zr/Hf $\approx$ 47, Bea *et al.*, 2006) has a Zr concentration of ~487,000 ppm, the concentration of Zr in a zircon-saturated melt is:

ppm Zr in the melt =
$$EXP(13.096 + (3.8 + (0.85*(M - 1)) - 12900/T)) \qquad (2)$$

or, in an approximate, reduced form:

ppm Zr in the melt =
$$EXP(17.1544 + 1.35235/ASI - 0.017SiO_2 - 12900/T) \qquad (3)$$

This expression indicates that zircon solubility increases with T but decreases with SiO_2 and ASI, with T as the most influential parameter.

After more than 25 years of intensive work on zircon stability, the only serious objection raised against the Watson and Harrison model is that the voluminous zircon overgrowths in migmatites apparently indicate a higher-than-expected solubility of zircon in low-temperature felsic melts. Zircon overgrowths in migmatites, however, have been explained convincingly by Nemchin *et al.* (2001) as a consequence of Ostwald ripening. That is to say, they do not involve higher zircon solubility, and that the Watson and Harrison model remains valid. The fact that this model has not been improved upon, over more than 25 years of intensive research, represents a rare case among the body of work in experimental petrology. Furthermore, as with most major advances in science, the Watson and Harrison model has led to the development of new application fields, in this case (1) zircon saturation thermometry (Hanchar and Watson, 2003) and (2) zircon solution kinetics (Watson, 1996).

The zircon saturation temperature of a rock (*T*-Zr) is the temperature required for a melt with the same major-element composition to reach its equilibrium Zr concentration, given that the source contained an excess of zircon. *T*-Zr is calculated from expression 1 and provides an estimate of the maximum temperature attained by the melt at its source. The topology of Zr concentration versus temperature curves obtained with expression 1 is such that relatively major fluctuations in Zr concentration have little influence on the estimated temperature. For example, for a rock (melt) composition with $M = 1.2$, 250 ppm Zr yields a *T*-Zr = 820°C whereas 300 ppm yields *T*-Zr = 840°C (Fig. 1). This indicates that the estimation of *T*-Zr is robust, scarcely affected by analytical uncertainty in Zr determination.

Despite objections, e.g. that the composition of a rock may not match the composition of the melt, and that some zircon might be entrained as suspended crystals, the concept of zircon saturation temperature has proved its usefulness. It has led Miller *et al.* (2003) to classify North American granites in two broad groups, "hot" and "cold" granites. Whereas the former have a mean T-Zr $\approx 837°C$ and are inheritance-poor, the latter have a mean T-Zr $\mathring{A} \approx 766°C$ and are inheritance-rich. With minor variations, this twofold division of granitoids seems to occur in most orogenic belts and, as suggested by these authors, probably reflects fundamental differences in the mechanisms of magma generation, transport and emplacement.

The development of the concept of zircon saturation temperature and the closely-related concepts of apatite saturation (Harrison and Watson, 1984; Montel *et al.*, 1988; Hanchar and Watson, 1993; Wolf and London, 1993) and monazite saturation (Rapp and Watson, 1986; Montel, 1993), which control the concentration of P and *REE* in the melt, opened the door to consideration of equilibrium-disequilibrium processes during melt extraction, based on the chemical compositions of crustal melts. Lower-than-expected concentrations of Zr, P or *REE* were first interpreted as resulting from rapid melt extraction from the source, faster than required for the accessory minerals to dissolve until saturation is reached, due to their slow solution kinetics (e.g. Watt *et al.*, 1996). However, undersaturation in Zr, P or *REE* can be equally well explained by the textural positions of the accessories which, in common crustal rocks, are commonly included in major minerals, especially in biotite and garnet (e.g. Bea, 1996b). Included accessories can be partially shielded from the melt, especially at low melt-fractions, thus leading to melt undersaturation in the elements that the accessory phases contain – exactly the same effect as for rapid melt extraction.

Though Zr undersaturation does not yield reliable information about the kinetics of melt extraction, the presence of pre-magmatic zircons in an igneous rock, especially in the cases where magma temperature was equal to or greater than T-Zr, allows us to infer the heating/cooling gradients of the magma. In a paper published in 1996, Watson developed a 3D model for the instantaneous dissolution rate of a spherical zircon in a felsic melt. This model allows us to calculate the change in the radius of a zircon crystal suspended in a melt, as a function of heating and/or cooling rate (Watson, 1996). The equation is

$$dr/dt \times 1017 = -U((1.25 \times 1010/r)\exp(-28380/T) + 7.24 \times 108 \exp(-23280/T)) \quad (4)$$

where dr/dt is the instant dissolution rate (cm/s), r is the radius of an spherical zircon crystal (cm), T is the absolute temperature (K), U is the difference between the current Zr concentration of the melt and the concentration required for zircon saturation according to equation 1.

This equation is a simplification based on numerous physical and numerical experiments on Zr diffusion in melts. According to the Watson, however, it deviates by <10% from the results of more rigorous, moving boundary, finite-difference methods (Watson, 1996). Thus, it can be used safely for most purposes related to modelling the genesis of granitic rocks.

Equation 4 relies on the 1983 zircon solubility model expressed in equation 1 and on Zr diffusion in the melt, which depends additionally on the H_2O content of the melt. The equation treats temperature and melt composition as independent variables (the latter for calculating zircon solubility), but it assumes a constant melt H_2O content of 3 wt.%. In principle, this assumption might appear to be a serious limitation. In practice, however, it does not critically affect the model because crustal magmas rarely have <2 to 3 wt.% of dissolved H_2O (e.g. Clemens, 1984; Scaillet *et al.*, 1998) and the effects of dissolved H_2O on Zr diffusion mostly occur in the first $2-3$ wt.% (Harrison and Watson, 1983). This model has been used successfully by Bea *et al.*, (2007) to explain the abnormally elevated zircon inheritance in the Cambrian-Ordovician granitic magmas of Iberia.

Thus, this paper by Watson and Harrison (1983) deserves a place in this compilation of landmark papers because of its stability as a piece of work, because it has generated such a wide spectrum of useful concepts and because it has provided important constraints on processes in high-grade metamorphic rocks and granitic magmas, from their anatectic birth to their final emplacement and crystallization.

Earth and Planetary Science Letters, 64 (1983) 295–304
Elsevier Science Publishers B.V., Amsterdam – Printed in The Netherlands

295

[6]

Zircon saturation revisited: temperature and composition effects in a variety of crustal magma types

E. Bruce Watson [1] and T. Mark Harrison [2]

[1] *Department of Geology, Rensselaer Polytechnic Institute, Troy, NY 12181 (U.S.A.)*
[2] *Department of Geological Sciences, State University of New York at Albany, Albany, NY 12222 (U.S.A.)*

Received November 19, 1982
Revised version received April 25, 1983

Hydrothermal experiments in the temperature range 750–1020°C have defined the saturation behavior of zircon in crustal anatectic melts as a function of both temperature and composition. The results provide a model of zircon solubility given by:

$$\ln D_{Zr}^{zircon/melt} = \{-3.80 - [0.85(M-1)]\} + 12900/T$$

where $D_{Zr}^{zircon/melt}$ is the concentration ratio of Zr in the stoichiometric zircon to that in the melt, T is the absolute temperature, and M is the cation ratio $(Na + K + 2Ca)/(Al \cdot Si)$. This solubility model is based principally upon experiments at 860°, 930°, and 1020°C, but has also been confirmed at temperatures up to 1500°C for $M = 1.3$. The lowest temperature experiments (750° and 800°C) yielded relatively imprecise, low solubilities, but the measured values (with assigned errors) are nevertheless in agreement with the predictions of the model.

For $M = 1.3$ (a normal peraluminous granite), these results predict zircon solubilities ranging from ~100 ppm dissolved Zr at 750°C to 1330 ppm at 1020°C. Thus, in view of the substantial range of bulk Zr concentrations observed in crustal granitoids (~50–350 ppm), it is clear that anatectic magmas can show contrasting behavior toward zircon in the source rock. Those melts containing insufficient Zr for saturation in zircon during melting can have achieved that condition only by consuming all zircon in the source. On the other hand, melts with higher Zr contents (appropriate to saturation in zircon) must be regarded as incapable of dissolving additional zircon, whether it be located in the residual rocks or as crystals entrained in the departing melt fraction. This latter possibility is particularly interesting, inasmuch as the inability of a melt to consume zircon means that critical geochemical "indicators" contained in the undissolved zircon (e.g. heavy rare earths, Hf, U, Th, and radiogenic Pb) can equilibrate with the contacting melt only by solid-state diffusion, which may be slow relative to the time scale of the melting event.

1. Introduction

The behavior of zircon during crustal magma production and evolution is of major concern to geochemists for several reasons: not only is this ubiquitous accessory mineral a significant host for several widely used trace elements (notably Y and the heavy rare earths [1–3], and of course Zr), but it also plays a primary role in U-Pb geochronology.

Existing knowledge of zircon stability in the presence of magmatic liquids comes principally from two sources, the more important of which may be Pb isotope studies of zircons in natural rocks. These investigations have shown that zircon *can* survive crustal melting events, as evidenced by the preservation of billion-year-old radiogenic Pb (Pb*) zircon ages in igneous rocks of much younger age (e.g. [4]). Additional information on zircon behavior comes from laboratory studies of its saturation systematics in various magmatic liquid analogs. In principal, such studies enable a direct evaluation not only of the timing and amount of zircon precipitation from a cooling magma, but also of the tendency for zircon to remain in the residue or be consumed during anatexis. Where rocks of the continental crust are concerned, the main limitation of existing experimental data is that they apply mainly to peralkaline felsic compositions [5] and lunar magmas at temperatures in excess of 1070°C [6]; both of these show rather high zircon solubilities relative to those implied for low-temperature crustal melts by Pb isotope studies and by sparse experimental data available for non-peralkaline compositions at temperatures below 900°C [5,7].

The present study was undertaken with a view toward providing detailed knowledge of zircon saturation systematics in a variety of hydrous, low-temperature, intermediate to felsic magma types. This information can in turn be used to better understand the redistribution of geochemically important elements initially contained in zircon (e.g. REE, Hf, U, Th, and Pb*) during partial fusion events in the crust.

2. Experimental and analytical methods

2.1. General approach

As in the previous studies by Dickinson and Hess [6] and Watson [5], the overall objective in this work was to determine the amount of dissolved Zr required to saturate the melts of interest in zircon. On the basis of previous low-temperature (700–800°C) results for peraluminous and metaluminous melts, we anticipated very low zircon solubilities in crustal compositions and consequent difficulties with in situ measurement of dissolved Zr in the quenched glasses. For this reason, we planned the majority of runs at temperatures higher than those normally prevailing during anatexis. This approach yielded sufficiently high zircon solubilities that good precision in the Zr analysis could be achieved with the electron microprobe, and thus provided a firm basis for down-temperature extrapolation to conditions at which zircon solubility could be measured with only marginal precision.

TABLE 1

Compositions of starting glasses (in wt.%)

	AA	LKA	HKA	BTC	Obsidian [a]
SiO_2	59.6	60.7	61.9	65.1	76.1
TiO_2	0.6	0.7	0.8	0.8	0.1
Al_2O_3	18.1	17.5	17.1	18.4	13.0
FeO	5.2	4.6	4.0	4.8	0.7
MgO	2.6	2.4	1.6	1.6	0.1
CaO	9.1	7.3	5.7	2.8	0.5
Na_2O	2.7	3.8	4.2	1.7	3.7
K_2O	0.7	1.6	3.3	3.8	4.8
ZrO_2	1.5	1.5	1.5	1.0	

[a] Crystal-free obsidian from Lake County, Oregon. This glass was mixed with other starting glasses to obtain intermediate compositions (see text and Table 2).

The study was based on four primary starting compositions, three of which are synthetic, siliceous "andesites" (compositions AA, LKA, and HKA in Table 1) varying mainly in their CaO and alkali content, and one of which is a synthetic pelite (BTC in Table 1). Through crystallization of amphibole and/or plagioclase at run conditions, these four compositions produced a spectrum of melt types ranging in SiO_2 content from 61 to ~ 75 wt.% (anhydrous basis). When necessitated by compositional gaps in the generally systematic coverage, the primary compositions were mixed with each other or with a natural, crystal-free obsidian (Table 1) to obtain intermediate liquids.

2.2. Experimental details

The starting compositions were prepared from reagent-grade oxides and carbonates and were subjected to three cycles of fusion at 1450°C followed by grinding to a glass powder. Included in the "andesite" and "pelite" glasses were 1.5 and 1.0 wt.% dissolved ZrO_2, respectively; these levels were sufficient to cause crystallization of at least 1.2% zircon in all experiments except four of those in which a primary starting composition was "diluted" with the natural obsidian. The runs were made in sealed gold capsules initially containing 15 mg of glass powder and 3–25 wt.% distilled H_2O. (See Table 2 for specific information on added H_2O for each experiment, and note that

TABLE 2

Summary of run information

Run No.	T (°C)	P (kbar)	Duration (hours)	Starting glass [a]	H_2O [b] (wt.%)	Phases in addition to glass and zircon [c]
1	930	6	44	AA	11.6	hbl(5), vapor
2	930	6	44	LKA	11.2	hbl(5), vapor
3	930	6	44	HKA	11.8	hbl(< 5), vapor
4	930	6	44	BTC	11.5	vapor
5	1020	6	19	AA	6.0	none
6	1020	6	19	LKA	6.1	none
7	1020	6	19	HKA	5.9	none
8	1020	6	19	BTC	6.1	none
9	930	6	50	HKA : obsidian = 1 : 2	11.1	vapor
10	930	6	50	HKA : obsidian = 1 : 1	10.6	vapor
11	930	6	50	AA	4.7	hbl(8), plag(8)
12	930	6	50	LKA	4.9	hbl(6), plag(4)
13	930	6	50	HKA	5.4	hbl(5)
14	930	6	50	BTC	3.1	plag(2)
15	800	1.7	284	HKA : obsidian = 1 : 2	25	hbl(5), opaque(< 1), vapor
16	800	1.7	284	HKA : obsidian = 1 : 1	25	hbl(10), opaque(< 1), vapor
19	800	1.7	284	HKA	25	hbl(20), plag(20), opaque(< 1), vapor
20	800	1.7	284	BTC	25	hbl(10), plag(10), vapor
21	860	1.2	240	HKA : obsidian = 1 : 2	25	hbl(15), plag(5), vapor
22	860	1.2	240	HKA : obsidian = 1 : 1	25	hbl(< 10), plag(< 10), vapor
23	860	1.2	240	HKA	25	hbl(10), plag(< 30), vapor
24	860	1.2	240	BTC	25	hbl(10), plag(10), vapor
25	750	2.1	240	HKA : obsidian = 1 : 2	25	hbl(10), plag(20), vapor
26	750	2.1	240	HKA : obsidian = 1 : 1	25	hbl(15), plag(20), vapor
41	1020	6	17	BTC : obsidian = 1 : 1	6.1	none
42	1020	6	17	HKA : obsidian = 1 : 1	6.2	none
43	1020	6	17	LKA : BTC = 1 : 1	6.4	none
44	1020	6	17	HKA	11.4	vapor
45	1020	6	17	BTC	11.6	vapor

[a] See Table 1 for starting glass compositions.

[b] Bulk H_2O sealed in Au capsule.

[c] Numbers in parentheses are visual estimates of phase abundances (vol.%).

some runs were H_2O-saturated and some were not. The maximum amount of dissolved H_2O in any quenched glass is ~ 12 wt.%) Runs at 930° and 1020°C were made in a piston-cylinder apparatus at approximately 6 kbar, using 3/4″ diameter assemblies consisting of NaCl and Pyrex sleeves with internal filler pieces of unfired pyrophyllite, Pyrex, and crushable Al_2O_3. Tungsten 3% rhenium vs. W24%Re thermocouples were used to monitor and control temperature over run durations of 16-50 hours. The remaining experiments (860°, 800°, and 750°C) were made in cold-seal pressure vessels at 1.2, 1.7, and 2.1 kbar, respectively, for durations of 10-12 days. All pertinent run information is summarized in Table 2.

2.3. Analysis

The Zr contents of the quenched glasses in all run products were determined with a MAC electron microprobe, using 15 kV accelerating potential and a sample current of ~ 0.065 μA. In most cases, it was possible to broaden the beam spot to ~ 10-12 μm diameter, thus preventing excessive damage to the H_2O-bearing glasses. Zirconium L_{α_1} X-rays were collected through a PET analyzing crystal, which effectively separates the Zr signal from nearby interfering peaks, and also gives low background count rates (~ 2 cps; these were determined at offset positions on either side of the Zr peak). Standardization was achieved prior to and at the end of each measurement session by counting on homogeneous glasses (starting mixtures LKA, HKA and BTC (Table 1) containing 1 or 1.5 wt.% dissolved ZrO_2. Unknown samples were analyzed at 10-15 spots for total count times of 300-900 second on the peak and at each background position. Uncertainties in Zr concentration given in Table 3 are based on counting statistics and are in the 2σ level.

The major elements (Table 3) were measured on an average of 6 spots for each sample using an energy-dispersive microprobe system. Special care was taken to deal with the problem of Na (and K) loss during analysis of the hydrous glasses, and alkali concentrations obtained were checked against values estimated by mass balance of the bulk charge. (For cases in which sodium loss during analysis was obvious, the reported Na values are estimates based on the Na contents of the starting glasses, and are identified as such in Table 3.) It should also be noted that some iron loss to the Au container was experienced by charges run in the piston-cylinder apparatus; this problem is considered unimportant in the present study.

3. Results and discussion

The zirconium contents of 29 zircon-saturated melts are summarized in Table 3 and presented graphically in Fig. 1 as a plot of dissolved Zr vs. the cation ratio $M = (Na + K + 2Ca)/(Al \cdot Si)$ of the melts. This ratio is a somewhat complex melt composition parameter, and its use in systematizing the data requires some explanation and justification. Previous work [5,6] has clearly demonstrated the dependence of zircon solubility on melt composition, and has shown that basic melts in general can dissolve more zircon at a given temperature than acidic melts. Although identification of the key compositional variable with which to define the solubility systematics is largely empirical, there are several criteria with which to restrict choices and anticipate a functional form.

Our choice of the cation ratio, M, reflects prior observations [5] of increasing zircon solubility with increasing $(Na + K)/Al$ above unity in simple Na_2O-K_2O-Al_2O_3-SiO_2 melts (we now include Ca

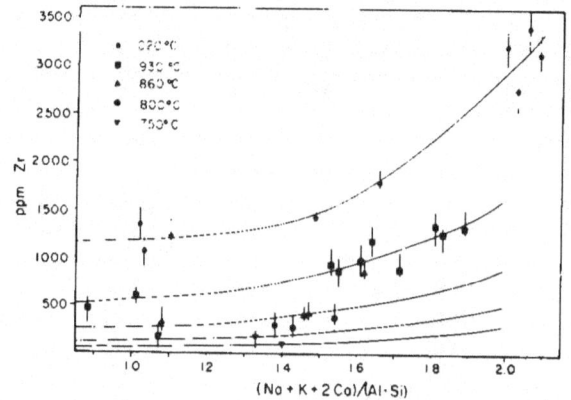

Fig. 1. Results of hydrothermal zircon saturation/solubility experiments at temperatures of 1020°, 930°, 860° and 750°C. The residual zirconium concentration in the glass following zircon crystallization is plotted here against a measure of the melt basicity, the cation ratio $M = (Na + K + 2Ca)/(Al \cdot Si)$. The justification for the fitted curves is discussed in detail in the text. The 1020°, 930° and 860°C curves are visual fits to the experimental data, whereas the 800° and 750°C curves are derivative from a model illustrated in Fig. 2. Error bars shown are at the level of $\pm 2\sigma$.

is the numerator for the complex melts studied here), and also takes into account the presence of the SiO_2 activity term in the solubility product of $ZrSiO_4$, which predicts a dependence of zircon solubility upon $1/SiO_2$. The general form of a plot of Zr concentration in zircon-saturated melts against a composition parameter such as M is constrained by simple thermodynamics in the following ways.

First, it is clear that a saturation isotherm cannot intersect the composition axis—rather, it must show upward concavity, approaching the composition axis in asymptotic fashion. (This is so because the required equivalence of chemical potentials of ZrO_2 in melt and zircon precludes zero concentration in the melt.) Secondly, if we express the crystallization of zircon from a melt by the equilibrium:

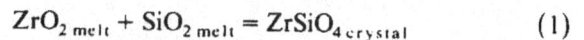

$$ZrO_{2\,melt} + SiO_{2\,melt} = ZrSiO_{4\,crystal} \qquad (1)$$

and formulate an equilibrium constant:

$$K_{eq(1)} = \frac{[ZrSiO_4]_{zircon}}{[ZrO_2]_{melt}[SiO_2]_{melt}}$$

TABLE 3

Major element (wt.%) and Zr (ppm) concentrations in zircon-saturated melts. Numbers in parentheses following major element values are 1 standard deviation on 5–7 analysis spots. Estimated errors associated with Zr values are at the 2σ level (see text)

Run No.:	1	2	3	4	5	6	7	8	9	10
SiO₂	54.2(0.3)	56.2(0.4)	57.4(0.4)	58.8(0.4)	56.5(0.3)	58.2(0.4)	58.4(0.6)	61.6(0.4)	66.6(0.4)	63.8(0.5)
TiO₂	0.5(0.1)	0.6(0.1)	0.7(0.1)	0.6(0.2)	0.5(0.1)	0.7(0.1)	0.7(0.10)	0.8(0.1)	0.4(0.1)	0.4(0.1)
Al₂O₃	16.9(0.2)	16.8(0.1)	16.8(0.2)	17.1(0.3)	16.8(0.3)	16.6(0.2)	16.1(0.2)	17.7(0.3)	13.6(0.2)	14.2(0.1)
FeO	4.1(0.2)	3.1(0.2)	2.6(0.2)	3.6(0.2)	3.4(0.1)	2.6(0.2)	2.2(0.1)	2.8(0.2)	1.2(0.1)	1.6(0.1)
MgO	2.0(0.1)	1.6(0.2)	1.2(0.1)	1.4(0.2)	2.4(0.2)	2.4(0.2)	1.5(0.1)	1.4(0.1)	0.6(0.1)	0.8(0.1)
CaO	7.6(0.1)	5.9(0.2)	4.8(0.1)	2.6(0.1)	8.4(0.1)	6.8(0.2)	5.6(0.2)	2.8(0.2)	2.1(0.1)	3.0(0.1)
Na₂O	2.3 [a]	3.3 [a]	3.7 [a]	1.5 [a]	2.5(0.2)	3.4 [a]	3.7 [a]	1.4(0.1)	3.5 [a]	3.5 [a]
K₂O	0.8(0.1)	1.4(0.1)	3.0(0.1)	3.3(0.1)	0.8(0.1)	1.5(0.1)	3.0(0.1)	3.6(0.2)	4.0(0.1)	3.6(0.1)
Zr	1320±200	1340±200	1260±200	600±150	3150±250	2780±250	3420±250	1070±200	940±170	1180±170
Total	88.5	89.0	90.3	89.0	91.6	92.5	91.5	92.2	92.1	91.4
M [b]	1.89	1.81	1.83	1.01	2.09	2.03	2.06	1.03	1.53	1.64

Run. No.:	11	12	13	14	15	16	19	20	21	22
SiO₂	58.9(0.5)	59.4(0.4)	59.4(0.3)	64.2(1.0)	68.1(0.5)	68.1(0.3)	67.8(0.2)	69.5(0.1)	66.6(0.3)	66.4(0.5)
TiO₂	0.7(0.1)	0.7(0.1)	0.6(0.1)	0.6(0.2)	0.1(0.1)	0.2(0.1)	0.2(0.1)	0.2(0.2)	0.2(0.1)	0.2(0.1)
Al₂O₃	15.8(0.2)	16.5(0.3)	17.2(0.1)	17.8(0.3)	13.0(0.2)	13.0(0.3)	13.0(0.1)	13.2(0.1)	13.7(0.2)	14.2(0.2)
FeO	4.1(0.2)	3.2(0.3)	2.3(0.2)	3.7(0.5)	1.1(0.2)	1.0(0.1)	0.8(0.3)	1.4(0.1)	1.4(0.1)	1.2(0.2)
MgO	1.4(0.1)	1.5(0.4)	0.8(0.1)	1.6(0.1)	0.1(0.1)	0.1(0.1)	0.2(0.1)	0.7(0.1)	n.d.	n.d.
CaO	5.6(0.2)	4.7(0.5)	4.4(0.2)	1.8(0.1)	1.2(0.1)	1.2(0.1)	1.3(0.1)	1.5(0.1)	0.8(0.1)	1.4(0.1)
Na₂O	2.2(0.2)	3.4 [a]	3.6(0.1)	1.6(0.2)	3.4 [a]	3.6 [a]	3.8 [a]	1.6 [a]	4.6 [a]	3.8(0.2)
K₂O	0.9(0.1)	1.8(0.1)	3.3(0.1)	3.8(0.1)	4.2(0.1)	4.5(0.1)	4.8(0.2)	4.2(0.1)	5.2(0.1)	4.8(0.2)
Zr	870±170	980±170	880±170	460±160	300±140	270±130	390±140	180±140	860±170	400±140
Total	89.7	91.3	91.7	95.1	91.2	91.7	91.9	92.3	92.6	92.0
M [b]	1.55	1.61	1.72	0.88	1.38	1.43	1.54	1.07	1.62	1.46

Run No.:	23	24	25	26	41	42	43	44	45	
SiO₂	66.9(0.7)	68.5(1.5)	68.9(0.4)	69.0(0.6)	67.4(0.3)	67.0(0.4)	61.4(0.2)	57.9(0.5)	60.3(0.2)	
TiO₂	0.4(0.1)	0.4(0.3)	0.1(0.1)	0.1(0.1)	0.5(0.1)	0.4(0.1)	0.7(0.1)	0.7(0.1)	0.7(0.1)	
Al₂O₃	13.8(0.2)	13.3(1.4)	12.4(0.3)	12.6(0.1)	15.2(0.2)	14.7(0.2)	17.7(0.1)	16.3(0.3)	17.3(0.2)	
FeO	1.4(0.2)	1.9(0.6)	0.9(0.1)	1.0(0.1)	2.3(0.1)	2.0(0.2)	3.4(0.1)	2.4(0.1)	3.1(0.2)	
MgO	0.2(0.1)	0.8(0.1)	n.d.	n.d.	0.7(0.1)	0.8(0.1)	1.9(0.1)	1.6(0.1)	1.5(0.1)	
CaO	1.4(0.1)	1.6(0.6)	1.0(0.1)	1.0(0.1)	1.6(0.1)	3.1(0.1)	4.9(0.1)	5.4(0.1)	2.7(0.1)	
Na₂O	3.5 [a]	1.7(0.2)	3.5 [a]	3 [a]	2.4 [a]	3.6 [a]	2.6 [a]	3.7 [a]	1.5 [a]	
K₂O	4.9(0.1)	4.2(0.2)	4.4(0.1)	4.6(0.1)	4.0 [a]	3.9 [a]	2.4(0.1)	3.0 [a]	3.4 [a]	
Zr	410±140	320±160	<100	180±140	1230±190	1800±180	1430±170	3210±200	1360±160	
Total	92.5	92.4	91.2	91.3	94.2	95.6	95.2	91.3	90.5	
M [b]	1.47	1.08	1.40	1.33	1.10	1.66	1.49	2.00	1.20	

[a] Value based on approximate mass balance of bulk charge.
[b] M = cation ratio $(Na + K + 2Ca)/(Al \cdot Si)$ (see text).
n.d. = not detected.

we see that because of the expected dependence of $\ln K_{eq(1)}$ upon $1/T$, we should also anticipate such a dependence for $\ln[ZrO_2]_{melt}^{-1}$. In a series of melts having similar composition parameters, it is reasonable to assume that $[ZrO_2]_{melt}$ is proportional to dissolved zirconia concentration. If the com-

position parameter is "correctly" chosen, then, the measured Zr contents of zircon-saturated melts should increase in regular fashion $[\ln(1/Zr) \propto 1/T]$ for melts in which the composition variable is similar.

For experiments made over a range in tempera-

300

ture on a variety of melt compositions, the constraints noted above predict a saturation diagram (Zr concentration in the melt vs. melt composition) consisting of a family of "nested", concave-up, isothermal solubility curves. A judiciously chosen melt composition variable should yield a saturation diagram displaying these characteristics, and in addition, of course, should yield a coherent variation in dissolved zirconium concentration with composition. The cation ratio, M, plotted in Fig. 1, is successful in all respects: with the exception of run 21, which will be discussed later, all data plot close to or on the curves shown, and the curves have all the desired and expected characteristics. The diagram is especially convincing when it is noted that a plot of $(Na + K + 2Ca)/(Al \cdot Si)$, for instance, does not reveal the same agreement but instead shows considerable scatter, underscoring the importance of accounting for the divalent nature of calcium in the ratio.

It is apparent in Fig. 1 that the imprecise results of the lower-temperature runs (750° and 800°C) do not allow independent definition of the curves shown. However, the solubility isotherms at 1020°, 930°, and (to some extent) 860°C are sufficiently well defined to provide a basis for extrapolation to lower temperatures. This extrapolation was done by drawing smooth curves through the 1020°, 930°, and 860°C data, and recasting the apparent solubilities at M values of 1.3, 1.6 and 1.9 (1.3 and 1.6 only for 860°C) as the concentration ratio of zirconium in a stoichiometric zircon to that in the melt ($D_{Zr}^{zircon/melt}$). The natural logarithm of these $D_{Zr}^{zircon/melt}$ values was then plotted against reciprocal absolute temperature (Fig. 2), and the down-temperature extensions of the defined lines were used to construct the 800° and 750°C curves shown in Fig. 1. It is clear that although the lower-temperature data were insufficient in themselves in defining saturation surfaces, they do agree, within error, with the curves predicted from solubility behavior at higher temperatures. As noted previously, the error bars in Fig. 1 represent X-ray counting uncertainty at the $\pm 2\sigma$ level. Attempts were made to improve upon these uncertainties in the low-temperature runs by re-analysis for dissolved Zr with a newly-acquired TlAP crystal, which has a much higher sensitivity for Zr-L$_\alpha$

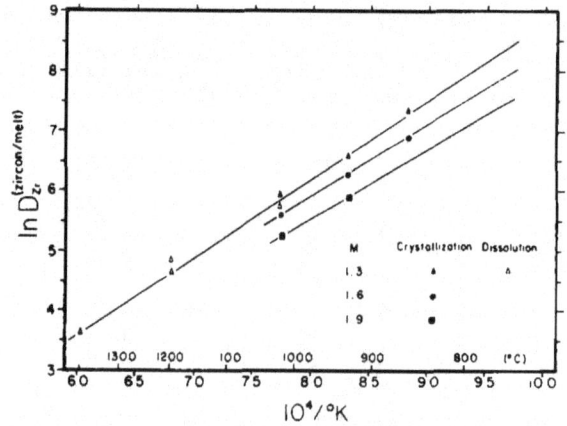

Fig. 2. Plot of $\ln D_{Zr}^{zircon/melt}$ for three M values (1.3, 1.6, 1.9) versus the reciprocal absolute temperature of the experiment. For $M = 1.6$ and 1.9, the lines are based solely upon crystallization experiments at 1020°, 930° and 860°C; each solid symbol represents the intersection of a vertical line of appropriate M value in Fig. 1 with the indicated solubility isotherm. For $M = 1.3$, the solid symbols have the significance just noted; the open symbols represent zirconium concentration in the melt at the surface of a dissolving zircon crystal, and thus are higher temperature "reversals" of the crystallization experiments. The down-temperature extensions of the lines in this diagram enabled solubility isotherms at 750° and 800°C to be drawn in Fig. 1 (see text for further explanation).

X-rays than does PET. These attempts yielded zircon solubilities indistinguishable from those previously determined. They also had equally high uncertainty, which in this case is attributable either to difficulty in completely avoiding tiny zircons in the charges or to heterogeneity of the quenched glasses. (We have in fact confirmed the expectation of zome Zr heterogeneity in the low-temperature melts: measured Zr diffusivities indicate an average transport distance of only about 10 μm in 10 days at 750°C [8].)

The general solubility model illustrated by the curves in Fig. 1 is given, for $M = 0.9$ to 1.7, by:

$$\ln D_{Zr}^{zircon/melt} = \{-3.80 - [0.85(M - 1)]\} + 12900/T \qquad (2)$$

As shown in Fig. 2, this model has been further substantiated, specifically at $M = 1.3$, by a series of experiments carried out in a companion study of the kinetics of zircon dissolution in crustal melts [8]. These experiments are, in effect, higher temperature (1020–1500°C) "reversals" of those in the present work, because equilibrium at the zircon/melt interface was achieved by dissolution of zircon rather than by crystallization. The consistency of the two approaches is striking, as is the large temperature interval over which ln $D_{Zr}^{zircon/melt}$ is linear in $1/T$. The slope of this line yields an enthalpy of zircon dissolution of 25.6 ± 0.2 kcal/mole; this value is affected only slightly by changes in M over the range 0.9 to ~ 1.7, which encompasses the variety of magma compositions normally derivable by crustal anatexis ($M = 1.9$—a broadly andesitic composition—ΔH has dropped to 23.0 kcal/mole).

Despite the encouraging consistency of the zircon solubility model embodied in equation (2), we caution that is it not completely general, inasmuch as previous simple system results [5,7], most notably for peralkaline melts, are not well described by equation (2). Watson [5] did note, however, that the remarkably high solubility of zircon in peralkaline melts is due specifically to complexing of dissolved Zr^{4+} with "free" alkalies not associated with Al in the melt. Such complexing is precluded in non-peralkaline compositions, so it makes sense that zircon solubility in melts of the present study should be governed by different factors. In this regard, it is interesting to note that the only solubility value obtained in this study that is inconsistent with our model (i.e., run 21) is for a highly alkaline composition. The glass analysis given in Table 3 is not actually peralkaline [(Na + K)/Al = 0.97], but the estimated errors in the microprobe analyses for Al and K, along with the considerable uncertainty in the estimated Na content, easily allow (Na + K)/Al to exceed 1, in which case the apparently anomalous high zircon solubility would be expected.

In view of the demonstrated success of our empirical model in describing zircon saturation systematics for a range of crustal melt compositions and temperatures, we are confident in applying equation (2) to the examples of zircon behavior during crustal melting discussed in the next section. Before doing so, however, it may be worthwhile to note that two variables of general importance in crustal melting—namely, H_2O content and pressure—probably are not of great

significance in determining zircon solubilities. A considerable range in dissolved H_2O content is represented by the data plotted in Fig. 1, but there is no indication that this variation introduces any scatter into the plot. Note in particular that runs 44 and 45 are "wetter" duplicates of runs 7 and 8, respectively; the later runs contain considerably more H_2O (Tables 2 and 3), but the zircon solubilities are not significantly different. (The dissolved H_2O contents of the quenched glasses are best represented by the amounts of added H_2O and H_2O-undersaturated runs, and are given very approximately by the difference from a 100% microprobe analysis total for the H_2O-saturated runs—see Tables 2 and 3.) In our previously-cited [8] companion study of zircon dissolution kinetics, we have confirmed the general insensitivity of zircon solubility to dissolved H_2O content, but we have also learned that this lack of dependence applies only at H_2O concentrations in excess of 1.5–2 wt.%. At 1200–1500°C, a granitic melt containing $\sim 0.2\%$ H_2O will dissolve 30–40% less zircon than the same melt with $\sim 2\%$ H_2O, yet further increases in dissolved H_2O have little or no effect on zircon solubility. Although similar systematics probably apply at crustal melting temperatures, experimental confirmation is precluded by sluggish kinetics. In any case, crustal anatectic magmas containing < 1.5–2 wt.% H_2O are probably relatively uncommon.

No explicit test of pressure effects on zircon solubility was made, but because of the internal consistency of our data, we infer that the ~ 5 kbar maximum difference in pressure among our experiments has a minimal influence on zircon saturation. This conclusion is consistent with the observation of Green and Watson [9] that pressure has only a minor influence in apatite saturation systematics.

4. Areas of application

The zircon saturation model embodied in equation (2) constitutes valuable input into any assessment of the role of zircon in the geochemistry of crustal magmatic systems. For illustrative purposes, we will consider in varying detail situations in which meaningful constraints can be placed on the behavior of trace elements and isotope systems during melting and crystallization. In many of these examples, knowledge of kinetic factors is needed before a completely rigorous evaluation is possible, but the general principles of zircon behavior should nevertheless become clear.

4.1. Zircon saturation in fractional crystallization series

Because Zr is a generally incompatible trace element of relatively high abundance in mafic and intermediate rocks, its concentration is likely to rise to sufficient levels in felsic derivatives that zircon saturation will be encountered. This conclusion has of course been reached previously without the benefit of solubility data simply on the basis of zirconium behavior in consanguineous rock series and from observation of apparent zircon "phenocrysts" (e.q. [10,11]). The contribution of the present saturation model is to enable fairly precise calculation of the timing of initial zircon precipitation and of the proportion of bulk zirconium in the magma that is bound in zircon at any subsequent time. Thus, while gravitational removal of zircons from a differentiating magma might be revealed by a negative slope on a Zr variation diagram (bulk Zr vs. differentiation index), the solubility model can identify instances of zircon saturation even when the crystals remain suspended in the magma. Given knowledge of zircon/liquid partition coefficients for elements concentrated in zircon (e.g. heavy rare earths, Y, Hf, U and Th), it is also possible to deduce the proportion of these elements that is in zircon relative to that in the rest of the magma at any point in a fractional crystallization sequence.

4.2. Redistribution of zircon-hosted trace elements during partial melting

As discussed in some detail by Watson [5], residual zircon in the source regions of crustal magmas can have a significant influence on the concentrations of REE (and other trace elements) in the derived melt [1–3]. The primary input needed to model such effects is of course zircon solubility

data and trace element partition coefficients. Provided that a good estimate of the temperature of melting can be made, our saturation model can be used to predict, for any given melt composition, the proportion (if any) of zircon originally present in the source that will tend to remain unmelted. A simple mass balance reveals, for example, that a source containing 50 ppm Zr as zircon is unlikely to retain zircon beyond ~ 50% melting at 750°C, whereas sources with ≥ 100 ppm Zr will contain residual zircon at all degrees of melting. The extent to which REE, Y, U, Th, and Hf are withheld from the melt is determined by the absolute amount of undissolved zircon, which is obviously controlled by the original zirconium (zircon) content of the source.

4.3. Inheritance of radiogenic Pb (Pb*)

Several recent studies of U-Pb systematics in zircons from Phanerozoic S-type granitoids [12] have clearly demonstrated that undissolved zircon can carry significant portions of accumulated Pb* into anatectic magmas [4,13–16]. This realization is accompanied by the observation that the cores of zircons containing inherited Pb* are often subrounded, in contrast with the younger magmatic rims, which well-defined terminations. The subrounded cores have been interpreted as detrital zircons, rounded by abrasion at the earth's surface. While this may sometimes be the case, it is equally plausible that the rounding represents partial resorption of pre-existing, well-formed zircons of the source rock during peak melting conditions. In either case, the cores may contain inherited Pb: however, an origin by partial resorption may be more consistent with the zircon behavior predicted by our saturation model.

Zircon solubility data have a direct bearing on the problem of inherited Pb* because they can be used to predict magmatic conditions under which zircon will tend to dissolve and thereby allow its radiogenic Pb to mix with common Pb of a newly-formed melt. For example, a detailed U-Pb zircon study of the southeast Australian Berridale Batholith revealed that the S-type granitoids contained "the large components of older Pb" [14] although inherited Pb* was observed in one I-type.

As the S-types of the Berridale Batholith apparently contain only marginally more Zr than the I-types (150 and 120 ppm, respectively) [17], it seems unlikely that the contrast in concentration can explain the disproportionate solubility of the I-type zircons. However, the generation of I-types requires substantially greater temperatures (800–900°C) than that for S-types granitoids ($\leq$ 700°C) [18], and produces somewhat more mafic melts (i.e. larger M value). Though both these effects will result in enhanced zircon dissolution in the I-type melt, for illustrative purposes we shall consider the former parameter in isolation. For an M value of 1.4, a normal peraluminous granite, the I-type magma at 850°C would not become saturated until a concentration of 340 ppm Zr was reached. In contrast, zircons would be soluble in the S-type magma below a melt concentration of about 50 ppm. This immediately provides a basis for interpreting the observed general absence of inherited zircons from the Berridale I-types [14].

Although we recognize that virtually all of the zirconium in a slowly cooling granitoid magma will eventually be incorporated into crystalline zircon ‡, our concern here is with the solubility behavior at peak conditions—if the inherited zircon is at any time soluble in the contacting melt, the Pb* will be homogenized throughout the entire system. In the case of the S-type granitoids, only a small fraction of the recycled (entrained) zircon must dissolve in the melt to reach saturation. This would result in preservation of most of the zircon and thus prevent extensive redistribution of Pb*, provided that the solid state transport of Pb* within the zircon was relatively sluggish. For the low-Zr, high-temperature I-type granitoids, on the other hand, the likely dissolution of most re-cycled zircon before attainment of saturation would result in complete loss of chronological information on the source. Although Compston and Chappell [20] have shown that the source of

the southeast Australian I-type granitoids, including the Berridale Batholith, is at least 1 Ga old, the Berridale I-types show little evidence of an inherited zircon component—a phenomenon consistent with complete dissolution of all zircons in the source region.

Other isotopic systems will be equivalently affected by the solubility behavior of zircon during anatexis. As a result of the nearly identical geochemical behavior of Hf and Zr, zircon preferentially partitions Hf into the lattice with respect to Lu, which gives this phase the potential to be a useful indicator of the initial ^{176}Hf/^{177}Hf ratio of a rock system [21]. However, this condition will not be met if much of the Hf is incorporated into insoluble zircon. In a similar fashion, although the HREE-enriched nature of the rare earth pattern of zircon [1–3] makes it an attractive candidate for ^{147}Sm/^{143}Nd geochronology, the assumption of an initially uniform ^{143}Nd/^{144}Nd composition requires that any pre-existing zircon, as well as other accessory phases, have dissolved in or equilibrated with the melt.

5. Concluding remarks

Although we believe in the general validity of the types of applications discussed above, it is prudent to accompany the conclusions with several qualifications concerning the attainment of equilibrium in natural melting situations: (1) it has not been demonstrated that pre-existing zircons will dissolve fast enough to force saturation on a contacting melt over the time interval of the melting event; (2) if zircons in the source do dissolve to the point of saturating a partial melt, the extent to which trace element equilibrium is approached via diffusion in the zircon is largely unknown (as suggested in the discussion of inherited Pb* in "recycled" zircons, Pb diffusion may be very slow; diffusivities of other elements are completely uncharacterized); and (3) if zircon in the source is isolated from the partial melt by inclusion in major mineral phases, it will not interact with the melt in any way. It would probably be overly pessimistic to anticipate untractable behavior in all three (or even one) of the above considerations, but the

‡ Of the common rock-forming minerals in non-alkaline rocks, only hornblende has been suggested as a significant host for Zr (e.g. [19]). However, in a companion study of Zr partitioning between hornblende and melt, we have determined a crystal/liquid partition coefficient of only 0.31 ± 0.05 for dacitic melt at 950°C.

304

likely existence of some complications does bring into question the validity of developing detailed models of the role of zircon in crustal geochemistry at the present time. Our continuing research on zircon is centered on kinetics of dissolution in felsic melts and on trace element diffusion—the forthcoming information should expedite more rigorous treatment of the applications outlined above at a later date.

Acknowledgements

This research was supported through National Science Foundation Grants EAR-8121275, EAR-8025587, and EAR-8212453.

References

1 H. Nagasawa, Rare earth concentrations in zircons and apatites and their host dacites and granites, Earth Planet. Sci. Lett. 9, 359–364, 1970.

2 J.G. Arth, Behavior of trace elements during magmatic process— a summary of theoretical models and their applications, J. Res. U.S. Geol. Surv. 4, 41–47, 1976.

3 E.B. Watson, Some experimentally determined zircon/liquid partition coefficients for the rare earth elements, Geochim. Cosmochim. Acta 44, 895–897, 1980.

4 R.T. Pidgeon and M. Aftalion, Cogenetic and inherited zircon U-Pb systems in granites: Palaeozoic granites of Scotland, in: Crustal Evolution in Northwestern Britain and Adjacent Regions, Geol. J. Spec. Paper 10, 183–220, 1978.

5 E.B. Watson, Zircon saturation in felsic liquids: experimental data and applications to trace element geochemistry, Contrib. Mineral. Petrol. 70, 407–419, 1979.

6 J.E. Dickson and P.C. Hess, Zircon saturation in lunar basalts and granites, Earth Planet. Sci. Lett. 57, 336–344, 1982.

7 L. Larsen, Measurement of solubility of zircon (ZrSiO$_4$) in synthetic granitic melts, EOS 54, 479, 1973.

8 T.M. Harrison and E.B. Watson, kinetics of zircon dissolution and diffusion of zirconium in granitic melts, Contrib. Mineral. Petrol. (in press).

9 T.H. Green and E.B. Watson, Crystallization of apatite in natural magmas under high pressure, hydrous conditions, with particular reference to "orogenic" rock series, Contrib. Mineral. Petrol. 79, 96–105, 1982.

10 L.R. Wager and R.L. Mitchell, The distribution of trace elements during strong fractionation of basic magmas: a further study of the Skaergaard intrusion, East Greenland, Geochim. Cosmochim. Acta 1, 129–209, 1951.

11 C.K. Brooks, On the distribution of zirconium and hafnium in the Skaergaard intrusion, East Greenland, Geochim. Cosmochim. Acta 33, 357–374, 1969.

12 B.W. Chappell and A.J.R. White, Two contrasting granite types, Pacific Geol. 8, 173–179, 1974.

13 I.S. Williams, The Berridale Batholith: a lead and strontium isotopic study of its age and origin, Ph.D. Thesis, Australian National University, Canberra, A.C.T., 1977.

14 I.S. Williams, U-Pb evidence for the pre-emplacement history of granitic magmas, Berridale Batholith, southeastern Australia, U.S. Geol. Surv. Open-file Rep. 78–701, 455–457, 1978.

15 W. Compston and I.S. Williams, Protolith ages from inherited zircon cores measured by a high mass-resolution ion microprobe, Abstr. 5th Int. Conf. on Geochronology, Cosmochronology and Isotope Geology, 63–64, 1982.

16 T.C. Liew, M.T. McCulloh, R.W. Page, B.W. Chappell and I. McDougall, A Nd, Sr and U-Pb zircon isotopic study of the granitoid batholiths of Penninsular Malaysia, Abstr. 5th Int. Conf. on Geochronology, Cormochronology and Isotope Geology, 215–216, 1982.

17 A.J.R. White, I.S. Williams and B.W. Chappell, Geology of the Berridale 1 : 100,000 sheet (8625), Geological Survey of New South Wales, 1977.

18 T.C. Liew, A Nd, Sr, and U-Pb zircon isotopic study of the granitoid batholiths of Penninsular Malaysia, Ph.D. Thesis, Australian National University, Canberra, A.C.T. (in preparation).

19 J.A. Pearce and M.J. Norry, Petrogenetic implications of Ti, Zr, Y and Nd variations in volcanic rocks, Contrib. Mineral. Petrol. 69, 33–47, 1979.

20 W. Compston and B.W. Chappell, Sr-isotope evolution of granitoid source rocks, The Earth: Its origin, structure and evolution, M.W. McElhinny, ed., pp. 377–426, Academic Press, London, 1979.

21 P.J. Patchett, O. Kouvo, C.E. Hedge and M. Tatsumoto, Evolution of continental crust and mantle heterogeneity: Evidence from Hf isotopes, Contrib. Mineral. Petrol. 78, 279–297, 1981.

CHAPTER 14

Wall, V.J., Clemens, J.D. and Clarke, D.B. (1987) Models for granitoid evolution and source compositions. *Journal of Geology*, **95**, 731–749.

During the late 1970s and 1980s, once it had been accepted that most granites were derived through melting of the continental crust, granite petrologists gave increasing attention to the processes responsible for the chemical-mineralogical diversity of granitoids. White and Chappell (1977) proposed the restite hypothesis, which was further refined by Chappell *et al.* (1987). This hypothesis proposed that granitic magmas are a suspension of restitic crystals in a melt phase, and that most variability in granitic plutons resulted from different degrees of unmixing between the restitic minerals and the melt. Therefore, the study of granite suites can allow us to estimate the compositions of their sources, thus opening a panoply of exciting possibilities and awakening an interest in granites among those working in other branches of the geosciences. During the 1980s, the restite hypothesis was accepted by many authors as capable of explaining most aspects of granite petrogenesis. For example Wyborn *et al.* (1981) proposed that phenocrysts in a highly porphyritic (~60 vol.%) dacite are in fact restitic crystals, and Chen *et al.* (1990) concluded that mafic inclusions in I-type granites represent fragments of more refractory source materials carried up in the magma.

The paper of Wall *et al.*, based on experimental, textural, geochemical and isotopic evidence, dismantled the arguments in support of the restite hypothesis, and critically reviewed various models for the chemical evolution of granitoid suites. The authors did not deny that some granitic magmas might have contained large fractions of suspended restitic crystals but presented compelling evidence that neither restite unmixing nor restite incorporation is essentially responsible for the chemical variability found in most granitic suites. One the most important contributions of this paper is, perhaps, the establishment of a link between the fraction of crystals in suspension and the mode of emplacement of a granitic magma. They wrote that the retention of a significant fraction of restite would require ascent to high crustal levels by a diapiric mechanism, which

in turn would require very high magma temperatures to partially melt (that is, to soften) the envelope rocks. Magma temperatures well above the solidus and high restite fraction are, therefore, a *contradictio in terminis* and, as such, highly improbable. Melt extraction and transport from the source channelled through narrow conduits would inexorably filter the magma, especially if melt segregation and transport were fast. If so, magmas would necessarily reach high crustal levels almost devoid of suspended restitic crystals.

It is important to emphasize that the occurrence of restitic zircon cannot be invoked as a proof of the restite unmixing hypothesis because the grain sizes of zircons in the source rocks are always very small, commonly <100 μm equivalent spherical diameter. Crystals of that size cannot be segregated easily from a viscous, probably thixotropic fluid. This is not the case for the major minerals because their grain sizes, at the source, are usually no smaller than 3–4 mm. Thus, if they were to decrease to a size analogous to zircon, they would occupy a negligible volume fraction (< 0.05 vol.%). Assuming that all minerals melt equally, the persistence of 40% of restitic crystals restricts the diminution of their crystal radii to 70% of the initial value. If some minerals do not melt at all, their radii will not decrease. In either case, restite crystals of major minerals will be much larger than zircon crystals and, therefore, they will be segregated much more easily than zircons or other accessory minerals.

It seems, therefore, that granites are not simple mirrors of their sources. Their mineralogy and geochemistry can certainly provide us with information about the source rocks, but it is not as direct a relationship as the restite hypothesis implies. The ideas and questions posed by Wall *et al.* (1987) are at the foundation of many aspects of modern granite science. This paper marked a point of inflection in the conception about how granitic magmas evolve and has influenced many subsequent works, among them, the seminal paper of Clemens and Mawer (1992).

VOLUME 95

NUMBER 6

THE JOURNAL OF GEOLOGY

November 1987

MODELS FOR GRANITOID EVOLUTION AND SOURCE COMPOSITIONS[1]

V. J. WALL, J. D. CLEMENS, AND D. B. CLARKE[2]

Department of Earth Sciences, Monash University, Clayton, Victoria, 3168, Australia
C.N.R.S., U.A.10 and O.P.G.C., 5 rue Kessler, 63038 Clermont-Ferrand Cedex, France
Department of Geology, Dalhousie University, Halifax, Nova Scotia, B3H 3J5, Canada

ABSTRACT

This paper critically examines the bases of various models for the chemical evolution of granitoid suites. The proposed mineralogical, textural, and chemical criteria for the recognition of restite components are found to be equivocal and can be plausibly explained by crystal fractionation and accumulation, as well as by magma mixing and assimilation of non-restitic solids. These processes demonstrably account for the bulk of the chemical variation in a variety of granitic suites, although restite-controlled variation may have played a role in some types of metasediment-derived suites. Granitoid magmas with "non-minimum" melt chemistry are probably rather common. The petrographic, geochemical, and isotopic features of individual plutons should be assessed carefully to elucidate the relative importance of various mechanisms for chemical evolution. Conclusions as to the character of magma source materials and magma evolution paths that are based on the restite model should be treated with caution.

INTRODUCTION

White and Chappell (1977) suggested an hypothesis (the restite unmixing model) to explain chemical variations in granitoid suites (see also Piwinskii 1968; Zeck 1970; Presnall and Bateman 1973; Price 1983; Clark and Lyons 1986; Scambos et al. 1986). According to their model, many granitoid magmas are comprised of variable mixtures of liquid (granitic melt) and solid (restite: melt-depleted source material) components which separate to varying degrees to produce the chemical variation observed in plutonic bodies.

The restite model is primarily based on two sets of observations/interpretations. White and Chappell noted that major and trace element contents of individual granitoid suites commonly define linear trends on Harker variation diagrams. They interpreted these trends as the results of unmixing of SiO_2-poor restite and SiO_2-rich melt fractions. They also observed that xenolith or inclusion suites contain mineral assemblages which are broadly compatible with their granitic hosts and that inclusions (enclaves) are more abundant in the mafic members of granitoid suites. The inclusions are interpreted as restite materials. Thus, an implication of the restite model is that granitoids "image" their sources (Chappell 1979) in a simple manner, allowing calculation of the composition and age of source materials (e.g., Compston and Chappell 1979; McCulloch and Chappell 1982; Price 1983; Chappell 1984). Kay (1984, p. 545) points out that "the compositional (and isotopic) diversity of continental crustal magma sources lends a particularly *ad hoc* flavor to the discussions that are based purely on element systematics in the melt products themselves . . . The residual mineralogy of the "restite" has been paramaterized almost at will by some workers to fit geochemical models."

---

[1] Manuscript received December 26, 1986; accepted June 1, 1986.
[2] Please direct all reprint requests to D. B. Clarke.

[JOURNAL OF GEOLOGY, 1987, vol. 95, p. 731–749]
0022-1376/87/9506-0003$1.00

731

From the outset we wish to state that we share Chappell and White's views that *some* features of granitoid chemistry (particularly isotopes and some trace elements) may reflect source chemistry, and that restite unmixing may be an important mechanism in some granitoid suites. However, we believe that the evidence for restite-controlled chemical variation should be critically examined. Is the restite model a *Revolution In Petrology* or should it be left to *Rest In Peace*?

MINERALS OF GRANITIC ROCKS

Restites, Xenocrysts, or Magmatic Phases?—Materials commonly held to be of restite origin take a variety of forms; single grains (although no convincing argument has been advanced to explain how the refractory residuum disintegrates to single grains), "clots" of one or more phases, and also texturally and mineralogically complex aggregates (xenoliths, enclaves, or inclusions). In general terms, residual (restite) phases should: (1) be in equilibrium with granitic melt at the temperatures and pressures of the anatectic environment; and (2) exhibit textural features consonant with growth in this high-grade metamorphic setting.

Restite phases must be the earliest phases in inferred paragenetic or crystallization sequences. The lower temperatures and pressures, higher $X_{H_2O}^{melt}$ and changes in chemical character of magmas late in their history should result in restite phases becoming unstable and reacting to produce other phases, a sort of fossil restite. However, many putative "restite" phases: (1) have euhedral grain shapes; (2) exhibit grain-size variations concordant with the magmatically crystallized phases of their hosts; and (3) are stable at low pressures under magmatic conditions. These features suggest growth of these materials from melts during ascent and emplacement. Some examples may help clarify these points.

Garnet, Cordierite, and Orthopyroxene.—Cordierite and garnet in peraluminous granitoids have commonly been interpreted as restites (e.g., Birch and Gleadow 1974; Flood and Shaw 1975; Wyborn and Chappell 1979; Hamer and Moyes 1982; Pattison et al. 1982; Clark and Lyons 1986). The (rare) presence of sillimanite inclusions in these phases and compositional zoning patterns in garnet (resembling those of metamorphic origin) are

the main lines of evidence for their residual character. Phillips et al. (1981) and Clemens and Wall (1981, 1984) regard the bulk of garnet and cordierite in several granitic and volcanic suites as products of magmatic crystallization. Taken together, the euhedral grain shapes, the paucity of inclusions, and their chemical variations in differentiated suites constitute firm grounds for this interpretation. Phillips et al. (1981) describe granites in which the grain-size of cordierite varies concordantly with that of the host rock, indicating that most of the cordierite crystallized at the emplacement level. Clemens and Wall (1984) document similar relations among garnets in a volcanic suite. Clemens et al. (1979) and Clemens and Wall (1984) demonstrated that cordierite developed relatively late in the crystallization sequences of some peraluminous volcanic suites. However, they also recognized a minor component of garnets with anomalous compositions, zoning patterns, and abundant inclusions. These could have a restite origin. Also, Allan and Clarke (1981) and Maillet and Clarke (1985) have recognized xenocrystic and magmatic garnets and both magmatic and hydrothermal cordierites in the South Mountain batholith, Nova Scotia. There is no clear evidence that any of these minerals are restites.

Theoretical studies (e.g., Abbott and Clarke 1979) and experimental investigations (Clemens and Wall 1981) of phase relations in peraluminous systems have demonstrated that garnet and cordierite are normal products of the crystallization of peraluminous melts, especially at conditions outside the muscovite stability field. Clemens and Wall showed that early crystallization of cordierite is restricted to fairly low pressures. The results of their multiple saturation experiments indicate that liquids initially coexisting with a range of assemblages appropriate to magma generation from peraluminous source rocks must, even after complete separation from the restite, crystallize highly aluminous phases.

In peraluminous compositions, orthopyroxene exhibits a wide crystallization field (generally above 750–780°C) compatible with its magmatic habit in volcanics and some granites (e.g., Clemens and Wall 1981, 1984; cf. Wyborn and Chappell 1979). On the basis of geobarometry applied to pyroxene- and olivine-bearing assemblages, Clemens (1982) showed that crystallization of some orthopyroxene occurred at low-to-moderate pressures unlikely to be representative of magma source regions. Cordierite-, garnet-, and orthopyroxene-bearing granitoids need not contain much residual refractory component.

Clinopyroxene and Hornblende.—Pyroxene- and amphibole-bearing mafic "clots," and/or clinopyroxenes rimmed by amphibole, are common in metaluminous granitoids. For these, a restite or "re-equilibrated restite" origin has been put forward by Presnall and Bateman (1973, p. 3197), White and Chappell (1977, p. 11), and Chappell (1978, p. 274).

Pyroxenes rimmed by amphibole could represent material crystallized from the melt at high T and lower f_{H_2O} which has partially reacted with the magma later in its crystallization history. Indeed, clinopyroxenes have broad P-T-f_{H_2O} stability fields in intermediate to silicic magmas (e.g., Eggler 1972; Green 1972; Eggler and Burnham 1973; Naney and Swanson 1980; Wang et al. 1980; Thompson 1983; Rutherford 1985; Clemens et al. 1986). It should be noted that the early, crystalline phases (that may control chemical variation) may not be the same as the observed minerals in the granites. False assumptions regarding the nature of the early crystals could have detrimental effects on modeling of the chemical variation.

The Ca-Tschermakite (CATS) contents of clinopyroxene in equilibrium with plagioclase and quartz can be used as a geobarometer (e.g., Wells 1980). The low CATS content of clinopyroxenes from granitic rocks indicates relatively low crystallization pressures, incompatible with restite origin. As an example, Higgins et al. (1986) demonstrated that clinopyroxene phenocrysts in the metaluminous St. Mary's rhyodacite crystallized at P <3 kb. If the magmas concerned were not saturated with quartz during clinopyroxene crystallization, this conclusion is further reinforced.

Mafic clots (which commonly contain plagioclase; see below) may represent accumulations of relatively early-crystallized magmatic material. Such crystals will tend to cluster in order to minimize surface energy (E. B. Watson pers. comm. 1987). Minerals in the clots commonly have compositions similar to those of isolated crystals. These same relationships are seen in andesites and other volcanic rocks for which there is no compelling evidence for the presence of restite. The fine-grained, quench-like textures of some mafic aggregates in metaluminous granitoids is incompatible with a restite origin.

Plagioclase.—Major and trace element abundances in granitoid rocks are significantly influenced by variations in the composition and abundance of plagioclase. Petrographic and experimental data (Clemens and Wall 1981; Phillips et al. 1981; Clemens and Wall 1984; Clemens et al. 1986) indicate that relatively calcic plagioclase develops early in the crystallization sequence of most granitic magmas. It is possible that some plagioclase in these rocks has a restite origin. White and Chappell (1977, p. 17) suggest that complexly zoned and twinned plagioclase is characteristic of modified restite in granitoids. They and coworkers further interpret the irregular, patchilly zoned ("mottled"), often corroded, calcic cores to be definitive evidence of restite origin. There are compelling arguments against these hypotheses.

Plagioclase with patchy zonation and commonly with corroded (partially resorbed) cores is common in basalts and gabbros (Vernon 1983). In view of the instability of plagioclase in the mantle peridotite sources, such occurrences cannot be interpreted as restites. By inference, similar textures in granitic rocks can be explained without recourse to the restite hypothesis.

Plagioclase crystals and their internal compositional zones are commonly subhedral to euhedral, even when in complex aggregates. The zoning is mostly normal with overprinted, fine-scale oscillations. Zones commonly surround point nucleii (e.g., Phillips et al. 1981; Clemens and Wall 1984). Even the corroded cores of plagioclase crystals commonly display complex zoning patterns (Vance 1962; Vernon 1983). Such microstructures are exceedingly uncommon in plagioclases from high-grade metamorphic rocks believed to be the sources of most granitoid magmas. In some migmatites the melanosomes are demonstrably restitic in character and do not generally contain plagioclases with complex zoning patterns (Johannes

1985). Conversely, Vernon (1983) points out that *unzoned* cores do not necessarily have a restite origin (c.f. Wyborn 1983), since crystallization from granitic melts at small degrees of undercooling can produce large, homogeneous crystals (Swanson 1977).

There seems little reason to depart from petrological orthodoxy in ascribing plagioclase zoning patterns to crystal-liquid processes. For water-undersaturated conditions, plagioclase saturation surfaces in granitic compositions have steep dP/dT slopes (e.g., Whitney 1975; Burnham 1979; Naney and Swanson 1980; Clemens and Wall 1981; Naney 1983). The composition of plagioclase in equilibrium with granitic melt is therefore sensitive to variations in P, T, and f_{H_2O} during magma ascent and emplacement. Varying degrees of undercooling or departure from saturation of the melt with particular plagioclase compositions, coupled with the slow diffusion rates of Al and Si in the melts and plagioclase crystals, can produce a variety of zoning patterns and resorption textures (e.g., Vance 1962 and references therein; Swanson 1977; Johannes 1978). Magma mixing (or mingling), with its combined thermal and compositional effects, can lead to complex resorption and zoning patterns also (e.g., Hibbard 1981; Tsuchiyama 1985; Cocirta 1986). Crustal assimilation could presumably have similar effects.

The composition of plagioclase cores is not always "anomalously calcic" as is sometimes inferred by proponents of the restite model. The cores of plagioclases in the Violet Town Volcanics (An$_{50}$; Clemens and Wall 1984) have compositions close to those of near-liquidus plagioclases formed in crystallization experiments on a natural rhyodacite composition (e.g., 5 kb, 850°C, melt H_2O content = 2.5 wt %, An$_{49}$; Clemens 1981), Burnham's (1979, 1981) models for the phase relations of hydrous felsic magmas suggest that, especially at deep crustal pressures and moderate melt water contents, the composition of near-liquidus plagioclase should commonly be rather calcic. Given better phase equilibrium data for plagioclase-melt relations, useful tests of the restite origin of plagioclase cores could be made.

Zircon.—Abundant evidence has been presented (e.g., Silver and Deutsch 1963; Pidgeon and Aftalion 1978) that many granitic rocks contain old, crustally recycled zircons. Textural evidence of rounded zircon cores with euhedral overgrowth supports the Pb isotope data. Watson and Harrison (1983) have shown that the solubility of Zr in granitic melts is low and that many granites have much higher Zr contents than can have been dissolved in the melts. It is clear that many granites (particularly the peraluminous, metasediment-derived types) contain refractory zircons. In metamorphic rocks, zircons occur mostly as inclusions in biotite crystals. During anatexis the biotite breaks down and contributes components to the melt. It is not surprising, therefore, that refractory zircons could be derived from the source rocks or from wall rocks assimilated during ascent and emplacement of a pluton.

The abundance of Zr in granitic magmas is largely controlled by the behavior of zircon and in many granites, some fraction of the zircon may have a restite origin. Thus, the variations in concentration of Zr may be difficult to ascribe uniquely to any particular process. Similar arguments may hold for some other trace elements (e.g., Y, Ce, La) that are mainly concentrated in highly refractory accessory phases, such as monazite and xenotime. At the same time, the other trace element and major element variations may be controlled by processes such as crystal fractionation. Such mixed behavior can greatly complicate geochemically based differentiation models.

ENCLAVES IN GRANITIC ROCKS

Xenoliths, Magma Mixing Products, Cognate Cumulates, or Restites?—White and Chappell (1977) have noted that many of the more mafic granitoids contain abundant enclaves (also see Didier 1973) that are largely unrelated to their wall rocks and are broadly compatible with their inferred source materials (e.g., metasedimentary inclusions in some peraluminous granites). In the latest version of the restite model (A. J. R. White pers. comm. 1985), the enclaves are not considered to be derived from the principal source rocks and do not play a dominant role in the restite unmixing process. Plagioclase cores and mafic "clots" are the major restite components, while the enclaves represent minor source lithologies that may not have undergone melting. If this is correct, then at least some of the enclaves should have metamorphic fabrics and parageneses.

Enclaves can have a variety of origins, and we suggest that most plutons contain a number of different types. Below, we examine the characteristics of enclaves of various provenances (see also Vernon 1983).

Accidental Inclusions (Xenoliths).—This group includes crustal materials derived from the wall rocks near the emplacement level, and those derived at deeper levels and related to neither the present wall rocks nor the primary source region.

In many high-level intrusions, xenoliths of both these origins are readily recognizable. Inclusions derived from shallow depth are fine-grained and may exhibit hornfelsic textures. Mineral assemblages and chemistry may be incompatible with the magma (e.g., pelitic metasedimentary xenoliths in hornblende-bearing, metaluminous granites). Xenoliths derived from deeper levels may contain relicts of high-pressure mineralogy, and there may be evidence of reaction with the enclosing magma.

The problem of identifying accidental inclusions in synmetamorphic or "regional aureole" granites (White et al. 1974) is more difficult, as foliated types could have either restite or xenolithic origins. In these cases, chemical incompatibility (with magma) would be the major criterion for recognizing true xenolithic inclusions. The presence of such xenoliths suggests the possibility of contamination by partial assimilation of these xenoliths.

Evidence for assimilation would include the presence of xenoliths in various stages of digestion and the existence of isotopic mixing curves (Taylor 1980; DePaolo 1981; Clarke et al. 1987). Existing numerical analyses (e.g., Shaw 1974; Marsh and Kantha 1978) do not favor assimilation as a major process in granitic magmas. Vernon (1983) notes that no clear textural evidence of xenolith assimilation has been presented in the literature. However, White et al. (1986) and Ague and Brimhall (1987) note that the chemical features of some batholiths in California result from assimilation of pelitic wall rocks (c.f. Todd and Shaw, 1985). Muecke and Clarke (1981) and Clarke and Muecke (1985) present mineralogical, chemical, and isotopic evidence that the South Mountain batholith

magma assimilated some of its low-grade metasedimentary wall rocks near the margins of the pluton. In summary, the available evidence suggests that assimilation is a measurable process in the evolution of granitic magmas.

Magma Mixing Products.—Larsen et al. (1938), Eichelberger and Gooley (1977) and Eichelberger (1978, 1980) have shown that partial mixing of mafic and silicic magmas is widespread in calc-alkaline volcanic associations. Clear examples of the mingling of mafic and silicic magmas also occur in plutonic suites (e.g., Reid et al. 1980, 1983; Cantagrel et al. 1984; Furman and Spera 1985; Kistler et al. 1986).

The field evidence for magma mixing is mainly the occurrence of chilled mafic fragments in a silicic matrix. In the more slowly cooled intrusive environment, quench textures are partly obscured by recrystallization effects. In some metaluminous granites the presence of fine-grained, mafic inclusions may indicate rapid recrystallization of largely liquid "blobs." Such inclusions are commonly rounded, have crenulated margins, and are flow layered, compatible with their being in a partially liquid state. Some contain enclaves themselves (e.g., Didier et al., 1982). Mixing may occur at deep, near-source conditions or in shallow, composite magma chambers.

Mixing is the classic cause of linear variation in major and trace element Harker diagrams. Isotopic data for a suite of rocks influenced by high-level magma mixing should also reveal pseudo-isochrons or mixing lines. If a granitic pluton has these field, geochemical and isotopic features it is reasonable to conclude that magma mixing has played a major role in its chemical evolution. For some metaluminous granites particularly, magma mixing or mingling seems to have been important. Vernon (1983) and Didier and co-workers (Didier 1982; Didier et al. 1982; Cantagrel et al. 1984) believe it is the dominant process.

Cognate Cumulates.—In some silicic volcanic suites and granites, cognate cumulates are common and may dominate the enclave populations. Such cognate inclusions are well documented among suites of mafic igneous rocks. Phillips et al. (1981) and Vernon and Flood (1982) believe that many microgranitic

736 V. J. WALL, J. D. CLEMENS, AND D. B. CLARKE

enclaves with igneous textures represent accumulations of near-liquidus phases formed on the cool walls of magma chambers and in volcanic conduits. Didier (1973) cites numerous examples of such mafic-enriched chilled margins and enclaves derived from them. The fine grain-size is attributed to high degrees of undercooling, abundant nucleation, and rapid crystal growth.

Such cognate cumulates are characteristically enriched in early-crystallized ferromagnesian phases and calcic plagioclase (Pabst 1928; Bateman and Chappell 1979; Phillips et al. 1981) with variable amounts of intercumulus quartz and feldspar. Igneous textures such as euhedral crystal shapes and flow fabrics are common (Didier et al. 1982; Vernon 1983). However, there is usually evidence of some solid state textural modification. Vernon (1983) notes that some textures in these enclaves resemble those of high-grade metamorphic rocks.

On mineralogical grounds alone, these cognate inclusions could be confused with restites, since early-crystallized minerals may resemble restite phases. However, igneous microstructures and isotopic coherence and chemical compatibility with the magma (lack of evidence for reaction) would be characteristic features of cognate inclusions. Distinguishing between a cognate comagmatic origin and one due to magma mixing presents more difficulties. However, true linear variation in Harker plots of major and trace elements, possible isotopic incoherence between large enclaves and matrix, and the textural features noted above should serve to distinguish inclusions of magma mixing origin.

If these criteria are applied to the inclusions in many granitoid suites commonly held to be restite controlled (e.g., White and Chappell 1977; Chappell 1984) the inclusions are best interpreted as cognate cumulates and magma mixing products.

Restites.—The source regions of crustally derived granites should be high-grade, regionally metamorphosed terranes of upper amphibolite to granulite facies (Fyfe 1973*a*, 1973*b*; White and Chappell 1977; Clemens and Wall 1981; Clemens 1984). Thus, the most compelling mineralogical and textural evidence for the restite character of an inclusion would be: (1) granoblastic grain shapes; (2)

foliated fabrics; (3) high-temperature, relatively anhydrous, refractory mineral assemblages; (4) assemblages consistent with pressures in excess of emplacement level; (5) alkali- and silica-depleted compositions; and (6) a degree of chemical compatibility with the enclosing granitic rock depending on the extent of re-equilibration of melt and solids en route to the surface. Note also, that criteria (3) and (4) may be complicated by the occurrence of retrograde reactions occurring in response to the lower P and T and higher f_{H_2O} conditions that develop as a magma separates from the source, rises, and evolves chemically.

In metasediment-derived granitoids, restite inclusions would contain assemblages involving quartz, garnet, cordierite, biotite, kyanite, sillimanite, hercynite, calcic plagioclase, and K-feldspar, with hypersthene in less aluminous types and corundum in SiO_2-poor rocks. For metaluminous rocks, derived from mafic to intermediate igneous sources, restite inclusions would contain clinopyroxene and calcic plagioclase. Orthopyroxene and garnet may be present, but K-feldspar and quartz should be absent because these phases crystallize only late in the histories of such magmas. Scapolite could be a residual phase in some meta-igneous sources also. Restite inclusions should commonly show evidence of mineralogical and textural readjustment in the lower-T, lower-P, more hydrous environment near magma emplacement level. Retrograde cordierite and biotite may form in metasediment-derived magmas, whereas in metaluminous rocks, hornblende may replace high-grade pyroxenes. Such reactions may be facilitated and complicated by infiltration of the inclusions by magma.

Isolated restite crystals will commonly be anhedral and may have euhedral, magmatic overgrowths. They may poikilitically enclose ovoid inclusions of other minerals. Mica, sillimanite or amphibole inclusions in restite grains may form foliaceous aggregates and trains. Except in magmatic overgrowths, restite plagioclase should not show complex oscillatory zoning but may be unzoned or reversely zoned.

As with magma mixing, chemical variation controlled by a single restite component should produce linear arrays on Harker plots. However, the heterogeneity of crustal sources (especially those dominated by meta-sediments) should give scattered arrays in such diagrams. Isotopic variation should indicate general coherency between restite inclusions and enclosing rocks. Again, the probable variety among source rock types for large granitic bodies should result in some degree of scatter, reflecting the possibility of more than one restite and melt composition and incomplete homogenization during melting.

Many peraluminous suites derived from metasedimentary sources contain material of probable restite origin (e.g., Phillips et al. 1981; Price 1983; Clemens and Wall 1984). In these cases, however, restite enclaves and individual restitic mineral grains may be distinguished on textural and chemical grounds from magmatic minerals and cognate magmatic enclaves. Where the proportion of restite has been estimated (e.g., Clemens and Wall 1984), it usually comprises only a few percent of the magma; cognate microgranitic enclaves far outnumber those of probable restite or other origins.

In metaluminous rocks, enclaves showing the textural features expected of restites are virtually unknown. Textural, geochemical, and isotopic characteristics favor magma mixing or cognate magmatic origins for essentially all microgranitic enclaves in such rocks.

SUMMARY OF THE MINERALOGICAL AND TEXTURAL RELATIONS

Inclusions in granitoids and silicic volcanics have predominantly non-restite origins. This is especially true of metaluminous rocks. Igneous inclusions are the most abundant type, although some granites contain mixed populations. Even inclusions derived from the source region are not necessarily representative of the dominant source lithology. Individual mineral grains and clots also show features typical of magmatic origins. Rocks containing relatively abundant restite are also enriched in early magmatic, cumulate material.

EXPERIMENTAL CONSTRAINTS

Corroded Plagioclase Cores and Clinopyroxene Inclusions.—It was concluded above that the corroded cores in plagioclase crystals from most granites were almost entirely formed by magmatic processes. The recent experiments of Ellis and Thompson (1986) in the system $CaO-MgO-Al_2O_3-SiO_2-H_2O$ shed further light on the formation of these cores in metaluminous magmas.

Ellis and Thompson showed that amphibole stability in the high-T part of the system is controlled by reactions such as:

$$Amp \pm Fl = An + Opx + Cpx \pm Qtz + M$$

where Fl and M denote aqueous fluid and hydrous melt, respectively. Such equilibria have steep, positive dP/dT slopes with Amp on the low-T sides. Ellis and Thompson (1986, fig. 9) show that cooling paths of magmas, in equilibrium with calcic plagioclase + pyroxenes $\pm$ quartz, will commonly intersect these retrograde reactions. When this happens, early magmatic pyroxene and calcic plagioclase are resorbed and amphibole is precipitated. In systems containing Na_2O, such as natural magmas, a more sodic plagioclase may be precipitated on the corroded remnant of the early, high-T feldspar. Inclusions of clinopyroxene in the corroded cores of these feldspars and amphiboles may represent armored relics of the high-T magmatic phases.

Liquidus Temperatures of Granitic Compositions.—Piwinskii and Wyllie (1968) note that the liquidus temperatures of the more mafic members of granitic suites are too high for these compositions to have been liquids. However, Clemens and Wall (1981) pointed out that it is difficult to identify parent magma compositions. Some experiments have been carried out on rocks bearing significant accumulations of crystals. For example, Clemens and Wall (1981) presented the crystal-liquid phase equilibria for Strathbogie granite 889. This is the most mafic sample from the batholith and is interpreted as a crystal cumulate because it has 50% phenocrysts in a microgranitic groundmass. Its liquidus (875–930°C at 5 kb) cannot be taken to be that of the parent magma. An example of how experiments on such cumulate rocks can be used to constrain models for differentiation is given below.

Figure 1 shows the results of visual estimates of the amounts of crystals in the run products of experiments on 889. Geothermometry on the Strathbogie rocks indicates

738   V. J. WALL, J. D. CLEMENS, AND D. B. CLARKE

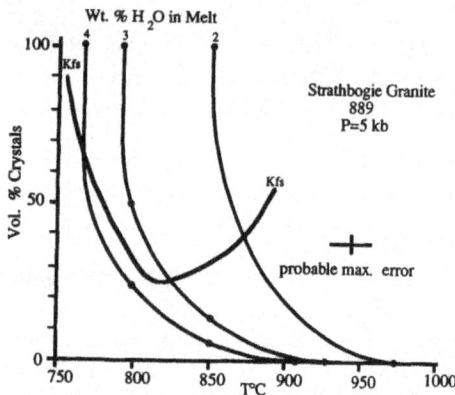

FIG. 1.—Volume percent crystals (visual estimates) as a function of melt water content and T for Strathbogie granite 889 at 5 kb. Data points are shown as dots. The position of the K-feldspar saturation boundary is shown also. Data are from Clemens (1981).

TABLE 1

POSSIBLE CRYSTAL CONTENTS FOR
STRATHBOGIE GRANITE

Water Content (wt %)	$T(°C)$		
	800	850	900
3	50	10	2
4	30	5	1

early crystallization at $T = 810$–$890°C$ (Grt-Crd, Wells 1979; Grt-Bt, Ferry and Spear 1978). Clemens and Wall (1981) suggest 5 kb phase relations as the best model for early crystallization of the Strathbogie magma, and Clemens (1981) concluded that the crystallization sequence was best modelled for melt water contents of 2.8–4 wt %. Figure 1 was used to calculate the amounts of crystals that the Strathbogie magma may have contained for a range of temperatures and melt water contents (table 1). The results range from <1 to 50 vol %. Since 889 contains almost no K-feldspar phenocrysts, and K-feldspar saturation occurs only late in the crystallization history of the Strathbogie magma (Phillips et al. 1981), figure 1 can be used to limit the maximum crystal content to ~60% for a melt water content of 4 wt % and ~30% if the melt had 3 wt % H_2O. The most reasonable conditions for early crystallization are: 850°C and 3–4 wt % H_2O in the melt. Thus, granite 889 may have contained only 5 to 10% crystals during its early magmatic history. Even for this cumulate rock, the amount of crystalline material (restite and/or magmatic) must have been small. For the bulk magma, the amount would probably have been as low as 1–2%. Even if all this were restite, it could not control differentiation as ~17% plagioclase, quartz, and biotite need to be separated in order to account for the observed chemical variations (Phillips et al. 1981).

Compositions of Granitic Liquids.—Some recent work with simplified granitic compositions (Miller et al. 1985) has suggested that granitic melts saturated with biotite or hornblende can contain no more than 2.5 wt % total FeO + MgO at temperatures up to 1050°C. These results have been interpreted as evidence that rock compositions in the granodiorite-tonalite range cannot exist as liquids at the T-f_{H_2O} conditions generally inferred for generation and emplacement of such magmas.

As we stated above, we believe that many of the more mafic granitic rocks represent accumulations of crystalline material. However, we would reject these experiments as evidence of the abundance of restite on two grounds:

1. The experiments described by Miller et al. (1985) involve seeding haplogranitic glasses with crystals of natural micas and amphiboles. Watson and Miller (written comm. 1987) inform us that all runs were carried out with an initial, high-T step in which all seed crystals were destroyed. Following this, T was lowered to final run conditions. Naney and Swanson (1980) report that such experiments result in metastable persistence of ferromagnesian phases to temperatures above their stability limits and that nucleation of tectosilicate phases is strongly inhibited. Both these factors will result in anomalously felsic quench glasses.

It seems better to nucleate and grow ferromagnesian phases from an amorphous starting material in which both stable and metastable nucleii form during the low-T history of the run. Once final run conditions are attained the stable phases grow while others disappear (Naney and Swanson 1980; Clemens 1981). Clemens and Wall (1984) reported the results of such experiments on a peraluminous, rhyodacitic glass. Analysis of the glass coexisting with Qtz, Pl, Bt, Opx, Grt, and Ilm (produced in a 136 hr experiment at 5 kb, 850°C and with 3 wt % H_2O in the melt) shows 2.9 wt % total FeO + MgO on an anhydrous basis. This result was obtained by broad-beam microprobe analysis and may suffer from counting losses on Mg, common in analyses of hydrous glasses. Naney (1983) reports quench glass compositions for 8 kb experiments on synthetic granitic rocks. For T in the range of 650–950°C, these glasses have FeO + MgO in the range 0.82 to 3.16 (wt %, anhydrous basis). An interesting set of experiments was carried out by Rutherford et al. (1985) on crushed pumice from the May 18, 1980 eruption of Mount St. Helens. The starting material was only about 60% glass, but at P = 1–3.2 kb, T = 875–1090°C and in run times of 5–55 hrs, they obtained homogeneous glasses with 62–76% SiO_2, 1–8% H_2O, and 2.22–6.05% FeO + MgO. Further corroborating evidence can be found in the experiments of Vielzeuf and Holloway (1985), who studied fluid-absent melting in a pelitic composition at 10 kb. At temperatures of 875–1050°C they produced quench glasses with 66.6–73.2% SiO_2, 3.22 to 3.74% H_2O, and 1.4–7.1 wt % FeO + MgO. Also, Grapes (1986) reports the compositions of melts formed from pelitic xenoliths in a trachyte magma at 2–3 kb and 873–1034°C. The quench glasses in these xenoliths have SiO_2 between 6.35 and 67.6%, estimated water contents around 8%, and between 1.31 and 5.88% total FeO + MgO. Note also that Brown and Fyfe (1970) produced partial melts at 1–10.5 kb and 730–900°C from quartz-dioritic rocks and found quench glasses with 1.17 to 5.20 wt % FeO + MgO. These results show that granitic to tonalitic compositions *can* exist as liquids at reasonable T and f_{H_2O} conditions.

2. Some granodioritic to tonalitic magmas undoubtedly contained substantial amounts of entrained crystals. However, this crystalline component cannot be *assumed* to be of restitic origin. Indeed, the weight of evidence suggests that most of this material consists of accumulated, early magmatic phases.

RESTITE UNMIXING AND THE
GEOCHEMICAL EVIDENCE

Harker Diagrams and Linear Element Variations.—Linear trends in Harker variation diagrams form the primary basis for postulating that restite unmixing was the cause of the variation within some granitoid suites. For many suites the variation of element concentrations with SiO_2 is decidedly *non*-linear and, for these, there is general agreement that any two-end-member mixing process is inadequate to explain the data. Most such suites can be well modeled by combinations of Rayleigh fractionation and variable retention of the refractory crystals (e.g., McCarthy and Groves 1979; Bellieni et al. 1981; Tindle and Pearce 1981; Miller and Mittlefehldt 1982; Lee and Christiansen 1983; Clemens and Wall 1984; Baker and McBirney 1985; Bateman 1985; Clarke and Muecke 1985; and Higgins et al. 1986). Chappell (1984) and White and Chappell (1983) argue convincingly that groups of granitic plutons can be divided into suites, each rock within a suite being derived from a similar source material. This partitioning into suites forms the linchpin of the restite model, because it is within suites that linear element variation trends may be discerned. There is, however, an unsettling aspect to this. Whether a given rock "belongs" with a particular suite depends, largely, on whether it fits on the linear chemical variation trend defined by rocks of that suite. In other words, recognition of suites is partly on the basis of linear trends of Harker plots. (It is obvious that if a model guides the sampling, the sampling will confirm the model. This is as true for these suites as it would have been if random amounts of mafic enclaves had been included in analyzed granites; the chemical data would "confirm" that the variation was controlled by varying amounts of "restite".) There are rocks whose geographic, geological, and petrographic characteristics suggest kinship with a particular suite but which have not been included, primarily because they plot off the linear trends defined by the rest of the rocks. These "orphan" rocks are invariably at either the extreme mafic or felsic end of the compositional spectrum in the suite concerned. However, there are many suites of granitic rocks and silicic volcanics in which element variations do closely approximate linear trends. Thus, it is worthwhile investigating the causes of linearity on Harker diagrams.

True linear variation will result in cases of

simple mixing involving two end members (e.g., magma mixing or restite unmixing). This is only true for the restite model if all the restite fragments are chemically homogeneous. If not, the large and/or mafic ones will unmix more readily than the small and/or felsic ones, resulting in decidedly non-linear trends on Harker variation diagrams. Also, if restite can separate, so can nucleated minerals, making recognition of the restite effect doubly difficult.

Under certain circumstances, even pure Rayleigh fractionation can yield sensibly linear trends. For most elements and oxides, in most granitic magmas, the *bulk* crystal-liquid distribution coefficients (D) will fall in the range 0.3–3.0. Figure 2 illustrates the implications for Rayleigh fractionation trends. The first relevant feature of the variation in the figure is that, for "incompatible" elements ($D < 1$), the variation will never deviate greatly from linearity. The second important thing of note is that the degree of curvature displayed on a Harker plot for a "compatible" element ($D > 1$) is not only determined by bulk D for the element concerned, but also is strongly influenced by the bulk D value for SiO_2.

In many metaluminous granitoids, quartz is late and interstitial; the early-crystallizing phases (which may control differentiation) include plagioclase, pyroxenes, and minor accessories. In such cases the bulk D_{SiO_2} will be near 0.75–0.85; depending on the plagioclase composition and the proportions of fractionating phases. In contrast, quartz is almost always a near-liquidus phase in peraluminous magmas and is accompanied by plagioclase, minor biotite, and accessory phases. The result is that, for peraluminous types, bulk D_{SiO_2} is commonly 0.9. Even though parental peraluminous magmas commonly have higher SiO_2 contents than metaluminous types, considerably greater fractionation is necessary to produce high-silica rocks (76% SiO_2). Figure 2 shows that, for a peraluminous magma with an initial SiO_2 content of 70%, crystallization of an assemblage with bulk $D = 3$ will produce a clearly curved variation if it fractionates to a composition with 76% SiO_2. A model metaluminous melt with an initial SiO_2 content of 65% will fractionate toward the high silica composition while producing only a very gently curved trend.

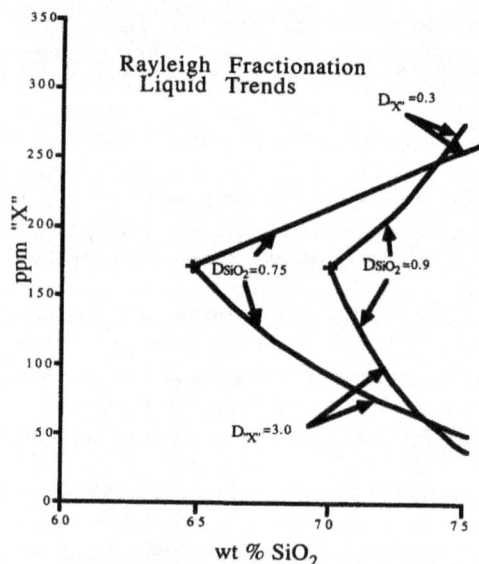

FIG. 2.—Model liquid fractionation trends on Harker diagrams: *a.* for an incompatible element ($D = 0.3$) in a magma where the fractionating crystalline assemblage has $D_{SiO_2} = 0.75$; *b.* for the same element in a magma in which the fractionating assemblage has $D_{SiO_2} = 0.9$; *c.* for a compatible element ($D = 3.0$) and $D_{SiO_2} = 0.75$; *d.* for the same compatible element but with $D_{SiO_2} = 0.9$. Initial concentration of both trace elements is assumed to be 170 ppm. At 76% SiO_2 the magma with $D_{SiO_2} = 0.75$ is 46% crystalline while that with $D_{SiO_2} = 0.9$ is 56% crystallized.

It is worth noting that the SiO_2 scale in figure 2 is rather compressed and the vertical scale is somewhat expanded, accentuating the curvature of the fractionation paths. Further effects that enhance linearity in variation diagrams for rock groups whose internal chemical variations are controlled by crystal fractionation include accumulation of crystals in the more mafic liquids, incomplete sampling of rock units (segments of curved trends appear linear), and scatter in the data (obscuring subtle features of the variation). Last, a small amount of equilibrium crystallization may be followed by variable mixing of the solid and liquid phases to produce genuine straight-line mixing trends. A possible example of this is the granite described by Sultan et al. (1986).

The Jindabyne Suite.—The preceding analysis indicates that the simple observation of linear trends in Harker plots does not favor any particular process for producing the ob-

2

2

served variation. Here we will examine the geochemistry of a specific group of rocks that has been cited as a classic example of restite-controlled variation, the Jindabyne suite in the Kosciusko batholith of southeastern Australia (Hine et al. 1978; Compston and Chappell 1979; Chappell 1984).

The metaluminous Jindabyne suite contains eight separate plutons that have a total outcrop area of 120 km^2. Rocks range in composition from high-alumina gabbro and gabbroic diorite to tonalite. The chemical variation is roughly continuous from 60–67% SiO$_2$. Data for the more mafic part of the suite are scattered, reflecting the small proportions of these rocks exposed and probably their low abundance in the suite. The gabbro and gabbroic diorite occur as a small, isolated, roughly circular stock and a finger-like appendage on one of the larger tonalite plutons respectively.

Both major and trace element variation diagrams for the Jindabyne suite exhibit near straight-line trends (see e.g., Hine et al. 1978, figs. 7 and 8) and White and Chappell (1983) state that chemical variation in the suite is controlled by restite unmixing. Figure 3 shows the variations in Rb, Sr, Ba, and Ni plotted against SiO$_2$. If such trends represent simple mixing of a restite and a melt component, any intermediate rock composition can be expressed as some fixed proportion of restite and melt. Even if absolute end members (pure restite and pure melt) cannot be identified among the sampled rocks, intermediate compositions can be described in terms of fixed proportions of the extreme high- and low-SiO$_2$ rocks. This implies that the calculated proportion of restite in a given rock must be the same for all analyzed elements. This property should provide a sensitive test of the validity of the restite model.

For the Jindabyne suite it is assumed that the rock containing 47.25% SiO$_2$ (KB53) represents pure restite while pure melt is represented by KB7 at 67.15% SiO$_2$. Using these end members we have calculated the proportions of restite and melt needed to form KB5 at 60.18% SiO$_2$. The results for major elements (Hine et al. 1978) are shown graphically in figure 4. The SiO$_2$ contents of the

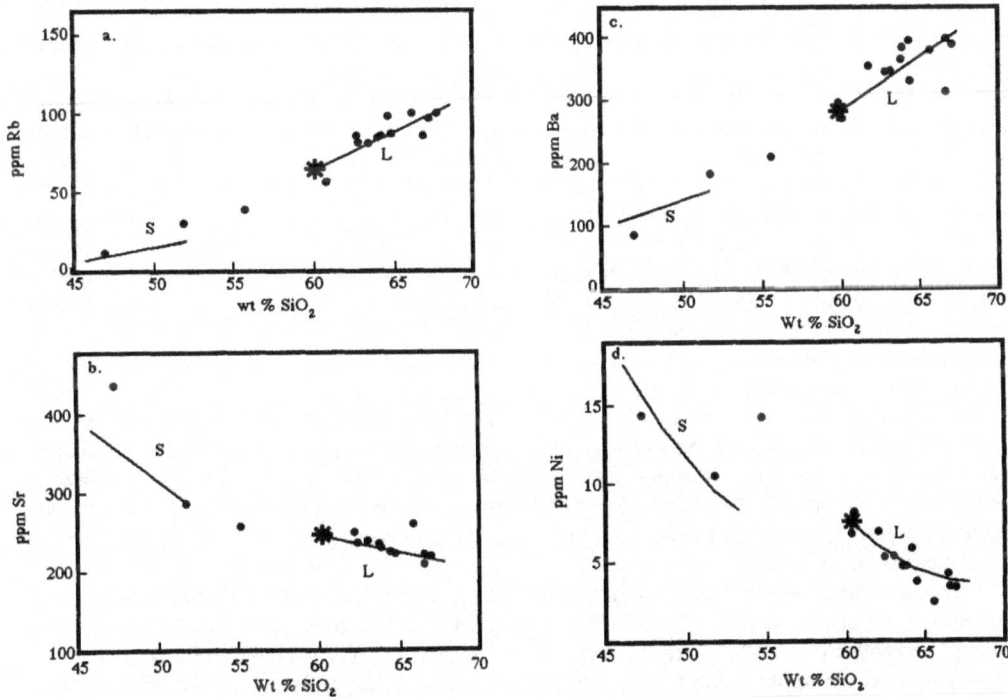

FIG. 3.—Trace element Harker plots for rocks of the Jindabyne suite: a. Rb; b. Sr; c. Ba; d. Ni. Data from Hine et al. (1978) and Chappell (1984). The predicted solid (S) and liquid evolution trends (L) for 40% pure Rayleigh fractionation of parent magma (*) are shown for comparison. See table 2 for model details.

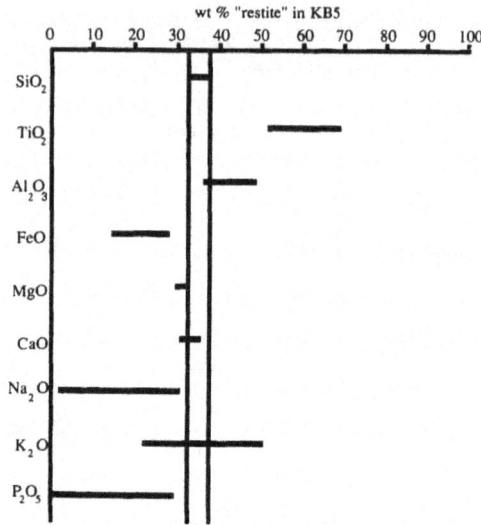

Fig. 4.—Amounts of restite (KB53) necessary to be mixed with melt (KB7) in order to form rock KB5 (in the Jindabyne suite), for several analyzed elements. Primary analytical data from Hine et al. (1978) and Chappell (1984). Bar widths represent estimates of the errors associated with the calculations.

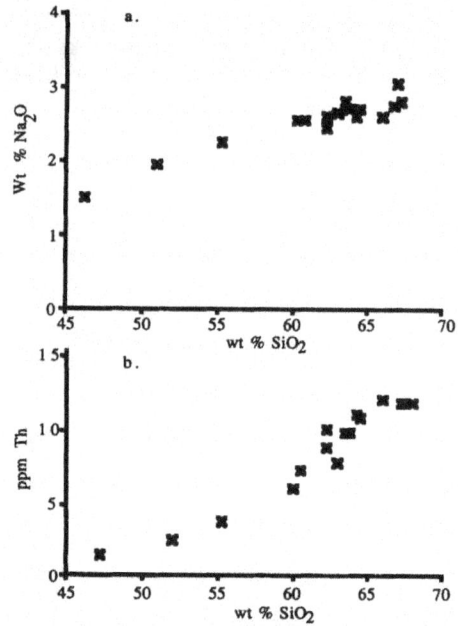

Fig. 5.—Examples of non-linear variations in Harker plots for rocks of the Jindabyne suite: a. Na_2O; b. Th. Data from Hine et al. (1978) and Chappell (1984).

chosen end members and target composition fix the required restite content of KB5 at $35 \pm 2\%$. The scatter in the analytical data for some elements results in large uncertainties in the calculated proportions of restite. However, the plot shows that the results for Na, P, Ti, and Fe deviate unacceptably from the SiO_2 value. Results for the trace elements analyzed by Hine et al. (1978) show also that Sr, Y, Zr, and Th variations do not easily fit with simple mixing lines (e.g., fig. 5). This suggests that restite unmixing may not have been the mechanism controlling the chemical variation in the Jindabyne suite.

To illustrate the viability of an alternative differentiation scheme we have modeled the variations in Rb, Sr, Ba, and Ni in terms of crystal-liquid fractionation. The results are shown in figure 3, where the solid and liquid evolution trends can be compared with the actual geochemical data. For the model we have assumed that the gabbroic rock (KB53) is a nearly pure crystal cumulate, that the initial melt had 60% SiO_2, and that trace element concentrations are appropriate to this silica content. Hine et al. (1978) give a modal analysis of KB53: 60% plagioclase cores of An_{80}, 12% clinopyroxene, 10% orthopyroxene, and

2% magnetite. We have assumed that these phases, in these proportions, crystallized from the magma throughout the differentiation process. Table 2 shows the parameters used in the modeling. The calculations were performed using the computer program "CRYSTALLIZATION" (Petitpierre and Boivin 1983).

From the results in figure 3 it is apparent that most of the suite could be interpreted as a succession of nearly crystal-free liquids derived by fractionation of 0–40% of the chosen crystalline assemblage. The more mafic rocks in the suite plot close to the compositional trend for the solids and could be interpreted as accumulations of precipitated crystals mixed with varying amounts (0–70%) of intercumulus liquid. The apparent straight-line variations in the chemistry of the suite can be viewed as artifacts of the low bulk D_{SiO_2} value (0.76), the low bulk D values for the modeled trace elements (0.15–2.7), and the effects of variable mixing between accumulated crystals and derivative liquids. This model can also explain the apparent absence of rocks with SiO_2 between 55.5 and 60.2%. The low-SiO_2 rocks may contain

MODELS FOR GRANITOID EVOLUTION   743

TABLE 2

PARAMETERS USED IN TRACE ELEMENT MODELING

| Fractioning Phase: | Crystal-Liquid Distribution Coefficients[a] Used | | | | Composition of Initial Liquid |
	Plagioclase An$_{80}$	Clinopyroxene	Orthopyroxene	Magnetite	
SiO$_2$[b]	.77	.79	.79	.000001[c]	60 wt %
Rb	.2	.02	.02	.000001[c]	70 ppm
Sr	2.0	.08	.02	.000001[c]	255 ppm
Ba	.5	.02	.02	.000001[c]	300 ppm
Ni	.000001[c]	8.0	9.0	20.0	8 ppm
Proportion in Solid:	.74	.143	.119	.024	

[a] Gill (1978).
[b] Calculated from probable SiO$_2$ contents of phases.
[c] Assumed values.

mostly crystals, and there may have been near-perfect crystal-liquid separation during the fractionation process.

In Chappell (1984) the Jindabyne suite is modified to exclude the high: alumina gabbro and gabbroic tonalite samples (47.25% and 52.14% SiO$_2$ respectively). Chappell reports that the most mafic member of the suite is now considered to be a new sample (KB139, 55.50% SiO$_2$) from a low-SiO$_2$ portion of the pluton known as the Grosses Plain Tonalite (Hine et al. 1978). Chappell (1984, p. 260) also states that, "in any suite, the most mafic granite composition can be taken as the *model source* composition"; i.e., melt plus maximum amount of restite.

We have thus undertaken modeling of the diminished Jindabyne suite as a Rayleigh fractionation series using a composition close to KB139 as the initial *liquid*. The results are nearly as acceptable as those shown in figure 3. In figure 6 we show the results for Ba as an example. The mean deviation of the Ba concentrations from those predicted by the Rayleigh curve is 25.71 ppm compared to 21.64 ppm for the linear regression, also shown in figure 6. Statistical tests at the 0.01 level of significance show that it is impossible to choose between the Rayleigh and the linear regression. The fits for Rb, Sr, and Ni are as good as, or better than, that for Ba. So even the "*model source*" composition could have been a liquid parent magma.

We may perform a test on *this* data set, similar to that used previously to construct figure 4. Thus, using KB139 to represent one mixing end member and KB7 as the other, we

calculate the amounts that must be mixed to form KB5. Note that these rock analyses all fall close to the linear fits to the analytical data on Harker plots. As before, there are unacceptable departures from the constraining SiO$_2$ result. Results for TiO$_2$, MgO, Na$_2$O, Sr, and Zr suggest that, in the case of the Jindabyne suite, geochemistry alone cannot provide definitive evidence of the causes of the observed elemental variations.

The major conclusion from the foregoing is that, although linear trends on Harker variation diagrams are permissive evidence for a mixing process, there are other mechanisms that can produce similar patterns of variation. It has already been demonstrated that the

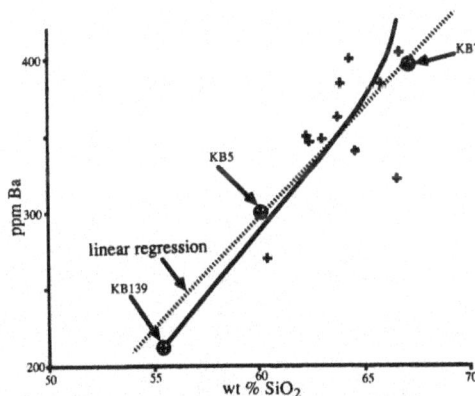

FIG. 6.—Harker plot for Ba in the Jindabyne suite. Points for rocks KB139, KB5, and KB7 are indicated. The dashed line is a linear fit to the data. The solid line is the predicted Rayleigh fractionation trend using KB139 as the initial liquid composition. Other model parameters are the same as in table 1 except that $D_{Ba}^{Pl-L} = 0.16$ and $D_{Ba}^{Opx-L} = 0.03$.

textural and mineralogical evidence cited for the presence of abundant restite in granitoid suites is equivocal at best. The same may be said of the geochemical evidence, even in the case of a classic "restite-controlled" suite such as the Jindabyne.

Suites and Magmas.—There is one further question arising here that needs attention. On what scale should we examine granitic rocks for evidence of differentiation mechanisms? Are suites the proper units to work with?

Chappell (1984) defines a suite as a group of rocks with strong petrographic, chemical, and isotopic similarities. The rocks may come from a single pluton, or a group of plutons, or a group of parts of plutons. Chappell (p. 695) then goes on to state that suites "are the basic unit(s) in discussing most chemical and isotopic features." If we are to use suites in discussing differentiation we must assume that a suite represents the products of solidification of a single magma. There are reasons, however, to doubt the validity of this assumption.

Chappell (1984, p. 696) writes that "different suites result from different sources rather than different processes of crystallization or solidification." Also, "A single suite is derived from *effectively* uniform source rocks" (italics occurs). Thus, there is no requirement that the rocks grouped as a suite be comagmatic, only that the source compositions be similar. Clearly, a given suite may contain rocks crystallized from several separate magmas.

There is no doubt that the concept of granitoid (and silicic volcanic) suites is useful as a means of grouping rocks that may have had similar source regions. Such suites may not, however, represent comagmatic lineages and should not generally be used as the basic units for discussing differentiation or differentiation mechanisms. It is important to study the chemical variations in individual plutons at a much finer resolution than that used to characterize suites.

Isotopic Evidence.—As discussed earlier, the crystals and lithic inclusions in granitic rocks and silicic volcanics may have at least four distinct origins: accidental xenolithic/xenocrystic, restitic, cumulate (from one liquid), and magma mixing (from two or more liquids). Isotopic studies of granites *and* their enclaves are very scarce, as are studies of the isotope chemistry of phenocrysts and enclosing glass in silicic volcanic rocks. However, these lines of investigation are potentially useful for discerning the provenance of the enclaves and crystals in granitic rocks.

Accidental xenoliths should commonly be out of isotopic equilibrium with the enclosing magma. It seems probable that such isotopic contrasts will be preserved in some types of xenoliths for some isotopic systems. The controlling factors include residence time in the magma, size of xenolith, chemical and isotopic contrast with the magma, and the presence or absence of fluid interactions (J-L. Duthou, C. Pin and P. Vidal pers. comm. 1987). In cases where assimilation of xenoliths has occurred, isotopic mixing lines should be evident (e.g., Clarke et al. 1987), and the data may show considerable scatter (Scambos et al. 1986). There may be discordance in apparent rock ages derived from different isotopic systems (e.g., Rb-Sr vs. K-Ar), and there may be strong correlations between isotopic ratios (e.g., $^{87}Sr/^{86}Sr_i$ and $\delta^{18}O$) and SiO_2. Products of magma mixing should show similar relations, with inclusions and some crystals out of isotopic equilibrium with the enclosing rock. Both restitic and cumulate material should be in isotopic equilibrium with the magma. Rock series derived by restite unmixing or fractional crystallization should show concordance in ages derived from different isotopic systems and lack of correlation between SiO_2 content and isotopic composition (unless the rocks are hydrothermally altered).

Detailed stable (e.g., O and S) and radiogenic isotope studies (e.g., Rb-Sr, Sm-Nd, U-Pb) could thus provide useful constraints on models for the origin of enclaves and crystals based on textural and geochemical evidence. We would encourage more such integrated, highly detailed studies of individual plutons and comagmatic series.

MELT SEGREGATION, ASCENT, AND
RESTITE UNMIXING

The question of whether restite is entrained during segregation of the melt from the residual source rocks and, if entrained, what factors contribute to its removal, are intimately concerned with degree of partial melting, magma temperature, water content and viscosity, ascent rates, segregation mechanism, and whether granitic melts behave as Newtonian or Bingham substances. In general, the most favorable conditions for a magma to clear itself of restite exist at or near the conditions of magma production.

Barr (1985) studied migmatites in the Moine terrane of Scotland. He described melt that had migrated into low stress parts of dilatant shear zones. In Barr's model, as the melt proportion increases, a point is reached where the strength of the material in the shear zone is drastically lowered (Arzi 1978; Van der Molen and Paterson 1979). At this point, melt begins to be expelled by filter pressing while the shear zone maintains an equilibrium melt content. Barr gives an example of such a shear zone in which 70% melting occurred with the extraction of 93% of the melt. In this case the shear zone is restite-rich but separation of mobile magma from the restite was inefficient, and the mass did not move far from the site of partial fusion.

An example of "hydraulic" fracturing as a melt segregation mechanism is given by Nicolas and Jackson (1982). These authors describe melt segregation into two sets of dykes in rocks undergoing partial melting and plastic flow. Melt is observed to collect progressively in interconnecting veinlets and to be eventually evacuated by "hydrofracturing" in larger, extensional dykes. Such fractures may be self-propagating by virtue of the differences in compressibility and thermal expansion of the magma and host rocks (Knapp and Knight 1977; Walther and Orville 1982).

In the case of melt localization in ductile shears, it seems likely that the magmas would retain significant amounts of restite. However, as a result of filtering, flow segregation, and gravity effects, melt collection and transport via veins and fractures would favor a low restite content from the outset. The faster and further a magma travels and the narrower the conduits, the less restite (or any kind of crystalline and lithic material) it will contain (Shaw 1965; Brown and Fyfe 1970; Fyfe 1973a, 1973b; Sleep 1974; Sparks 1977; Marsh 1982).

Retention of a large proportion of restite in a rising magma would seem to require ascent by a diapiric mechanism. Just as mafic diapirs cannot ascend through unmelted mantle rocks (Turcotte and Ahern 1978), diapiric ascent of granitic magmas in the crust requires partially molten envelope rocks (e.g., Bateman 1984). Thus, if significant restite is to be retained in high-level plutons the magmas would need to have sufficient heat to partially melt their cool wall rocks at medium to high crustal levels and still remain fluid themselves. For this to occur, magma temperatures would need to substantially exceed those of their solidi.

For metasediment-derived, peraluminous granitic suites, the parent magmas would commonly need to have contained 40% restite in order that the observed chemical variation be controlled by restite unmixing. The experiments of Clemens (1981) and Clemens and Wall (1981, 1984) show that, for a variety of peraluminous granite compositions and over a broad range of water contents, such magmas are 40% crystalline at, or close to, K-feldspar saturation. At 2–5 kb, K-feldspar saturation occurs only 15–25°C above the solidus, and 100–130°C below the silicate liquidus (see e.g., fig. 1). Thus any such magma, with 40% crystals, would be close to its solidus and have little ability to ascend diapirically. Such a magma would be emplaced at a deep level, not far from its source. High-level and volcanic magmas are therefore unlikely to move as crystal-rich mushes and hence are unlikely to differentiate by restite unmixing.

CONCLUSIONS

On mineralogical, textural, and chemical grounds, neither restite incorporation nor unmixing can be considered as major factors contributing to the chemical variation of many (?most) granitoids and silicic volcanics. This conclusion applies especially to high-level or epizonal and volcanic suites. Some mesozonal plutons (regional aureole granites; White et al. 1974), particularly mesozonal metasediment-derived types, do contain appreciable amounts of restite. Even in these instances, mechanisms of fractionation, accumulation, and mixing of magmas and solid materials may contribute to the observed chemical variations.

"Non-minimum melt" granitic magmas may be the rule rather than the exception. Such melts are generated at relatively high temperatures and have low-to-moderate water contents. This implies that crustal anatexis leading to large-volume granitic magmatism results from extreme thermal

746   V. J. WALL, J. D. CLEMENS, AND D. B. CLARKE

conditions in the crust. Higher temperature granitoid magmas have greatly enhanced potential for fractionation, assimilation, and mixing processes during ascent and emplacement. Restite-poor granitoids may result from melt extraction and ascent processes that differ significantly from those operating in restite-rich magmas. The latter may require relatively high degrees of partial melting and diapiric ascent. "Minimum" or "near-minimum"melts are generally unlikely to be emplaced at high crustal levels.

Conclusions as to the mineralogy and composition of the source regions of the granitoid magmas, based on the restite model, should be viewed with great caution. In specific cases where the presence of restite can be demonstrated, the roles of processes other than restite unmixing should be carefully examined. We urge petrologists and geochemists to consider *all* lines of evidence regarding chemical and mineralogical variations within granitic suites and especially in individual plutons. Groups of granitic rocks or silicic volcanics should be studied from a broad-based perspective, integrating field relations, petrology, major and trace element geochemistry, and if possible, both stable and radiogenic isotopic studies. If this is not done, then the full potential of granitic rocks for clarifying the nature of materials, processes, and conditions involved in magma generation, ascent, and emplacement may not be realized. Restite cannot be left to *Rest In Peace* but the model is not always *Requisite In Petrology.*

ACKNOWLEDGMENTS.—We wish to thank E. B. Watson, C. F. Miller, and an anonymous reviewer for their thoughtful comments. D. C. Merrett produced the figures on a Macintosh computer.

REFERENCES CITED

ABBOTT, R. N., JR., and CLARKE, D. B., 1979, Hypothetical liquidus relationships in the subsystem Al_2O_3-FeO-MgO projected from quartz, alkali feldspar, and plagioclase for a $(H_2O) < 1$: Can. Mineral., v. 17, p. 549–560.

AGUE, J. J., and BRIMHALL, G. H., 1987, Granites and batholiths of California: products of local assimilation and regional-scale contamination: Geology, v. 15, p. 63–66.

ALLAN, B. D., and CLARKE, D. B., 1981, Occurrence and origin of garnets in the South Mountain batholith, Nova Scotia: Can. Mineral., v. 19, p. 19–24.

ARZI, A. A., 1978, Critical phenomena in the rheology of partially melted rocks: Tectonophysics, v. 44, p. 173–184.

BAKER, B. H., and McBIRNEY, A. R., 1985, Liquid fractionation. Part III: geochemistry of zoned magmas and the compositional effects of liquid fractionation: Jour. Vol. Geotherm. Res., v. 24, p. 55–81.

BARR, D., 1985, Migmatites in the Moines, *in* ASHWORTH, J. R., ed., Migmatites: Glasgow, Blackie & Son Ltd., p. 225–264.

BATEMAN, P. C., and CHAPPELL, B. W., 1979, Crystallization, fractionation, and solidification of the Tuolumne Intrusive Series, Yosemite National Park, California: Geol. Soc. America Bull., v. 90, p. 465–482.

BATEMAN, R., 1984, On the role of diapirism in the segregation, ascent, and final emplacement of granitoid magmas: Tectonophysics, v. 110, p. 211–231.

———, 1985, Progressive crystallization of a granitoid diapir and its relationship to stages of emplacement: Jour. Geol., v. 93, p. 645–662.

BELLIENI, G.; MOLIN, G. M.; and VISONA, D., 1979, The petrogenetic significance of the garnets in the intrusive massifs of Bressanone and Vedrette di Ries (Eastern Alps-Italy): N. Jb. Miner. Abh., v. 136, p. 238–253.

BIRCH, W. D., AND GLEADOW, A. J. W., 1974, The genesis of garnet and cordierite in acid volcanic rocks: evidence from the Cerberean Cauldron, Central Victoria, Australia: Contrib. Mineral. Petrol., v. 45, p. 1–13.

BROWN, G. C., and FYFE, W. S., 1970, The production of granitic melts during ultrametamorphism: Contrib. Mineral. Petrol. v. 28, p. 310–318.

BURNHAM, C. W., 1979, The importance of volatile constituents, *in* YODER, H. S., ed., The Evolution of the Igneous Rocks—Fiftieth Anniversary Perspectives: Princeton N.J., Princeton Univ. Press, p. 439–482.

———, 1981, The nature of multicomponent aluminosilicate melts, *in* RICHARD, D. T., and WICKMAN, F. E., eds., Chemistry and geochemistry of solutions at high temperatures and pressures: Phys. Chem. Earth, v. 13, p. 197–229.

CANTAGREL, J-M.; DIDIER, J.; and GOURGAUD, A., 1984, Magma Mixing: origin of intermediate rocks and "enclaves" from volcanism to plutonism: Phys. Earth Planet. Inter., v. 35, p. 63–76.

CHAPPELL, B. W., 1978, Granitoids from the Moonbi district, New England Batholith, eastern Australia: Jour. Geol. Soc. Australia, v. 25, p. 267–284.

———, 1984, Source rocks of I- and S-type granites in the Lachlan fold Belt, southeastern Australia: Royal Soc. (London) Philos. Trans., v. A310, p. 693–707.

CLARK, R. G., JR., and LYONS, S. B., 1986, Petrogenesis of the Kinsman intrusive Suite:

peraluminous granites of western New Hampshire: Jour. Petrol., v. 27, p. 1365–1393.

CLARKE, D. B.; HALLIDAY, A. N.; and HAMILTON, P. J., 1987, Neodymium isotopic constraints on the origin of peraluminous granitoids of the South Mountain batholith, Nova Scotia: (In prep.).

——, and MUECKE, G. K., 1985, Review of the petrochemistry and origin of the South Mountain Batholith and associated plutons, Nova Scotia, Canada, in HALL, C., ed., High Heat Production (HHP) Granites, Hydrothermal Circulation, and Ore Genesis: London, Inst. Mining and Metal., p. 41–54.

CLEMENS, J. D., 1981, The origin and evolution of some peraluminous acid magmas (experimental, geochemical, and petrological investigations): Unpub. Ph.D. thesis Monash University, Australia, 577 p.

——, 1982, the Tolmie Igneous Complex, Australia: high-T, S-type rhyolites with polybaric crystallization histories: Geol. Soc. America Abs. with Prog., v. 14, p. 464–465.

——, 1984, Water contents of intermediate to silicic magmas: Lithos, v. 17, p. 273–287.

——; HOLLOWAY, J. R.; and WHITE, A. J. R., 1986, Origin of an A-type granite: experimental constraints: Am. Mineral., v. 71, p. 317–324.

——, and WALL, V. J., 1979, Crystallization and origin of some "S-type" granitic magmas, in Symposium on crust and upper mantle of southeastern Australia: Bur. Mineral Res. Rec. 1979/2, p. 10–11.

——, and ——, 1981, Crystallization and origin of some peraluminous (S-type) granitic magmas: Can. Mineral., v. 19, p. 111–132.

——, and ——, 1984, Origin and evolution of a peraluminous silicic ignimbrite suite: the Violet Town Volcanics: Contrib. Mineral. Petrol., v. 88, p. 354–371.

COCERTA, C., 1986, Les enclaves microgrenues sombre du massif de Bono (Sardaigne septentrionale). Signification petrogenetique des plagioclases complex et de leur inclusions: C. R. Acad. Sc. Paris, v. 302, ser. II, no. 7, p. 441–446.

COMPSTON, W., and CHAPPELL, B. W., 1979, Sr-isotope evolution of granitoid source rocks, in McELHINNY, M. W., ed., The Earth: Its Origin, Structure, and Evolution: London, Academic Press, p. 377–426.

DePAOLO, D. J., 1981, Trace element and isotope effects of combined wallrock assimilation and fractional crystallization: Earth Planet. Sci. Letters, v. 53, p. 189–202.

DIDIER, J., 1973, Granites and Their Enclaves: Amsterdam, Elsevier, 393 p.

——, 1982, The problem of enclaves in granitic rocks, a review of recent ideas on their origin, in Geology of granites and their metallogenic relations: Proc. Int. Sym. Nanjing, China, p. 137–144.

——; DOUTHOU, J. L.; and LAMEYRE, J., 1982, Mantle and crustal granites and the nature of their enclaves: Jour. Vol. Geotherm. Res., v. 14, p. 125–132.

EGGLER, D. H., 1972, Water-saturated and under-saturated melting relations in a Paracutin andesite and an estimate of water content in the natural magma: Contrib. Mineral. Petrol., v. 34, p. 261–271.

——, and BURNHAM, C. W., 1973, Crystallization and fractionation trends in the system andesite – H₂O – CO₂ – O₂ at pressures to 10 kb: Geol. Soc. America Bull., v. 84, p. 2517–2532.

EICHELBERGER, J. C., 1978, Andesitic volcanism and crustal evolution: Nature, v. 275, p. 21–27.

——, 1980, Vesiculation of mafic magma during replenishment of silicic magma reservoirs: Nature, v. 288, p. 446–450.

——, and GOOLEY, R., 1977, Evolution of silicic magma chambers and their relationship to basaltic volcanism, in The earth's crust: Am. Geophys. U. Mon., v. 20, p. 57–78.

ELLIS, D. J., and THOMPSON, A. B., 1986, Subsolidus and partial melting reactions in the quartz-excess CaO + MgO + Al₂O₃ + SiO₂ + H₂O system under water-excess and water-deficient conditions to 10 kb: some implications for the origin of peraluminous melts from mafic rocks: Jour. Petrol., v. 27, p. 91–121.

FLOOD, R. H., and SHAW, S. E., 1975, A cordierite-bearing granite suite from the New England Batholith, NSW, Australia: Contrib. Mineral. Petrol., v. 52, p. 157–164.

FURMAN, T., and SPERA, F. J., 1985, Co-mingling of acid and basic magma with implication for the origin of mafic I-type xenoliths: field and petrochemical relations of an unusual dike complex at eagle lake, Sequoia National Park, California, USA: Jour Vol. Geotherm. Res., v. 24, p. 151–178.

FYFE, W. S., 1973a, The generation of batholiths: Tectonophysics, v. 17, p. 273–283.

——, 1973b, The granulite facies, partial melting, and Archean crust: Phil. Royal Soc. (London) Philos. Trans., v. A273, p. 457–461.

GILL, J. B., 1978, Role of trace element partition coefficients in models of andesite genesis: Geochim. Cosmochim. Acta, v. 42, p. 709–724.

GRAPES, R. H., 1986, Melting and thermal reconstitution of pelitic xenoliths, Wehr volcano, East Eifel, West Germany: Jour. Petrol., v. 27, p. 343–396.

GREEN, T. H., 1972, Crystallization of calc-alkaline andesite under controlled high-pressure hydrous conditions: Contrib. Mineral. Petrol., v. 34, p. 150–166.

HAMER, R. D., and MOYES, A. B., 1982, Composition and origin of garnet from the Antarctic Peninsula Volcanic Group of Trinity Peninsula; Jour. Geol. Soc. London, v. 139, p. 713–720.

HIBBARD, M. J., 1981, The magma mixing origin of mantled feldspars: Contrib. Mineral. Petrol., v. 76, p. 158–170.

HIGGINS, N. C.; TURNER, N. J.; and BLACK, L. P., 1986, The petrogenesis of an I-type volcanic-plutonic suite: the St. Marys Porphyrite, Tasmania: Contrib. Mineral. Petrol., v. 92, p. 248–259.

HINE, R.; WILLIAMS, I. S.; CHAPPELL, B. W.; and WHITE, A. J. R., 1978, Geochemical contrasts between I- and S-type granitoids of the Kosciusko Batholith: Jour. Geol. Soc. Australia, v. 25, p. 219–234.

JOHANNES, W., 1978, Melting of plagioclase in the system Ab – An – H_2O and Qz – Ab – An – H_2O at P_{H_2O} = 5 kbars, an equilibrium problem: Contrib. Mineral. Petrol., v. 66, p. 295–303.

———, 1985, The significance of experimental studies for the formation of migmatites, in ASHWORTH, J. R., ed., Migmatites: Glasgow, Blackie & Son Ltd, p. 36–85.

KAY, R. W., 1984, Elemental abundances relevant to identification of magma sources: Royal Soc. (London) Philos. Trans., v. A310, p. 535–547.

KISTLER, R. W.; CHAPPELL, B. W.; and PECK, D. L., 1986, Isotopic variations in the Tuolumne Intrusive Suite, central Sierra Nevada, California: Contrib. Mineral Petrol., v. 94, p. 205–220.

KNAPP, R. B., and KNIGHT, J. E., 1977, Differential thermal expansion of pore fluids: fracture propagation and microearthquake production in hot pluton environments: Jour. Geophys. Res., v. 82, p. 2515–2522.

LEE, D. E., and CHRISTIANSEN, E. H., 1983, The granite problem as exposed in the southern Snake Range, Nevada: Contrib. Mineral. Petrol., v. 83, p. 99–116.

LARSEN, E. S.; IRVING, J.; GONYER, F. A.; and LARSEN E. S., III, 1938, Petrologic results of a study of the minerals from the Tertiary volcanics rocks of the San Juan region, Colorado: Am. Mineral., v. 23, p. 227–257.

MAILLET, L. A., and CLARKE, D. B., 1985, Cordierite in the peraluminous granites of the Meguma zone, Nova Scotia, Canada; Min. Mag., v. 49, p. 695–702.

MARSH, B. D., 1982, On the mechanics of igneous diapirism, stoping, and zone melting: Am. Jour. Sci., v. 282, p. 808–853.

———, and KANTHA, L. H., 1978, On the heat and mass transfer from an ascending magma: Earth Planet. Sci. Letters, v. 39, p. 435–443.

McCARTHY, T. S., and GROVES, D. I., 1979, The Blue Tier Batholith, northeastern Tasmania—a cumulate-like product of fractional crystallization: Contrib. Mineral. Petrol., v. 71, p. 193–209.

McCULLOCH, M. T., and CHAPPELL, B. W., 1982, Nd isotopic characteristics of S- and I-type granites: Earth Planet. Sci. Letters, v. 58, p. 51–64.

MILLER, C. F., and MITTLEFEHLDT, D. W., 1982, Crystal fractionation: controlling factor of differentiation in highly felsic magma chambers?: Geol. Soc. America Abs. with Prog., v. 14, p. 565.

———; WATSON, E. B.; and RAPP, R. P., 1985, Experimental investigation of mafic mineral-felsic liquid equilibria: preliminary results and petrogenetic implications (abs.): EOS, v. 66, p. 1130.

MUECKE, G. K., and CLARKE, D. B., 1981, Geochemical evolution of the South Mountain batholith, Nova Scotia: Can. Mineral., v. 19, p. 133–146.

NANEY, M. T., 1983, Phase equilibria of rock-forming ferromagnesian silicates in granitic systems: Am. Jour. Sci., v. 283, p. 993–1033.

———, and SWANSON, S. E., 1980, The effect of Fe and Mg on crystallization in granitic systems: Am. Mineral., v. 65, p. 639–653.

NICOLAS, A., and JACKSON, M., 1982, High temperature dikes in peridotites: origin by hydraulic fracturing: Jour. Petrol., v. 23, p. 568–582.

PABST, A., 1928, Observations on inclusions in the granitic rocks of the Sierra Nevada: Univ. Calif. Pub., Dept. Geol. Sci. Bull., v. 17, p. 325–386.

PATTISON, D. R. M.; CARMICHAEL, D. M.; and ST-ONGE, M. R., 1982, Geothermometry and geobarometry applied to early Proterozoic "S-type" granitoid plutons, Wopmay Orogen, Northwest Territories, Canada: Contrib. Mineral. Petrol., v. 79, p. 394–404.

PETITPIERRE, E., and BOIVIN, P., 1983, CRYSTALLIZATION: a computer program for modeling the crystallization of a magmatic liquid: Computers and Geosciences, v. 9, p. 455–461.

PHILLIPS, G. N.; WALL, V. J.; and CLEMENS, J. D., 1981, Petrology of the Strathbogie batholith: a cordierite-bearing granite: Can. Mineral., v. 19, p. 47–64.

PIDGEON, R. T., and AFTALION, M., 1978, Cogenetic and inherited zircon U-Pb systems in granites: Paleozoic granites of Scotland and England, in BOWES, D. R., and LEAKE, B. E., eds., Crustal evolution in northwestern Britain and adjacent regions: Geol. Jour. Spec. Issue, v. 10, p. 123–142.

PIWINSKII, A. J., 1968, Studies of batholithic feldspars: Sierra Nevada, California: Contrib. Mineral. Petrol., v. 17, p. 204–223.

———, and WYLLIE, P. J., 1968, Experimental studies of igneous rock series: a zoned pluton in the Wallowa Batholith, Oregon: Jour. Geol., v. 76, p. 205–234.

PRESNALL, D. C., and BATEMAN, P. C., 1973, Fusion relations in the System $NaAlSi_3O_8$ – $CaAl_2Si_2O_8$ – $KAlSi_3O_8$ – SiO_2 – H_2O and generation of granitic magmas in the Sierra Nevada Batholith: Geol. Soc. Am. Bull., v. 84, p. 3182–3202.

PRICE, R. C., 1983, Geochemistry of a peraluminous granitoid suite from Northeastern Victoria, southeastern Australia: Geochim. Cosmochim. Acta, v. 47, p. 31–42.

REID, J. B. JR.; EVANS, O. C.; and FATES, D. G., 1980, Quenched droplets of basaltic magma in granites of the Pliny Range, N.H.: Geol. Soc. America Abs. with Prog., v. 12, p. 507.

———; ———; and ———, 1983, Magma mixing in granitic rocks of the central Sierra Nevada, California: Earth Planet. Sci. Letters, v. 66, p. 243–261.

ROBIN, P-Y. F., 1979, Theory of metamorphic segregation and related processes: Geochim. Cosmochim. Acta, v. 43, p. 1587–1600.

RUTHERFORD, M. J.; SIGURDSSON, H.; CAREY, S.; and DAVIS, A., 1985, The May 18, 1980 eruption of Mount St. Helens 1. Melt composition and experimental phase equilibria: Jour. Geophys. Res., v. 90, p. 2929–2947.

SCAMBOS, T. A.; LOISELLE, M. C.; and WONES, D. R., 1986, The Center Pond pluton: the restite of the story (phase separation and melt evolution in granitoid genesis): Am. Jour. Sci., v. 286, p. 241–280.

SHAW, H. R., 1965, Comments on viscosity, crystal settling, and convection in granitic magmas: Am. Jour. Sci., v. 263, p. 120–152.

———, 1974, Diffusion of H₂O in granitic liquids: part I. Experimental data; part II. Mass transfer in magma chambers, in HOFFMAN, A. W.; GILETTI, B. J.; YODER, H. S., JR., and YUND, R. A., eds., Geochemical transport and kinetics: Carnegie Inst. Washington Pub. 634, p. 139–170.

SILVER, L. T., and DEUTSCH, S., 1963, Uranium-lead isotopic variation in zircons: a case study: Jour. Geol., v. 71, p. 721–758.

SLEEP, N. H., 1974, Segregation of magma from a mostly crystalline mush: Geol. Soc. America Bull, v. 85, p. 1225–1232.

SPARKS, R. S. J.; PINKERTON, H.; and MACDONALD, R., 1977, The transport of xenoliths in magmas: Earth Planet. Sci. Letters, v. 35, p. 234–238.

SULTAN, M.; BATIZA, R.; and STURCHIO, N. C., 1986, The origin of small-scale geochemical and mineralogic variations in a granite intrusion. A crystallization and mixing model: Contrib. Mineral. Petrol., v. 93, p. 513–523.

SWANSON, S. E. 1977, Relation of nucleation and crystal growth rate to the development of granitic textures: Am. Mineral., v. 72, p. 966–978.

TAYLOR, H. P., JR., 1980, The effects of assimilation of country rocks by magmas on ¹⁸O/¹⁶O and ⁸⁷Sr/⁸⁶Sr systematics: Earth Planet. Sci. Letters, v. 47, p. 243–254.

THOMPSON, R. N., 1983, Thermal aspects of the origin of Hebridean Tertiary acid magmas. II. Experimental melting behavior of the granites at 1 kbar P_{H₂O}: Mineral. Mag., v. 47, p.111–120.

TINDLE, A. G., and PEARCE, J. A., 1981, Petrogenetic modeling of in situ fractional crystallization in the zoned Loch Doon pluton, Scotland: Contrib. Mineral. Petrol., v. 78, p.196–207.

TODD, V. R., and SHAW, S. E., 1985, S-type granitoids and an I-S line in the Peninsular Ranges batholith, southern California: Geology, v. 13, p. 231–233.

TSUCHIYAMA, A., 1985, Dissolution kinetics of plagioclase in the melt of the system: diopside - albite - anorthite, and the origin of dusty plagioclase in andesites: Contrib. Mineral. Petrol., v. 89, p. 1–16.

TURCOTTE, D. L., and AHERN, J. L., 1978, A porous flow model for magma migration in the asthenosphere: Jour. Geophys. Res., v. 83, p. 767–772.

VAN DER MOLEN, I., 1985, Interlayer material transport during layer-normal shortening. Part 1. The model: Tectonophysics, v. 115, p. 275–295.

———, and PATERSON, M. S., 1979, Experimental deformation of partially melted granite: Contrib. Mineral. Petrol., v. 70, p. 229–318.

VANCE, J. A., 1962, Zoning in igneous plagioclase: normal and oscillatory zoning: Am. Jour. Sci., v. 260, p. 746–760.

VERNON, R. H., 1983, Restite, xenoliths, and microgranitoid enclaves in granites: Jour. Proc. Royal Soc. NSW, v. 116, p. 77–103.

———, and FLOOD, R. H., 1982. Some problems in the interpretation of microstructures in granitoid rocks, in RUNNEGAR, B., and FLOOD, P., eds., New England Geology: Armidale, NSW (Australia), Univ. New England and AHV Club, p. 201–210.

VIELZEUF, D., and HOLLOWAY, J. R., 1985, Fluid absent melting of pelitic rocks at 10 kbar. A theoretical and experimental approach (abs.): EOS, v. 66, p. 1149.

WALTHER, J. V., and ORVILLE, P. N., 1982, Volatile production and transport in regional metamorphism: Contrib. Mineral. Petrol., v. 79, p. 252–257.

WANG, L-K.; ZHAO, B.; ZHU, W-F.; CAI, Y-J.; and LI, T-J., 1980, Characteristics and melting experiments of granites in southern China: Mining Geol. Spec. Issue, v. 8, p. 29–38.

WATSON, E. B., and HARRISON, T. M., 1983, Zircon saturation revisited: temperature and composition effects in a variety of crustal magma types: Earth Planet. Sci. Letters, v. 64, p. 295–304.

WELLS, P. R. A., 1980, Thermal models for the magmatic accretion and subsequent metamorphism of continental crust: Earth Planet. Sci. Letters, v. 46, p. 253–265.

WHITE, A. J. R., and CHAPPELL, B. W., 1977, Ultrametamorphism and granitoid genesis: Tectonophysics, v. 43, p. 7–22.

———, and ———, 1983, Granitoid types and their distribution in the Lachlan Fold Belt, southeastern Australia, in RODDICK, J. A., ed., Circum-Pacific plutonic terranes: Geol. Soc. America Mem. 159, p. 21–34.

———, ———, and CLEARY, J. R., 1974, Geologic setting and emplacement of some Australian Palaeozoic batholiths and implications for intrusive mechanisms: Pacific Geol., v. 8, p. 159–171.

———; CLEMENS, J. D.; HOLLOWAY, J. R.; SILVER, L. T.; CHAPPELL, B. W., and WALL, V. J., 1986, S-type granites and their probable absence in southwestern North America: Geology, v. 14, p. 115–118.

WHITNEY, J. A., 1975, The effects of pressure, temperature, and X_{H₂O} on phase assemblage in four synthetic rock compositions: Jour. Geol., v. 83, p. 1–31.

WYBORN, D., 1983, Fractionation processes in the Boggy Plain zoned pluton: Unpub. Ph.D. thesis, The Australian National University, Camberra.

WYBORN, L. A. I., and CHAPPELL, B. W., 1979, Geochemical evidence for the existence of a pre-Ordovician sedimentary layer in southeastern Australia, in Symposium on crust and upper mantle of southeastern Australia: Bur. Min. Res. Rec. 1979/2, p. 104.

ZECK, H. P., 1970, An erupted migmatite from Cerrodel Hoyazo, SE Spain: Contrib. Mineral Petrol., v. 26, p. 225–246.

CHAPTER 15

Goranson, R.W. (1932) Some notes on the melting of granite. *American Journal of Science*, **23**, 227–236.

When people think of early experimental work on granites, the giants whose names initially come to mind are invariably Tuttle and Bowen. These workers are rightly credited with the first systematic and exhaustive exploration of the H_2O-saturated haplogranite system and the formulation of many important ideas on the origins of granitic magmas. However, they were not the first to produce valuable insights into the behaviour of granitic systems using experimental techniques and not the first to draw inferences from the behaviour of related experimental systems.

Roy W. Goranson was a petrologist, experimental chemist, thermodynamicist and physicist. It was he who performed the first systematic series of H_2O-saturated and -undersaturated partial melting and crystallization experiments on natural granitic materials. Prior to this, he had already published on the solubility of H_2O in granitic melts (Goranson, 1931). Working in the Geophysical Laboratory of the Carnegie Institute of Washington, and using quite primitive experimental apparatus, he overcame considerable technical difficulties in carrying out experiments that lasted hundreds of hours, at high temperatures and pressures. Goranson managed to make some highly important observations, in a sense extending the much earlier work of the chemist Bunsen (also included in this volume). However, few people have ever heard of Goranson's work, enshrined in this 10-page paper, eclipsed as it was by later and related work carried out in the same laboratory.

The experiments that he carried out involved partial melting of natural, finely powdered granitic rock, and also the crystallization of natural granitic glass (obsidian), both under hydrous conditions, at pressures up to ~100 MPa. To illustrate the importance of this work, here is a list of what we believe to be his major finings. Some were based on direct observation, some on the results of earlier work in related chemical systems and some on intuition and general petrological knowledge.

- determination of the solidus and liquidus temperatures of granite, and recognition that the crystallization interval can be several hundreds of degrees wide
- recognition that solid granitic rocks will commonly not represent the compositions of the magmas from which they began to crystallize
- recognition that the volatiles originally present, dissolved in a granitic magma, will mostly be lost to the system, and will not be represented by the hydrous phases (e.g. micas) that may be present in the solid rocks
- recognition that quartz is a natural product of crystallization of granitic magmas, at relatively low temperatures, and that it will crystallize as the β polymorph, except in some pegmatites
- recognition that high-temperature, aqueous fluids are capable of dissolving substantial amounts of silicate material derived from co-existing granitic melts
- determination that the solubility of H_2O in granitic melt at 100 MPa will be 6.5 wt.%
- observation that Fe^{3+} has a very low solubility in granitic melts
- recognition that, on crystallization, hydrous granitic magmas will concentrate H_2O into the residual melt and eventually evolve a fluid phase (i.e. undergo second boiling)
- prediction that this volatile release can lead to solidification of the magma
- recognition that granites, aplites, pegmatites and quartz veins represent stages in the crystallization of a granitic magma, with the pegmatites straddling the gap between the magmatic and hydrothermal regimes

This represents quite an impressive list of significant observations and inferences that were subsequently built upon and refined by many later experimental workers. We do not seek to detract from those later achievements but rather to appreciate the insights of this pioneer in experimental granite petrology.

SOME NOTES ON THE MELTING OF GRANITE

ROY W. GORANSON.

There has been considerable discussion concerning the melting interval of granite,[1] but up to the present time there seems to be no unanimity of opinion as to the actual values of these temperatures. Some of this uncertainty can probably be traced to the fact that there is no one clean-cut answer to the problem. For the term granite is used not for a definite composition but to classify a rather indefinite range of compositions. Variations in one or more of the components of granite, within this composition range may alter appreciably the value of the liquidus temperature.

There are also experimental difficulties with which to contend, and it is perhaps owing to the high viscosity of dry silicate melts with consequent difficulty of attaining equilibrium within the time available for a laboratory measurement that some writers have assumed liquidus temperatures for granite to lie above those of gabbros.

It may also be possible that the high melting temperatures of crystalline silica (quartz, tridymite, and cristobalite), have influenced opinions on melting temperatures, although geologic evidence points to the conclusion that granite represents a late stage in fractionation by crystallization from a cooling magma originally more femic.

There is still another consideration to be introduced in any attempt at evaluating the crystallization temperature interval of a granite magma. For all magmas contain volatile components, of which perhaps only a small part becomes bound up in the crystalline phases (granite), the greater portion of these volatiles playing a complex rôle in what are commonly called post-magmatic processes. Of these volatile constituents water is the most abundant, and in this paper is presented an account of some results obtained by melting Stone Mountain granite in the presence of water.

---

[1] Granite, like any other mixture of more than one component except a eutectic composition, does not melt or solidify at a point but over a temperature interval. The temperature at which the last crystals disappear or the first ones appear, T_l, is called the liquidus temperature; the temperature at which the first liquid appears or last liquid disappears, T_s, is called the solidus temperature. In the case of granite the temperature at which the last crystals formed is not a true solidus temperature but represents the point at which the residual liquid became separated from the crystalline phases. $T_L - T_S$ is the melting interval for the mixture.

227

228 *Goranson—Some Notes on the Melting of Granite.*

PART I. EXPERIMENTAL RESULTS.

DESCRIPTION OF MATERIALS USED.

The granite used in these melting determinations came from Stone Mountain near Atlanta, Georgia. It is a normal biotite-bearing muscovite granite of light grey color. The granite glass, used in some experiments and listed in the experimental results as such, was obtained by melting this granite in a platinum-rhodium resistance furnace until the glass obtained was essentially homogeneous. The obsidian listed in the experimental results occurs at Cerro Noagua, New Mexico. The composition of the granite and obsidian are given below in columns 1 and 2 respectively. The analyses were made by Dr. E. S. Shepherd of this Laboratory.

	1.	2.
SiO_2	73.39	76.35
Al_2O_3	14.41	12.34
Fe_2O_3	0.09	0.47
FeO	0.70	0.60
MgO	0.27	0.03
CaO	1.05	0.38
Na_2O	3.96	4.29
K_2O	5.07	4.39
H_2O+	0.39	0.91
H_2O-	0.05	0.00
TiO_2	0.27	0.09
ZrO_2	tr.	tr.
P_2O_5	0.57	0.02
Cl	0.03	0.06
S	0.02	tr.
BaO	0.06	tr.
Sum	100.34	99.85
Density	2.66 at 20°C	2.33 at 25°C*

* The density of the normal obsidian is 2.3297 at 25° C, of the ignited obsidian is 2.3372 at 25° C as determined by a pycnometer method using a water thermostat.

EXPERIMENTAL DETAIL.

The granite, ground extremely fine (less than 200 mesh), and water were inserted in a platinum capsule which was then sealed. Sufficient hydrostatic pressure had to be put on these capsules to compensate for the water vapor pressure inside them. The method of loading and sealing the capsules, and description of the bomb used for obtaining the necessary pressure and temperature are given in an earlier paper.[2] When

[2] Goranson, Roy W., this Journal, 22, 481, 1931.

obsidian or granite glass was used the procedure was identical with that for granite.

The experimental results are presented in Tables A and B, the experiment number being listed at the left for reference. The temperature, pressure, and time of run are given in degrees Centigrade, bars, and hours respectively. Temperature was controlled by a regulator operated on the Wheatstone bridge principle using the furnace winding as one arm of the bridge. The regulator controlled the temperature within ± 5° C. Pressure was read on a Bourdon-type gauge which was calibrated by a dead-weight gauge of the Amagat type. The dial is calibrated in kilograms per sq. cm. and can be read to ± 10 kilograms per cm.^{2}, which is sufficiently accurate for the present purpose.

Runs longer than 10 hours were discontinuous, as it was not safe to leave the bomb unattended.

Results obtained on Stone Mountain granite are given in Table A, those obtained on Stone Mountain granite glass and Cerro Noagua obsidian are given in Table B.

TABLE A.

Experimental Results on Stone Mountain Granite.

T denotes the temperature and *P* the pressure during the run.

A1-30   *T = 1000° C; P = 950 bars; 1 hour run.*
     Granite + no water: no glass observed. The biotite is darkened.
     Granite + 2.7% water: about 40% glass.
     Granite + 15% water: about 80% glass. Crystalline grains are all rounded and corroded. Hematite plates have formed.

A1-32   *T = 900° C; P = 1500 bars; 3 hour run.*
     Granite + no water: a few small isotropic patches of index about 1.485. Biotite completely altered.
     Granite + 4.8% water (water however leaked out sometime during the run); patches of glass of index about 1.498 enclosing branches and scales of hematite. The residual quartz grains are all rounded.
     Granite + 3.8% water: all glass of index about 1.488. The glass is sprinkled abundantly with tiny bubbles.

2A-12   *T = 888° C; P = 960 bars; 3 hour run.*
     Granite + 5.3% water: chiefly glass. A few rounded quartz grains are present. Small hexagonal plates of hematite have developed.

2A-8   *T = 862° C; P = 590 bars; 4 hour run.*
     Granite + 8.8% water: Light grey cylinder of glass interspersed with crystalline grains. Index of glass about 1.488.

2A-12   *T = 821° C; P = 960 bars; 3 hour run.*
     Granite + 5% water: about 30% glass. Quartz grains are all corroded and rounded. Hematite and magnetite were observed.

2A-18c   *T = 816° C; P = 960 bars; 21 hour run.*
     Granite + 2% water (lost all the water some time during the run): nearly all glass. Only a few rounded crystalline grains present.

230 *Goranson—Some Notes on the Melting of Granite.*

2A-12   *T = 778° C; P = 960 bars; 3 hour run.*
      Granite + 5% water: The crystalline grains are cemented with glass. Hematite has separated out.

2A-18a   *T = 778° C; P = 960 bars; 4 hour run.*
      Granite + 50% water: about 85% glass. The crystalline grains present are all rounded.

2A-20   *T = 723° C; P = 960 bars; 33 hour run.*
      Granite + 17% water (lost all water some time during run): about 90% glass. Some recrystallization seems to have taken place in the glass, perhaps after the water had seeped out. The original crystals present are all badly corroded.

2A-20   *T = 723° C; P = 960 bars; 55 hour run.*
      Granite + 20% water: Some portions of the glass seem to be devoid of any crystalline material. Other portions have a few shreds of crystalline material together with development of hematite flakes. Probably less than 0.1% crystalline material.

2A-22   *T = 704° C; P = 960 bars; 47.5 hour run.*
      Granite + 9.5% water: 50% or more of glass. The crystalline material present consists of rounded feldspar and quartz together with recrystallized hematite. Tiny bubbles are sprinkled through the glass.

2A-23   *T = 704° C; P = 960 bars; 102 hour run.*
      Granite + 4.4% water: about 70% or more of glass. All but the larger quartz grains have disappeared. There is some recrystallization of hematite flakes and small needles of rather high index. There are also tiny bubbles sprinkled through the glass.

2A-26   *T = 704° C; P = 960 bars; 207.5 hour run.*
      Granite + 4.2% water: 75% or more of glass. The crystalline grains are chiefly highly corroded quartz. Recrystallization of hematite flakes and small needles of rather high index. Tiny bubbles are sprinkled throughout the glass.

2A-26   *T = 704° C; P = 960 bars; 262.5 hour run.*
      Granite + 7% water (lost all water some time after 55 hours): about 70% glass. The crystalline grains are chiefly corroded quartz. There is recrystallization of hematite and some colorless needles which perhaps took place after the water had escaped.

1B-24   *T = 600° C; P = 385 bars; 460 hour run.*
      Granite + water: a small amount of glass was observed. The biotite is completely altered to magnetite and chlorite; the oligoclase has been attacked and probably partially dissolved by the water. Needles similar to those described in the 460 hour run of Table B occur in void spaces.

TABLE B.

Experimental Results on Cerro Noagua Obsidian and Stone Mountain Granite Glass.

3A-28   *T = 750° C; P = 980 bars; 4¾ hour run.*
      Stone Mt. granite glass + 6.2% water: Recrystallization of hematite and of a few small, high index rods has taken place.

3A-30   *T = 900° C; P = 490 bars; 4 hour run.*
      Stone Mt. granite glass + 3.7% water: A few scattered rod-like crystals have formed. They are extremely small and probably make up only a very small fraction of a per cent of the whole amount.

2A-2   *T = 815° C; P = 570 bars; 4 hour run.*
      Obsidian + no water: sintered to a long rod. Some crystallization has taken place, the glass being liberally spattered with tiny birefracting grains.

2A-18b   *T = 751° C; P = 960 bars; 6 hour run.*
Obsidian + 30% water: a few tiny rods have crystallized out.

1B-24   *T = 600° C; P = 385 bars; 460 hour run.*
Stone Mt. granite glass + 2.6% water: The powder sintered to a dark grey rod during the run. It appears to be a mass of birefracting grains too small to be clearly resolved and it is impossible to estimate the proportion of glass to crystals. Growing out from the glass are long needles or rods which in places form a radiating fibrous mass. These needles extinguish parallel to the elongation or close to it. Elongation is negative. Index 1.64 ± 0.01. In places V-shaped ends were observed. These crystals probably grew from the water-rich solution which had leached the components of the crystals from the glass.

3A-18   *T = 600° C; P = 980 bars; 4 hour run.*
Stone Mt. granite glass + 7% water: The glass is speckled with crystalline material. There are at least three kinds of crystals present: hematite. laths of plagioclase (?) and needles or rods of index higher than the laths and of negative elongation. In some places the laths and rods radiate from a central crystalline grain and resemble spherulitic growths. In places crystalline grains are arranged in parallel and concentric bands; these bands have a microfelsitic texture. This banding may be a kind of flow structure caused by kneading of the capsule due to pressure changes.

DISCUSSION OF RESULTS.

In the presence of water vapor at 960 bars pressure Stone Mountain granite became completely molten in 3 hours at 900° C; in 55 hours it was more than 99% glass at 723° C with only a few corroded remnants of the larger quartz grains remaining. In 207 hours at 704° C there remained only 20 per cent or less of crystalline material; since the only original material left which could be identified was quartz, this estimate of crystals is probably high. After 4 hours at 600° C Stone Mountain granite glass with 6% or more of water in solution contained numerous tiny crystals. Two distinct species, besides hematite, were observed but could not be identified. One type consisted of plates with the same sign and general appearance as feldspar. The other type was of rods or needles of higher index than the laths and had a negative elongation. In a 460-hour run at 600° C and 385 bars pressure on Stone Mountain granite glass and water, the glass appeared to be essentially crystalline. In this preparation hexagonal outlines were observed which may be quartz, but in none of the other glasses was any material observed which might have been quartz. In order to determine whether quartz would crystallize readily a 3-hour run was made on silica glass plus water at 900° C and 1000 bars. On examination the material was found to have crystallized completely to quartz. Hence, if conditions are suitable, quartz should undoubtedly crystallize from the granite melt.

A 460-hour run was made on granite plus water at 600° C and 385 bars pressure. After the run a small amount of glass, perhaps less than 1%, was observed cementing some of the grains. The biotite had been completely altered to magnetite and chlorite, and the oligoclase grains appeared to be somewhat leached by the water.

The presence of needle-like rods in the glass even up to 900° was rather puzzling. They have indices of about 1.64 ± .01 and negative elongation. Some of these rods had V-shaped ends and in places branched out in a radiating fibrous pattern. One of the runs was of sufficient duration (460 hours), to allow them to grow large enough to permit determination of the foregoing optical data. In this preparation the needles grew out from the glass into the void spaces. (This run differed from the other runs in that a special bomb was built such that the water generated all the pressure and hence allowed voids to remain intact in and around the glass.) The inference is that these needles must have crystallized from the water-rich solution which leached the constituents of this mineral from the glass. There are two non-crystalline phases present in the system; one consisting of water plus material leached from the granite or granite glass, the other consisting of granite glass plus water. The presence of considerable material soluble in the water-rich phase is apparent on evaporating off the excess water after a run, as there was always a white residue left. Whether the rods present in the other glasses had a similar origin is not known, because they were completely enclosed in the glass. They could, however, have grown from the surface of the glass grains and become enclosed on coalescence and flow of the glass.

CONCLUSIONS.

It is concluded that at 700 ± 50° C and under a water vapor pressure of 1000 bars Stone Mountain granite will become, except for hematite, completely liquid. The melt will have 6.5% of water in solution.[3]

Considerable crystalline material formed in granite glass subjected to a temperature of 600° and a water vapor pressure of about 385 bars for 460 hours. A small amount of

[3] Goranson, Roy W., op. cit.

glass, probably less than 1%, was present in granite subjected to the same procedure. If equilibrium were attained in this time then at about $575 \pm 25°$ and a pressure of 385 bars 99% or more of the silicate portion of the system granite-water will be crystalline.

In these experiments the iron present seemed to be almost entirely oxidized by the water vapor and, as it is very insoluble in the granite-water melt, much of it crystallized out as hematite at temperatures below 1200°. At 1200° the glass contained no hematite and was of a light straw color.

PART II. GENERAL DISCUSSION.

There are still insufficient data to give a coherent picture, and, to obtain it, hypotheses which are not fully substantiated must be introduced. Therefore the following discussion will be subject to change as more data become available.

Here the hypothesis to be introduced is an expression of the relation between concentration of water and liquidus temperature of granite.

The initial slope of the melting temperature lowering is given by the equation

$$\left(\frac{\delta \theta}{\delta m_1}\right)_P = \frac{1.9864\theta^2}{18\Delta x_1\, m^2_1}$$

where θ denotes the absolute thermodynamic temperature, m_1, the mass fraction of granite, and Δx_1, the heat of solution of the last crystalline phase of granite to go into solution in gram calories per gram. This equation, if we assume the heat of fusion to be 100 calories per gram, gives us an initial lowering of 10.5° per 1% of water. It is not applicable, however, except for very low water concentrations. The rigorous general equation cannot be used through lack of data.

The results of Morey[4] on the systems, H_2O-K_2SiO_3 and H_2O-$K_2Si_2O_5$, should give a closer approximation at the higher water concentrations. The initial slope for the system, H_2O-K_2SiO_3, is, according to the foregoing equation, about 10° C per 1% of water. Morey's observed temperature lowerings for the system, H_2O-K_2SiO_3, are: 35.8° C for a

[4] Morey, G. W., Jour. Am. Chem. Soc., **39**, 1207, 1917.

234 *Goranson—Some Notes on the Melting of Granite.*

2% water solution, 376° C for a 6.3% water solution. The slope of the liquidus for this system thus steepens considerably with increasing water content. Observed temperature lowerings for the system, $H_2O-K_2Si_2O_5$, are: 8° for a 0.6% water solution, 441° for an 8.3% water solution.

From these values we can now calculate a liquidus temperature for the dry granite, i.e. the temperature at which it will be completely molten. The calculated value is $T_L = 1050 \pm 50°$ C.

The calculated liquidus temperatures in degrees Centigrade (and pressures in bars) of Stone Mountain granite are: 1035 (140); 1005 (260); 963 (400); 908 (550); 840 (730); 750 (930); and 600 (1100) for 1, 2, 3, 4, 5, 6 and 7 weight per cent of water respectively.

As will be observed from the above computation, the liquidus does not have a maximum pressure in the crystallization temperature range of a granite magma. Furthermore, the temperature-pressure-water solubility relations are such that an error of 100° in the calculated temperatures will not affect the validity of the foregoing statement.

The end temperature in the crystallization process of a granite will not be a solidus temperature but the temperature at which this normal process becomes interrupted by a separation of the residual solution from the coexisting crystallized phases of granite.

Wright and Larsen[5] have evolved a set of criteria to determine whether quartz crystallized as the high or low temperature modification, and used it in a study of quartz from granites, dikes, pegmatites, and veins. These criteria have also been used by others for this purpose. The data indicate that quartz in granite, granite porphyry, and aplite dikes crystallized as the high temperature form, and that pegmatite quartz straddled the inversion, i.e. in the initial stages quartz crystallized as the high temperature form and in the final stages as the low temperature form. This would place the initial stage of pegmatite formation as about 700°, the final temperature as perhaps 550° C.

Let us assume we have a granite magma containing 1% of water in solution and at a depth of, say, 10 kilometers. It will begin to crystallize at about 1025° C. Crystallization, with corresponding increase in concentration of water in the residual liquid, will proceed with continued cooling. When the tem-

[5] Wright, F. E., and Larsen, E. S., this Journal, **27**, 421-47, 1909.

perature reaches 700° C about 85% of the original magma will have crystallized, the residual liquid containing about 6.5% of water in solution.

The residual solution, 15% of the original magma, would be available for the formation of aplites, pegmatites, and quartz veins; of this amount ⅔ would crystallize between 700° and 500°. Of the original magma containing 1% water in solution we would have as final products: 85% of granite (1035°-700°), 10% of aplite and pegmatite dikes (700°-550°), and 5% of quartz veins (500° and lower).

If this granite magma with 1% of water in solution lies at a depth of 4 kilometers crystallization will begin, as before, at about 1025° and continue normally until the temperature drops to 950° when about ⅔ of the original magma will have crystallized. At this point the pressure necessary to hold the increased concentration of water in solution will become equal to the hydrostatic head at this depth. But we know that the pressure-solubility curve is considerably steeper than the temperature-solubility curve at this pressure,[6] hence further crystallization will be accompanied by an ebullition of water.

There are then two coexistent fluid phases—one predominating in silicates and the other in water. Water will continue to boil out of the magma as silicates crystallize. In a 460-hour run at 600° and a vapor pressure of 385 bars less than 1% of the granite had melted. If in this time an equilibrium was reached then there would be little opportunity here for the formation of aplite and pegmatite dikes.

If the original granite magma contained 3% of water in solution at 10 kilometers depth crystallization would begin at about 965° and continue quietly until a temperature of about 700° was reached; at this temperature about half of the system would be crystalline, the solution containing about 6.5% water in solution. The pressure necessary to hold this amount of water in solution at 700° will be equivalent to the hydrostatic head at this depth and thus further crystallization would be accompanied by an ebullition of water with formation of two coexistent fluid phases. This would continue until the temperature-pressure curve dropped below the external load pressure.

[6] Goranson, Roy W., op. cit.

236 *Goranson—Some Notes on the Melting of Granite.*

SUMMARY.

In any attempt at evaluating the fusion conditions of granite there is a multiplicity of variables to consider. For a granite may vary in composition within rather hazily defined limits. Furthermore, granite magmas also contain other constituents, such as volatile material, which are only partially bound up in the minerals of granite. Of these volatile constituents water is the most abundant. In this paper is presented an account of some results obtained by melting Stone Mountain granite in the presence of water.

Part I. Experimental Results.

The granite, ground fine, and water were sealed up in platinum capsules and brought to the required temperature and pressure in a bomb. Pressure was obtained from the expansion of carbon dioxide, and temperature by means of a platinum resistance furnace enclosed in the bomb.

Equilibrium was difficult to obtain at temperatures below 800° C: for example, 50 hours was required to fuse Stone Mountain granite at a temperature of 723° C and under a water vapor pressure of 960 bars.

It is concluded that at 700 ± 50° C and under a water vapor pressure of 980 bars Stone Mountain granite will become essentially completely liquid, the resulting magma containing 6.5% water in solution; that at 575 ± 50° C and under a water vapor pressure of 385 bars the granite will be 99% + crystalline.

Part II. General Discussion.

After assuming a certain relation between liquidus temperatures of granite and solubility of water the liquidus of dry Stone Mountain granite is calculated to be 1050 ± 50°. The general course of crystallization of a granite magma containing water in solution is then discussed.

GEOPHYSICAL LABORATORY,
   CARNEGIE INSTITUTION OF WASHINGTON.

CHAPTER 16

Maaløe, S. and Wyllie, P.J. (1975) Water content of a granite magma deduced from the sequence of crystallization determined experimentally with water-undersaturated conditions. *Contributions to Mineralogy and Petrology*, **52**, 175–191.

Experimental petrology, on both synthetic and natural rock compositions, has contributed a great deal to our knowledge of the phase relations in granitic magmas. The real, ultimate aim of all this work has been to constrain the pressure-temperature conditions of formation and final crystallization of granitic magmas, the volatile contents of the melts and the likely protoliths from and melting reactions through which granitic magmas have formed. The classic experimental work of Tuttle and Bowen (1958) in the "haplogranite" system (quartz–albite–orthoclase–H_2O) was followed by many determinations of H_2O-saturated phase equilibria in SiO_2–$NaAlSi_3O_8$–$KAlSi_3O_8$–$CaAl_2Si_2O_8$–H_2O and its subsystems (e.g. James and Hamilton, 1969; Luth, 1969; Huang and Wyllie, 1975; Turner *et al.*, 1975; Johannes, 1978), their equivalents with Fe and Mg added (e.g. Naney and Swanson, 1980), and numerous experiments to determine the P-T phase relations in H_2O-saturated natural granitic rocks (e.g. Piwinskii, 1968; Piwinskii and Wyllie, 1970; Huang and Wyllie, 1973). Nevertheless, Tuttle and Bowen had already commented that most granitic magmas were probably not saturated in "water", and this was later echoed in the conclusions of a number of prominent workers (e.g. Burnham, 1967; Luth, 1969; Piwinskii, 1968; Piwinskii and Wyllie, 1970).

Thus, in the 1970s there came a growing realization that most granitic magmas that formed by partial melting of crustal protoliths began their existence as highly H_2O-undersaturated melts, and only evolved toward H_2O saturation late in their crystallization histories. In this case, the initial temperatures of granitic magmas might be considerably higher than was commonly assumed; many igneous geologists saw granites as relatively cool magmas, formed at temperatures near the H_2O-saturated solidus in the haplogranite system – somewhere around 650°C or so. The importance of f_{O_2}, and its control over both oxide and mafic mineral stability also became apparent. Thus there arose a need for careful experimental work that controlled not only P and T, but also f_{O_2}. Furthermore, if the H_2O contents of granitic magmas were to be properly constrained it would be necessary to carry out experiments at a range of H_2O contents in the region where the product melts would be H_2O-undersaturated. The idea was that, if you could constrain the pressure at which a granitic magma had crystallized (by geology, stratigraphy or geobarometry), you could carry out experiments designed to

determine the magma's phase relations in a T–XH_2O section. Then, by comparing the petrographically determined mineral crystallization sequence in the rock with the order of crystallization in the T-X section, you might be able to determine both the minimum temperature of initial crystallization and the range of possible initial melt H_2O contents. The groundwork for this leap forward was laid by Robertson and Wyllie (1971) who demonstrated the principles by which such isobaric T-X sections could be constructed, and also proposed probable T-X phase relations in some natural felsic rock compositions. Indeed, it could be argued, with some justice, that this work is the real landmark. However, we have chosen to celebrate the results of the first comprehensive, combined petrographic and experimental investigation in a granitic system – Maaløe and Wyllie (1975).

Maaløe and Wyllie recognized that, to make any sort of accurate determination of initial melt H_2O content of a granitic magma, they needed this sort of detailed H_2O-undersaturated work, to map out phase saturation boundaries in a T-X section. However, they also carefully controlled f_{O_2}, since Robertson and Wyllie (1971) had anticipated that this parameter could have strong effects on the stabilities of ferromagnesian phases. Furthermore, they thought long and hard about the issue of whether their experiments would come close to equilibrium conditions. In their study, the then standard approach of using finely crushed natural rock powder as a starting material was supplemented through the use of powdered glassy starting material made by fusion of the rock powder. The thinking was probably that, at H_2O-undersaturated conditions, attainment of equilibrium would take longer than in the H_2O-saturated experiments of previous workers. Using their technique, they could check whether both sorts of starting material produced the same phase assemblages and the same phase proportions at the same experimental conditions. This would not prove attainment of equilibrium but it would be a lot better than hope! In the event, however, it was clear that the crystalline starting material did not produce the same mineral assemblage as the glassy starting material. Maaløe and Wyllie seem to have concluded that the glassy starting material yielded results much closer to equilibrium because it was the results of this work that they used to construct their T-X section. This seems reasonable because, unless you include a superliquidus step to destroy the metastable nuclei formed during the run up to final experiment temperature, it will be far easier to resorb these tiny unstable nuclei than to resorb 10 μm-sized crystal fragments.

The inferences derived from the Maaløe and Wyllie experiments can be questioned because they chose an arbitrary pressure of 200 MPa for their runs. This may have been because they had no concrete geological information on which to base their choice, and because the only large-volume, low- to medium-pressure experimental apparatus that they had at their disposal were TZM cold-seal vessels; at the necessary temperatures, these could not be pushed to higher pressure without rupturing. Nevertheless this was pioneering work that showed how data on the temperature and volatile content of a real granitic magma could be obtained. Although this requires long series of somewhat difficult experiments, numerous subsequent studies used very similar techniques and produced a mass of valuable data. Together, these data have been used (among other things) to demonstrate that crustally-derived granitic magmas were indeed initially H_2O-undersaturated, that they were generated principally by fluid-absent partial melting reactions that occurred during amphibolite- to granulite-facies metamorphism of a variety of protoliths, and that the magmas had relatively low viscosities (see e.g. Scaillet *et al.*, 1998; Clemens and Watkins, 2001, and reference therein).

Contrib. Mineral. Petrol. 52, 175—191 (1975) — © by Springer-Verlag 1975

Water Content of a Granite Magma Deduced from the Sequence of Crystallization Determined Experimentally with Water-Undersaturated Conditions

Sven Maaløe and Peter J. Wyllie

Institut for Almen Geologi, Copenhagen, Denmark
Department of the Geophysical Sciences, University of Chicago, Chicago, Illinois 60637

Abstract. The sequence of crystallization in a biotite-granite from the Bohus batholith of Norway and Sweden, deduced from its texture, was magnetite, plagioclase, microcline, quartz, and finally biotite. Several sequences of crystallization were determined experimentally at 2 kb in the presence of varying H_2O contents, and this deduced sequence was reproduced only for H_2O contents below 1.2% by weight. The rock was fused to a homogeneous glass, and each experiment included samples of finely crushed rock and glass. The samples were reacted in Ag—Pd capsules with measured H_2O content in coldseal pressure vessels with NNO buffer. With excess H_2O (more than 6.5%) the crystallization interval extends from 865° C to 705° C. In the H_2O-deficient region, the solidus temperature remains unchanged as long as a trace of vapor is present, but the liquidus temperature increases as H_2O content decreases; with 0.8% H_2O the liquidus temperature is 1 125° C, the crystallization interval is 420° C, and a separate aqueous vapor phase is evolved only a few degrees above the solidus at 705° C. The biotite phase boundary increases slightly from 845° C with excess H_2O to 875° C with 1% H_2O, and it intersects the steep phase boundaries for quartz and feldspars; the sequence of crystallization changes at each intersection point. Similar diagrams at various pressures for related rock compositions involving muscovite, biotite and amphibole will provide grids useful in defining limits for the water content of granitic and dioritic magmas. Applications are considered for the Bohus batholith, other granitic rocks, and rhyolites. The Bohus magma could have been formed by crustal anatexis as a mobile assemblage of H_2O-undersaturated liquid and residual crystals with initial total H_2O content less than 1.2%, or it could have been derived by fractionation of a more basic parent with low H_2O content from mantle or subduction zone, but it could not have been derived from a primary andesite generated from mantle peridotite. We consider it unlikely that the H_2O content of large granitic magma bodies exceeds about 1.5% H_2O; these magmas are H_2O-undersaturated through most of their histories. Uprise and progressive crystallization of magma bodies produces H_2O-saturation around margins and in the upper regions of magma chambers. H_2O-saturated rhyolitic and dacitic magmas with phenocrysts can be tapped from the upper parts of the magma chambers.

S. Maaløe and P.J. Wyllie

Introduction

The significant role of water in the genesis, mobility, and crystallization of granitic magmas is well established, but there are few quantitative estimates of their water contents. According to Carmichael, Turner and Verhoogen (1974, p. 324): "So many experimental determinations of liquid-solid equilibria in silicate systems are made in the presence of a steam phase that the geologist can become lulled into accepting water saturation as a common state in nature ... there is really no good evidence to establish that, in the general case, it was saturated as it started to crystallize at depth." In fact, geologists have been explicitly warned in many experimental papers that the excess-water results represent a limiting condition that cannot be applied directly to natural magmas.

In their pioneering experimental study on the system $NaAlSi_3O_8$—$KAlSi_3O_8$—SiO_2—H_2O, Tuttle and Bowen (1958, p. 78) concluded that: "The fact that few granites have compositions near the "ternary" eutectic at 4000 kg/cm^2 water-vapor pressure, or near the minimum at 3000 kg/cm^2, suggests that granitic magmas are probably rarely saturated with water" ... "the water content of the liquids ... probably less than 2 weight per cent." They also illustrated (p. 122) the sensitivity of the granite liquidus temperature to water content. Burnham (1967, p. 72) noted that "a majority of shallow intrusions of felsic magmas initially were undersaturated with water for the prevailing load pressure at their source $(P_{e_{H_2O}} < P_l)$", and he formulated a model for the emplacement and crystallization of a water-undersaturated granitic magma using data from water-excess experimental results. Piwinskii and Wyllie (1968, 1970) determined the phase relationships with excess water for a series of granitic rocks from the Wallowa Batholith, stating that: "Caution is required in extrapolation from the experiments to the earth's crust because (1) the samples were run in the presence of excess water vapor, whereas natural magmas probably develop a free vapor phase only during the late stages of crystallization" (1968, p. 228), and "There are several factors suggesting that the normal product of partial fusion of a wide range of crustal rock types is a H_2O-undersaturated granite liquid;" (1970, p. 73). Similarly, Luth (1969) emphasized that phase relationships in the water-undersaturated region of granitic systems differ from those with excess water at the same total pressure, and Presnall and Bateman (1973) concluded that water-undersaturated phase relationships in the system Ab—An—Or—Qz—H_2O are "most relevant to fusion and crystallization processes that have taken place" in the Sierra Nevada Batholith.

Reaction rates in granitic systems are notoriously sluggish (Piwinskii, 1967, 1968; Piwinskii and Martin, 1970), and this retarded experimental progress with low water contents. Several experimental studies with water-undersaturated granitic systems at pressures below 10 kb have now been published, however, including those of Brown and Fyfe (1970), Robertson and Wyllie (1971a, 1971b), Eggler (1972), Eggler and Burnham (1973), Whitney (1975a), and Steiner, Jahns and Luth (1975). Results for similar systems have been extended to pressures of 30 kb or more by Green (1972), Huang and Wyllie (1973), and Stern and Wyllie (1973). The thermodynamic basis and theoretical treatment of silicate-water systems have been presented by Burnham and Davis (1971, 1974).

In this experimental study we have determined the sequence of crystallization of a granite from Norway as a function of water content, and compared this with the sequence of crystallization deduced from the texture of the original rock. This places limits on the possible water contents of the magma from which the granite crystallized. The original magma contained less than 1.2% H_2O, which is consistent with Eggler's (1972) experimentally based estimate of the water content of the Paricutin andesite before eruption.

Monzogranite (MG-1) from Norway and Its Crystallization Sequence

The granitic rock sample is a monzogranite (Streckeisen, 1967) from the Norwegian part of the Pre-cambrian Bohus Batholith in southeastern Norway and western Sweden. The batholith, 120 km long and 20–30 km wide, comprises about twenty discrete granite plutons, emplaced into metamorphic rocks of middle amphibolite facies. The granites display only a very small decrease in grain size towards their sharp contacts, indicating relatively high temperature for the country rocks during the intrusion. The depth of intrusion might therefore have been within the depth range of the amphibolite facies, 5–30 km.

The granite selected for this study (type MG-1) is fresh, with biotite and felsic minerals showing no sign of alteration, but with about 10% of the magnetite oxidized to hematite. The granite was intruded into another granite of similar composition, which has retained its composition to the very contact. The granite is devoid of pegmatitic schlieren, and pegmatites are extremely rare within the batholith as a whole.

The modal analysis, chemical analysis, and mineral chemistry are given in Tables 1 and 2. The granite is homogeneous, medium-to fine-grained, consisting of zoned plagioclase, perthitic microcline, quartz and biotite, with accessory minerals sphene, apatite, magnetite and amphibole. The tabular crystals of plagioclase and microcline may display a subparallel orientation, but the other minerals exhibit no systematic orientation.

The plagioclase crystals have an oscillatory zoned core with one to six reverse zones, and a normally zoned margin with steadily decreasing anorthite content. The radii of the cores are less than half the radii of the crystals. Small euhedral magnetite crystals occur throughout the plagioclase crystals, but microcline, quartz and biotite are included in the margins only. Quartz and biotite also exist as marginal inclusions in the microcline.

The textural features thus suggest the early nucleation of magnetite followed successively by plagioclase, microcline, and quartz and biotite. From the positions of the innermost inclusions of quartz and biotite in plagioclase we conclude that they began to crystallize at about the same time, with quartz probably preceding biotite, and this is the sequence illustrated in Fig. 2.

The composition and norm of a synthetic granite (R-1) studied experimentally by Whitney (1975a) are listed for comparison. This haplogranite is very similar to granite MG-1 in terms of normative feldspars and quartz.

Table 1. Chemical analyses, CIPW weight norms, and modal analysis of the Iddefjord granite (MG-1), compared with the synthetic granite (R-1) studied by Whitney (1957a)

Analysis	MG-1	R-1	Mode of MG-1	
SiO_2	71.30	73.98	Plagioclase	28.4
TiO_2	0.40	–	Microcline	39.1
Al_2O_3	14.44	15.07	Quartz	26.1
Fe_2O_3	1.00	–	Biotite	5.2
FeO	1.41	–	Muscovite	0.07
MnO	0.03	–	Amphibole	0.07
MgO	0.40	–	Magnetite	0.65
CaO	1.69	1.50	Sphene	0.26
Na_2O	3.44	3.75	Apatite	0.13
K_2O	5.45	5.70	Points counted	1533
H_2O^+	0.45	–		
H_2O^-	0.06	–		
P_2O_5	0.10	–		
Total	100.17	100.00		
Norm				
Q	26.1	26.5		
Or	32.3	34.0		
Ab	29.2	32.0		
An	7.8	7.5		
Hy	2.2	–		
Mt	1.5	–		
Il	0.76	–		
Ap	0.23	–		
Cc	0.05	–		

Analyst for MG-1: J.H. Scoon.

Table 2. Microprobe analyses of minerals in the Iddefjord granite (MG-1)

	Pl_c	Pl_i	Pl_o	Mi	Q	Bi	Mag
SiO_2	59.88	62.27	68.56	64.93	100.60	36.03	0.00
TiO_2	0.03	0.01	0.01	0.02	0.05	2.96	0.08
Al_2O_3	25.17	23.48	19.69	18.47	0.01	15.04	0.00
FeO	0.17	0.18	0.14	0.06	0.03	22.45	30.81
Fe_2O_3	–	–	–	–	–	–	68.48
MnO	–	–	–	–	–	0.50	0.00
MgO	–	–	–	–	–	8.51	0.00
CaO	6.82	4.87	0.18	0.01	0.02	0.01	0.00
Na_2O	7.76	8.82	11.56	1.02	0.09	0.13	0.00
K_2O	0.23	0.30	0.15	15.33	0.01	9.61	0.00
H_2O^+	–	–	–	–	–	3.83	–
H_2O^-	–	–	–	–	–	0.22	–
Total	100.06	99.93	100.29	99.84	100.81	99.29	100.11

Pl_c, plagioclase core; Pl_i, inner margin of plagioclase; Pl_o, outer margin of plagioclase; Mi, microcline; Q, quartz; Bi, biotite; Mag, magnetite. Water determined by the Penfield method.

Experimental Methods

Apparatus and Starting Material

The experiments were made in sealed Ag—Pd capsules within Rene 41 cold-seal pressure vessels with internal thermocouples and filler rods, using distilled water or argon as pressure media. Attempts to extend the temperature range in TZM pressure vessels were unsuccessful due to rupture of the vessels. Normal procedures were used for calibration of thermocouples and measurements of pressure and temperature (Piwinskii and Wyllie, 1968). Temperatures are believed accurate to $\pm 10°$ C and pressures to ± 100 bars. Other experimental details are given by Robertson and Wyllie (1971a).

Starting materials were the crushed rock and homogeneous glass prepared by three successive fusions of the rock. The grain size of the powders averaged 10 micron, with maximum dimensions reaching 50 micron. Each experiment contained both starting materials in separate capsules.

Oxygen fugacity was controlled using the NNO buffer (Eugster, 1957). About 50 mg of rock and glass powder were sealed into separate Ag—Pd capsules with the required proportions of distilled water, and the capsules were sealed within a larger gold capsule along with the buffer mixture. Many granitic rocks apparently crystallized at an oxygen fugacity near that of the NNO-buffer (Buddington and Lindsley, 1964; Eggler and Burnham, 1973), which is fortunate because the NNO-buffer is the only one with a reasonable life-time in the Rene 41 vessels; this varied from 7 days at 850° C to 30 days at 700° C. Nickel from the NNO-buffer alloyed with the Ag—Pd capsules, but only few runs were lost when leaks developed because of this.

The double-capsule buffer technique was designed for systems with an aqueous vapor phase. In charges where the water content is insufficient to saturate the liquid, the buffering effect is modified, as described by Whitney (1974). For an all liquid charge with the NNO-buffer at 800° C, $\log (f_{O_2})$ decreases from 10^{-14} at water-saturated conditions to 10^{-16} with 1% water.

Identification of Phases

The crushed run products were examined by X-ray and optical methods. Plagioclase formed tabular faceted crystals, alkali feldspar formed irregular crystals, and biotite formed hexagonal, pleochroic plates. Quartz did not develop crystal faces. A trace of opaque oxide was present in all runs, and we call this magnetite, because of its morphology, although ilmenite is possible (Eggler and Burnham, 1973). Acicular prisms with oblique extinction (clinopyroxene?) were observed in most runs above 700° C, but these were insufficient to register on the X-ray diffraction patterns.

In the vapor-absent region of two syenites, Robertson and Wyllie (1971a) were unable to attain equilibrium in runs of duration as long as the strength of Rene 41 pressure vessels would permit; runs at 2 kbar were held for 30 days at temperatures above 775° C, and for 6 months or more at lower temperatures. Whitney (1975a) studied granitic compositions using fired gels ground to particle size averaging about 5 micron, with runs varying from 3 days to 1 month. The results using gel were duplicated using crystallized gel, in both vapor-present and vapor-absent regions, and Whitney concluded that this evidence of equilibrium was due to the fine particle size.

Our run durations, using starting material with grain size near 10 microns, were as long or longer than those of Whitney (1975a). They varied from 2 months near the solidus (705° C) to 6 hours at temperatures above 900° C. Each experiment included a test for equilibrium by comparison of the results obtained from crystalline and glass starting materials. Plagioclase and biotite nucleated rapidly in the glass, and these two minerals occurred in approximately the same proportions in both glass and crystalline starting materials in most runs above 800° C. The amount of alkali feldspar produced in the glass did not reach the amount present in the crystalline starting material. Quartz did not nucleate at all in the glass starting materials, even in runs of 60 days duration, near the solidus. Lack of equilibrium in near-solidus runs was also reported by Piwinskii (1967).

The phase boundaries for plagioclase and biotite were located from the presence of these minerals in both glass and crystalline starting material, according to the X-ray powder patterns, while the phase boundaries for alkali feldspar and quartz were located by their disappearance from the crystalline starting material. Two of the runs defining the biotite phase boundary in the vapor-absent region

S. Maaløe and P.J. Wyllie

were discontinuous reversal runs. They were held first in the liquid field using TZM pressure vessels, then quenched and transferred to the standard vessels. The starting material for the second stage of each run was thus a glass with dissolved water.

Robertson and Wyllie (1971a) illustrated a reaction rim around charges in water-deficient runs. The rim contained gold blebs and a higher proportion of glass, probably due to the inward diffusion of water from the pressure medium outside of the gold capsule. Within our Ag—Pd capsules, a narrower reaction rim was observed without metallic blebs.

Experimental Results

Phase Fields Intersected

Experimental results for the granite MG-1 are plotted in Fig. 1 and some critical runs are listed in Table 3. Temperatures for the dehydrated rock composition are assumed to be the same as results presented for Whitney's (1975a) synthetic granite (R-1 in Table 1). We have also adopted the positions of Whitney's solidus, and water-saturation boundary through the crystallization interval. The saturated liquid contains 6.5 ± 0.5 weight per cent H_2O. Our runs are consistent with these, and with other measured or calculated values for similar rocks (Piwinskii, 1968; Robertson and Wyllie, 1971b, Fig. 7). The occurrence of trace amounts of magnetite and possible clinopyroxene have not been plotted on the diagram.

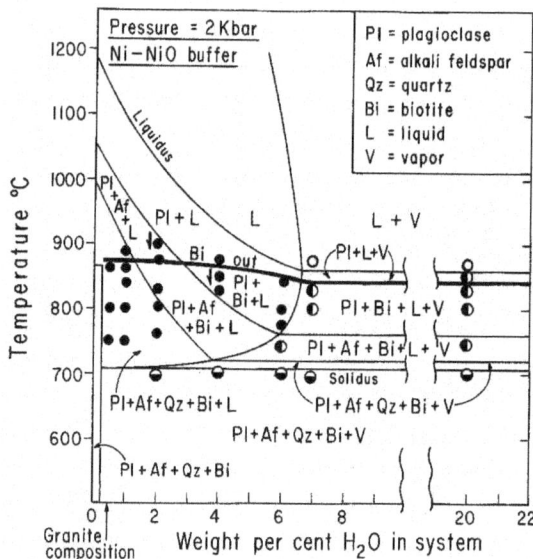

Fig. 1. Experimental results for granite MG-1. For chemical and mineralogical data see Tables 1 and 2. For critical experimental runs see Table 3. Each experiment included two samples, one with crushed rock, and the other with crushed glass. Note the boundary between phase fields with vapor (run circles open or partly open) and those without vapor (run circles closed). The heavy line is the upper stability limit of biotite: we have not determined its position for less than 1% H_2O, nor how it reaches the subsolidus vapor-absent assemblage. The vertical arrows identify runs held first in the field of liquid. The subsolidus assemblage, listed as Pl+Af+Qz+Bi, must include as well the dehydration products of biotite, which decreases from 5.2 modal % to 0% as H_2O in the system decreases from 0.2% to 0%

Table 3. Selected runs for the phase relationships of granite MG-1 at 2 kilobars pressure, with NNO buffer

Starting material	Weight (% H$_2$O)	Temp. (°C)	Time[a]	Crystalline phases[b]
cryst	0.45	860	17 d	Qz, Af, Pl, Bi
glass	0.45	860	17 d	Af, Pl
cryst	1	860	17 d	Qz, Af, Pl, Bi
glass	1	860	17 d	Af, Pl, Bi
cryst	1	887	6 h	Qz, Af, Pl
glass,	1	887	6 h	Af, Pl
cryst	2	827	14 d	Af, Pl, Bi
glass	2	827	14 d	Pl, Bi
cryst	2	875	1 d	Af, Pl
glass	2	875	1 d	Pl
cryst	2	901	6 h	Af, Pl
Rglass	2	901	6 h	Pl
cryst	4	852	2 d	Pl, Bi
Rglass	4	852	2 d	Pl, Bi
cryst	4	876	2 d	Pl
glass	4	876	2 d	Pl
cryst	6	840	7 d	Pl, Bi
glass	6	840	7 d	Pl, Bi
cryst	7	827	14 d	Pl, Bi
glass	7	827	14 d	Pl, Bi
cryst	7	875	1 d	—
glass	7	875	1 d	—
cryst	20	827	7 d	Pl, Bi
glass	20	827	7 d	Pl, Bi
cryst	20	850	7 d	Pl
glass	20	850	7 d	Pl

[a] d, days; h, hours.
[b] For abbreviations see Fig. 1. Runs Rglass with 2% and 4% H$_2$O were melted first at 1100° C and 1000° C (respectively), in their sealed capsules, quenched, and run again at the lower temperature listed.

The horizontal axis shows total H$_2$O content of the system. The rock contains 5.2 modal per cent biotite, with 3.8% H$_2$O (Table 2); this indicates that the rock is saturated with 0.2% H$_2$O under subsolidus conditions. The vertical boundary for 0.2% H$_2$O thus separates the vapor-absent from the vapor-present assemplages under subsolidus conditions. The rock analysis gives 0.45% H$_2$O + (Table 1), and the difference presumably represents H$_2$O firmly trapped within fluid inclusions or along intergranular boundaries.

With excess H$_2$O, the crystallization interval extends through 150° C, from the liquidus at 865° C to the solidus at 705° C. In the water-deficient region, where the H$_2$O content is less than that required to saturate the liquid, temperatures of the liquidus and the phase boundaries for the feldspars and quartz increase significantly, with little change in their positions relative to each other.

The solidus temperature remains unchanged in the presence of even a small amount of vapor. The crystallization interval therefore increases as H_2O content decreases.

The general arrangement of the phase boundaries for feldspars and quartz in the vapor-absent region is consistent with that deduced or measured experimentally by Robertson and Wyllie (1971a, 1971b), Huang and Wyllie (1973), and Whitney (1975a). Shimada's (1969) results for albite-H_2O suggested to Robertson and Wyllie (1971a, b) that the curves would be convex-upwards, but a concave-upwards curvature for these and other anhydrous minerals is now established in many rock-water systems (Huang and Wyllie, 1973; Stern and Wyllie, 1973; Whitney, 1975a). Fig. 1 is very similar to Whitney's (1975a) results for synthetic granite R-1 (Table 1); the liquidus temperature is higher by about 25° C.

The phase boundary limiting the stability of biotite cuts across the other phase boundaries in the vapor-absent region. It increases slightly from 845° C with excess H_2O to 875° C with 1% H_2O. For the rock composition itself, with only 0.45% H_2O, biotite persisted in the crystalline starting material at 860° C, but did not grow in the glass starting material. This increase in stability temperature of a hydrous mineral with decreasing H_2O content and a H_2O-undersaturated liquid appears to be well established (see Brown and Fyfe, 1970), and is consistent with experiments on phlogopite (Yoder and Kushiro, 1969), amphibole (Holloway, 1973; Eggler, 1972; Eggler and Burnham, 1973), and muscovite (Huang, Robertson and Wyllie, 1973; Huang and Wyllie, 1974). Carmichael et al. (1974, p. 121–122) consider these results to be "apparently in contradiction to the dictates of thermodynamics", and discuss reconciliation. The temperature difference between the maximum stability in the vapor-absent region and the stability limit with excess H_2O varies considerably from one system to another and from one total pressure to another. The maximum on the biotite curve estimated for biotite-granodiorite at 2 kbar by Robertson and Wyllie (1971a, 1971b) on the basis of results in amphibole eclogites (Millhollen and Wyllie, 1970, 1974; Merrill and Wyllie, 1972, 1975) was too high in temperature, according to the results in Fig. 1.

The effect of oxygen fugacity on the stability of biotite at 800° C and total H_2O content of 4.5% was tested in three runs with MW, NNO, and HM buffers. All runs contained biotite, with similar intensities for X-ray peaks in powder diffraction patterns.

Sequence of Crystallization

The sequence of crystallization of the felsic minerals from the MG-1 granite liquid is plagioclase, alkali feldspar, and quartz. The phase boundary for biotite intersects those for quartz and alkali feldspar in the vapor-absent region, and therefore biotite appears at different positions within the crystallization sequence, depending upon the H_2O content. The changes occur at H_2O contents of 1.2% and 2.5%, as shown in Fig. 2. For low H_2O contents biotite is the last mineral to crystallize, and for higher H_2O contents crystallization of biotite precedes that of alkali feldspar and quartz.

For compositions with more H_2O than 6.5%, vapor is present throughout the crystallization interval, and the H_2O content of the saturated liquid changes only slightly. With progressive crystallization of liquids containing less than 6.5% H_2O, the H_2O content of the liquid increases with progressive crystallization until it reaches saturation at the temperature given by the saturation boundary; note that this occurs closer to the solidus for mixtures with lower H_2O contents. The H_2O contents of liquids during crystallization of H_2O-undersaturated granitic liquids can be determined from the diagrams of Jahns and Burnham (1969).

Comparison of Phase Diagrams with Crystallization Sequence in Granite

Fig. 2 compares the inferred crystallization sequence in the natural granite, MG-1, with the sequences of crystallization determined for the granite with various H_2O contents, in a closed system at 2 kbar. If we assume arbitrarily that the natural granite was emplaced as a liquid magma at a depth of 7–8 km, and crystallized under equilibrium conditions as a closed system with H_2O as the only volatile component, we can conclude from Figs. 1 and 2 that the H_2O content of the initial liquid was between 0.2% and 1.2%, the minimum amount corresponding to that now stored in the biotite. According to the textural evidence, quartz began to crystallize just before biotite, which indicates a H_2O content closer to 1.2%. If the granite was emplaced as a mixture of liquid and crystals, the H_2O content of the magma would have the same limits, but it would be concentrated in the H_2O-undersaturated liquid portion of the magma, with the actual concentration depending on the percentage of crystals and liquid.

Evaluation of available but limited data at other pressures suggests that these H_2O limits vary little through the range of middle and upper crustal pressures corresponding to the probable depth of emplacement of the granite. The data include Whitney's (1975a) diagram for synthetic granite R-1 corresponding to Fig. 1 at 8 kbar, without the biotite phase boundary; Piwinskii's (1973a, 1973b)

Fig. 2. Sequence of crystallization in granite MG-1 deduced from its texture, compared with the experimentally determined crystallization sequences at 2 kb for different ranges of H_2O content. These sequences are taken from Fig. 1

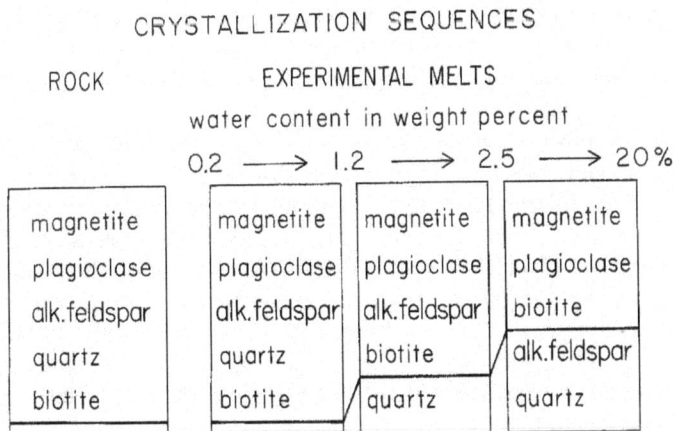

CRYSTALLIZATION SEQUENCES

ROCK   EXPERIMENTAL MELTS

water content in weight percent

0.2 $\longrightarrow$ 1.2 $\longrightarrow$ 2.5 $\longrightarrow$ 20%

ROCK			
magnetite	magnetite	magnetite	magnetite
plagioclase	plagioclase	plagioclase	plagioclase
alk.feldspar	alk.feldspar	alk.feldspar	biotite
quartz	quartz	biotite	alk.feldspar
biotite	biotite	quartz	quartz

diagrams for biotite in granitic rocks with excess water to 10 kbar; Holloway's (1973) study on the effect of pressure on the stability of amphibole in the water-deficient region; and Huang and Wyllie's (1973) results for muscovite granite.

The assumption of equilibrium crystallization is surely not valid. The plagioclase feldspars are zoned. Presnall and Bateman (1973) demonstrated convincingly that granitic rocks of the Sierra Nevada batholith were produced by fractional crystallization of a crystal-liquid mush. But fractional crystallization in a closed system requires a lesser H_2O content at higher temperatures than the 1.2% maximum indicated by the intersection of the quartz and biotite phase boundaries.

The field evidence suggests that these plutons crystallized under conditions that approximated closed systems, as described above. The absence of pegmatitic schlieren and rarity of pegmatites in the batholith suggest a low H_2O content. Figs. 1 and 3 show that only when crystallization is almost completed do magmas with low H_2O content release the aqueous vapor phase required for pegmatite formation (Jahns and Burnham, 1969). The presence of other volatile components such as CO_2 in amounts considered reasonable for granitic magmas could yield a vapor phase at higher temperatures, but the relative positions of phase boundaries are unlikely to change significantly (Wyllie and Tuttle, 1960; Eggler and Burnham, 1973). If the granite lost H_2O during crystallization as an open system, the amount lost was probably trivial until the closing stages of crystallization. The maximum H_2O content indicated by the intersection of the quartz and biotite phase boundaries in Fig. 1 probably remains valid even for the real crystallizing granite magma.

Petrological Applications

Granite MG-1

We have concluded that the texture of the granite MG-1 is consistent with its crystallization under closed conditions at 2 kbars from a magma with H_2O content between 0.2 and 1.2%. Let us assume an intermediate value, 0.8% H_2O; then the crystallization sequence is as shown in Fig. 3. Plagioclase, the liquidus mineral at 1125° C, crystallizes alone through a temperature interval of 145° C until it is joined by alkali feldspar at 980° C. Quartz begins to crystallize at 915° C, and biotite appears at 882° C. All four minerals crystallize together through a temperature interval of 177° C until solidification is completed at 705° C. A separate aqueous vapor phase is produced only a few degrees above the solidus.

Notice in Fig. 3 the high liquidus temperature, the wide crystallization interval of 420° C, and the fact that a vapor phase is released only during the closing stages of crystallization. The H_2O content of the liquid increases from 0.8% at the liquidus to about 6.5% just above 705° C where the vapor is released (Jahns and Burnham, 1969; Robertson and Wyllie, 1971b, Fig. 6).

The high liquidus temperature indicates that the rock MG-1 did not crystallize from a completely liquid magma generated by crustal anatexis, unless crustal temperatures have been considerably higher than most recent thermal models

Fig. 3. Crystallization interval and crystallization sequence for granite MG-1 with 0.8% H_2O at 2 kb pressure, taken from Fig. 1. Plagioclase feldspar is the liquidus phase, and a separate aqueous vapor phase appears only a few degrees above the solidus. The crystallization interval is 420° C

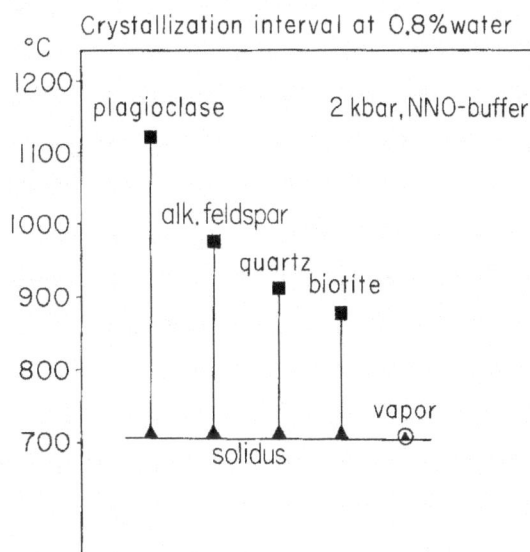

Crystallization interval at 0.8% water

would permit. If the magma was formed by crustal anatexis, it is more likely that the mobile mass produced consisted of an assemblage of H_2O-undersaturated liquid with suspended residual crystals (Piwinskii and Wyllie, 1968; Brown and Fyfe, 1970; Presnall and Bateman, 1973). The cores of the plagioclase crystals (Table 2) could readily be interpreted as unmelted metamorphic minerals. At 950° C, the magma could consist of about 50% liquid with suspended plagioclase and sanidine; with our assumed value of 0.8% H_2O for the magma, the liquid would contain 1.6% dissolved H_2O.

If the granite liquid was derived at 1120° C through fractionation of a more basic magma from mantle depths, the parent must have contained less than 0.8% H_2O, because crystallization of the parent magma normally increases the H_2O content of the derivative liquids. If this granite and the other plutons in the batholith were derived by fractionation of a parent andesitic liquid, we conclude that the primary andesite had a low H_2O content. This is consistent with calculations of H_2O contents of andesites generated in subduction zones according to Marsh and Carmichael (1974), 0.75%, and with Eggler's (1972) experimentally based estimate of less than 2.2% H_2O for the Paricutin andesite. Yoder (1969) noted that small amounts of water are required to match the experimental melting relations of many andesite suites with the petrography of the rocks. Anderson (1973) found that inclusions of andesitic glasses in olivine from tephra-rich eruptions contained less than 2% H_2O.

Primary andesites derived from mantle peridotite could not conceivably be parents for the low-H_2O granite magma, because such magmas require H_2O contents of 10 to 20% (Mysen et al., 1974; Nicholls, 1974). Even if primary andesite magmas are generated in the mantle, it seems unlikely that they could reach the surface for eruption under normal conditions (Nicholls, 1974; Stern, Huang and Wyllie, 1975).

Other Granitic Rocks

The experimental results in Fig. 1 are consistent with conclusions reached by Barker et al. (1975) for other granitic rocks in a detailed review of the Pikes Peak batholith. From a combination of mineral stability and melting phase diagrams, they concluded that the magmas of the batholith crystallized at a range of total pressures between 1 and 2 kbar, that the magmas were H_2O-undersaturated until 90–99% crystallized, and that crystallization was completed at temperatures near 700° C. They presented the sequence of crystallization for two syenites and four granites. In five of the six rocks, the crystallization of biotite was preceded by plagioclase, quartz, and alkali feldspar. Despite the variety of rock types involved, it is clear from the geometry of Fig. 1 that this sequence of crystallization can only occur for low H_2O contents, less than 1 or 2% in the magmas. The sequences of crystallization for various H_2O contents in Figs. 1 and 2 may thus be approximately applicable to other granitic magmas.

It is now possible to sketch the phase relationships corresponding to Fig. 1 for many rock types for any crustal pressure, using excess-H_2O results and the method described by Robertson and Wyllie (1971b), and using the results in Fig. 1 and those of Whitney (1975a) as a guide for the shapes of phase boundaries. Available excess-H_2O results were summarized by Robertson and Wyllie (1971a), and additional results to 10 kbar have been published by Piwinskii (1973a, 1973b, 1974, 1975).

Diagrams similar to Fig. 1 at various pressures for related rock compositions involving muscovite (Huang and Wyllie, 1973a), biotite, and amphibole (Eggler and Burnham, 1973) will provide a series of grids useful in defining limits for the water content of granitic and dioritic magmas. In the water-deficient region, the steep phase boundaries for the anhydrous silicates cross the more nearly isothermal phase boundaries for the hydrous silicates, with each intersection causing a change in the sequence of crystallization. It should be possible, eventually, to contour the vapor-absent space in terms of f_{H_2O} (Burnham and Davis, 1974). Values of f_{H_2O} can be calculated from data on coexisting biotite, sanidine and magnetite (Wones and Eugster, 1965), but there are uncertainties related to the mixing properties of Mg and Fe in biotite, and estimated values for the activity of Fe in biotite (Mueller, 1972; Wones, 1972).

We cited in the introduction several experimentally-based papers concluding that granitic magmas are rarely saturated with H_2O, and many others have reached the same conclusion for specific intrusions (e.g. Presnall and Bateman, 1973, Putman and Alfors, 1965, for the Sierra Nevada batholith; Barker et al., 1975, for the Pikes Peak batholith; Upton, 1960, for the Kungnat syenite complex of southwest Greenland). In contrast, we are not aware of any definitive evidence for granitic magma bodies containing high proportions of dissolved H_2O, except for strongly fractionated liquids corresponding to the closing stages of crystallization of a parent magma. Jahns and Burnham (1969, Fig. 2) illustrated the crystallization interval for granitic liquids at 5 kbar with various initial H_2O contents. There is complete gradation between granites and pegmatites, but on their diagram the name "granite" covers liquids with initially less than 2% H_2O, and "pegmatite" covers liquids with initially more than 5% H_2O.

A magma initially undersaturated with H_2O evolves a separate aqueous vapor phase during crystallization, and it is convenient to consider two limiting processes that cause this. First, there is isobaric cooling, represented by Fig. 3. Around the margins of a static magma body, the aqueous vapor is evolved as the magma cools and solidifies. Secondly, there is isothermal decrease in pressure, represented approximately by rapid uprise of magma. The magma will eventually reach a level where vapor is evolved from its upper parts; this can be seen from the diagrams of Robertson and Wyllie (1971b, Fig. 10B, across the line *d–e*) and Whitney (1975b, Fig. 3). Withney (1975b) calculated the temperature distribution in a simple quartz monzonite stock intruded to shallow levels, and plotted the equilibrium phase assemblages appropriate for the derived pressure-temperature distribution. This illustrated the formation around the intrusion of an envelope of magma evolving vapor. In this envelope, the residual liquid portion of the magma contains high H_2O contents, corresponding to saturation values at the varied pressures and temperatures.

Although the Sierra Nevada batholith crystallized from relatively dry magmas, there is evidence for the migration of large volumes of aqueous fluids along contacts between magma and wallrock or along the interface between magma and previously solidified magma. Moore and Lockwood (1973) explained "comb layering" and associated orbicular rocks as the product of deposition from the aqueous vapor, either on the solid walls of fluid-filled channels, or on inclusions suspended within the upward flowing fluid. This is consistent with the requirement that a vapor phase must be evolved at the crystallizing margin of a magma chamber, although some of the water could have been driven out of the neighboring wallrock during metamorphism by the pluton.

Rhyolites and Dacites

Although we believe that granitic magmas are H_2O-undersaturated through most of their history, there is evidence that many rhyolitic or dacitic obsidians and pumices involved H_2O-saturated liquids (Aramaki, 1971; Wood and Carmichael, 1973). If these were erupted from the upper portion of a magma chamber, we can anticipate from the preceding review of granitic intrusions that the liquid portion of the eruption could be H_2O-saturated, despite the probability that the magma body as a whole was H_2O-undersaturated. Liquid erupted from deeper portions of a magma chamber could contain less dissolved H_2O.

Calculations of f_{H_2O} and H_2O content of acid eruptions appear to involve many uncertainties. For a biotite rhyolite from the San Juan Mountains, Colorado, Wones and Eugster (1965) calculated f_{H_2O} between 125 and 600 bars. Following a discussion by Mueller (1972), the calculated range was increased to 200–2000 bars (Wones, 1972).

For amphibole-bearing rhyolitic pumices, lavas, and ignimbrites of New Zealand with phenocryst content ranging from 0–40%, Ewart *et al.* (1971) used various methods to calculate equilibration temperatures of 735–780° C, and f_{H_2O}

of 1.1–1.3 kbar, or P_{H_2O} of 1.9–2.3 kbar, depending on the method used. These estimates were refined by Wood and Carmichael (1973) using different mixing models for the mineral solid solutions. They calculated independent values for P_{total} and P_{H_2O}, and found that the amphibole equilibrated with the liquid under conditions approximating $P_{H_2O} = P_{total}$. Values for P_{total} are 3.98–4.06 kbar or 6.76–8.17 kbar, depending on the mixing model used, and corresponding values for P_{H_2O} are 4.45–5.1 kbar, or 7.2–9.0 kbar. This implies eruption of a saturated rhyolitic liquid with 9–12% dissolved H_2O from a minimum depth of 16 km. Comparison of diagrams such as Fig. 1 with phenocryst assemblages and sequences of crystallization in rhyolitic rocks might provide independent estimates of H_2O content that could place constraints on mixing models and thermodynamic calculations.

Stormer and Carmichael (1970) used the iron-titanium oxide geothermometer and the Kudo-Weill plagioclase geothermometer in a variety of rhyolitic obsidians and pumices to derive equilibration temperatures with the liquids formerly enclosing the phenocrysts. They concluded that the outer zones of the plagioclase phenocrysts equilibrated with $P_{H_2O} = 0.5$ kbar or less, which is consistent with H_2O fugacities deduced from the coexisting biotite-magnetite-sanidine assemblages, and that liquidus temperatures exceeded 900° C. The cores of plagioclase phenocrysts indicate temperatures more than 100° C higher than the margins, implying either a cooling interval in excess of 100° C, or a xenocrystic origin for the cores. $P_{H_2O} = 0.5$ kbar corresponds to H_2O content of about 2.5%, and for this H_2O content in Fig. 1 plagioclase crystallizes alone through a temperature interval near 1000–900° C. These results support the conclusion of Stormer and Carmichael (1970) that some rhyolitic liquids are erupted from fairly high-temperature H_2O-undersaturated granitic magma bodies.

Stormer and Carmichael (1970) noted that granites and rhyolites might have wider crystallization intervals than the synthetic equivalents represented by a narrow range of compositions in the Residua System (Tuttle and Bowen, 1958). It is clear from Figs. 1 and 3 that the crystallization interval for H_2O-undersaturated granitic liquids is several hundred degrees; this is true also even for the "ideal" granite minimum composition.

It has been proposed that an initially H_2O-undersaturated granitic magma could become saturated by interaction with meteoric H_2O when the magma body arrived at a high enough level in the crust. Friedman et al. (1974) and Lipman and Friedman (1975) reported that changes in oxygen isotope ratios for phenocrysts from rhyolitic sequences in Nevada, Colorado, and the Yellowstone plateau indicate a progressive depletion of the magmas in $\delta^{18}O$ with time. They explained this as reflecting a gradually increasing component of meteoric H_2O in batholithic-sized bodies of granitic magma. This requires diffusion of supercritical solutions into a H_2O-undersaturated magma, or bubbling of vapor through the melt after it becomes saturated. Taylor (1974) outlined other mechanisms for producing low-^{18}O magmas that he considered more plausible; these include melting, assimilation and dissolution of roof-rocks above the magma chamber, rocks that had been previously depleted in ^{18}O by the meteoric-hydrothermal circulation system associated with intrusive bodies.

S. Maaløe, S. and P.J. Wyllie, P.J.: Experimentally deduced water content of a granite magma

Water Content of a Granite Magma 189

Conclusion

We consider it unlikely that the H_2O content of large bodies of batholithic granitic magma exceeds about 1.5% (compare Fyfe, 1973). The conclusion that most granitic liquids are H_2O-undersaturated through most of their histories is not inconsistent with the conclusion that many rhyolitic eruptions are H_2O-saturated. Uprise and progressive crystallization of magma bodies produces H_2O-saturation in the upper regions of magma chambers, and at high levels in the crust interaction with meteoric fluids apparently can increase H_2O content. Rhyolitic eruptions with phenocrysts can be tapped from the H_2O-saturated tops of magma chambers. However, the values of 9–12% H_2O calculated by Wood and Carmichael (1973) are surprisingly high.

Acknowledgments. We thank W.-L. Huang and A. Berthelsen for their assistance and guidance in the laboratory and field, respectively; J.C. Scoon for analysing the granite MG-1, and T. Svane for calculating the norm; and A.T. Anderson for manuscript review. This research was supported by the Earth Sciences Section, National Science Foundation, NSF Grant DES 73-00191 A01. We wish to acknowledge also the general support of the Materials Research Laboratory by the National Science Foundation.

References

Anderson, A.T.: The before-eruption water content of some high-alumina magmas. Bull. Volcan. 37, 530–552 (1973)

Aramaki, S.: Hydrothermal determination of temperature and water pressure of the magma of Aira Caldera, Japan. Am. Mineralogist 56, 1760–1768 (1971)

Barker, F., Wones, D.R., Sharp, W.N., Desborough, G.A.: The Pikes Peak batholith, Colorado Front Range, and a model for the origin of the gabbro-anorthosite-syenite-potassic granite suite. Precambrian Research 2, 97–160 (1975)

Brown, G.C., Fyfe, W.S.: The production of granitic melts during ultrametamorphism. Contrib. Mineral. Petrol. 28, 310–318 (1970)

Buddington, A.F., Lindsley, D.H.: Iron-titanium oxide minerals and synthetic equivalents. J. Petrol. 5, 310–357 (1964)

Burnham, C.W.: Hydrothermal fluids at the magmatic stage. In: Geochemistry of hydrothermal ore deposits. (H.L. Barnes ed.), p. 34–76. New York: Holt, Rinehart and Winston 1967

Burnham, C.W., Davis, N.F.: The role of H_2O in silicate melts: I, P-V-T relations in the system $NaAlSi_3O_8$—H_2O to 10 kilobars and 1000° C. Am. J. Sci. 270, 54–79 (1971)

Burnham, C.W., Davis, N.F.: The role of H_2O in silicate melts: II. Thermodynamic and phase relations in the system $NaAlSi_3O_8$—H_2O to 10 kilobars, 700 to 1100° C. Am. J. Sci. 274, 902–940 (1974)

Carmichael, I.S.E., Turner, F.J., Verhoogen, J.: Igneous petrology. 739 p. New York: McGraw-Hill 1974

Eggler, D.H.: Water-saturated and undersaturated melting relations in a Paricutin andesite and an estimate of water content in the natural magma. Contrib. Mineral. Petrol. 34, 261–271 (1972)

Eggler, D.H., Burnham, C.W.: Crystallization and fractionation trends in the system andesite-H_2O—CO_2—O_2 at pressures to 10 kb. Geol. Soc. Am. Bull. 84, 2517–2532 (1973)

Eugster, H.P.: Heterogeneous reactions involving oxidation and reduction at high pressures and temperatures. J. Chem. Physics 26, 1160 (1957)

Ewart, A., Green, D.C., Carmichael, I.S.E., Brown, F.H.: Voluminous low temperature rhyolitic magmas in New Zealand. Contrib. Mineral. Petrol. 33, 128–144 (1971)

Friedman, I., Lipman, P.W., Obradovich, J.D., Gleason, J.D.: Meteoric water in magmas. Science 184, 1069–1072 (1974)

L223

Fyfe, W.S.: The generation of batholiths. Tectonophysics **17**, 273–283 (1973)

Green, T.H.: Crystallization of calc-alkaline andesite under controlled high-pressure conditions. Contrib. Mineral. Petrol. **34**, 150–166 (1972)

Holloway, J.R.: The system pargasite-H_2O—CO_2: a model for melting of a hydrous mineral with a mixed-volatile fluid – I. Experimental results to 8 kbar. Geochim. Cosmochim. Acta **37**, 651–666 (1973)

Huang, W., Wyllie, P.J.: Melting relations of muscovite-granite to 35 kbar as a model for fusion of metamorphosed subducted oceanic sediments. Contrib. Mineral. Petrol. **42**, 1–14 (1973)

Huang, W.L., Wyllie, P.J.: Melting relations of muscovite with quartz and sanidine in the K_2O—Al_2O_3—SiO_2—H_2O system to 30 kilobars and an outline of paragonite melting relations. Am. J. Sci. **274**, 378–395 (1974)

Huang, W.L., Robertson, J.K., Wyllie, P.J.: Melting relations of muscovite to 30 kilobars in the system $KAlSi_3O_8$—Al_2O_3—H_2O. Am. J. Sci. **273**, 415–427 (1973)

Jahns, R.H., Burnham, C.W.: Experimental studies of pegmatite genesis: I. A model for the derivation and crystallization of granitic pegmatites. Econ. Geol. **64**, 843–864 (1969)

Lipman, P.W., Friedman, I.: Interaction of meteoric water with magma: an oxygen-isotope study of ash-flow sheets from Southern Nevada. Geol. Soc. Am. Bull. **86**, 695–702 (1975)

Luth, W.C.: The systems $NaAlSi_3O_8$—SiO_2 and $KAlSi_3O_8$—SiO_2 to 20 kb and the relationship between H_2O content, P_{H_2O}, and P_{total} in granitic magmas. Am. J. Sci. **267-A**, 325–341 (1969)

Marsh, B.D., Carmichael, I.S.E.: Benioff zone magmatism. J. Geophys. Res. **79**, 1196–1206 (1974)

Merrill, R.B., Wyllie, P.J.: Hydrous upper mantle: water-excess and water-deficient melting relations of hornblende eclogite (abstract). Trans. Am. Geophys. Union **53**, 552 (1972)

Merrill, R.B., Wyllie, P.J.: Kaersutite and kaersutite eclogite from Kakanui, New Zealand – Water-excess and water-deficient melting at 30 kilobars. Geol. Soc. Am. Bull. **86**, 555–570 (1975)

Millhollen, G.L., Wyllie, P.J.: Relationship of brown hornblende mylonite to spinel peridotite mylonite at St. Paul's rocks. Abstracts with programs. Geol. Soc. Am., Denver **2**, 625 (1970)

Millhollen, G., Wyllie, P.J.: Melting of brown-hornblende mylonite from St. Paul's rocks under water-saturated and water-deficient conditions to 30 kilobars. J. Geol. **82**, 589–606 (1974)

Moore, J.G., Lockwood, J.P.: Origin of comb layering and orbicular structure, Sierra Nevada batholith, California. Geol. Soc. Am. Bull **84**, 1–20 (1973)

Mueller, R.F.: Stability of biotite: a discussion. Am. Mineralogist **57**, 300–316 (1972)

Mysen, B.O., Kushiro, I., Nicholls, I.A., Ringwood, A.E.: A possible mantle origin for andesitic magmas. Discussion of a paper by Nicholls and Ringwood. 1. Opening discussion, 2. Reply to opening discussion, 3. Comments on the reply of Nicholls and Ringwood, and 4. Final reply. Earth Planet. Sci. Lett. **21**, 221–229 (1974)

Nicholls, I.A.: Liquids in equilibrium with peridotitic mineral assemblages at high water pressures. Contrib. Mineral. Petrol. **45**, 289–316 (1974)

Piwinskii, A.J.: The attainment of equilibrium in hydrothermal experiments with "granitic rocks". Earth Planet. Sci. Lett. **2**, 161–162 (1967)

Piwinskii, A.J.: Experimental studies of igneous rock series: Central Sierra Nevada batholith, California. J. Geol. **76**, 548–570 (1968)

Piwinskii, A.J.: Experimental studies of granitoids from the Central and Southern Coast Ranges, California. Tschermaks Mineral. Petrogr. Mitt. **20**, 107–130 (1973a)

Piwinskii, A.J.: Experimental studies of igneous rock series, central Sierra Nevada batholith, California: Part II. Neues Jahrb. Mineral. Monatsh. H. 5, 193–215 (1973b)

Piwinskii, A.J.: Experimentelle Untersuchungen an granitischen Gesteinen von den südlichen Coast-Ranges, Transverse-Ranges und der Mojave-Wüste, Kalifornien. Fortschr. Mineral. **51**, 240–255 (1974)

Piwinskii, A.J.: Experimental studies of granitoid rocks near the San Andreas fault zone in the Coast and Transverse ranges and Mojave Desert, California. Tectonophysics **25**, 217–231 (1975)

Piwinskii, A.J., Martin, R.F.: An experimental study of equilibrium with granitic rocks at 10 kb. Contrib. Mineral. Petrol. **29**, 1–10 (1970)

Piwinskii, A.J., Wyllie, P.J.: Experimental studies of igneous rock series: a zoned pluton in the Wallowa batholith, Oregon. J. Geol. **76**, 205–234 (1968)

Piwinskii, A.J., Wyllie, P.J.: Experimental studies of igneous rock series: "Felsic Body Suite" from the Needle Point pluton, Wallowa batholith, Oregon. J. Geol. **78**, 52–76 (1970)

Presnall, D.C., Bateman, P.C.: Fusion relations in the system $NaAlSi_3O_8$—$CaAl_2Si_2O_8$—

$KAlSi_3O_8$—SiO_2—H_2O and generation of granite magmas in the Sierra Nevada batholith. Geol. Soc. Am. Bull. **84**, 3181–3202 (1973)

Putnam, G.W., Alfors, J.T.: Depth of intrusion and age of the Rocky Hill stock, Tulare County, California. Geol. Soc. Am. Bull. **76**, 357–364 (1965)

Robertson, J.K., Wyllie, P.J.: Experimental studies on rocks from the Deboullie stock, northern Maine, including melting relations in the water-deficient environment. J. Geol. **79**, 549–571 (1971a)

Robertson, J.K., Wyllie, P.J.: Rock-water systems, with special reference to the water-deficient region. Am. J. Sci. **271**, 252–277 (1971b)

Shimada, M.: Melting of albite at high pressures in the presence of water. Earth Planet. Sci. Lett. **6**, 447–450 (1969)

Steiner, J.C., Jahns, R.H., Luth, C.W.: Crystallization of alkali feldspar and quartz in the haplogranite system $NaAlSi_3O_8$—$KAlSi_3O_8$—SiO_2—H_2O at 4 kb. Geol. Soc. Am. Bull. **86**, 83–98 (1975)

Stern, C.R., Wyllie, P.J.: Water-saturated and undersaturated melting relations of a granite to 35 kilobars. Earth Planet.Sci. Lett. **18**, 163–167 (1973)

Stern, C.R., Huang, W.L., Wyllie, P.J.: Basalt-andesite-rhyolite-H_2O: crystallization intervals with excess H_2O and H_2O-undersaturated liquidus surfaces to 35 kilobars, with implications for magma genesis. Earth Planet. Sci. Lett. manuscript submitted

Stormer, J.C., Carmichael, I.S.E.: The Kudo-Weill plagioclase geothermometer and porphyritic acid glasses. Contrib. Mineral. Petrol. **28**, 306–309 (1970)

Streckeisen, A.L.: Classification and nomenclature of igneous rocks. Neues Jahrb. Mineral. Abhandl. **107**, 144–240 (1967)

Taylor, H.P.: The application of oxygen and hydrogen isotope studies to problems of hydrothermal alteration and ore deposition. Econ. Geol. **69**, 843–883 (1974)

Tuttle, O.F., Bowen, N.L.: Origin of granite in the light of experimental studies in the system $NaAlSi_3O_8$—$KAlSi_3O_8$—SiO_2—H_2O. Geol. Soc. Am. Mem. **74**, 153 p. (1958)

Upton, B.G.J.: The alkaline igneous complex of Kungnat Fjeld, South Greenland. Medd. Groenland 123, No. 4, 1–145 (1960)

Whitney, J.A.: The effect of reduced H_2O fugacity on the buffering of oxygen fugacity in hydrothermal experiments. Am. Mineralogist **57**, 1902–1908 (1974)

Whitney, J.A.: The effects of pressure, temperature and X_{H2O} on phase assemblages in four synthetic rock compositions. J. Geol. **83**, 1–31 (1975a)

Whitney, J.A.: Vapor generation in a quartz monzonite magma: a synthetic model with application to porphyry copper deposits. Econ. Geol. **70**, 346–358 (1975b)

Wones, D.R.: Stability of biotite: a reply. Am. Mineralogist **57**, 316–317 (1972)

Wones, D.R., Eugster, H.P.: Stability of biotite: experiment, theory, and application. Am. Mineralogist **50**, 1228–1272 (1965)

Wood, B.J., Carmichael, I.S.E.: P_{total}, P_{H_2O} and the occurrence of cummingtonite in volcanic rocks. Contrib. Mineral. Petrol. **40**, 149–158 (1973)

Wyllie, P.J., Tuttle, O.F.: Experimental investigation of silicate systems containing two volatile components. Part I, Geometrical considerations. Am. J. Sci. **258**, 498–517 (1960)

Yoder, H.S.: Calkalkalic andesites: experimental data bearing on the origin of their assumed characteristics. In: Proceedings of the Andesite Conference, A.R. McBirney ed. Oregon Dep. Geol. Mineral. Ind. Bull **65**, 77–89 (1969)

Yoder, H.S., Kushiro, I.: Melting of a hydrous phase: phlogopite. Am. J. Sci. **267A**, 558–582 (1969)

Prof. Dr. P.J. Wyllie
Department of Geological Sciences
University of Chicago
Chicago, Illinois 60637, U.S.A.

Accepted June 24, 1975

CHAPTER 17

Conrad, W.K., Nicholls, I.A. and Wall, V.J. (1988) Water-saturated and -undersaturated melting of metaluminous and peraluminous crustal compositions at 10 kb: evidence for the origin of silicic magmas in the Taupo Volcanic Zone, New Zealand, and other occurrences. *Journal of Petrology*, **29**, 765–803.

In 1976, P.J. Wyllie and co-workers published a compendium of their experimental work, in the gabbro-tonalite-granite-H_2O system, that placed specific constraints on magmatic sources, temperatures and processes that could generate granitic magmas. These authors emphasized that granites cannot represent primary melts of the mantle or subducted slabs, but can easily be generated in the crust at temperatures attained during regional metamorphism. Shortly before this, Chappell and White (1974) had introduced the concept of S- and I-type granites, which emphasizes the influence of the crustal source rock on the mineralogy and chemical composition of the resulting magmas. The S- and I-type duality that, with minor modifications and additions, can be recognized in most orogenic belts reflects the fact that two main types of materials, metasedimentary and meta-igneous, dominate the source regions of granitic magmas. Whereas magmas formed from metasedimentary sources tend to be markedly peraluminous, those derived from meta-igneous protoliths are commonly metaluminous or slightly peraluminous. Thus, the Alumina Saturation Index (ASI = mol. Al_2O_3/ $[CaO + Na_2O + K_2O]$) is a key geochemical parameter that reflects this primary contrast in source chemistry.

The paper of Conrad, Nicholls and Wall (1988) summarizes the first experimental work specifically designed to compare the melting behaviour and the compositions of partial melts of these two contrasting crustal sources. Using finely powdered natural rocks, instead of synthetic materials, the experiments were carried out at 10 kb (1 GPa), with varying degrees of melt H_2O saturation, achieved through the use of fluid phases with varying H_2O-CO_2 ratios. As the representative of metasedimentary sources they selected an aluminous greywacke with an ASI = 1.19, Mg# = 0.35, and as the representative of meta-igneous sources, they selected a dacite with ASI = 0.90 and Mg# = 0.50. Both starting materials contained ~30% of latent haplogranitic component (= 3.33 times the fraction of the least abundant CIPW-normative mineral among Q, Or and Ab). The results were of great importance for the state of knowledge of granite petrogenesis at that time. They shed light on many controversial aspects, such as the possible generation, by partial melting, of trondhjemitic migmatites less potassic than their sources. The results also showed that partial melts from metaluminous sources can be either metaluminous or moderately peraluminous, but that strongly peraluminous melts can only be derived from strongly peraluminous sources. They also confirmed previous results from Clemens and Wall (1981) with natural and model granitic compositions at 5 kb and over a wide range of H_2O activity, indicating that primary granitic liquids, saturated with quartz in their source regions, should contain ~70–73 wt.% SiO_2. This finding represented a serious objection against the requirements of the restite model of White and Chappell (1977), in which the melt component of the granitic magma is viewed as having highly felsic compositions (>73 wt.% SiO_2), consistent with their formation by low-temperature, near-minimum melting. The observed variations in the stability fields of quartz, plagioclase and biotite with water activity were of particular interest, as these can cause systematic changes in the proportions of felsic components of the melts produced from the same source, in response to melting at different (progressively higher?) temperatures. Thus, their work illustrated how a wide variety of granitoid magma compositions could be produced from similar source rocks depend-ing on P-T-a_{H_2O} conditions at which partial melting took place.

This paper is included in the set because it represents the first systematic study of the effects of bulk protolith composition on melt compositions formed in the crust, at a_{H_2O} values consistent with granulite-facies metamorphism and anatexis. Following on from this there were numerous partial melting studies that further investigated mainly fluid-absent partial melting in a wide variety of protolith materials, both natural and synthetic. Thus, the publication of Conrad, Nicholls and Wall (1988) heralded an era of investigation that has resulted in us knowing far better what kinds of materials and processes are involved in the genesis of granitic magmas. Most of the subsequent work used the fluid-absent, rather than the fluid-present, reduced aH_2O experimental strategy. Indeed, the differences in findings between these two supposedly equivalent approaches suggest that a re-examination of their work might pay dividends in understanding how fluid conditions influence the production and composition of granitic magmas.

Conrad, W.K., Nicholls, I.A. and Wall, V.J., Water-saturated and -undersaturated melting of metaluminous and peraluminous crustal compositions at 10 kb: evidence for the origin of silicic magmas in the Taupo Volcanic Zone, New Zealand, and other occurrences. *Journal of Petrology*, 1988, **29**, pp. 765–803, reproduced with the kind permission of Oxford University Press (http://petrology.oxfordjournals.org/).

0022-3530/88 $3.00

Water-Saturated and -Undersaturated Melting of Metaluminous and Peraluminous Crustal Compositions at 10 kb: Evidence for the Origin of Silicic Magmas in the Taupo Volcanic Zone, New Zealand, and Other Occurrences

by WALTER K. CONRAD[1], I. A. NICHOLLS[2] AND V. J. WALL[2]

[1]*Department of Geology, SUNY College at New Paltz, New Paltz, New York, 12561*
[2]*Department of Earth Sciences, Monash University, Clayton, Victoria 3168, Australia*
(*Received 12 December 1986; revised typescript accepted 8 February 1988*)

ABSTRACT

The melting relations of two proposed crustal source compositions for rhyolitic magmas of the Taupo Volcanic Zone (TVZ), New Zealand, have been studied in a piston–cylinder apparatus at 10 kb total pressure and a range of water activities generated by H_2O–CO_2 vapour. Starting materials were glasses of intermediate composition (65 wt.% SiO_2), representing a metaluminous 'I-type' dacite and a peraluminous 'S-type' greywacke. Crystallization experiments were carried out over the temperature range 675 to 975 °C, with aH_2O values of approximately 1·0, 0·75, 0·5, and 0·25. Talc–pyrex furnace assemblies imposed oxygen fugacities close to quartz-fayalite–magnetite buffer conditions.

Assemblages in both compositions remain saturated with quartz and plagioclase through 675–700 °C at high aH_2O, 725–750 °C at $aH_2O \approx 0.5$, and 800–875 °C at $aH_2O \approx 0.25$, corresponding to <60–70% melting. Concentrations of refractory mineral components (Fe, Mg, Mn, P, Ti) in liquids increase throughout this melting interval with increasing temperature and decreasing aH_2O. Biotite and hornblende are the only mafic phases present near the solidus in the dacite, compared with biotite, garnet, gedritic orthoamphibole, and tschermakitic clinoamphibole in the greywacke. Near-solidus melting reactions are of the type: biotite + quartz + plagioclase = amphibole ± garnet, potentially releasing H_2O for dehydration melting in the greywacke, but producing larger amounts of hornblende and releasing little H_2O in the dacite. At $aH_2O \approx 0.25$ and temperatures ≥825–850 °C, amphibole dehydration produces anhydrous mineral phases typical of granulite facies assemblages (clinopyroxene, orthopyroxene, plagioclase ± quartz in the dacite; garnet, orthopyroxene, plagioclase ± quartz in the greywacke) coexisting with melt proportions as low as 40%. Hornblende-saturated liquids in the dacite are weakly peraluminous (0·3–1·6 wt.% normative C—within the range of peraluminous TVZ rhyolites), whereas, at $aH_2O \approx 0.25$ and temperatures ≥925 °C, metaluminous partial melt compositions (up to 1·8 wt.% normative Di) coexist with plagioclase, orthopyroxene, and clinopyroxene. At all water activities, partial melts of the greywacke are uniformly more peraluminous (1·5–2·6 wt.% normative C), reflecting their saturation in the components of more aluminous mafic minerals, particularly garnet and Al-rich orthopyroxene. A metaluminous source for the predominantly Di-normative TVZ rhyolites is therefore indicated.

With decreasing aH_2O the stability fields of plagioclase and quartz expand, whereas that of biotite contracts. These changes are reflected in the proportions of normative salic components in partial melts of both the dacite and greywacke. At high aH_2O, partial melts are rich in An and Ab and poor in Or (trondhjemitic-tonalitic); with decreasing aH_2O they become notably poorer in An and richer in Or (granodioritic–granitic). These systematic variations in salic components observed in experimental metaluminous to strongly peraluminous melts demonstrate that a wide variety of granitoid magmas may be produced from similar source rocks depending upon P–T–aH_2O conditions attending partial melting. Some peraluminous granitoids, notably trondhjemitic leucosomes in migmatites, and sodic

[*Journal of Petrology*, Vol. 29, No. 4, pp. 765–803, 1988]

granodiorites and granites emplaced at deep crustal levels, have bulk compositions similar to near-solidus melt compositions in both the dacite and greywacke, indicating possible derivation by anatexis without the involvement of a significant restite component.

INTRODUCTION

Most experimental studies investigating the origins of granitic magmas have focussed either on melt compositions in model systems of salic components (e.g., Tuttle & Bowen, 1958; Luth *et al.*, 1964; James & Hamilton, 1969; Winkler, 1979), or on the phase relations of a variety of natural granitoid compositions (e.g., Wyllie and coworkers—see Wyllie *et al.*, 1976). The model system experiments first established that crustal anatexis may be important in producing granites, because low temperature water-saturated melts are analogous, in terms of salic components, to granites occurring worldwide. However, the condition of water-saturation is generally recognized as being reached only late in the magmatic history of many granites (e.g., Luth *et al.*, 1964; see, however, Winkler, 1979). Phase relations in natural granitoids have provided a guide to mineral assemblages which may have been present during anatexis or early magmatic crystallization (e.g., Clemens & Wall, 1981), but there are presently few compositional data for water-undersaturated silicic partial melts at pressures of the deep crust–conditions under which anatexis should be common (Tuttle & Bowen, 1958).

Fewer experimental studies have taken the more direct approach of examining silicic partial melts of natural compositions proposed to be melting sources for granitic magmas (e.g., Kilinc, 1969; Brown & Fyfe, 1970; Green, 1976; Thompson, 1981). In addition to the numerous experimental and analytical difficulties usually encountered in working with assemblages containing small proportions of silica-rich melt, this approach requires choices of source compositions to be made from amongst the diversity of possible magma sources in the crust.

The concept of I- and S-type granites Chappell & White, 1974; White & Chappell, 1977) is an important step towards placing broad, yet critical, constraints upon the nature of granite source regions. The degree of alumina-saturation in granites is considered to reflect two fundamentally contrasting types of source regions: metaluminous (dominantly igneous protoliths) and peraluminous (dominantly sedimentary protoliths). Within this context, we wish to address experimentally several questions which are applicable to the genesis of a wide variety of granitoids, including: (1) whether equilibrium partial melts of metaluminous and peraluminous bulk compositions are correspondingly distinctive; (2) to what extent do water-undersaturated partial melt compositions, particularly those in equilibrium with important Fe–Mg silicates such as biotite, hornblende and pyroxenes, differ from those in water-saturated model systems; and (3) the general, and controversial, question of whether natural granitoids resemble silicic partial melts of likely crustal sources, or if they represent mixtures of liquid and refractory crystalline material (i.e. restite).

Geological constraints upon crustal magma sources

The metaluminous and peraluminous crustal source compositions studied here are based upon previous models for rhyolite petrogenesis in the Taupo Volcanic Zone (TVZ), North Island, New Zealand. Sedimentary rocks exposed in the TVZ have been proposed to be analogues of deep-level basement rocks which underwent anatexis to produce rhyolitic magmas (Ewart, 1969; Ewart & Stipp, 1968). The zone is also part of a presently active volcanic arc (Cole, 1979), raising the possibility that the lower crust formed by underplating and intrusion of mafic or intermediate magmas (e.g., Conrad & Kay, 1984). This crust could then have acted as a source for the acid magmas (e.g., White, 1979).

A primary reason for using the TVZ as a reference for the source compositions is that several recent studies have thoroughly documented the underlying Mesozoic basement rocks. Figure 1 illustrates in a simplified ACF projection the compositions of 75 petrographically characterized greywackes and argillites from the Mesozoic Torlesse Supergroup flanking and underlying the TVZ (Reid, 1982, 1983). As discussed by Blattner & Reid (1982) and Reid (1983), the Torlesse rocks may be divided into two chemically distinct groups based on field relations and provenance: a eugeosynclinal 'western basement' facies and a miogeosynclinal 'eastern basement' facies.

The western basement facies is dominated by rocks which constitute a 'first generation S-type' source (Chappell and White, 1974), produced by rapid weathering, erosion and sedimentation of material derived from oceanic island arc sequences. Western basement greywackes contain up to 50% volcanolithic fragments, with the remainder mainly plagioclase, quartz, and minor K-feldspar. By contrast, sandstones from the more mature and pelitic eastern basement facies (see Fig. 1) have abundant metamorphic and granitic lithic fragments indicating derivation from a continental terrain (Reid, 1982, 1983).

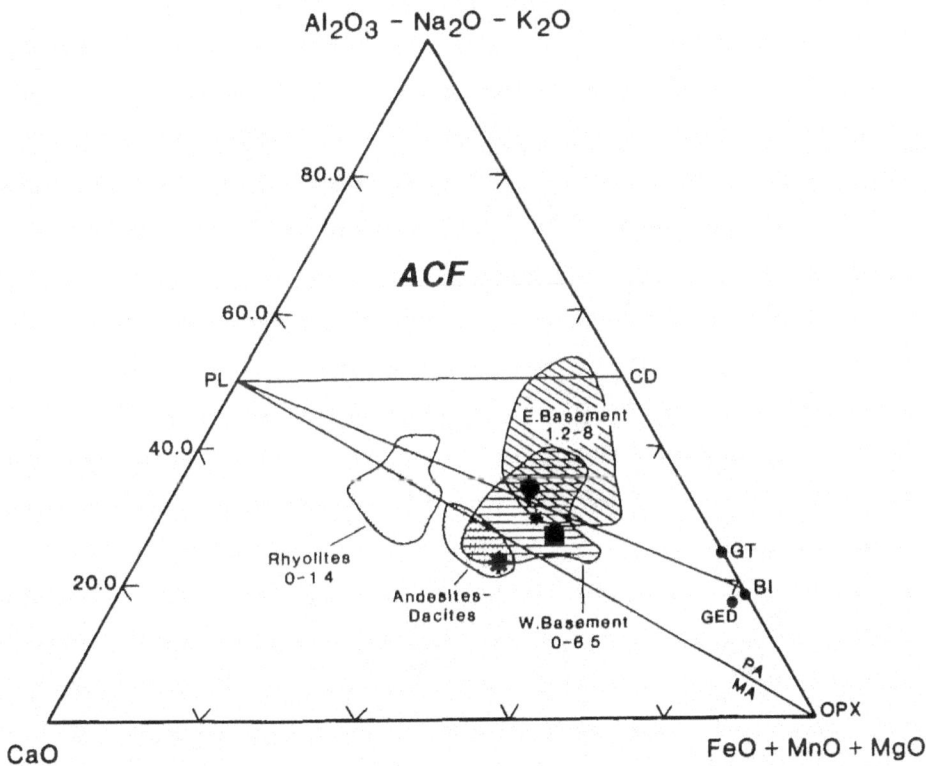

FIG. 1. Taupo Zone volcanic (Ewart, 1969) and sedimentary basement rocks (Reid, 1982) in ACF projection (total Fe as FeO). The plagioclase (PL)-orthopyroxene (OPX) join marks an approximate boundary between metaluminous (MA) and peraluminous (PA) fields. The boundary between I- and S-type granites, as defined by Chappell and White (1974) lies slightly above this join (1% normative C). The ranges of normative C contents within each field are also indicated. Symbols representing average TVZ compositions (Table 1) are: circle and square for average Eastern and Western basement, including argillite interbeds (Reid, 1983); small star and large asterisk for average TVZ greywacke and dacite experimental compositions. Compositions of garnet (GT), biotite (BI), cordierite (CD) and gedrite (GED) are also indicated.

WALTER K. CONRAD *ET AL.*

Western basement greywackes range widely from metaluminous to peraluminous compositions (Fig. 1), while interbedded argillites are all strongly peraluminous (see Table 1, Analysis 6). According to Reid (1982, 1983), the overall average composition of the western basement facies is mildly peraluminous, and quite similar to averages of the greywackes alone (Table 1, Analyses 1, 3, and 5). The greywackes have also been the focus of geochemical studies which tested possible genetic relationships between TVZ rhyolite magmas and sedimentary sources (Ewart & Stipp, 1968; Ewart, 1969; Blattner & Reid, 1982; Reid, 1983). An average greywacke composition derived from analyses of borehole material (Table 1, Analysis 1) was therefore chosen as a general model for immature sedimentary sources, and to directly test the sediment melting hypothesis for TVZ rhyolitic magmas.

A metaluminous average TVZ dacite composition (Fig. 1, Table 1) may be used to model a much wider variety of possible deep crustal 'I-type' sources, ranging from Q-normative gabbro through granodiorite. At high degrees of melting, residual mineral assemblages are those typical of more mafic rocks (i.e. pyroxenes, An-rich plagioclase). High temperature melts of the dacite should therefore be similar to those derived from more mafic bulk compositions at similar temperatures, but smaller degrees of melting, provided crystalline phases of similar composition are present. These liquids should also be similar to those produced by fractional crystallization of andesitic or dacitic magmas associated with island arc volcanism in the TVZ (e.g., Reid, 1983).

EXPERIMENTAL AND ANALYTICAL METHODS

Most experiments used glass starting materials in the presence of excess vapour. This approach commonly yields equilibrium or near-equilibrium crystal–liquid assemblages whose interpretation is not complicated by the presence of refractory minerals (e.g., plagioclase) as in some experiments using crystalline starting assemblages (Clemens & Wall, 1981; see, however, Johannes, 1980).

TABLE 1

Starting materials and comparative model source compositions

	1	2	3	4	5	6
SiO$_2$	65·0	65·2	64·79	71·36	64·68	67·05
TiO$_2$	0·73	0·60	0·80	0·54	0·85	0·74
Al$_2$O$_3$	16·5	15·7	16·61	14·98	16·29	18·63
FeO*	7·0	5·0	5·66	3·42	5·72	4·26
MnO	0·1	0·1	0·11	0·04	0·12	0·03
MgO	1·9	2·5	2·47	1·27	2·45	1·44
CaO	3·5	5·3	3·17	1·93	3·19	0·52
Na$_2$O	3·4	3·6	4·00	3·81	4·28	3·25
K$_2$O	1·9	1·7	2·21	2·52	2·18	3·84
P$_2$O$_5$	0·15	0·10	0·16	0·11	0·14	0·14
Total	100·18	99·80	99·98	99·98	99·90	99·90
Wt.% C	2·8	—	2·3	2·7	1·4	8·5
Wt.% Di	—	3·3	—	—	—	—

(1) Average Kawerau Geothermal Area greywacke starting composition, after Reid (1982- boreholes only).
(2) Average TVZ dacite starting composition, after Cole (1979).
(3) Average Western Basement, greywackes and argillites, volatile-free (Reid, 1983).
(4) Average Eastern basement, greywackes and argillites, volatile-free (Reid, 1983).
(5) Experimental TVZ greywacke composition of Reid (1982).
(6) Experimental TVZ argillite composition of Reid (1982).
FeO* = total Fe as FeO.

The dacite and greywacke starting compositions were prepared from oxide–carbonate mixes sintered at 1100 °C, fused twice on an Ir-strip heater to produce uniform glasses (Nicholls, 1974), and checked by microprobe analysis of selected glass fragments. 10 mg aliquots of finely ground greywacke and dacite glasses were run in paired 2·5 mm $Ag_{70}Pd_{30}$ capsules. The capsules were sealed with excess H_2O or H_2O plus $Ag_2C_2O_4$ to generate H_2O–CO_2 vapour with $X_{H_2O}^{vap}$ of approximately 0·75, 0·5, and 0·25 (cf. Eggler, 1972). Molar rock/vapour ratios used ranged from 0·1–0·3 for water-undersaturated to 0·3–0·5 for water-saturated conditions. Using microprobe analytical data for mineral assemblages, vapour compositions were calculated assuming: (i) ideal mixing of H_2O and CO_2; (ii) H_2O solubility in coexisting melts according to the model of Burnham (1979, 1981); and (iii) melt proportions derived from calculated modes (Tables A1, A2). Allowing for uncertainties in (ii) and (iii) and possible weighing errors, $X_{H_2O}^{vap}$ is considered accurate to $\pm 0\cdot05$. Calculated melt H_2O contents range from 4 wt.% H_2O at $aH_2O = 0\cdot25$ to 13–17 wt.% H_2O at $aH_2O \simeq 1\cdot0$. Samples were run for 24 to 384 h at 10 kb pressure in a 0·5 in (12·7 mm) piston–cylinder apparatus (Table 2), applying a 10% pressure correction for piston-in technique and talc-pyrex furnace assemblies. To prevent H_2O–CO_2 vapour from precipitating graphite during the initial stages of experiments at $X \leq 0\cdot50$ (due to H_2 liberated by reaction between the inner talc assembly and the graphite furnace), the capsules were surrounded with a 'hydrogen absorber' of powdered crystalline fayalite, magnetite and quartz. As shown by mineral compositions (see Amphibole section), resulting hydrogen fugacities within sample capsules were near QFM buffer conditions. Temperatures were controlled with a Pt/Pt–10%Rh thermocouple. Maximum temperature variations attributable to furnace design (Gust & Hibberson, 1979) and controller fluctuations are within ± 15 °C of the set temperature.

All phases were identified optically in grain mounts and confirmed by energy dispersive microprobe (EDS) analysis. The typical grainsize range was 5–20 μm. Stability fields of the major mineral phases throughout the melting interval are plotted on T–X (i.e. $X_{H_2O}^{vap}$) sections (Figs. 2 and 3). Minor phases not indicated on Figs. 2 and 3 are ilmenite and apatite in both the dacite and greywacke and also rutile in the dacite (Table 2).

Filled points in Figs. 2 and 3 indicate charges in which all phases have been analysed and modal abundances in wt.% calculated by mass balance (see Appendix, Tables A1, A2). Calculated percentages of glass in near-solidus charges were constrained by modal estimates from microprobe backscattered electron (BSE) images. Glasses at high aH_2O in the dacite contained variable amounts of quench amphibole (Table 2), but quench phases were not a significant problem in either the dacite or greywacke at reduced aH_2O, excluding some near-liquidus charges at very high ($>85\%$) degrees of melting. Iron-loss to sample capsules was estimated by using FeO as an additional phase during least-squares mass balance calculations on analysed assemblages. The amounts required to produce satisfactory calculated bulk charge compositions were large ($>30\%$ of original FeO*) only in assemblages with $\geq 80\%$ melt, particularly at high aH_2O (Tables A1, A2).

Microprobe operating conditions for EDS analysis were 15 kV, 5 nA, and 30 s counting(live) time, with on-line ZAF corrections. The hydrous glasses were analysed in two stages:

(1) *EDS analysis*, from which the major elements Si, Al, Ca, Na, and K were retained. To minimize the effects of Na-mobility, which also effects Si, Al and possibly Ca counting rates (cf. Borom & Hanneman, 1967; Rutherford *et al.*, 1985), each EDS glass analysis incorporates data accumulated from 5 to 50 individual 3–50 μm^2 areas (using SEM imaging and the beam scan facility), counted for 0·6–6 s (live time). The maximum counting times which could be used for a given scanned area without decrease in Na count rate were determined on large glass regions in near-liquidus charges; generally, glasses with higher water contents were more susceptible to Na-mobility and required shorter counting times. Where possible, the multiple spot analyses of water-rich glasses were checked by EDS analysis at 10 kV, using 50 μm^2 areas scanned without affecting Na count rate for up to several minutes.

(2) *Wavelength dispersive (WDS) analysis* for the minor elements Ti, Fe, Mn, Mg, and P. These elements were not affected by Na-mobility and were reproducibly analyzed using crystal spectrometers at beam currents of 30–50 nA. Concentrations were determined directly from calibration curves for fused rock standards analyzed during the same session. Under WDS analysis conditions, Fe gave the best counting statistics (limits of detection as defined by Reed (1975), of 0·01–0·04 %) and hence the best measure of glass homogeneity. Representative data are given in Table 3. Observed variations for FeO are typically 1–5 relative % (1σ), rising to ~ 10% for concentrations below 1% FeO. According to a simple index of homogeneity suggested by Reed (1975), of the analysed glasses listed in Table 9, 18 are homogeneous for all elements, 5 are possibly homogeneous, and only one glass (305A) is inhomogeneous (e.g., FeO = $0\cdot92 \pm 0\cdot08$ %, 1σ), probably due to minor quench amphibole.

TABLE 2

Experimental run products

T (°C)	Time (h)	Run #	Dacite	Run #	Greywacke
			$X_{H_2O}^{vap} = 0.25 \pm 0.05$		
800	74	614A	Q, P, Bi, Hbl, I, Kfs[1] (solidus)	614B	Q, P, Bi, Ged, Gt, I, trace Gl
825	120	624A	Q, P, Hbl, Opx, I, Kfs[1], Gl	624B	Q, P, Bi, Opx, Gt, I, Gl, Ap
850	90	617A	Q, P, Cpx, Opx, I, Gl	622B	Q, P, Opx, Gt, I, Gl
875	72	619A	Q?, P, Cpx, Opx, I, Gl	619B	Q, P, Opx, Gt, I, Gl
900	64	621A	Q, P, Cpx, Opx, I, Gl	621B	P, Opx, Gt, I, Gl
925	48	623A	P, Cpx, Opx, I, Gl	623B	P, Opx, Gt, I, Gl
950	48	625A	P, Cpx, Opx, Gl	625B	Gt, Gl
975	24	626A1	Cpx, Opx, Gl		
975	24	626A2*	Cpx, Opx, Gl		
			$X_{H_2O}^{vap} = 0.50 \pm 0.05$		
700	384	329A*[2]	Q, P, Bi, Hbl (solidus)	329B[2]	Q, P, Bi?, Gt?, I (solidus)
725	336	323A*	Q, P, Bi, Hbl, Gl	323B	Q, P, Bi, Ged, Gt, I, Gl
750	316	321A*	Q, P, Bi, Hbl, I, Gl, Ap	321B	Q, P, Bi, Ged, Gt, I, Gl, Ap
775	150	332A	P, Hbl, I, Gl	330B	P, Hbl, Ged, Gt, I, Gl, Ap
775	150	333A*	P, Hbl, I, Gl		
800	94	320A	P, Hbl, I, Gl, Ru	320B	Hbl, Gt, I, Gl, Ap
825	144	325A	Hbl, I	325B[3]	Hbl, Gt, I, Gl, Ap
825	72			332B[2]	Hbl, Gt, I, Gl
850	96			322B	Opx, Gt, I, Gl
875	72	324A*	Hbl, Cpx, Gl	324B[3]	Opx, Gt, I, Gl
900	72	326A*	Opx, Cpx, Gl	326B	P[4], Opx[4], Gt[4], Gl
			$X_{H_2O}^{vap} = 0.75 \pm 0.05$		
675	310	311A	Q, P, Bi, Hbl, I (solidus)	311B	Q, P, Bi, Hbl, Ged, I, Gl
700	219	310A	Q, P, Bi, Hbl, I, Gl, Ru		
725	169	308A	P, Hbl, I, Gl, Ru	308B	Q, Bi, Hbl, Ged, I, Gl
750	192	307A	Hbl[4], I, Gl, Ru	307B	Bi, Hbl, Ged, Gt, I, Gl
800	96	306A	Hbl, I, Gl, Ru	306B	Bi, Hbl, Ged, Gt, I, Gl
850	120	309A	Hbl, I, Gl	309B	Opx, Gt[2], I, Gl
875	120	327A*	Opx[4], Hbl, Gl	327B	Opx, Gt, I, Gl
			$X_{H_2O}^{vap} \simeq 1.00$		
675	336	305A	Q[2], P, Bi, Hbl[4], I, Gl, Ap	305B	Q, P, Bi, Hbl, Ged, I, Gl
700	190	304A	Hbl[4], I, Gl	304B	Q, Bi, Hbl, I, Gl
725	72	301A	Hbl[4], I, Gl	301B	Bi, Hbl, I, Gl
750	264	302A	Hbl, I, Gl	302B	Bi, Hbl, Gt, I, Gl, Ap
800	72	303A	Hbl, I, Gl, Ap	303B	Bi, Ged, Gt, I, Gl
825	76			312B	Opx, I, Gl
850	30	314A	Hbl[4], Gl		
925	30	315A	Opx[4], Gl		

Key to phases: Q = quartz; P = plagioclase; Kfs = alkali feldspar; Bi = biotite; Hbl = hornblende or Ca-amphibole; Ged = gedrite or Ca-poor amphibole; Gt = garnet; Cpx = Ca-clinopyroxene; Opx = orthopyroxene; I = ilmenite; Gl = glass; Ap = apatite; Ru = rutile.

Notes: [1] = presence confirmed by X-ray diffraction only; [2] = optical identification only; [3] = graphite present; [4] = skeletal quench phases.

Unless otherwise indicated, all phases confirmed by SEM imaging and EDS analysis.

* = Sintered oxide-mix starting materials.

PHASE RELATIONS AND MINERAL CHEMISTRY

Although the phase relations presented were derived from synthesis (unreversed) data, several observations suggest that they represent a close approach to equilibrium. Phase assemblages and compositions in two dacite charges near the quartz and plagioclase stability limits were duplicated using both glass and sintered oxide mix starting materials (Tables 2 and 3), suggesting that all phases (and especially quartz and feldspars) were able to nucleate from glass without major undercooling (cf. Edgar, 1973). The boundaries of stability fields and the modal proportions and compositions of phases

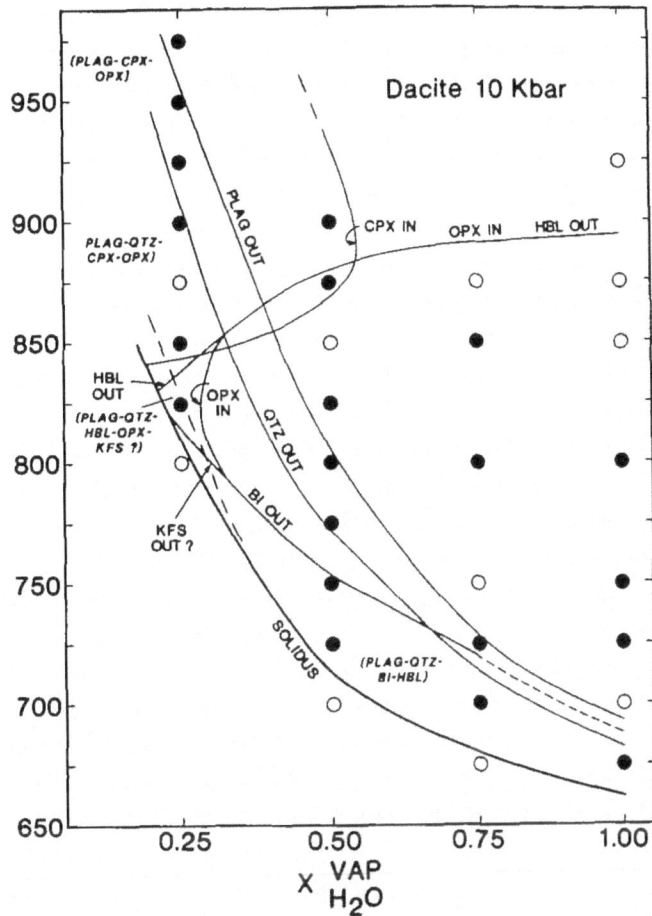

FIG. 2. $T°C$ vs. $X_{H_2O}^{vap}$ phase diagram for the experimental dacite composition at 10 kb. Phase boundaries marked 'In' and 'Out' refer to the sequence during melting. Important near-solidus and near-liquidus assemblages are indicated in brackets. Additional phases not represented are given in Table 2. Filled points indicate charges in which modal abundances have been calculated by mass balance (Appendix 1).

vary systematically, except in the case of amphiboles in some greywacke charges. The only clear evidence of disequilibrium is the presence of zonation in some crystalline phases (garnet, amphibole).

Solidus relations

From an extrapolation of the observed relationship between temperature and melt percentages (Fig. 4), the greywacke solidus is estimated to lie near 650 °C at $X \approx 1·0$ ($X_{H_2O}^{vap}$ hereafter abbreviated as X). Quench amphibole in the H_2O-saturated 675 °C dacite charge prohibits a similar extrapolation to the solidus, but a temperature of 650–675 °C is indicated. These results closely match water-saturated solidus temperatures in the system Q–Ab–An–Or with plagioclase An_{40-50} at 10 kb (Johannes, 1985, fig. 2.10), and they are within the range observed in other studies on natural intermediate to felsic rocks at 10 kb (e.g., Piwinski, 1973; Stern *et al.*, 1975).

The dacite solidus was bracketed at 675–700 °C at $X = 0·75$, 700–725 °C at $X = 0·5$, and 800–825 °C at $X = 0·25$ (Fig. 2). The presence of even very small amounts of melt was detectable by changes in grain morphology, particularly of amphibole and biotite, in BSE images. The greywacke solidus was bracketed only at $X = 0·5$, 700–725 °C, although very small percentages of melt at $X = 0·25$, 800 °C

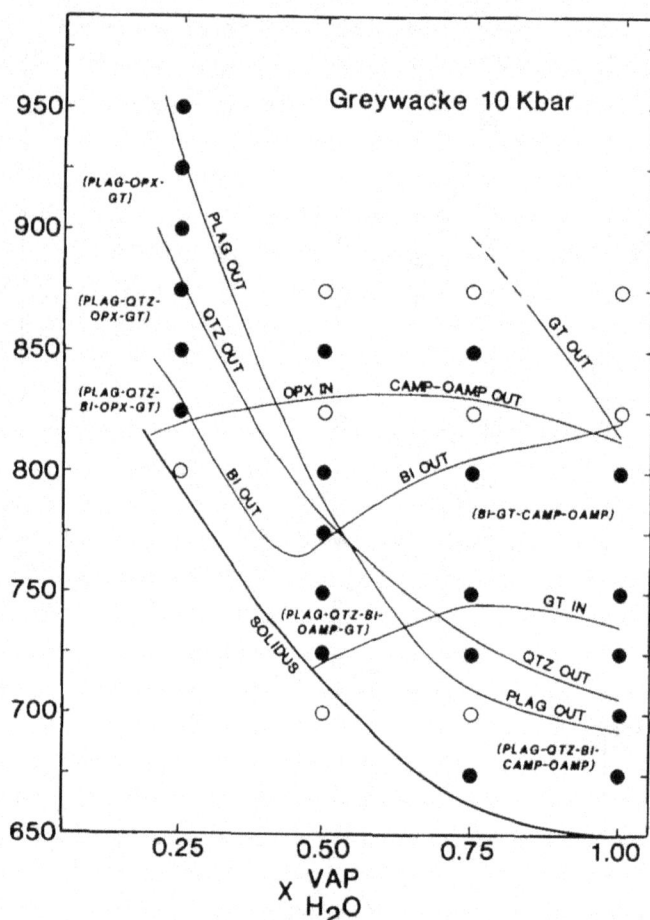

FIG 3 $T°C$ vs. $X_{H_2O}^{vap}$ phase diagram for the experimental greywacke composition at 10 kb. Symbols as in Fig. 2 and Table 2. The amphibole-out boundary indicates the maximum stability of both 1- and 2- amphibole assemblages as discussed in the text.

indicate close approach to the solidus. The greywacke appears to have consistently slightly lower solidus temperatures, compatible with its lower normative An and higher Q contents.

Quartz and plagioclase

The stability fields of quartz and plagioclase change considerably with aH_2O, profoundly influencing phase proportions and melt compositions. In both the dacite and greywacke, plagioclase and quartz stability limits diverge from the solidus toward higher temperatures, and a given melt proportion corresponds to progressively larger temperature intervals above the solidus, as aH_2O decreases (Figs. 2 and 3). In the dacite at high aH_2O, nearly 80% plagioclase-saturated melt is produced within 50°C of the solidus, while at $X = 0.25$ this melt percentage is not reached until 125–150°C above the solidus. Figure 4 demonstrates similar $T–X$-% melt relationships for the greywacke.

In the greywacke at high aH_2O quartz is stable to higher temperatures than plagioclase. This behaviour has been observed at 5–8 kb in H_2O-saturated experiments on a number of acid-intermediate compositions (e.g., Clemens & Wall, 1981; Huang & Wyllie, 1981; Stern & Wyllie, 1981), and compositions as mafic as quartz diorite and tonalite at 10 kb (Piwinski, 1973; Stern *et al.*, 1975). The relative increase in plagioclase stability with decreasing aH_2O results in a distinctive crossover of

TABLE 3

Phase compositions in runs using glass and oxide mix starting materials

| | Glasses | | | | | | | |
| | 626A1 (25, 5), [3] | | 626A2 (10, 2), [*] | | 332A (25, 5), [2] | | 333A (10, 2), [*] | |
(#)		±1σ		±1σ		±1σ		+1σ
SiO_2	66·82	0·2	65·96	0·9	72·29	0·5	73·18	0·9
TiO_2	0·61	1·8	0·58	21	0·18	28	0·16	56
Al_2O_3	16·05	0·8	16·56	0·8	15·73	0·9	15·73	1·5
FeO	4·34	1·6	4·74	0·8	1·64	2·9	1·74	0·9
MnO	0·08	47	nd	—	0·07	24	nd	—
MgO	1·40	3·3	1·62	4·4	0·30	0·2	nd	—
CaO	5·10	1·6	5·41	1·0	3·31	4·3	3·44	0·7
Na_2O	3·74	3·1	3·53	1·5	4·33	6·8	3·81	2·6
K_2O	1·72	1·3	1·59	0·7	2·00	4·0	1·92	0·0
P_2O_5	0·13	18	nd	—	0·14	24	nd	—
Total	93·23		92·29		89·76		85·37	
mg-no.	36·5		37·8		24·6		—	

| | Minerals | | | | | | | | | | |
| | 626A1 (4) | | 626A2 (2) | | 332A (5) | | 333A (2) | | 332A (9) | | 333A (1) |
(#)	Opx	±1σ	Opx	±1σ	Hbl	±1σ	Hbl	±1σ	Pl	±1σ	Pl
SiO_2	51·09	0·9	52·02	0·1	44·00	0·9	43·41	1·0	57·41	1·0	56·19
TiO_2	0·29	36	0·18	16	1·33	10	1·18	16	0·11	27	nd
Al_2O_3	4·18	10	3·19	8·9	13·02	2·2	11·93	5·6	26·58	1·5	27·91
FeO	21·65	3·5	22·04	2·3	17·89	2·9	18·08	10	0·57	42	0·16
MnO	0·45	13	0·36	33	0·24	24	0·22	29			
MgO	20·94	2·5	22·02	2·6	9·90	3·7	10·32	7·7			
CaO	2·67	9·6	1·75	19	10·23	2·1	10·77	0·7	9·24	4·9	10·33
Na_2O					1·18	4·0	1·10	13	5·88	3·1	5·40
K_2O					0·26	13	0·21	6·7	0·20	23	0·13
Total	101·27		101·56		98·05		97·22		99·99		100·12
mg-no.	63·3		64·0		49·7		50·4	%An	63·8		68·2

Key to symbols:
Glasses: (x, y): EDS data for x scanned areas accumulated into y groups, averaged. [*]: EDS data used for all elements; [x]: number of spots analysed by WDS. ±1σ: standard deviation about mean (rel. %); nd: below limit of detection.
Minerals: Abbreviations as for Table 2. (x): number of grains analysed.
Run conditions:
  626A1, 626A2: 975°C, $X = 0.25$, glass, sintered oxide mix starting materials respectively
  332A, 333A: 775°C, $X = 0.50$, glass, sintered oxide mix respectively

the quartz and plagioclase limits at $X = 0.5$, and at $X = 0.25$ the plagioclase limit lies 50°C above that for quartz (Fig. 3). A less pronounced crossover for the dacite at high aH_2O is possible within the constraints of the 25°C run spacing used (Fig. 2).

Microprobe analyses of plagioclase are listed in Table 4. Due to their small grain size ($<10\ \mu m$) and thin tabular shape, some plagioclase crystals could not be completely resolved from glass. Reported average compositions were calculated from all available analyses with cation totals ≥ 4.95 on an 8 oxygen basis. A number of analyses contain appreciable Fe (up to 1 wt. % as FeO) and trace amounts of TiO_2. Plagioclase in both the dacite and greywacke increases regularly in An content with increasing temperature at constant aH_2O. At intermediate and high aH_2O, and low temperatures (675–725°C), plagioclase composition ranges are An_{35-42} for the dacite and An_{34-39} for the greywacke, becoming only slightly less An-rich with decreasing aH_2O. At $X = 0.25$ and high temperatures (825–975°C) plagioclases are generally appreciably more An-rich (An_{43-52} in the dacite and An_{39-51} in the

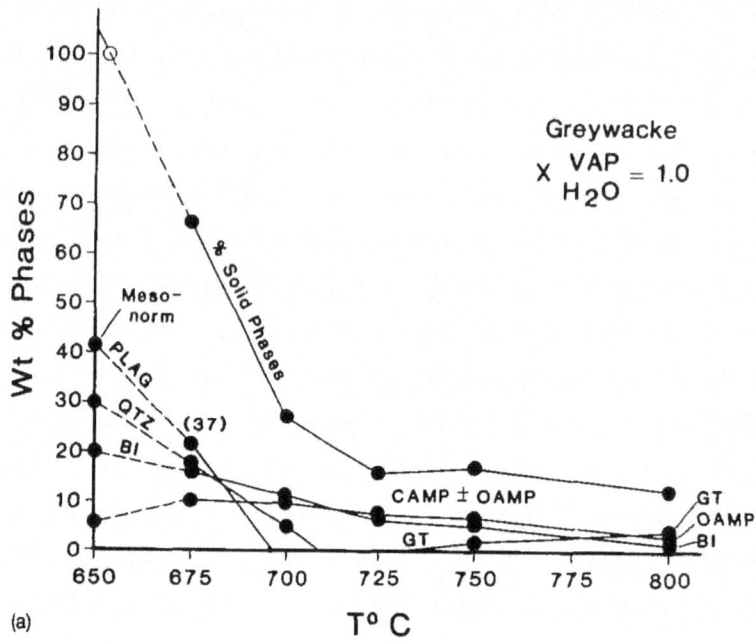

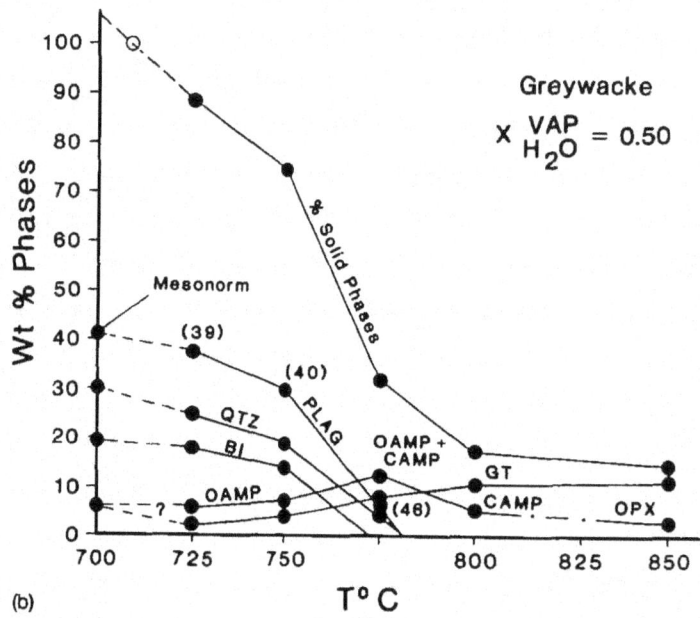

Fig. 4(a–b).

greywacke), suggesting that temperature influences plagioclase composition more strongly than water activity.

Alkali feldspar

Alkali feldspar was identified by X-ray diffractometry (but not confirmed by microprobe) only in near-solidus dacite charges at $X = 0.25$, 800 and 825°C. In the 800°C (subsolidus) assemblage it is

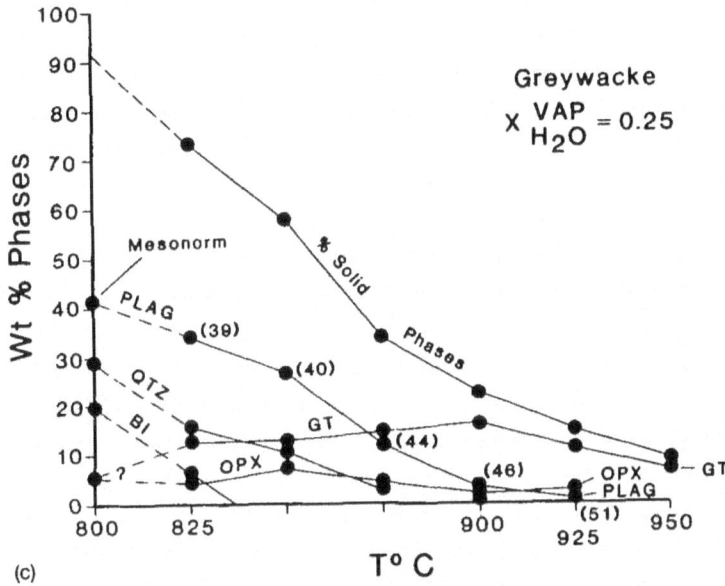

FIG. 4. Calculated modal abundances of phases (in wt.%) vs. $T°C$ in the greywacke composition at $X_{H_2O}^{vap} \approx 1\cdot0, 0\cdot5$, and $0\cdot25$. Dashed lines indicate an extrapolation of wt.% glass to the solidus temperature, and of mineral abundances to the bulk mesonorm (Mielke & Winkler, 1979). Average plagioclase An % (Table 4) is in parentheses.

accompanied by biotite. In the 825 °C assemblage, a $\bar{2}01$ alkali feldspar peak persists in spite of lowered X-ray line intensities due to the presence of glass, but biotite is replaced by minor orthopyroxene. Mass balance calculations for this assemblage, using a suitable K-rich alkali feldspar composition (Table 4), suggest that about 5 wt.% is present.

Pyroxenes

Average pyroxene compositions are given in Table 5. In the dacite, low cation totals for typically 5–10 μm Ca-clinopyroxenes (3·90–4·00, on a 6 oxygen basis) probably reflect incomplete resolution from glass. Two-pyroxene assemblages occur mainly at $X = 0.25$, with quartz and plagioclase to 900 °C and with plagioclase alone to 950 °C (Fig. 2). In these, composition ranges for the coexisting pyroxenes are similar to those obtained in experiments to define the pyroxene solvus by Lindsley (1983, Fig. 3), i.e. individual data points for the Ca-rich phase define a roughly linear trend at near-constant *mg*-number i.e. $100Mg/(Mg + Fe^{2+})$, whereas points for the Ca-poor phase show little variation in *ca*-number, i.e. $100Ca/(Ca + Mg + Fe^{2+})$, but a moderate spread in *mg*-number ($\pm 1\cdot2$–$2\cdot1$ mol %). When Ca-clinopyroxene compositions chosen for averaging are restricted to those with cation totals > 3·96, they group within relatively narrow ranges of *ca*-number (35–40 mol %—see Table 5). The limited number of near-stoichiometric analyses with *ca*-number of 20–35 mol % is then probably attributable to incomplete resolution of Ca-rich and Ca-poor phases ± glass.

Coexisting clino- and orthopyroxenes display systematic variations in major and minor element chemistry with increasing temperature: (1) *mg*-number increases, with clinopyroxene at a higher *mg*-number than coexisting orthopyroxene; (2) *Ca*-number values in clino- and orthopyroxene decrease and increase respectively, indicating a narrowing of the solvus with increasing temperature; (3) Al_2O_3 contents in clinopyroxene increase appreciably (from 4·4 to 5·4 wt.%), producing increased Ca-Tschermaks molecule, whereas there is little Al_2O_3 variation in orthopyroxene (3·6–3·8 wt.%); (4) MnO in clinopyroxene decreases and TiO_2 and Na_2O increase with increasing temperature. A single Ca–Cpx + Opx assemblage in the dacite at $X = 0.5$, 900 °C has pyroxenes notably less aluminous and more magnesian than any observed at $X = 0.25$ (Table 5). This primarily reflects the very high proportion of melt present (95%).

Orthopyroxenes have cation totals of 3·99–4·01, and corresponding calculated Fe_2O_3 contents range up to 1·5 wt.%. In the greywacke, variation in *mg*-number is most pronounced ($\pm 3\cdot7$–$4\cdot6$ mol %)

WALTER K. CONRAD ET AL.

TABLE 4

Plagioclase analyses

	624A	617A	621A	623A	625A	624B	622B	619B	623B	323A	321A
(#)	(6)	(4)	(6)	(7)	(7)	(5)	(6)	(7)	(8)	(4)	(6)
X	0·25	0·25	0·25	0·25	0·25	0·25	0·25	0·25	0·25	0·50	0·50
T°C	825	850	900	925	950	825	850	875	925	725	750
SiO_2	58·56	58·46	57·27	56·45	56·56	60·04	59·82	57·98	56·66	61·77	59·12
TiO_2	0·00	0·16	0·00	0·00	0·00	0·00	0·10	0·00	0·00	0·00	0·10
Al_2O_3	26·28	26·46	27·42	27·47	27·69	25·74	26·09	26·61	27·44	25·14	25·96
FeO	0·39	0·46	0·32	0·32	0·27	0·53	0·29	0·46	0·26	0·13	0·75
CaO	8·55	8·75	10·21	10·56	10·56	7·45	7·88	9·26	10·30	7·00	6·92
Na_2O	6·16	6·06	5·59	5·41	5·34	6·43	6·50	5·97	5·51	7·13	6·87
K_2O	0·46	0·52	0·22	0·21	0·19	0·60	0·62	0·33	0·23	0·69	0·66
Total	100·40	100·87	101 03	100·42	100·61	100·79	101·30	100·61	100·40	101·86	100·38
An %	43·4	44·4	50·2	51·9	52·2	39·0	40·1	46·2	50·8	35·2	35·8

	332A	320A	323B	321B	330B	310A	308A	311B	305A	305B	K F'spar
(#)	(9)	(9)	(7)	(7)	(6)	(5)	(5)	(5)	(5)	(7)	
X	0·50	0·50	0·50	0·50	0·50	0·75	0·75	0·75	1·00	1·00	
T°C	775	800	725	750	775	700	725	675	675	675	
SiO_2	57 41	56·24	57·82	58·80	58·25	59·73	57·63	60·53	59·38	60·00	64·20
TiO_2	0·11	0·00	0·16	0·00	0·16	0·12	0·00	0·12	0·07	0·00	0·00
Al_2O_3	26·58	27·37	25·94	25·59	26·68	24·98	26·73	25·05	25·27	25·68	19·10
FeO	0·57	0·25	0·43	0·33	0·72	0·46	0·00	0·45	0·19	0·47	0·40
CaO	9·24	10·21	7·99	7·82	9·16	7·16	8·55	6·90	7·52	7·62	0·34
Na_2O	5·88	5·58	6·86	6·53	5·94	6·79	6·51	7·39	6·78	7·14	2·60
K_2O	0·20	0·16	0·28	0·39	0·39	0·37	0·13	0·23	0·24	0·24	12·76
Total	99·99	99·81	99·48	99·46	101·30	99·61	99 55	100·67	99·45	101·15	99·40
An %	46·5	50·3	39·2	39·8	46·0	36·8	42 1	34·0	38·0	37·1	

A-series runs are for dacite composition; B-series runs are for greywacke composition. # = number of individual microprobe analyses averaged; $X = X_{H_2O}^{mp}$; An% = 100[Ca/(Ca + Na)]. K F'spar (Deer et al., 1972) used in mass balance calculations as discussed in the text.

in orthopyroxenes of some assemblages in which garnets also show variable mg-number, and often reverse Mg–Fe zoning (see below). These orthopyroxenes are much less calcic (0·6–1·0 vs. 1·5–2·3 wt.% CaO) and more aluminous (7·2–9·8 wt.% Al_2O_3) than any in the dacite, reflecting the respective bulk compositions. ca-number increases systematically with temperature at $X = 0·25$ (Table 5), but Al_2O_3 variations within individual grains in a single charge can be quite large, obscuring any temperature-related variation.

Garnet

In the greywacke at $X = 0·25$ and 0·5, garnet is present at the solidus and throughout the melting interval. Increasing aH_2O reduces its stability relative to amphibole at lower temperatures (see below), and it first appears well above the solidus at $X = 0·75$ and 1·0. Average garnet compositions are listed in Table 5.

Crystals are often reversely zoned from Fe-rich cores to Mg-rich rims, with increases in mg-number of 2–10 mol%. While this zoning is generally more pronounced at higher temperatures, it appears to be independent of the proportion of melt present and the estimated Fe-loss from the charge e.g., at 800–825°C, reverse zoning of 5–7 mol % mg-number occurs in garnet of assemblages at both $X = 0·5$ (80% melt, 0·5–1·2 wt.% Fe-loss) and $X = 0·25$ (25% melt, negligible Fe-loss). In spite of zoning, the averages given (which represent paired core and rim analyses from individual grains) show systematic

TABLE 5

Pyroxene and garnet analyses

	Pyroxenes										
	624A	617A*	617A*	621A*	621A*	623A*	623A*	625A*	625A*	326A*	326A*
(#)	(3)	(4)	(3)	(5)	(6)	(6)	(5)	(6)	(7)	(1)	(7)
X	0·25	0·25	0·25	0·25	0·25	0·25	0·25	0·25	0·25	0·50	0·50
T°C	825	850	850	900	900	925	925	950	950	900	900
SiO_2	49·68	49·99	50·70	50·57	50·37	50·27	50·12	50·90	50·08	52·77	55·24
TiO_2	0·34	0·24	0·66	0·29	0·62	0·30	0·79	0·32	0·77	0·24	0·26
Al_2O_3	3·85	3·56	4·68	3·58	4·76	3·82	4·90	3·58	5·43	1·77	1·85
FeO	29·69	28·75	16·04	26·77	15·38	25·68	14·26	24·24	13·63	21·06	6·26
MnO	0·75	0·60	0·41	0·45	0·32	0·43	0·31	0·41	0·28	0·39	0·14
MgO	15·07	16·20	11·18	17·30	11·50	18·10	12·03	19·14	12·85	22·17	14·90
CaO	1·50	1·56	17·42	1·89	17·45	1·96	17·28	2·28	16·91	1·80	21·83
Na_2O	0·00	0·00	0·30	0·00	0·25	0·00	0·32	0·00	0·35	0·00	0·69
Total	100·88	100·90	101·39	100·85	100·65	100·56	100·01	100·87	100·30	100·20	101·17
mg-no	47·5	50·3	55·4	53·5	57·1	56·2	60·1	59·0	62·7	65·2	80·9

	Pyroxenes				Garnets						
	624B	622B	619B	623B	624B	622B	619B	623B	323B	321B	330B
(#)	(7)	(4)	(5)	(5)	(10)	(6)	(6)	(8)	(6)	(6)	(6)
X	0·25	0·25	0·25	0·25	0·25	0·25	0·25	0·25	0·50	0·50	0·50
T°C	825	850	875	925	825	850	875	925	725	750	775
SiO_2	46·73	45·72	47·10	47·50	37·69	38·11	38·20	38·45	38·31	37·74	37·94
TiO_2	0·35	0·36	0·45	0·52	0·94	1·04	1·32	1·36	0·72	0·76	1·04
Al_2O_3	7·19	9·81	7·87	8·00	20·94	20·67	20·64	21·13	20·92	20·93	20·86
FeO	33·21	32·82	30·77	27·95	31·30	30·53	29·00	27·08	30·78	30·08	29·55
MnO	0·44	0·40	0·35	0·26	0·54	0·56	0·68	0·58	1·80	1·36	0·98
MgO	12·28	11·89	13·25	14·72	5·40	5·76	6·36	7·37	2·72	3·20	4·07
CaO	0·56	0·55	0·80	0·95	3·83	4·19	4·68	5·06	6·18	6·32	6·64
Na_2O	0·00	0·00	0·00	0·00	0·13	0·21	0·15	0·00	0·20	0·20	0·20
Total	100·76	101·55	100·59	99·90	100·77	101·07	101·03	101·03	101·63	100·59	101·28
mg-no.	39·7	39·2	43·4	48·4	23·5	25·2	28·1	32·7	13·6	15·5	19·7

Symbols as in Table 3. Other symbols are: * = two-pyroxene assemblage, *mg*-no = $100[Mg/(Mg + Fe^{2+})]$.

Fe^{2+}-Mg partitioning behaviour with respect to their host glasses. Garnet has *mg*-number approximately equal to that of coexisting glass in assemblages at 925–950 °C, but (in close agreement with the observations of Green, 1977) it is the phase of lower *mg*-number at lower temperatures. Because this relationship holds in assemblages with both large and small percentages of liquid, it is clearly not due to preferential Fe-loss from the melt.

Contents of P_2O_5 (0·1–0·2 wt.%) in the garnets are significant, and decrease with increasing temperature, while TiO_2 contents increase with increasing temperature. Calcium contents increase with increasing aH_2O, and decreasing temperature and *mg*-number (Table 5). Overall, CaO contents are higher than those of most garnets in silicic rocks (e.g., Green, 1977) and the 5 kb experimental garnets of Clemens & Wall (1981). This probably reflects the relatively high pressures of the present experiments.

Ilmenite

Ilmenite is present throughout almost the entire melting interval in both compositions, disappearing only at near-liquidus temperatures of 875–925 °C in the dacite and 900–950 °C in the greywacke. Representative ilmenite compositions are listed in Table 6. Microprobe analyses indicate appreciable SiO_2 and Al_2O_3, but this probably reflects incomplete resolution of the typically 1 to 2 µm grains from surrounding glass. Where analysed SiO_2 was high and cation totals low, ilmenite compositions were approximated by normalizing TiO_2, FeO, MnO, and MgO to 100 wt.%. Generally

TABLE 6

Ilmenite analyses

(#)	624A*	617A*	624B*	619B*	623B*	320A	321B	330B	311B	308B	305B*	304B*
	(2)	(2)	(2)	(4)	(2)	(3)	(1)	(2)	(1)	(1)	(1)	(3)
X	0·25	0·25	0·25	0·25	0·25	0·50	0·50	0·50	0·75	0·75	1·00	1·00
T°C	825	850	825	875	925	800	750	775	675	725	675	700
SiO_2	—	—	—	—	—	1·60	1·81	—	0·96	1·02	0·70	0·72
TiO_2	49·78	51·36	50·36	52·03	52·31	50·61	48·09	51·59	48·95	49·78	51·93	50·35
Al_2O_3	—	—	—	—	—	0·52	0·60	0·81	0·46	1·59	0·27	0·40
FeO	47·87	45·99	47·16	44·50	44·16	43·33	43·85	47·30	45·66	43·86	45·71	47·25
MnO	0·67	0·52	—	0·17	0·19	0·76	0·39	0·30	0·90	0·58	0·63	0·44
MgO	1·68	2·12	2·48	3·30	3·34	2·00	1·11	1·39	0·65	1·35	0·76	0·83
Total	—	—	—	—	—	98·82	95·85	101·39	97·58	99·18	—	—

* = Analysis normalized to 100 wt.%, omitting oxides other than those listed. All other symbols are as in Table 3.

high Ti/Fe ratios indicate stoichiometric compositions with little Fe_2O_3. Contents of MnO and MgO are appreciable, in general respectively decreasing and increasing with increasing temperature at constant aH_2O.

Amphiboles

Amphiboles are major phases over large $T-aH_2O$ ranges in both compositions e.g., from the solidus to 800-825 °C in the dacite. In both the dacite and greywacke they break down at higher temperatures to form orthopyroxene (joined by Ca-clinopyroxene at low aH_2O in the dacite). Amphibole phase relations in the greywacke are notable for assemblages with both hornblende and aluminous gedrite, one of the few occurrences of coexisting Ca- and Fe-Mg amphiboles reported from experiments within the melting interval of natural rock compositions (cf. Gilbert *et al.*, 1982; Green, 1982).

Average amphibole compositions are listed in Table 7. Ferric iron contents were calculated as average values by the methods of Spear & Kimball (1983) and site occupancies assigned according to the procedure outlined by Robinson *et al.* (1982). In the dacite, amphibole grains were generally compositionally uniform, except for slight reverse zonation to higher *mg*-numbers in higher temperature charges. In many greywacke charges at high aH_2O, compositional variability presented a more serious problem, similar to that encountered with Ca-rich pyroxenes. The compositions obtained at amphibole rims near contacts with the glass are the most systematic, defining the widest miscibility gaps, and the indicated number of analyses averaged (Table 7) is thus drawn (with some discretion) from a much larger pool, often 20 to 30 individual analyses from each charge.

The total range of $Fe^{3+}/(Fe^{3+}+Fe^{2+})$ ratios in hornblendes is between 0·10 and 0·33, although the majority have less than 0·20. These low calculated percentages of Fe^{3+} are close to the measured values of Spear (1981) at oxygen fugacities of the wüstite–magnetite (WM) and QFM oxygen buffers. Low and consistent oxygen fugacities in these experiments are also indicated by the occurrence of ilmenite in both compositions and low Fe^{3+} calculated from stoichiometry in ilmenite, garnet and pyroxenes. Gedrite and Ca-poor amphiboles generally have lower Fe^{3+} contents than hornblendes, with $Fe^{3+}/(Fe^{3+}+Fe^{2+})$ ratios from 0·04 to 0·15 (see also Oba & Nicholls, 1986).

Figure 5 illustrates M4 site occupancy in the experimental amphiboles. Also shown are the compositional limits of the three major amphibole groups separated by miscibility gaps at metamorphic temperatures: Fe-Mg-, Na-, and Ca-amphiboles (Robinson *et al.*, 1982).

Dacite hornblendes

At intermediate and high aH_2O a broad stability field of hornblende extends from the solidus to 875-900 °C, while at $X = 0.25$ hornblende is replaced by orthopyroxene and clinopyroxene above 850 °C. According to the classification scheme of Leake (1978), the dacite amphiboles range from hornblende to tschermakitic hornblende. They are intermediate between tschermakite and pargasite, with 0·16 to 0·29 cations present in the A site (23 oxygen basis). The total content of cations involved in Tschermaks substitutions (Al^{VI}, Fe^{3+}, Ti) is relatively high. These calcic hornblendes also have a substantial glaucophane component (0·21 to 0·32 Na cations in the M4 site—Fig. 5), possibly reflecting both high crystallization pressures and the intermediate bulk composition of the dacite (cf. Gilbert *et al.*, 1982; Arai & Hirai, 1985).

Systematic compositional variations in the dacite hornblendes with increasing temperature are exhibited by increasing Ti and Fe^{3+} contents and *mg*-number. Al^{IV} and Al^{VI} first increase, as coexisting biotite melts out, and then decrease. The decrease in Al^{IV} is less pronounced than that of Al^{VI}, which is consistent with a large decrease in Al-Tschermaks substitutions, as indicated in other experiments with amphibole-bearing assemblages (Helz, 1973; Spear, 1981).

Greywacke amphiboles

Amphiboles in the greywacke occur in three types of assemblages:

(1) *Single-amphibole assemblages.* Gedrite alone occurs in near-solidus assemblages at $X = 0.25$, 800 °C and $X = 0.5$, 725-750 °C, and in a near-liquidus assemblage at $X = 1.0$, 800 °C (Table 2). Ca-amphibole is the sole amphibole in several runs at $X = 1.0$ and also at $X = 0.5$, 800 °C.

(2) *Two discrete coexisting phases* (Type 1 two-amphibole assemblages). The most notable such assemblages are at $X = 0.5$, 775 °C and $X = 0.75$, 725 °C, where Ca-amphibole and gedrite are both abundant (approximately 5-10%). Rarer (1-2%) gedrite grains are also present with Ca-amphibole in several charges at $X = 0.75$.

TABLE 7

Amphibole analyses

	624A	323A	321A	332A	320A	323B*	321B*	330B†	330B†	320B	310A	308A
(#)	(8)	(4)	(3)	(5)	(3)	(4)	(5)	(7)	(9)	(6)	(5)	(5)
X	0·25	0·50	0·50	0·50	0·50	0·50	0·50	0·50	0·50	0·50	0·75	0·75
T°C	825	725	750	775	800	725	750	775	775	800	700	725
SiO_2	45·16	45·15	44·88	44·00	44·14	42·84	42·93	45·04	44·06	44·80	42·96	44·03
TiO_2	1·78	1·05	1·44	1·33	1·81	0·53	0·66	0·82	1·08	1·38	1·19	1·16
Al_2O_3	12·11	10·94	11·83	13·02	13·64	15·89	15·45	15·38	6·43	12·50	14·51	16·57
FeO	17·37	21·32	19·62	17·89	15·99	27·36	26·48	25·92	1·53	15·36	16·81	14·85
MnO	0·27	0·40	0·35	0·24	0·17	0·46	0·51	0·26	0·24	0·10	0·26	0·20
MgO	10·17	8·64	8·91	9·90	10·46	8·11	8·04	9·25	7·62	12·84	8·34	8·84
CaO	8·80	8·99	10·06	10·23	10·43	0·74	0·88	1·57	6·58	7·28	10·70	10·57
Na_2O	1·50	1·80	1·56	1·18	1·72	1·28	1·34	1·30	1·67	1·74	1·58	1·85
K_2O	0·44	0·40	0·72	0·26	0·32	0·00	0·00	0·10	0·28	0·11	0·79	0·51
Total	97·55	98·69	99·37	98·05	98·68	97·21	96·29	99·64	99·49	96·11	97·14	98·58
mg-no.	55·8	48·3	50·6	56·3	60·4	37·0	37·3	40·9	42·1	67·9	49·8	54·2

	311B†	311B†	308B‡	308B‡	306B†	306B†	305A	303A	305B‡	305B‡	304B	303B*
(#)	(2)	(7)	(1)	(7)	(3)	(4)	(1)	(4)	(2)	(5)	(7)	(5)
X	0.75	0·75	0·75	0·75	0·75	0·75	1·00	1·00	1·00	1·00	1·00	1·00
T°C	675	675	725	725	800	800	675	800	675	675	700	800
SiO_2	43·17	41·57	41·11	41·92	44·63	45·90	43·48	46·59	45·32	43·48	42·52	47·47
TiO_2	0·36	0·43	0·62	0·84	0·61	1·16	1·19	1·33	0·72	0·63	0·54	0·66
Al_2O_3	17·98	19·33	12·71	17·61	14·27	14·29	13·97	11·21	13·33	16·33	17·41	10·31
FeO	27·65	22·97	26·48	21·08	24·65	16·56	16·88	13·77	28·16	24·37	22·64	21·43
MnO	0·97	0·70	0·75	0·32	0·26	0·26	0·21	0·25	0·61	0·47	0·30	0·17
MgO	5·78	4·98	9·97	7·06	10·79	12·25	9·21	13·16	7·02	5·72	6·61	14·96
CaO	1·85	6·27	3·47	6·99	1·39	6·86	10·88	11·01	3·39	5·14	7·06	1·38
Na_2O	1·24	1·83	1·33	1·71	1·17	1·72	1·55	1·59	1·94	1·96	1·72	1·11
K_2O	0·00	0·10	0·00	0·17	0·00	0·17	0·32	0·36	0·00	0·18	0·17	0·00
Total	99·00	98·78	96·44	97·70	97·77	99·17	97·78	99·27	100·49	98·28	98·97	97·49
mg-no.	28·3	31·3	50·7	42·3	46·8	63·2	54·0	69·9	34·1	32·3	39·0	59·2

Unless otherwise indicated, all analyses are hornblendes or 'Ca-amphiboles' as discussed in the text. Symbols indicating amphibole assemblages discussed in the text are: * = gedrite-only, † = type I coexisting amphiboles, ‡ = type II coexisting amphiboles. All other symbols and captions are in Table 3

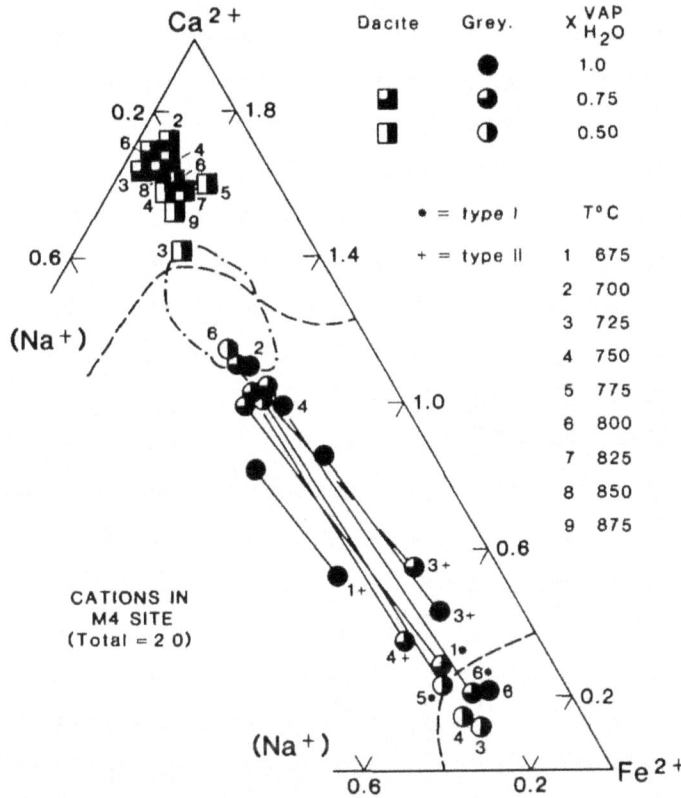

FIG. 5. Amphibole compositions in terms of proportions of Ca, Na, and $(Mg + Fe^{2+})$ cations in the M4-site. Fe^{3+} calculated by the methods of Spear & Kimball (1983). Dacite amphiboles at $X_{H_2O}^{vap} = 1.0$ and 0.25 are not shown for clarity. The dashed boundaries indicate the wide miscibility gap between natural coexisting metamorphic amphiboles (fig. 20 of Robinson *et al.*, 1982). The dot-dashed field encloses the range of subcalcic hornblendes synthesized in earlier 3 kb hydrothermal experiments (Reid, 1982).

(3) *Discrete compositions in composite grains* (Type II two-amphibole assemblages). Gedrite-like compositions are found as cores rimmed by Ca-amphibole in rare composite grains (< 1%) at $X = 0.75$ and 1.0.

'Ca-amphiboles' are optically monoclinic, and they have 1.1–1.4 (Ca + Na) in the M4 site, i.e. $(Ca + Na)_{M4}$. According to the classification of Leake (1978), they range through cummingtonite, hornblende, ferrotschermakitic hornblende, and ferrotschermakite, with no obvious miscibility gaps (Fig. 5). The H_2O-saturated experiments of Reid (1982) with a similar TVZ greywacke (Table 1) at 3 kb produced similar Fe- and Al-rich subcalcic hornblendes (Fig. 5). On the other hand, natural occurrences of highly sub-calcic hornblendes appear to be rare. Hawthorne, *et al.* (1980) described tschermakitic hornblendes, coexisting with gedrite and cummingtonite, with $(Ca + Na)_{M4}$ of 1.556; these would plot close to some of the experimental sub-calcic hornblendes of Fig. 5.

Gedrites in single amphibole assemblages are optically orthorhombic, and they have about 0.4 (Ca + Na)$_{M4}$. In type I two-amphibole assemblages they have 0.4–0.6 (Ca + Na)$_{M4}$, but their optical properties could not be determined. In type II two-amphibole assemblages, gedritic Ca-poor core compositions in composite grains have 0.6–0.9 (Ca + Na)$_{M4}$, and they are thought to represent either poorly-resolved two-phase material or more extensive, perhaps metastable, solid-solution towards Ca-amphibole compositions. Reid (1982) did not observe gedritic amphibole in his 3 kb experiments (although considerable gedrite component is present in solid-solution with hornblende—Fig. 5), suggesting that gedrite stability is restricted to higher pressures. However, gedrite crystallized in limited experiments with the present greywacke at 5 kb, $X = 1.0$ (Nicholls, 1983).

TABLE 8

Biotite analyses

(#)	624B (6)	323A* (5)	321A* (3)	323B* (4)	321B* (5)	310A* (3)	311B (3)	308B (4)	306B (5)	305A (4)	305B (5)	304B (5)	301B (5)	303B (5)
X	0·25	0·50	0·50	0·50	0·50	0·75	0·75	0·75	0·75	1·00	1·00	1·00	1·00	1·00
T°C	825	725	750	725	750	700	675	725	800	675	675	700	725	800
SiO_2	39·82	36·48	37·16	36·94	36·41	38·49	35·58	36·62	37·34	37·88	36·88	36·55	36·80	38·24
TiO_2	4·54	4·30	4·58	3·00	3·38	3·13	1·81	2·55	2·70	2·44	1·88	1·77	1·91	2·15
Al_2O_3	17·21	15·33	15·19	18·01	17·91	18·94	20·21	18·98	17·48	17·52	19·79	18·49	18·06	16·42
FeO	15·25	20·45	19·39	21·97	21·56	16·06	20·37	17·93	15·08	16·76	20·72	21·75	20·73	15·71
MnO	0·00	0·00	0·07	0·00	0·05	0·00	0·00	0·00	0·00	0·12	0·00	0·08	0·00	0·00
MgO	9·91	10·12	10·19	7·12	7·54	9·78	6·64	9·45	13·51	10·88	7·87	8·84	9·94	14·57
CaO	0·00	0·00	0·10	0·00	0·00	0·00	0·00	0·00	0·00	0·00	0·00	0·00	0·00	0·00
Na_2O	0·53	0·56	0·63	0·74	0·88	0·62	0·61	0·67	0·62	0·50	0·51	0·53	0·53	0·52
K_2O	8·06	8·77	8·68	8·21	8·31	8·99	8·02	8·30	8·15	8·62	8·16	8·03	7·99	7·88
Total	95·32	—	—	—	—	—	93·24	94·50	94·88	94·76	95·81	96·04	95·96	95·49
mg-no.	53·7	46·9	48·4	36·6	38·4	52·0	36·7	48·4	62·2	53·6	40·4	42·0	46·1	62·3

* = average microprobe analysis normalized to 96 wt %. All other symbols are as in Table 3.

In natural metamorphic assemblages with coexisting Ca-rich amphiboles, increasing pressure causes increased glaucophane substitution and this may be accompanied by a large cummingtonite component (Arai & Hirai, 1985). High tschermakite contents in hornblendes are also attributable to high pressures (e.g., Deer *et al.*, 1972; Gilbert *et al.*, 1982). In assemblages with both tschermakitic hornblende and gedrite, hornblende also contains a large cummingtonite component (Hawthorne *et al.*, 1980). Cameron (1975) and Oba & Nicholls (1986) have suggested from experimental studies at 3–5 kb, and pressure estimates for natural assemblages, that increasing PH_2O narrows the solvus between Fe–Mg and Ca-amphiboles. The very high degrees of Ca-(Mg + Fe)-Na miscibility observed in Ca-amphiboles and within two-amphibole assemblages in the greywacke are therefore qualitatively compatible with the 10 kb pressure of the present study. Also apparent in Fig. 5 is a progressive narrowing of the miscibility gap between Ca- and Mg–Fe amphiboles with increase in aH_2O from 0·5 to 1·0.

Most reported naturally-occurring hornblende–gedrite assemblages, used in part to define the miscibility gap shown in Fig. 5, relate to basic protoliths with *mg*-number for mafic minerals higher than those in the greywacke experimental assemblages (c.f. Robinson, *et al.*, 1982; Hawthorne, *et al.*, 1980). The relatively Fe-rich compositions of the greywacke amphiboles at high aH_2O may also favour extended Ca-miscibility, as in pyroxenes. The most Fe-rich (*mg*-number = 0·31–0·42), subcalcic and notably tschermakitic Ca-amphiboles (2·6–3·4 Al cations) are found in garnet-free assemblages in the greywacke at high aH_2O. In the presence of garnet, gedrite is the only amphibole present under near-solidus conditions, while under near-liquidus conditions Ca-amphiboles are generally more calcic, less aluminous (2·1–2·8 Al cations), and Mg-rich (*mg*-number = 0·42–0·68).

Biotite

Average biotite compositions are listed in Table 8. Extremely small biotite grains in some charges yielded stoichiometric analyses with low totals, which were normalized to 96 wt.%. Biotite compositions are plotted in Fig. 6, which illustrates octahedral site occupancy vs. total Al content. At temperatures of 750–800 °C the greywacke biotites plot close to ideal trioctahedral biotite–eastonite compositions, while at lower temperatures (675–750 °C) they contain substantial octahedral site

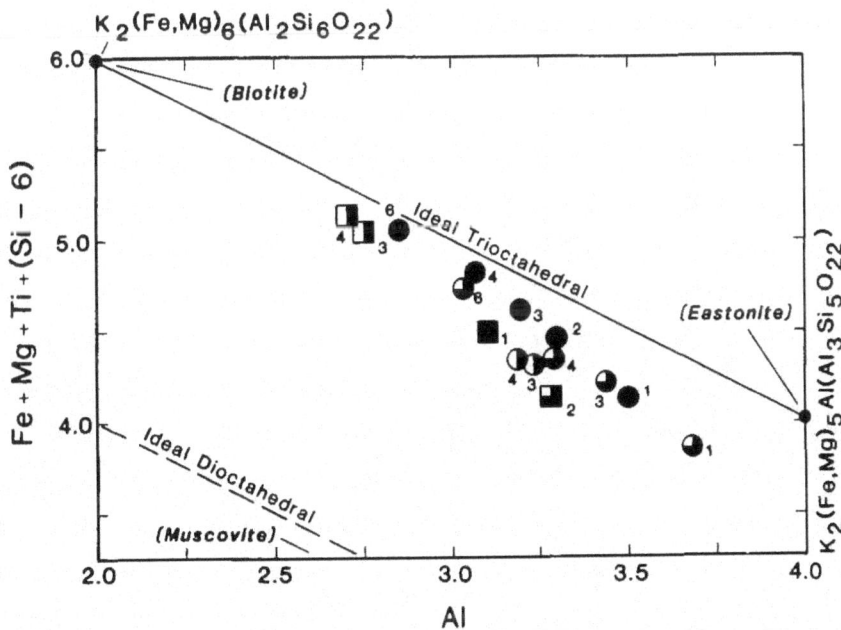

FIG. 6. Biotite compositions in terms of octahedral-site occupancy vs. total Al cations after Tracy (1978). Substitutions within ideal trioctahedral- and dioctahedral-mica compositions project as parallel lines. Symbols as in Fig. 5.

784   WALTER K. CONRAD *ET AL.*

vacancies, as do natural metamorphic biotites crystallized in this temperature range (Tracy, 1978) and equivalents crystallized experimentally at high pressure (Green, 1981).

At constant aH_2O, biotite compositions systematically increase in *mg*-number and decrease in Al with increasing temperature (Fig. 6, Table 8). Variations in (K + Na) cations are strongly correlated with Al^{IV}, and both are of approximately equal magnitude (≤ 1.5 atoms per formula). A much larger variation in Al^{VI} is positively correlated with Ti content, but the small absolute variation in Ti implies that more than one substitution involving Al^{VI} is operative. Figure 6 suggests that the primary substitution mechanism controlling biotite Al variation is:

$$3(R^{2+})^{VI} = 2(Al^{3+})^{VI} + \square^{VI} \qquad (1)$$

With decreasing aH_2O (at similar temperatures) biotites in both the dacite and greywacke show systematically increasing K, Na, and Ti. The large increase of Ti with decreasing aH_2O is particularly good evidence for the following substitution (cf. Bohlen *et al.*, 1980):

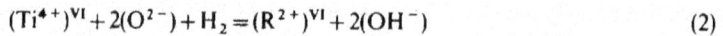

$$(Ti^{4+})^{VI} + 2(O^{2-}) + H_2 = (R^{2+})^{VI} + 2(OH^-) \qquad (2)$$

The available experimental data on coexisting hornblende and biotite at reduced aH_2O is sparse (e.g., Naney, 1983). In the present study, biotites with low *mg*-number (0·4) in the greywacke are appreciably more magnesian than coexisting hornblendes, but as biotite *mg*-number increases toward 0·6 in both the dacite and greywacke, values for hornblendes become almost equal. Similar dependence

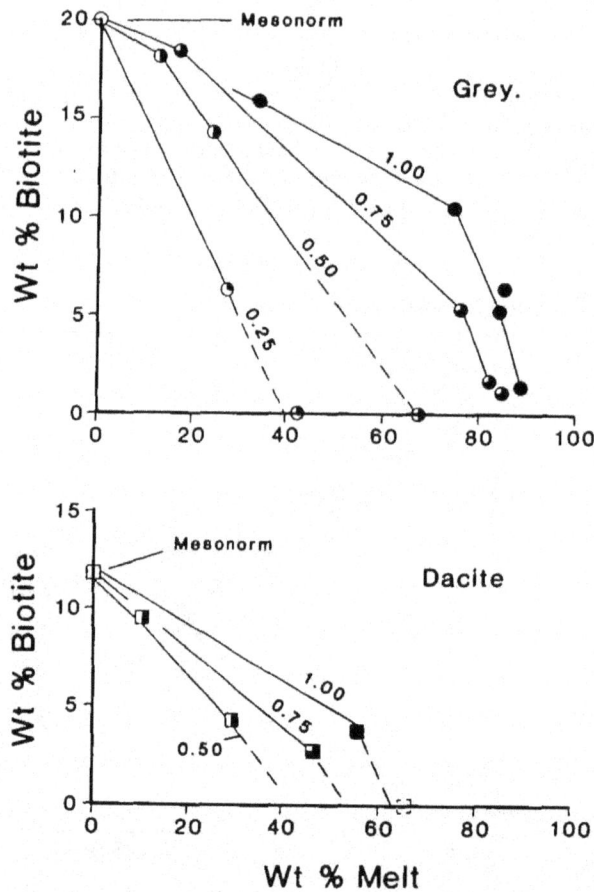

FIG. 7. Calculated wt.% biotite vs. wt.% glass in the indicated melting range of the greywacke (top) and dacite (bottom) compositions. Symbols for varying aH_2O as in Fig. 5, with all trends extrapolated to wt.% biotite in the bulk mesonorm (Mielke & Winkler, 1979).

of Fe^{2+}/Mg partitioning behaviour between biotite and hornblende on *mg*-number is observed in natural assemblages (e.g., Czamanske *et al.*, 1981; see also Speer, 1984).

MELTING REACTIONS INVOLVING BIOTITE AND AMPHIBOLE

Figure 7, 8, and 9 illustrate the changes in modal (wt.%) abundances of crystalline phases and liquid with increasing temperature. In both compositions, melting near the solidus is characterized by breakdown of biotite, plagioclase, and quartz to form residua with modally increasing amphibole $\pm$ garnet $\pm$ pyroxene. This indicates that incongruent melting reactions are forming amphibole from biotite. The sole exception is at $X = 0.25$ in the dacite, where breakdown of biotite to form K-feldspar may be completed before the onset of melting.

From these general modal relationships and simplified phase compositions based upon microprobe analyses, several biotite melting reactions analogous to those known from previous experiments (discussed in detail by Winkler, 1979) may be inferred. This treatment is necessarily simplified, since most phases are present over large temperature intervals and melt via sliding reactions involving solid solutions. In the greywacke, the following near-solidus melting reactions predominate (normalized to 1 biotite formula unit):

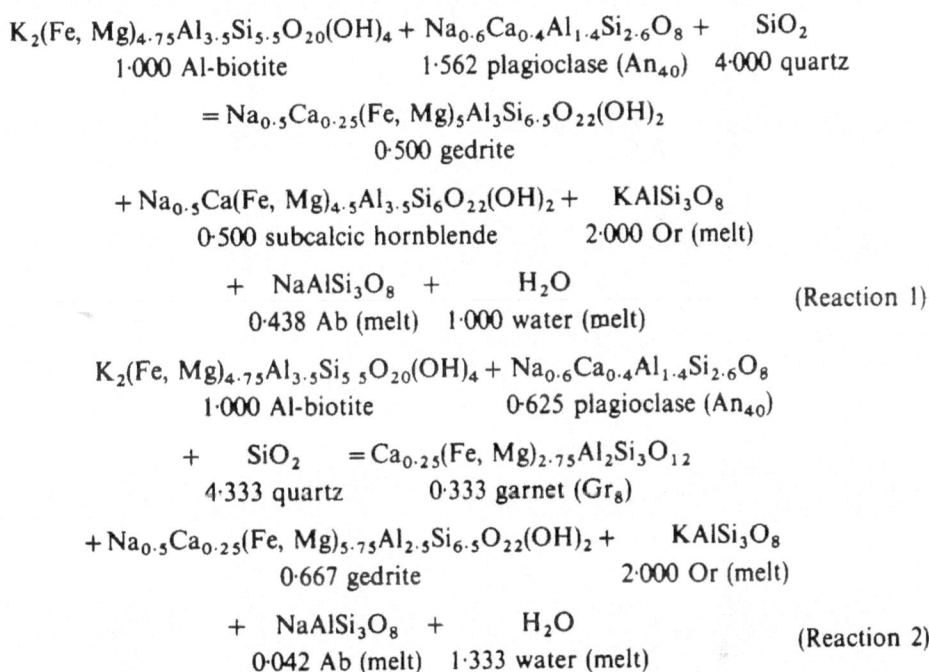

$$K_2(Fe, Mg)_{4.75}Al_{3.5}Si_{5.5}O_{20}(OH)_4 + Na_{0.6}Ca_{0.4}Al_{1.4}Si_{2.6}O_8 + \quad SiO_2$$
$$\text{1·000 Al-biotite} \qquad \text{1·562 plagioclase (An}_{40}) \quad \text{4·000 quartz}$$

$$= Na_{0.5}Ca_{0.25}(Fe, Mg)_5Al_3Si_{6.5}O_{22}(OH)_2$$
$$\text{0·500 gedrite}$$

$$+ Na_{0.5}Ca(Fe, Mg)_{4.5}Al_{3.5}Si_6O_{22}(OH)_2 + \quad KAlSi_3O_8$$
$$\text{0·500 subcalcic hornblende} \qquad \text{2·000 Or (melt)}$$

$$+ \quad NaAlSi_3O_8 \quad + \quad H_2O$$
$$\text{0·438 Ab (melt)} \quad \text{1·000 water (melt)} \qquad \text{(Reaction 1)}$$

$$K_2(Fe, Mg)_{4.75}Al_{3.5}Si_{5.5}O_{20}(OH)_4 + Na_{0.6}Ca_{0.4}Al_{1.4}Si_{2.6}O_8$$
$$\text{1·000 Al-biotite} \qquad \text{0·625 plagioclase (An}_{40})$$

$$+ \quad SiO_2 \quad = Ca_{0.25}(Fe, Mg)_{2.75}Al_2Si_3O_{12}$$
$$\text{4·333 quartz} \qquad \text{0·333 garnet (Gr}_8)$$

$$+ Na_{0.5}Ca_{0.25}(Fe, Mg)_{5.75}Al_{2.5}Si_{6.5}O_{22}(OH)_2 + \quad KAlSi_3O_8$$
$$\text{0·667 gedrite} \qquad \text{2·000 Or (melt)}$$

$$+ \quad NaAlSi_3O_8 \quad + \quad H_2O$$
$$\text{0·042 Ab (melt)} \quad \text{1·333 water (melt)} \qquad \text{(Reaction 2)}$$

Reaction 1 is important at $X = 0.75$ and 1.0, where near-solidus assemblages comprise plagioclase, quartz, biotite, ilmenite, tschermakitic hornblende, and often a minor proportion of gedrite (considerable gedrite is also present in solid solution with hornblende). Reaction 2 occurs at $X = 0.5$ and very near the solidus at $X = 0.25$, where the assemblage is plagioclase, quartz, biotite, garnet, gedrite, and ilmenite. At higher percentages of partial melting an additional reaction producing the assemblage garnet-hornblende from biotite–quartz–plagioclase also occurs. This is evident at $X = 0.5$, 775–800°C (Figs. 3 and 4) where biotite is replaced by hornblende, joining gedrite in an assemblage with both amphiboles abundant.

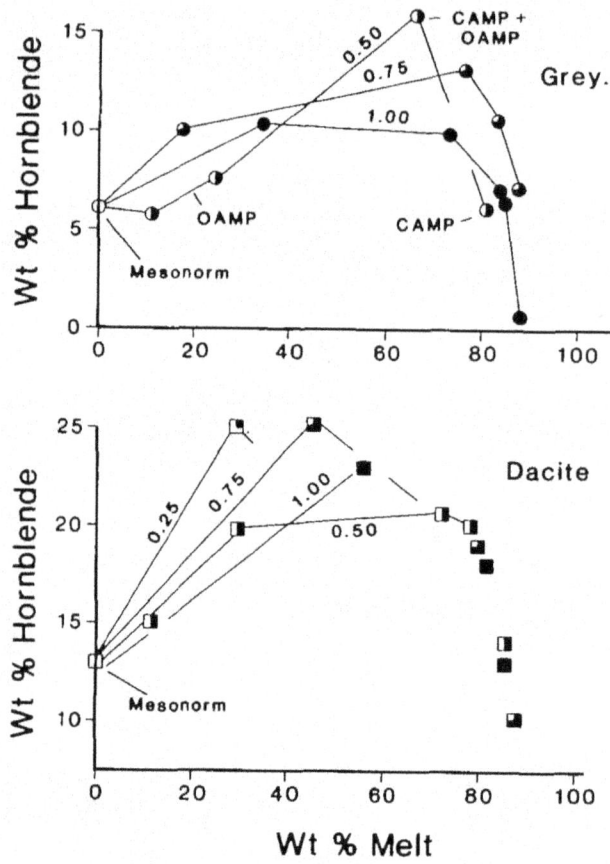

FIG. 8. Calculated wt % amphibole vs. wt.% glass in the indicated melting range in the greywacke (top) and dacite (bottom) compositions. Coexisting Ca-rich- and Ca-poor-amphiboles in the greywacke assemblages are summed. Symbols as in Fig. 5.

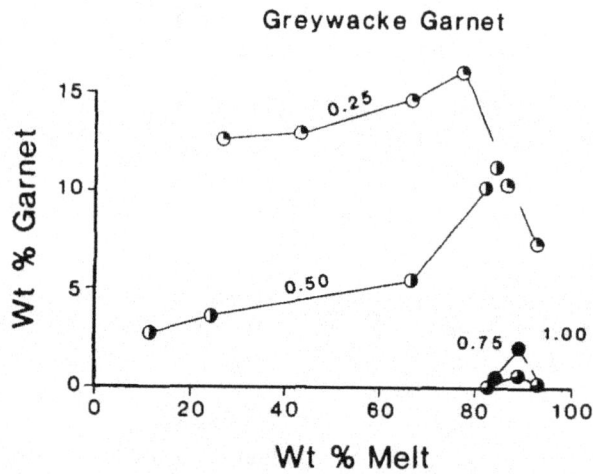

FIG. 9. Calculated wt.% garnet vs. wt.% glass thoroughout the melting range in the greywacke. Symbols as in Fig. 5.

Incongruent biotite melting reactions explain the unusual T–X stability field of biotite in the greywacke, compared to other systems where biotite remains a near-liquidus phase at intermediate aH_2O (e.g., Maaløe & Wyllie, 1975; Clemens & Wall, 1981). The steep drop of the biotite-out boundary from $X = 0.75$ to 0.5 (Fig. 3) coincides with increasing plagioclase stability, suggesting that increasing activity of An and Ab reduces biotite stability by the above reactions. Reactions 1 and 2 also illustrate how decreasing aH_2O: (i) increases the relative amounts of biotite and quartz consumed and K_2O released to melt; and (ii) increases the amount of H_2O released to the melt, as garnet replaces hornblende. The greywacke amphiboles are converted to garnet with decreasing aH_2O (Figs. 8, 9).

In the greywacke, both hornblende and biotite decrease in Al-Tschermaks components with increasing temperature throughout the near-solidus melting interval (Tables 7, 8). In contrast, melting reactions in the dacite are notable for producing relatively greater amounts of hornblende with increasing Al-contents (as biotite melts out):

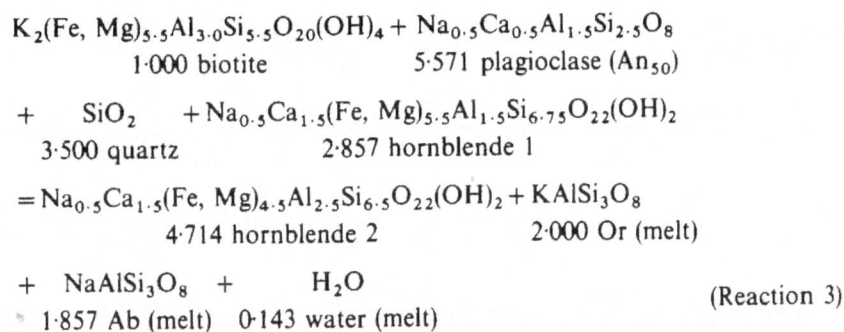

$$K_2(Fe, Mg)_{5.5}Al_{3.0}Si_{5.5}O_{20}(OH)_4 + Na_{0.5}Ca_{0.5}Al_{1.5}Si_{2.5}O_8$$
$$1.000 \text{ biotite} \qquad\qquad 5.571 \text{ plagioclase } (An_{50})$$

$$+ \quad SiO_2 \quad + Na_{0.5}Ca_{1.5}(Fe, Mg)_{5.5}Al_{1.5}Si_{6.75}O_{22}(OH)_2$$
$$3.500 \text{ quartz} \qquad\qquad 2.857 \text{ hornblende 1}$$

$$= Na_{0.5}Ca_{1.5}(Fe, Mg)_{4.5}Al_{2.5}Si_{6.5}O_{22}(OH)_2 + KAlSi_3O_8$$
$$4.714 \text{ hornblende 2} \qquad\qquad 2.000 \text{ Or (melt)}$$

$$+ \quad NaAlSi_3O_8 \quad + \quad H_2O$$
$$1.857 \text{ Ab (melt)} \quad 0.143 \text{ water (melt)}$$
(Reaction 3)

In the greywacke, breakdown of 20 wt.% biotite evolves 5–10 wt.% amphibole and substantial H_2O should therefore be liberated. Analogous reactions could initiate dehydration melting under vapour-absent or H_2O-deficient conditions in the crust (cf. Brown & Fyfe, 1970). In the dacite, initially smaller amounts (12 wt.%) of biotite are converted to relatively larger amounts (> 10 wt.%) of hornblende, and little or no H_2O is liberated; under H_2O-vapour-absent conditions, breakdown of biotite would perhaps produce assemblages with K-feldspar, plagioclase, and abundant amphibole, and not initiate substantial partial melting. These contrasting reactions illustrate how hydrated metaluminous assemblages (particularly those lacking quartz) may reach very high temperatures without undergoing anatexis, with the solidus determined by amphibole dehydration (up to 1050 °C in mafic rocks e.g., Allen & Boettcher, 1978), whereas mildly peraluminous assemblages, (particularly those containing quartz), begin to melt at much lower temperatures (< 850 °C) determined by biotite dehydration (e.g., Clemens & Wall, 1981).

EXPERIMENTAL GLASS COMPOSITIONS

Table 9 lists all glass compositions at $X = 1.0$, 0.75, 0.5, and 0.25 which are saturated in plagioclase and/or quartz. The glass analyses are normalized to 100 wt.% with un-normalized totals also given. Totals were highly variable, depending on microprobe spot size (probably reflecting volatilization and differing amounts of void space scanned), and estimates of melt H_2O contents by difference are not considered reliable.

Variations of normative salic components with aH_2O.

CIPW (wt. %) normative components of the glasses are projected into the ternary systems Q–Ab–Or and An–Ab–Or in Figs. 10 and 11. Decreasing aH_2O in both model source

TABLE 9
Glass analyses

| X | 624A 0·25 | 617A 0·25 | 621A 0·25 | 623A 0·25 | 625A 0·25 | 624B 0·25 | 622B 0·25 | 619B 0·25 | 623B 0·25 | 323A 0·50 | 321A 0·50 | 332A 0·50 | 320A 0·50 | 326A 0·50 |
T°C	825	850	900	925	950	825	850	875	925	725	750	775	800	900
SiO_2	72·81	71·96	71·03	68·90	67·92	73·66	72·78	72·19	69·16	71·98	72·16	72·29	70·93	65·76
TiO_2	0·18	0·29	0·44	0·65	0·66	0·19	0·26	0·38	0·61	0·13	0·13	0·18	0·28	0·66
Al_2O_3	15·39	15·29	15·18	15·58	15·80	15·02	15·10	15·26	16·24	16·05	15·99	15·73	16·02	16·57
FeO	1·78	2·12	2·84	3·39	3·90	1·48	1·62	2·11	3·39	0·86	0·95	1·64	2·12	4·46
MnO	0·04	0·05	0·07	0·06	0·08	0·01	0·04	0·02	0·02	0·05	0·07	0·07	0·07	0·10
MgO	0·30	0·49	0·59	0·95	1·12	0·31	0·44	0·60	0·90	0·15	0·14	0·30	0·42	1·52
CaO	2·37	2·62	3·38	4·09	4·30	1·37	1·63	2·32	3·30	2·38	2·41	3·31	3·92	5·13
Na_2O	3·63	3·59	4·01	4·24	4·24	3·87	4·07	4·19	3·93	4·49	4·61	4·33	4·10	3·88
K_2O	3·42	3·49	2·30	1·99	1·86	3·92	3·90	2·68	2·22	3·85	3·48	2·00	2·02	1·75
P_2O_5	0·09	0·10	0·16	0·13	0·12	0·16	0·15	0·25	0·25	0·05	0·06	0·14	0·12	0·16
Total	91·16	92·10	91·52	92·82	93·07	93·23	92·48	91·87	92·79	94·95	95·00	89·76	86·53	88·06
mg-no.	23·1	29·2	27·0	33·3	33·8	27·2	32·6	33·6	32·1	23·7	20·8	24·6	26·1	37·8
					Wt.% C.I.P.W. normative components									
Q	31·95	30·06	29·13	24·81	23·18	32·10	29·26	30·90	27·70	25·28	26·07	31·01	29·04	20·75
C	1·62	1·08	0·33	—	—	2·30	1·58	1·85	1·97	0·29	0·40	0·76	0·25	—
Ab	30·71	30·37	33·93	35·87	35·87	32·74	34·44	35·45	33·25	37·99	39·00	36·64	34·69	32·83
Or	20·21	20·63	13·59	11·76	10·99	23·17	23·05	15·84	13·12	22·76	20·57	11·82	11·94	10·34
An	11·17	12·35	15·72	17·61	18·59	5·75	7·11	9·88	14·74	11·48	11·56	15·51	18·66	22·63
Di	—	—	—	1·56	1·66	—	—	—	—	—	—	—	—	1·50
Hy	3·79	4·73	6·09	6·84	8·16	3·19	3·72	4·78	7·50	1·83	2·01	3·59	4·61	10·31

X	323B	321B	330B	310A	308A	311B	308B	305A	305B	304B
X	0·50	0·50	0·50	0·75	0·75	0·75	0·75	1·00	1·00	1·00
T°C	725	750	775	700	725	675	725	675	675	700
SiO_2	71·99	72·05	70·72	70·81	71·85	70·56	70·15	70·71	71·16	70·55
TiO_2	0·09	0·09	0·18	0·11	0·17	0·10	0·17	0·13	0·09	0·13
Al_2O_3	16·42	16·34	16·46	17·60	16·48	17·81	17·51	17·24	17·30	17·22
FeO	1·62	1·75	2·17	0·85	1·07	1·43	1·67	0·92	1·35	1·90
MnO	0·06	0·03	0·04	0·06	0·05	0·07	0·03	0·04	0·04	0·06
MgO	0·17	0·25	0·37	0·18	0·33	0·26	0·45	0·21	0·27	0·40
CaO	2·44	2·48	3·10	3·15	3·98	3·40	3·83	3·62	3·33	3·85
Na_2O	4·95	4·42	4·28	4·76	4·03	4·92	4·12	4·99	4·92	4·20
K_2O	2·18	2·44	2·50	2·40	1·97	1·33	1·86	2·08	1·41	1·53
P_2O_5	0·09	0·16	0·18	0·08	0·08	0·12	0·20	0·06	0·12	0·15
Total	88·69	90·35	89·88	88·36	92·62	88·81	86·91	92·24	87·99	86·91
mg-no.	15·8	20·3	23·3	27·4	35·5	24·5	32·4	28·9	26·3	27·3
				Wt.% C.I.P.W. normative components						
Q	28·29	30·34	27·86	26·46	31·26	28·34	29·46	25·12	28·85	30·30
C	1·70	2·30	1·51	1·64	0·67	2·38	2·24	0·34	2·91	2·02
Ab	41·88	37·40	36·21	40·27	34·10	41·63	34·86	42·22	41·63	35·54
Or	12·88	14·42	14·78	14·19	11·64	7·86	10·99	12·29	8·33	9·04
An	11·52	11·26	14·20	15·11	19·22	16·08	17·70	17·57	15·74	18·12
Di	—	—	—	—	—	—	—	—	—	—
Hy	3·36	3·74	4·68	1·94	2·60	3·24	3·96	2·07	3·08	4·38

All glass compositions are normalized to 100 wt.%.
Total = unnormalized total from combined EDS and WDS analyses as discussed in the text.

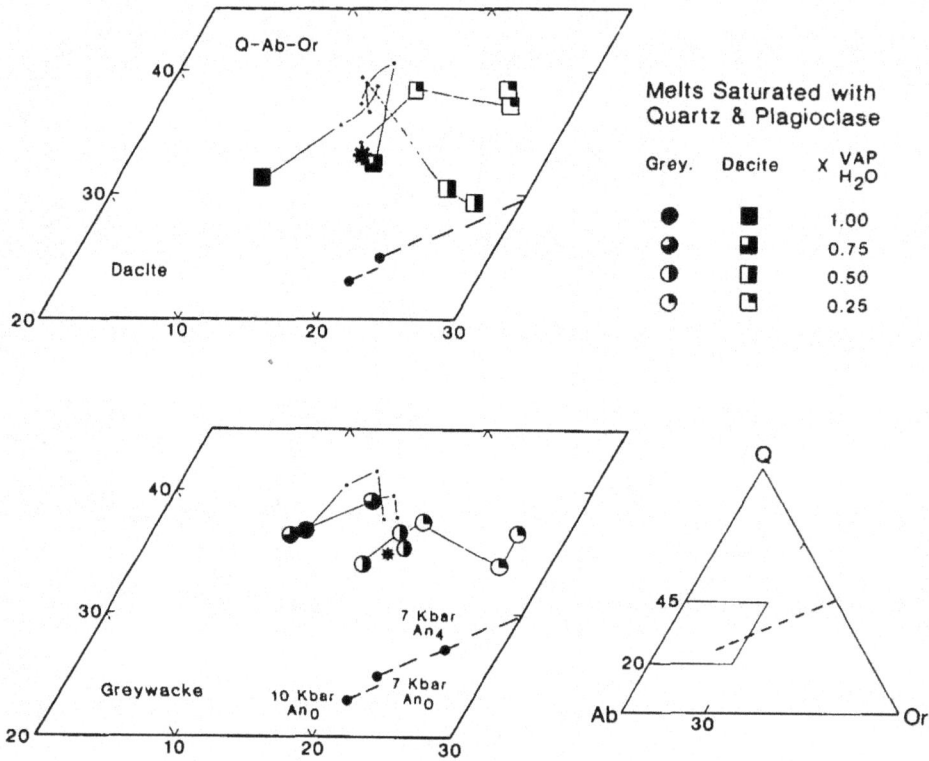

FIG. 10. Analyzed glass compositions throughout the melting range in the greywacke and dacite compositions projected in terms of CIPW normative quartz (Q), albite (Ab) and orthoclase (Or). Symbols as in Figs. 1 and 5. Large symbols indicate assemblages at varying aH_2O saturated in plagioclase+quartz. Smaller points indicate compositions at higher percentages of melting trending towards the respective bulk compositions near the liquidus. Locations of cotectic and minimum melting compositions at 10 and 7 kb taken from Winkler (1979).

compositions produces melts with progressively decreasing An and increasing Or. Near-solidus melts projected into the system Q–Ab–Or move towards Or, with constant or slightly increasing Q contents, and in the system An–Ab–Or move away from An towards the Ab–Or join. These pronounced compositional changes occur in quartz- and plagioclase-saturated melts at comparable percentages of partial melting. Once plagioclase and quartz are eliminated, all melt compositions converge towards the bulk dacite and greywacke compositions.

The relative changes in normative salic components in these quartz- and plagioclase-saturated melts bear a simple relationship to experimental data for model systems. It is well established that with increasing PH_2O eutectics in the systems Q–An, Q–Ab, and Q–Or shift toward lower temperatures, with the magnitude of the depression in the relative order Q–An > Q–Ab > Q–Or (see tables 9·4 and 9·5 of Burnham, 1981). For example, the Q–An eutectic is at 1369 °C at 1 atm. (dry) and 757 °C at $PH_2O = 10$ kb, a depression of 612 °C. The corresponding 1 atm. to 10 kb PH_2O depression in the system Q–Or is from 990 to 710 °C (280 °C). The system Q–Ab has an intermediate value of 412 °C. Comparison of anhydrous and water-saturated eutectic temperatures at 10 kb (Burnham, 1981; Boettcher et al., 1984) shows that increase in aH_2O at constant load pressure has similar effects on relative stabilities of feldspar components.

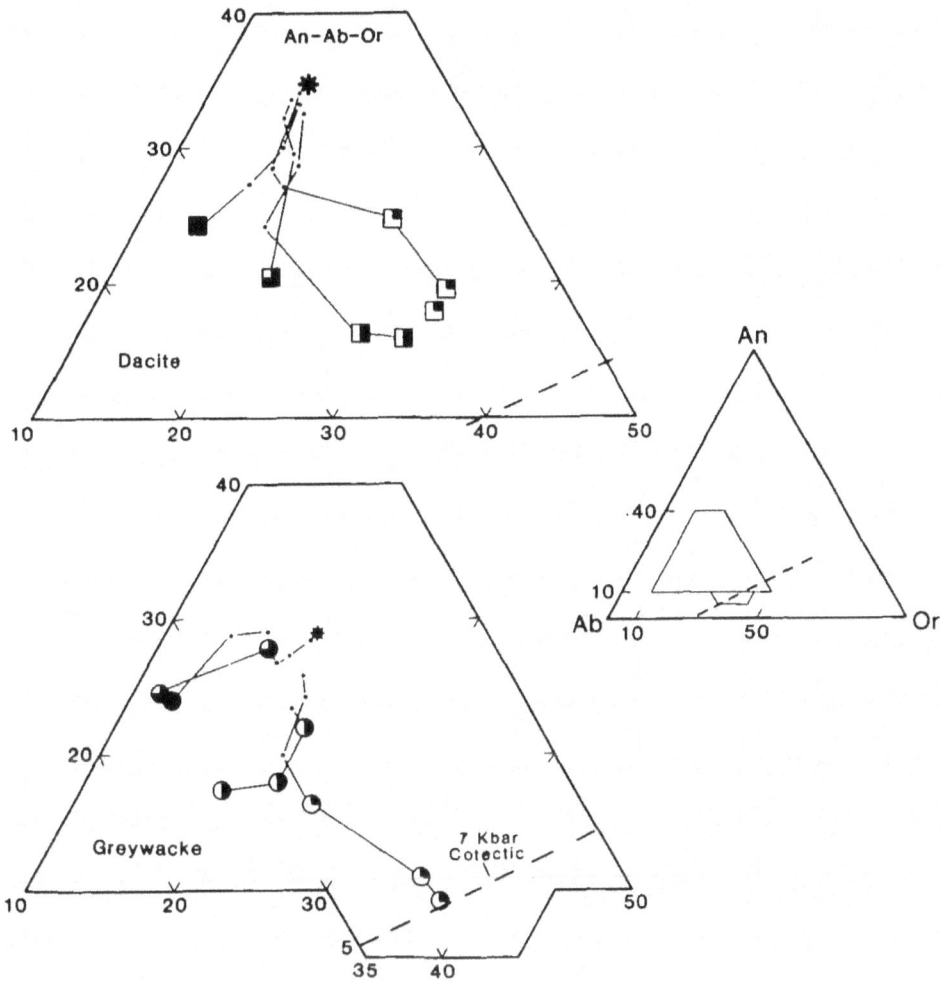

FIG. 11. Analyzed glass compositions throughout the melting range in the greywacke and dacite bulk compositions projected in terms of CIPW normative albite (Ab), anorthite (An) and orthoclase (Or). Symbols as in Fig. 10
Location of cotectic taken from Winkler (1979).

Clearly, increase in aH_2O (and melt water content) leads to preferential solution of the
components of plagioclase, especially An. Melts in equilibrium with both plagioclase and
alkali feldspar should therefore be relatively enriched in plagioclase components (especially
An) at high aH_2O and in alkali feldspar components (especially Or) at low aH_2O. In residues
of partial melting, the relative stability of alkali feldspar should be increased at high aH_2O
and that of plagioclase at low aH_2O.

With the exception of that in the dacite assemblage at $X = 0.25$, 825 °C, all melts studied
here are undersaturated in alkali feldspar. Hence it is difficult to make direct comparisons
with behaviour in model systems such as Q–An–Ab–Or–H_2O. Nevertheless, the biotite
breakdown reactions and T–X patterns of biotite stability discussed above (broadly biotite
breakdown, releasing Or component to melt, suppressed by increasing aH_2O) are also
consistent with a process in which increasing aH_2O increases the solubility of An and Ab in
melts relative to that of Or. The additional importance of biotite in retaining Or component

is particularly clear for greywacke assemblages at high aH_2O, in which biotite is abundant and melt K_2O contents are actually lower than in the bulk composition.

In systems of quartz and feldspar components, the relative stability of quartz, and its solubility in eutectic melts, are also sensitive to change in P_{Total} and aH_2O. The 10 kb data of Burnham (1981) indicate that increase in aH_2O tends to increase the normative Qz contents of eutectic melts. However, the glass compositions of the present study (Table 9) suggest that the increased solubility of An component at high aH_2O is dominant. The lowest SiO_2 contents of melts saturated in quartz and plagioclase (70–71 wt.%, on an anhydrous basis) are associated with the highest normative An contents at high aH_2O (1·0 and 0·75). Silica contents of melts nearer the solidus at low aH_2O are slightly higher (71–73 wt.%), but they decrease at higher temperatures and degrees of melting as levels of dissolved refractory components increase and limits of quartz saturation are exceeded.

Variation in alumina-saturation of silicate melts

Over the range of water activities studied, melts saturated with plagioclase and quartz are all peraluminous for both the dacite and greywacke. The relative degrees of Al-saturation of the two compositions are preserved (Figs. 1 and 12), with melts of the dacite having 0·2–1·6% normative C compared with 1·6–2·3% for the greywacke. The coexisting mafic assemblages include hornblende ± biotite in the dacite as opposed to garnet, gedrite, tschermakitic hornblende and biotite in the greywacke.

In earlier experiments at 5 kb, $X = 1·0$ and 0·75, 700–850 °C, Nicholls (1983) produced plagioclase ± quartz-bearing assemblages with low-Al hornblende ± Ca-clinopyroxene ± biotite in the dacite, and gedrite, biotite ± aluminous (~ 6 wt.% Al_2O_3) orthopyroxene in the greywacke. These assemblages were comparable in all but the absence of garnet in the greywacke to those formed at 10 kb, but their fine grainsize allowed reliable glass analyses to be obtained only at higher degrees of melting. In experiments at 3 kb, $X = 1·0$ on TVZ basement rock compositions, Reid (1982) crystallized tschermakitic hornblende (see Fig. 5) in a greywacke and cordierite in an argillite composition (Table 1). The six glass analyses derived from this study are all peraluminous, but they vary widely in normative C (up to 5%), and their reliability is difficult to assess.

The only Di-normative melts produced in this study are those in the dacite at low aH_2O (0·50 and 0·25), high temperatures (900–975 °C) and high melt proportions (> 75 wt.%; see Tables A1, 9). Melts in the dacite at $X = 0·25$ become metaluminous only at temperatures 50–75 °C above the incoming of Ca-clinopyroxene, which unexpectedly becomes progressively more aluminous (4·5–5·5% Al_2O_3; see Table 5) with increasing temperature.

Available data therefore confirm that, although peraluminous silica-rich melts may coexist with assemblages including metaluminous minerals (especially at high aH_2O and low temperatures), these melts are less strongly C-normative than those which coexist with only peraluminous minerals. Highly peraluminous minerals with limited Al solid solution, such as garnet or gedrite, buffer melts at highly peraluminous compositions, as demonstrated for assemblages including Al-silicates (Kilinc, 1969; Green, 1976; Clemens & Wall, 1981). The stoichiometries of biotites and amphiboles allow wide variation in ratios of Al to Mg, Fe and Ca, and these minerals can therefore coexist with melts ranging widely in degree of Al-saturation. Biotites buffer melts which also coexist with plagioclase and quartz at moderately to strongly peraluminous compositions, whereas common hornblendes, such as the pargasite-tschermakite of the experimental dacite, buffer at weakly to moderately peraluminous compositions (cf. conclusions from some previous experimental investi-

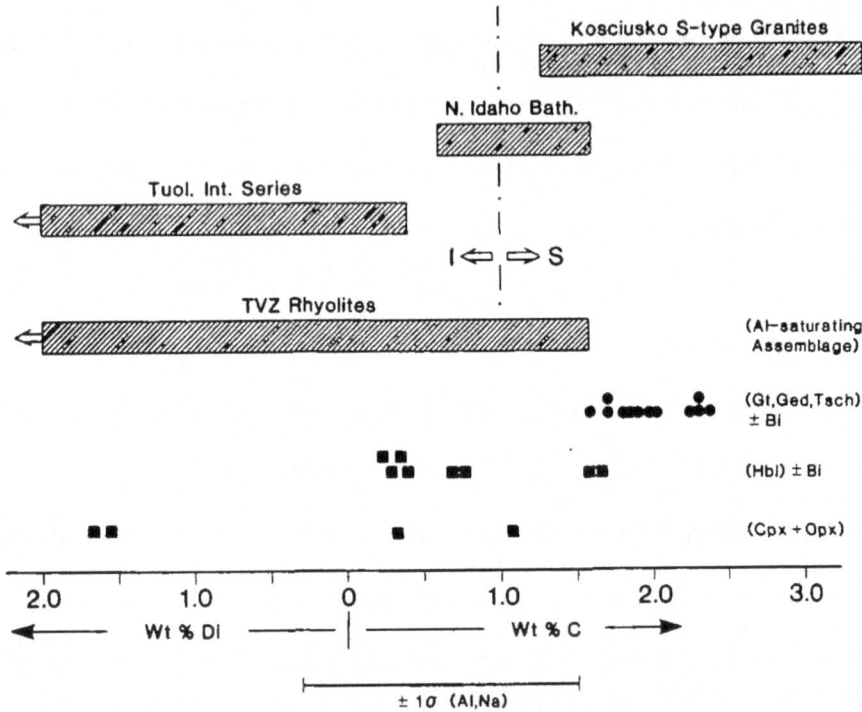

FIG. 12. Wt.% normative corundum (C) and diopside (Di) for the experimental melts and comparative ranges from natural granitoid suites. Fe–Mg–Al silicate phases which appear to define a level of Al-saturation in the assemblages and coexisting melts are indicated in parentheses. Circles are for greywacke melts in equilibrium with garnet, gedrite, or tschermakitic hornblende + quartz ± plagioclase ± biotite. Squares (upper row) are for dacite melts saturated with hornblende + plagioclase + quartz ± biotite. Lower row of squares indicates a transition at $X_{H_2O}^{m.p} \approx 0.25$ from peraluminous to metaluminous melts in equilibrium with clinopyroxene + orthopyroxene + plagioclase ± quartz. Field for N. Idaho Batholith taken from Hyndman (1984). The TVZ rhyolites have both Di- and C-normative compositions typical of the progressively fractionated Tuolumne Intrusive series, an I-type suite (Bateman & Chappell, 1979). Near-solidus dacite partial melts cluster between 0·2 and 1·6 wt.% C, spanning the upper range of normative C in TVZ rhyolites. Greywacke partial melts have uniformly higher normative C appropriate for classic S-type granitoids in the Kosciosko Batholith (Hine *et al.*, 1978).

gations—Helz, 1973, 1976; see also Zen, 1986). The relatively few data for assemblages including Ca-clinopyroxene indicate that restricted Al solid solution at crustal pressures and temperatures allows it to buffer coexisting silica-rich liquids only at metaluminous to weakly peraluminous compositions.

Results of experiments in the system CMASH closely parallel those described above. Di-normative compositions melt eutectically at 1 atm. to form metaluminous liquids, but with increasing PH_2O they begin to melt peritectically to form peraluminous liquids in equilibrium with hornblende (Kushiro & Yoder, 1972; Ellis & Thompson, 1986). Liquids in equilibrium with Al-rich Ca-clinopyroxene may be peraluminous at high PH_2O (e.g., Kushiro & Yoder, 1972), but at reduced aH_2O (2 wt.% H_2O in the charge), Ellis & Thompson (1986) produced metaluminous liquids in equilibrium with Ca clinopyroxene + orthopyroxene ± plagioclase at 5 and 10 kb.

Concentrations of refractory components in melts

Peraluminous melts produced by low to moderate degrees of melting of both the dacite and greywacke show significant increases in concentrations of Fe, Mg, Mn, Ti, and P with increasing temperature (cf. Miller *et al.*, 1985). Melts coexisting with plagioclase and quartz to 60–70% melting contain 2·3–3·7 wt.% normative (Hy + Il + Ap) at $X \geq 0.75$, 675–700 °C, 2·2–5·4% at $X = 0.50$, 725–775 °C and 3·9–7·3% at $X = 0.25$, 825–900 °C. At higher temperatures, levels of these components in high percentage melts approach those of the bulk compositions e.g., at $X = 0.25$, 70–85% melts in equilibrium with plagioclase, but not quartz, contain 9·2–11·0% normative (Hy + Il + Ap ± Di). The corresponding oxide concentrations are similar to those of many natural granitic compositions—FeO 3·4–3·9, MgO 0·9–1·1, TiO_2 0·6–0·7, P_2O_5 >0·1 wt.%.

COMPARISON OF MELT COMPOSITIONS WITH TVZ RHYOLITES AND OTHER GRANITOIDS

Comparison with and origin of Taupo Volcanic Zone rhyolites

In terms of normative salic components, lower percentage melts of both the dacite and greywacke at $X \leq 0.50$ are generally similar to averages of Taupo Zone rhyolites (Table 10). On this basis, deep crustal equivalents of either composition could be possible sources for rhyolite magmas. Comparisons of bulk compositions of TVZ rhyolites and basement greywackes, with emphasis on Na_2O/K_2O ratios, led Ewart (1969) and others to conclude that the greywackes were a suitable source for the rhyolite magmas, but this reasoning could also apply to equivalents of exposed andesites and dacites. Equivalence in terms of relative proportions of salic components between possible igneous and sedimentary source compositions simply reflects the oceanic arc provenance of the Western Basement greywackes, and their large component of volcanic lithic fragments (Table 1).

The TVZ rhyolites are generally of 'I-type' character. Normative C contents in the data set of Ewart (1969) reach 1·4%, slightly above the I/S boundary of Chappell & White (1974), but within the range of melts of the dacite (Fig. 12). From the data summarized in Fig. 12, and those of other experimental studies discussed above (Reid, 1982; Nicholls, 1983), we conclude that metaluminous to mildly peraluminous magmas represented by the TVZ rhyolites are unlikely to have equilibrated with strongly peraluminous residual phases such as garnet, cordierite and gedrite observed within the melting intervals of moderately to strongly peraluminous source compositions (Table 1) over a wide range of pressures, temperatures and water activities applicable to the crust. Therefore, equivalents of exposed TVZ basement sedimentary rocks do not appear to be suitable sources for all but a few members of the TVZ rhyolite suite (which may in fact represent differentiated magmas).

Comparisons with peraluminous granitoids

Melts in the dacite and greywacke are in equilibrium with mineral assemblages typical of the upper amphibolite to granulite facies in metaluminous and peraluminous compositions. Near-solidus melts saturated with plagioclase, quartz and one or more mafic minerals thus provide information on the composition of 'primary' crustal melts (cf. Presnall & Bateman, 1973; Wyllie *et al.*, 1976).

At $X = 1.0$ and 0·75, these melts have 70–71% SiO_2 and >17 wt.% Al_2O_3, reflecting high normative An contents. Those nearest the solidus are notably poor in potassium, with 2·1–2·4% and 1·3–1·9% K_2O in the dacite and greywacke respectively. In terms of normative An, Ab, and Or, these melts range from high-Al trondhjemites to tonalites (Barker, 1979).

TABLE 10

Comparative TVZ rhyolite and other granitoid compositions

	1	2	3	4	5	6	7	8	9
SiO_2	70·87	69·20	71·50	72·49	70·0 –73·0	72·5	73·73	71·65	67·32
TiO_2	0·23	0·25	0·27	0·14	0·18– 0·27	0·32	0·29	0·24	0·55
Al_2O_3	17·10	16·90	15·82	15·30	15·2 –16·1	13·62	13·30	14·87	15·68
FeO*	1·18	1·80	1·41	0·84	1·1 – 1·5	2·54	1·64	1·57	3·10
MnO	0·03	0·07	0·02	0·09	—	0·07	0·06	0·04	0·06
MgO	0·65	0·71	0·47	0·32	0·12– 0·5	0·49	0·36	0·38	1·10
CaO	2·78	2·93	2·48	1·99	1·5 – 2·7	1·95	1·67	1·87	3·59
Na_2O	5·35	5·10	5·29	3·85	4·0 – 4·8	4·41	4·28	3·98	4·17
K_2O	0·76	1·87	2·12	4·31	2·6 – 4·2	2·93	3·17	4·19	3·22
P_2O_5	0·05	0·14	—	—	—	0·08	0·06	0·08	0·19
LOI†	0·68	0·69	0·58	0·30	—	—	—	1·14	0·79
Total	99·68	99·66	99·96	99·63	—	98·91	98·56	100·01	99·77
Wt.% C.I.P.W. normative components									
Q	29·2	24·1	25·7	28·2	—	29·1	31·6	27·0	20·2
C	2·5	1·5	0·3	0·7	0·6 – 1·6	—	—	0·6	—
Or	4·5	11·0	12·4	25·5	—	17·3	18·7	24·8	19·0
Ab	4·3	43·2	44·8	32·6	—	37·3	36·2	33·7	35·3
An	13·5	13·6	12·3	9·9	—	8·7	7·7	8·8	14·6
Di	—	—	—	—	—	0·4	0·2	—	1·7

(1) Leuco-trondhjemitic, biotite–quartz–plagioclase gneiss in migmatite (Misch, 1968).
(2) *In situ* migmatite leucosome (Winkler & Breitbart, 1978).
(3) Two-mica granodiorite (Anderson & Rowley, 1981).
(4) Garnet–two-mica granodiorite (Anderson & Rowley, 1981).
(5) Range of 12 analyses, Northern Idaho Batholith (Hyndman, 1984).
(6) Average of 18 TVZ rhyolite analyses (Cole, 1979).
(7) Average TVZ rhyolite of Reid (1983).
(8) 'I-type granite'—Johnson granite porphyry, Tuolumne Intrusive Series (Bateman & Chappell, 1979).
(9) 'I-type mafic granodiorite'—Half-dome granodiorite, Tuolumne Intrusive Series (Bateman & Chappell, 1979).
    * FeO = total Fe as FeO
    † LOI = reported loss on ignition, total H_2O and CO_2

Kilinc (1969, 1972) produced normatively similar melts of a natural greywacke reacted at 2–8 kb, 700–750 °C, with KCl–NaCl solutions.

Generation of near-solidus melts relatively depleted in potassium has particular significance for an important group of migmatites in metasedimentary rocks referred to as 'trondhjemitoid' by Ashworth (1985). Firstly, the present experiments demonstrate that quartz–plagioclase–biotite assemblages can melt at similarly low temperatures to those with alkali feldspar (see Solidus Relations, above, and cf. Hoschek, 1976; Johannes, 1985). Secondly, under some conditions (e.g., in the presence of residual biotite in the greywacke at high aH_2O) melts may be less potassic than the parent composition, with K_2O contents increasing as biotite disappears with progressive melting. These relationships appear to invalidate the arguments of Misch (1968) and Yardley (1978) that trondhjemitic migmatites (see Table 10) less potassic than their host rocks could not be anatectic.

With decreasing aH_2O, quartz- and plagioclase-saturated melt compositions move toward higher SiO_2 (71–73%) and lower Al_2O_3 (~16 and 15% at $X = 0·50$ and 0·25 respectively), reflecting increasing Or and decreasing An contents. At $X \leq 0·50$, melts of the dacite have sufficient Or to be classed as granites. In the greywacke, the temperature stability of biotite decreases from $X = 0·50$ to 0·25, with lower temperature melts having higher K_2O

contents, and the melt composition range is from sodic granodiorite to granite (Table 9). These pronounced changes in melt compositions demonstrate clearly that source chemistry is only one of several factors which may influence the nature of primary granitic melts of quartzo-feldspathic material.

Clemens & Wall (1981), in melting experiments with natural and model granitic compositions at 5 kb and over a wide aH_2O range, produced glasses with SiO_2 contents similar to those of the present study. Their data therefore support the conclusion that primary granitic liquids *saturated with quartz* in their source regions should have ~ 70–73% SiO_2. Granitic rocks with lower SiO_2 contents (forming the bulk of many major batholiths) could then represent liquids derived from sources previously depleted in quartz, magmas carrying significant mafic restite or cumulate components, or the products of mixing between acid and mafic–intermediate magmas. Leucocratic granites with SiO_2 contents greater than about 73% SiO_2 could most simply represent magmas affected by crystal fractionation.

Granites and granodiorites in the northern Idaho Batholith (Hyndman, 1983, 1984) were emplaced at mid-crustal levels (15–20 km; $P \simeq 5$ kb) and appear to have neither moved far from the site of magma generation nor undergone extensive crystal fractionation. These granitoids show similarities to melts in both the dacite and greywacke at $X = 0.5$, 725–750 °C (Fig. 12, Table 10). Hyndman emphasized the transitional I/S-type compositional and mineralogical affinities of the Idaho Batholith, attributing these characteristics to derivation of magmas from a 'semi-pelitic, quartzofeldspathic metasedimentary' source, or a 'micaceous quartzofeldspathic orthogneiss' (Hyndman, 1983, p. 98). The present experiments, however, suggest that the Idaho Batholith magmas could represent melts of quartzofeldspathic sources lacking a major pelitic (i.e. aluminous) component, originating at relatively high water activities and low temperatures.

Anderson & Rowley (1981) describe mildly peraluminous granites and sodic granodiorites from the Whipple Mountains, California, associated with regional metamorphism and deformation at pressures around 3 kb (Table 10). These granitoids are similar in composition to near-solidus melts of the dacite over a wide range of water activities, and they could therefore have been derived from the melting of metaluminous sources (see also White et al., 1986).

At lower aH_2O ($X = 0.25$, 825–850 °C), near-solidus melts of the dacite are good analogues for mildly peraluminous granites in predominantly metaluminous I-type suites such as the Tuolumne intrusive series (Bateman & Chappell, 1979; see Table 10). Significantly, the experimental melts have comparable amounts of iron (1.8–2.1% total FeO), magnesium (0.3–0.5% MgO) and minor elements (0.05% MnO, 0.18–0.29% TiO_2, 0.10% P_2O_5). Granites such as those of the Tuolumne series could therefore represent 'primary' melts of metaluminous sources, or alternatively derivatives of more mafic magmas which evolved to produce mineral assemblages and water contents similar to those of the experimental melts (cf. Cawthorn & Brown, 1976; Green, 1978).

Comparisons with metaluminous granitoids

The Ca-clinopyroxene–orthopyroxene–plagioclase assemblage which coexists with met-aluminous melts in the dacite is typical of near-liquidus assemblages of basaltic–andesitic compositions at low aH_2O and high temperatures. These compositions may crystallize hornblende (and sometimes quartz) nearer the solidus (e.g., Eggler, 1972; Allen & Boettcher, 1978; Green, 1978). The experimental metaluminous melts have low SiO_2 (68–69%) and high concentrations of mafic and refractory components (see Glass compositions, above), similar to I-type granite compositions (Table 10, and cf. Hine *et al.*, 1978; Bateman & Chappell, 1979). The metaluminous and refractory nature of many felsic to mafic granodiorite magmas (if they are of crustal origin—see Wyllie *et al.* (1976)) therefore need not reflect the presence of a significant metaluminous restite component, but may simply be the consequence of melting of mafic–intermediate source rocks at low water activities and high temperatures.

On the other hand, the experimental metaluminous melts differ from many natural granites of similar SiO_2 content in their lower K_2O contents, which reflect high degrees of melting of the relatively K-poor dacite composition. Clearly, more potassic metaluminous melts could be produced from more potassic sources under similar conditions of melting (provided biotite was absent from the residues). Alternatively, high K_2O contents and K_2O/Na_2O ratios might be expected for smaller percentage melts of more mafic source compositions. This would be the case particularly if residual amphibole persisted to temperatures above the biotite stability limit at low aH_2O, thus causing K to behave much less compatibly than Na.

CONCLUSIONS FOR GRANITOID PETROGENESIS

(1) From O, Sr, and Nd isotopic studies, Perfit *et al.* (1981) and Blattner & Reid (1982) have proposed that magmas parental to basalts, andesites and rhyolites in the TVZ are derived from below the exposed Mesozoic greywacke–argillite sedimentary rocks and contaminated by interaction with these *en route* to the surface. The experimental results presented here are consistent with an origin for the mainly Di-normative TVZ rhyolite magmas from either melting of more strongly Di-normative lower crust or fractionation of arc-related mafic–intermediate magmas, in either case with the attendant possibility of contamination of magmas by shallower crustal material. The experimental data alone cannot distinguish between these two modes of origin. They are, however, not consistent with an origin mainly involving melting of the dominantly C-normative greywacke–argillite sedimentary pile.

(2) Incongruent melting reactions which consume biotite and produce amphibole (in the absence of alkali-feldspar) are dependent upon temperature and aH_2O and greatly influence the compositions of partial melts near the solidus. Granitoids with proportions of salic components deviating substantially from the ternary minimum or cotectic trough in the model systems Q–Ab–Or–H_2O or Q–An–Ab–Or–H_2O (as defined by Winkler & Breitbart, 1978; Winkler, 1979; see however, Johannes, 1985), such as trondhjemites, need not have accumulated plagioclase or quartz, or have been derived from extremely K_2O-depleted bulk compositions.

(3) The experimental melts nearer the solidi for both the metaluminous and peraluminous source compositions (Table 9) are similar in important respects (especially normative salic components and normative C, and also Fe and Mg contents) to bulk compositions in a range of peraluminous amphibolite facies migmatites, granodiorites and granites (Table 10). Their ranges of normative C contents are primarily determined by the aAl_2O_3 of the associated solid assemblages, temperature and aH_2O (see also Clemens & Wall, 1981), and they correspond closely to those proposed by Chappell & White (1974) as reflecting the contrasted source compositions and mineralogies of I- and S-type granites.

(4) An element of the restite model of White & Chappell (1977) which is difficult to reconcile with the experimental results is the requirement that high silica (76 wt.% SiO_2) 'minimum melts' form by anatexis under low temperature hydrous conditions within the biotite and/or hornblende stability fields. In fact, the present results show that near-solidus

melting in biotite and hornblende-bearing assemblages produces higher-silica melts only at low aH_2O (with normative An in the melts at a minimum and Or at a maximum), and hence at temperatures ($> 800°C$) sometimes considered to be too high for realistic crustal melting conditions. Within the 5–10 kb pressure range most typical of crustal anatexis, experimental melts of natural source compositions (this study; Clemens & Wall, 1981), even when saturated with quartz, have significantly less silica (70–73 wt.% SiO_2) than 'minimum melt' compositions observed in hydrous model systems of salic components. This evidence indicates that high silica granite compositions present in many granitic suites, which may be used to constrain the 'minimum melt' component for purposes of modelling of magma evolution by restite unmixing (see Compston & Chappell, 1979), may actually represent fractionated members of these suites, with SiO_2 contents higher than those of any primary partial melts of the associated magma source. This could lead to overestimation of the proportions of any restite component present in members of the suite.

(5) Experimental results for the dacite source show that Di-normative partial melts may form in equilibrium with granulite facies assemblages at low aH_2O and temperatures in excess of 900°C, as recently suggested by Ellis & Thompson (1986) from work on model systems. Melt compositions with the high FeO and MgO contents typical of mafic granodiorites, quartz diorites, and tonalites have been observed in the present study only at extremely high degrees of melting. The presence of restite components from mafic–intermediate sources (White & Chappell, 1977) or admixtures of mafic magmas (Reid *et al.*, 1983), have been invoked to account for the relatively high contents of refractory components in the more mafic members of granitic suites. However, some suites, especially in large calc-alkaline batholiths within orogenic belts and in other tectonic settings, probably originated with very low initial water contents (< 1–2 wt.%) at temperatures approaching those of basaltic magmas (e.g., Wyllie *et al.*, 1976). Further experiments, using more mafic source compositions under lower water activities ($aH_2O < 0.25$) and at higher temperatures ($> 950°C$) than those used in the present study, are required to assess whether highly mafic granitoids can also represent compositions which were once largely liquid.

ACKNOWLEDGEMENTS

Frank Reid kindly provided the data used in determining the experimental average Kawerau greywacke composition, and also a draft manuscript describing greywacke–argillite melting experiments at 3 kb PH_2O. A. Hohmann, R. Douglass, L. Jones, W. Manley, and R. Clarke provided analytical and technical assistance. D. Gelt drafted the figures, and P. Hermansen and J. Muir processed the manuscript. E. Mikucki, T. Oba, and T. Sekine are thanked for helpful discussions concerning granites in general and the specifics of experimental techniques. We thank M. Pichavant, T. H. Green, and W. Johannes for helpful comments and constructive criticism of the manuscript. W. K. C. gratefully acknowledges the support of a Monash University Vice-Chancellors Post-doctoral Fellowship.

REFERENCES

Allen, J. C., & Boettcher, A. L., 1978. Amphiboles in andesite and basalt: II. Stability as a function of P–T–f_{H_2O}–f_{O_2}. *Am. Miner.* **63**, 1074–87.

Anderson, J. L., & Rowley, M. C., 1981. Synkinematic intrusion of peraluminous and associated metaluminous granitic magmas, Whipple Mountains, California. *Can. Miner.* **19**, 83–101.

Arai, S., & Hirai, H., 1985. Compositional variation of calcic amphiboles in Mineoke metabasites, Japan, and its bearing on the actinolite–hornblende miscibility relationship. *Lithos* **18**, 187–99.

Ashworth, J. R., 1985. Introduction. In: Ashworth, J. R. (ed.) *Migmatites*. London: Blackie & Son Ltd., 1–35.

Barker, F., 1979. Trondhjemite: definition, environment and hypotheses of origin. In: Barker, F. (ed.) *Trondhjemites, Dacites and Related Rocks.* New York: Elsevier, 1–12.

Bateman, P. C., & Chappell, B. W., 1979. Crystallization, fractionation and solidification of the Tuolumne Intrusive Series, Yosemite National Park, California. *Bull. geol. Soc. Am.* **80**, 465–82.

Blattner, P., & Reid, F., 1982. The origin of lavas and ignimbrites of the Taupo Volcanic Zone, New Zealand, in the light of oxygen isotope data. *Geochim. cosmochim. Acta* **46**, 1417–29.

Boettcher, A. L., Guo, O., Bohlen, S. R., & Hanson, B., 1984. Melting in feldspar-bearing systems to high pressure and the structure of aluminosilicate liquids. *Geology* **12**, 202–4.

Bohlen, S. R., Peacor, D. R., & Essene, E. J., 1980. Crystal chemistry of a metamorphic biotite and its significance in water barometry. *Am. Miner.* **65**, 55–62.

Borom, M. P., & Hanneman, R. E., 1967. Local compositional changes in alkali silicate glasses during electron microprobe analysis. *J. appl. Phys.* **38**, 2407–9.

Brown, G. C., & Fyfe, W. S., 1970. The production of granitic melts during ultrametamorphism. *Contr. Miner. Petrol.* **28**, 310–8.

Burnham, C. W., 1979. The importance of volatile constituents. In: Yoder, H. S. (ed.) *The Evolution of the Igneous Rocks: Fiftieth Anniversary Perspectives.* Princeton, New Jersey: Princeton University Press, 439–82.

——1981. The nature of multicomponent aluminosilicate melts. In: Richards, D. T., & Wickham, F. Z. (ed.) *Chemistry and Geochemistry of Solutions at High Temperature and Pressure, Physics and Chemistry of the Earth.* New York: Pergamon Press, 197–229.

Cameron, K. L., 1975. An experimental study of actinolite-cummingtonite phase relations with notes on the synthesis of Fe-rich anthophyllite. *Am. Miner.* **60**, 375–90.

Cawthorn, R. G., & Brown, P. A., 1976. A model for the formation and crystallization of corundum-normative calc-alkaline magmas through amphibole fractionation. *J. Geol.* **84**, 467–76.

Chappell, B. W., & White, A. J. R., 1974. Two contrasting granite types. *Pacific Geol.* **8**, 173–4.

Clemens, J. D., & Wall, V. J., 1981. Origin and crystallization of some peraluminous (S-type) granitic magmas. *Can. Miner.* **19**, 111–31.

Cole, J. W., 1979 Structure, petrology, and genesis of Cenozoic volcanism, Taupo Volcanic Zone, New Zealand-a review. *N.Z. J. Geol, Geophys.* **22**, 631–57.

Compston, W., & Chappell, B. W., 1979. Sr-isotope evolution of granitoid source rocks. In: McElhinny, M. W. (ed.) *The Earth, its Origin, Structure and Evolution.* Academic Press: London, 377–426.

Conrad, W. K., & Kay, R. W., 1984. Ultramafic and mafic inclusions from Adak Island: crystallization history, and implications for the nature of primary magmas and crustal evolution in the Aleutian Arc. *J. Petrology* **25**, 88–125.

Czamanske, G. K., Ishihara, S., & Atkin, S. A., 1981. Chemistry of rock-forming minerals of the Cretaceous-Paleocene Batholith in Southwestern Japan and implications for magma genesis. *J. geophys. Res.* **86**, 10431–69.

Deer, W. A., Howie, R. A., & Zussman, J., 1972. *An Introduction to the Rock-forming Minerals.* London: Longman Group Ltd.

Edgar, A. D., 1973. *Experimental Petrology: Basic Principles and Techniques.* Oxford: Clarendon Press.

Eggler, D. H., 1972. Water-saturated and under-saturated melting relations in a Paricutin andesite and an estimate of water content in the natural magma. *Contr. Miner. Petrol.* **34**, 261–71.

Ellis, D. J., & Thompson, A. B., 1986. Subsolidus and partial melting reactions in the quartz-excess $CaO+MgO+Al_2O_3+SiO_2+H_2O$ system under water-excess and water deficient conditions to 10 kb: some implications for the origin of peraluminous melts from mafic rocks. *J. Petrology* **27**, 91–121.

Ewart, A., 1969. Petrochemistry and feldspar crystallization in the silicic volcanic rocks, central North Island, New Zealand. *Lithos* **2**, 371–88.

——Stipp, J. J., 1968. Petrogenesis of the volcanic rocks of the Central North Island, New Zealand, as indicated by a study of $^{87}Sr/^{86}Sr$ ratios, and Sr, Rb, K, U and Th abundances. *Geochim. cosmochim. Acta* **32**, 699–736.

Gilbert, M. C., Heltz, R. T., Popp, R. K., & Spear, F. S., 1982. Experimental studies of amphibole stability. In: Veblen, D. R., & Ribbe, P. H. (eds.) *Reviews in Mineralogy, volume 9b, Amphiboles: Petrology and Experimental Phase Relations.* Washington, D. C.: Mineral Soc. Am. 229–353.

Green, T. H., 1976. Experimental generation of cordierite- or garnet-bearing granitic liquids from a pelitic composition. *Geology* **4**, 85–8.

——1977. Garnet in silicic liquids and its possible use as a *P-T* indicator. *Contr. Miner. Petrol.* **65**, 59–67.

——1978. A model for the formation and crystallization of corundum-normative calc-alkaline magmas through amphibole fractionation: a discussion. *J. Geol.* **86**, 269–72.

——1981. Synthetic high-pressure micas compositionally intermediate between the dioctahedral and trioctahedral mica series. *Contr. Miner. Petrol.* **78**, 452–8.

——1982. Anatexis of mafic crust and high pressure crystallization of andesite. In: Thorpe, R. S. (ed.) *Andesites: Orogenic Andesites and Related Rocks.* Chichester: Wiley, 465–87.

——Ringwood, A. E., 1968. Genesis of the calc-alkaline igneous rock suite. *Contr. Miner. Petrol.* **18**, 105–62.

Grove, T. L., Gerlach, D. C., & Sando, T. W., 1982. Origin of calc-alkaline series lavas at Medicine Lake volcano by fractionation, assimilation and mixing. *Contr. Miner. Petrol.* **80**, 160–82.

Gust, D. A., & Hibberson, W., 1979. Temperature calibration of the 5/32" graphite furnace assembly. A. N. U. Res. School of Earth Sciences Ann. Rept., 126–8.

Hawthorne, F. C., Griep, J. L., & Curtis, L., 1980. A three amphibole assemblage from the Tallan Lake Sill, Peterborough County, Ontario. *Can. Miner.* **18**, 175–84.

Helz, R. T., 1973. Phase relations of basalts in their melting range at $PH_2O = 5$ kb as a function of oxygen fugacity. Part I. Mafic Phases. *J. Petrology* **14**, 249–302.

—— 1976. Phase relations of basalts in their melting ranges at $PH_2O = 5$ kb. Part II. Melt compositions. *Ibid* **17**, 139–93.

Hine, R., Williams, I. S., Chappell, B. W., & White, A. J. R., 1978. Contrasts between I- and S-type granitoids of the Kosciosko batholith. *J. geol Soc. Aust.* **25**, 219–34.

Holloway, J. R., & Burnham, C. W., 1972. Melting relations of basalt with equilibrium water pressure less than total pressure. *J. Petrology* **13**, 1–29.

Hoschek, G., 1976. Melting relations of biotite + plagioclase + quartz. *N. Jb. Miner. Mh.* **1976**, 79–83.

Huang, W. L., & Wyllie, P. J., 1981. Phase relationships of an S-type granite with H_2O to 35 kilobars: muscovite granite from Harney Peak, South Dakota, *J. geophys. Res.* **86**, 10515–29.

Hyndman, D. W., 1983. The Idaho Batholith and associated plutons, Idaho and western Montana. In: Roddick J. (ed.) *Circum-Pacific Plutonic Terrains.* Geol. Soc. Amer. Mem. **159**, 213–40.

—— 1984. A petrographic and chemical section through the northern Idaho Batholith. *J. Geol.* **92**, 83–102.

James, R. S., & Hamilton, D. L., 1969. Phase relations in the system $NaAlSi_3O_8$–$KAlSi_3O_8$–$CaAl_2Si_2O_8$–SiO_2 at 1 kilobar water vapour pressure. *Contr. Miner. Petrol.* **21**, 111–41.

Johannes, W., 1980. Metastable melting in the granite system Qz–Or–Ab–An–H_2O. *Ibid.* **72**, 73–80.

—— 1985. The significance of experimental studies for the formation of migmatites. In: Ashworth, J. R. (ed.) *Migmatites* London: Blackie & Son Ltd., 36–85.

Kilinc, I. A., 1969. Experimental metamorphism and anatexis of shales and greywackes. Pennsylvania State Univ. Ph.D. thesis, University Park, PA. 181 pp

—— 1972. Experimental study of partial melting of crustal rocks and formation of migmatites (abstract). *Intl. Geol Cong.*, Sec. 2, 109–13.

Kushiro, I., & Yoder, H. S., 1972. Origin of calc-alkaline peraluminous andesite and dacite. *Yb. Carnegie Inst., Wash.* **71**, 411–3.

Leake, B. E., 1978. Nomenclature of amphiboles. *Am. Mineral.* **63**, 1023–53.

Lindsley, D. H., 1983. Pyroxene thermometry. *Ibid.* **68**, 477–93.

Luth, W. C., Jahns, R. H., & Tuttle, O. F., 1964. The granite system at pressures of 4 to 10 kilobars. *J geophys. Res.* **69**, 759–73.

Maaloe, S., & Wyllie, P. J., 1975. Water content of a granite magma deduced from the sequence of crystallization determined experimentally with water-undersaturated conditions. *Contr. Miner. Petrol.* **52**, 175–91.

Mielke, P., & Winkler, H. G. F., 1979. Eine bessere Berechnung der Mesonorm für granitische gesteine. *N. Jb. Miner. Mh.* **1979**, 471–80.

Miller, C. F., Watson, E. B., & Rapp, R. P., 1985. Experimental investigation of mafic mineral-felsic liquid equilibria: preliminary results and petrogenetic implications (abstract). *Eos* **66**, 1130.

Misch, P., 1968. Plagioclase compositions and non-anatectic origin of migmatitic gneisses in the northern Cascade mountains of Washington State. *Contr. Miner. Petrol.* **17**, 1–70.

Naney, M. T., 1983. Phase equilibria of rock-forming ferromagnesian silicates in granitic systems. *Am. J. Sci.* **283**, 993–1033.

Nicholls, I. A., 1974. A direct fusion method of preparing silicate rock glasses for energy-dispersive electron microprobe analysis. *Chem. Geol.* **14**, 151–7.

—— 1983. Experimental studies on crustal source materials for rhyolitic magmas, Taupo Zone, New Zealand (abstract). *Abstracts, Sixth Australian Geol. Convention, Canberra.* 158–9.

Oba, T., & Nicholls, I. A., 1986. Experimental study of cummingtonite and Ca-Na amphibole in the system Cum–Act–Pl–Qz–H_2O. *Am. Miner.* **71**, 1354–65.

Perfit, M. R., McCulloch, M. T., & Froude, D., 1981. Sr and Nd isotopic variations in volcanic and plutonic rocks from the Aleutian Islands and the Taupo Volcanic Zone, New Zealand: Implications for island arc magma genesis (abstract). *IAVCEI 1981 Symposium: Arc Volcanism* **24**, 292–3.

Piwinski, A. J., 1973. Experimental studies of granitoids from the Central and Southern Coast Ranges, California. *Tschermaks Min. Petr. Mitt.* **20**, 131–54.

Presnall, D. C. & Bateman, P. C., 1973. Fusion relations in the system $NaAlSi_3O_8$–$CaAl_2Si_2O_8$ –$KAlSi_3O_8$–SiO_2–H_2O and generation of granitic magmas in the Sierra Nevada batholith. *Bull. geol. Soc. Am.* **84**, 3181–202.

Reed, S. J. B., 1975. *Electron Microprobe Analysis.* Cambridge University Press.

Reid, F. W., 1982. Geochemistry of Central North Island greywackes and genesis of silicic magmas. Ph.D. thesis, Victoria University, Wellington, New Zealand, 329pp.

—— 1983. Origin of the rhyolitic rocks of the Taupo Volcanic Zone, New Zealand. *J. Volcanol. geotherm. Res.* **15**, 315–38.

Reid, J. B., Evans, O. C., & Fates, D. G., 1983. Magma mixing in granitic rocks of the central Sierra Nevada, California. *Earth planet. Sci. Lett.* **66**, 243–61.

Robinson, P., Spear, F. S., Schumacher, J. C., Laird, J., Klein, C., Evans, B. W., & Doolan, B. L., 1982. Phase relations of metamorphic amphiboles: natural occurrence and theory. In: Veblen, D. R., & Ribbe, P. H. (eds.)

Reviews in Mineralogy, volume 9B, Amphiboles: Petrology and Experimental Phase Relations. Washington, D. C.: Mineral. Soc. Am., 1–227.

Rutherford, M. J., Sigurdsson, H., Carey, S., & Davis, A., 1985. The May 18, 1980, eruption of Mount St. Helens, I. Melt compositions and experimental phase equilibria. *J geophys. Res.* **90**, 2929–47.

Spear, F. S., 1981. An experimental study of hornblende stability and compositional variability in amphibolite. *Am. J. Sci.* **281**, 697–734.

—— Kimball, K. L., 1983. RECAMP—a Fortran IV program for estimating Fe^{3+} contents in amphiboles. *Computers Geosci.* **10**, 317–25.

Speer, J. A., 1984. Micas in igneous rocks. In: Bailey, S. W. (ed.) *Reviews in Mineralogy, volume* 13, *Micas.* Washington, D. C.: Mineral. Soc. Am., 299–355.

Spulber, S. D., & Rutherford, M. J., 1983. The origin of rhyolite and plagiogranite in oceanic crust: an experimental study. *J. Petrology* **24**, 1–25.

Stern, C. R., Huang, W. L., & Wyllie, P. J., 1975. Basalt–andesite–rhyolite–H_2O: crystallization intervals with excess H_2O and H_2O-undersaturated liquidus surfaces to 35 kilobars, with implications for magma genesis. *Earth planet. Sci. Lett.* **28**, 163–7.

—— Wyllie, P. J., 1981. Phase relationships of I-type granite with H_2O to 35 kilobars: the Dinkey Lakes biotite-granite from the Sierra Nevada batholith. *J. geophys. Res.* 10412–22.

Thompson, R. N., 1981. Thermal aspects of the origin of Hebridean Tertiary acid magmas. I. An experimental study of partial fusion of Lewisian gneisses and Torridonian sediments. *Miner. Mag.* **44**, 161–70.

Tracy, R. J., 1978. High grade metamorphic reactions and partial melting in pelitic schists, West-Central Massachussetts. *Am. J. Sci.* **278**, 150–78.

Tuttle, O. F., & Bowen, N. L., 1958. Origin of granite in the light of experimental studies in the system $NaAlSi_3O_8$–$KAlSi_3O_8$–SiO_2–H_2O. *Geol. Soc. Am. Mem.* **74**, 153 pp.

White, A. J. R., 1979. Sources of granite magmas (abstract). *Geol. Soc. Amer. Abstr. Prog.* **11**, 539.

—— Chappell, B. W., 1977. Ultrametamorphism and granitoid genesis. *Tectonophysics* **43**, 7–22.

—— Clemens, J. D., Holloway, J. R., Silver, L. T., Chappell, B. W., & Wall, V. J., 1986. S-type granites and their possible absence in southwestern North America. *Geology* **14**, 115–8.

Winkler, H. G. F., 1979. *Petrogenesis of Metamorphic Rocks. Fifth edition.* New York: Springer Verlag.

—— Breitbart, R., 1978. New aspects of granitic magmas. *N. Jb. Miner. Mh.* **1978**, 463–80.

Wyllie, P. J., Huang, W. L., Stern, C. R., & Maaløe, S., 1976. Granitic magmas: possible and impossible sources, water contents and crystallization sequences. *Can. J. Earth Sci.* **13**, 1007–19.

Yardley, B. W. D., 1978. Genesis of the Skajit Gneiss migmatites, Washington, and the distinction between possible mechanisms of migmatization. *Bull. geol. Soc. Am.* **98**, 941–51.

Zen, E.-A., 1986. Aluminum enrichment in silicate melts by fractional crystallization: some mineralogic and petrographic constraints. *J. Petrology* **27**, 1095–17.

WALTER K. CONRAD ET AL.

APPENDIX

TABLE A1

Calculated modal abundances: dacite

T°C	Run #	Glass	Qtz	Plag	Hbl	Bi (KFs)	Cpx	Opx	Ilm	Σr²	FeO
					$X_{H_2O}^{vap} = 0.25$						
825	624A	25·1	13·6	34·3	18·6	(4·7)	—	3·6	0·3	0·24	—
850	617A	44·5	6·6	31·7	—	—	8·8	8·3	0·2	0·15	—
875	619A	65·6	0·3	18·2	—	—	8·3	7·0	0·0	0·02	—
900	621A	66·8	0·2	18·8	—	—	6·2	8·2	0·0	0·28	—
925	623A	76·5	—	12·0	—	—	4·7	6·7	0·0	0·20	—
950	625A	82·5	—	7·8	—	—	4·9	4·8	—	0·18	—
975	626A1	94·4	—	—	—	—	2·3	3·4	—	0·82	—
					$X_{H_2O}^{vap} = 0.50$						
725	323A	11·0	20·2	44·0	15·1	9·5			0·0	0·84	—
750	321A	28·6	15·4	31·9	20·5	3·6			0·0	0·37	—
775	332A	72·2	—	5·1	22·9	—			0·0	0·27	—
800	320A	78·1	—	1·5	20·3	—			0·1	0·08	—
825	325A	84·0	—	—	15·9	—			0·1	0·11	—
875	324A	85·1	—	—	14·0	—	0·8		—	0·17	—
900	326A	95·4	—	—	—	—	1·2	3·4	—	0·10	—
					$X_{H_2O}^{vap} = 0.75$						
700	310A	48·1	11·9	13·0	25·1	1·9			0·0	0·33	—
725	308A	78·0	—	0·5	20·6	—			0·5	0·39	0·3
800	306A	80·8	—	—	17·7	—			0·2	0·73	1·2
850	309A	87·0		—	10·7	—			0·5	0·14	1·7
					$X_{H_2O}^{vap} = 1.00$						
675	305A*	55·5	9·2	8·9	22·9	3·6			0·0	0·08	—
750	302A	81·0	—	—	18·0	—			0·0	0·34	0·9
800	303A	85·5	—	—	13·2	—			0·0	0·27	1·2

Calculated modal abundances are given in wt.%. Representative mineral-glass compositions used in calculations are given in Tables 2–8. Σr² = sum of squares of residuals for 7 major elements (Si, Al, Fe, Mg, Ca, Na, K). FeO = calculated wt.% FeO lost from charge.

* Quench crystallization makes this calculation highly uncertain, from 16–55 wt.% glass.

TABLE A2

Calculated modal abundances: greywacke

T°C	Run #	Glass	Qtz	Plag	C-amp	O-amp	Bi	Opx	Gt	Ilm	Σr²	FeO
						$X_{H_2O}^{comp} = 0.25$						
825	624B	24.9	6.3	35.3			6.9	4.8	11.8	0.4	0.22	—
850	617B	41.8	0.3	27.5			—	8.3	12.1	0.0	0.23	—
875	619B	66.4	3.0	11.7			—	4.3	14.5	0.0	0.14	—
900	621B	77.4	1.1	3.4			—	1.9	16.2	0.0	0.07	—
925	623B	85.3	—	1.0			—	2.3	10.8	0.1	0.02	0.5
950	625B	91.9	—	—			—	—	7.2	—	0.16	0.8
						$X_{H_2O}^{comp} = 0.50$						
725	323B	12.6	24.1	36.7		5.8	17.6		2.9	0.1	0.01	—
750	321B	23.1	19.7	31.7		7.4	13.6		4.1	0.3	0.07	—
775	330B	67.0	4.4	7.1		11.4	—		5.4	0.0	0.05	—
800	320B	81.9	—	—	4.8	—	—		10.3	0.5	0.11	1.2
825	322B	84.4	—	—	6.2	—	—	3.4	11.4	0.4	0.03	0.5
						$X_{H_2O}^{comp} = 0.75$						
675	311B	16.7	24.1	30.0	10.0	0.0	18.5			0.6	0.08	—
725	308B	76.1	3.7	—	5.8	7.4	5.3		0.0	1.7	0.51	—
750	307B	82.8	—	—	1.7	10.6	2.3		1.7	1.3	0.30	1.2
800	306B	86.4	—	—	1.9	4.7	1.8		0.1	0.8	0.06	2.7
850	309B	91.9	—	—	—	—	—	4.2		0.6	0.26	3.2
						$X_{H_2O}^{comp} = 1.00$						
675	305B	34.0	17.3	21.4	10.3	0.0	16.0			0.7	0.08	—
700	304B	73.7	4.6	—	9.9	—	10.5			1.3	0.13	—
725	301B	84.9	—	—	6.6	0.0	6.4			1.0	0.20	1.0
750	302B	83.4	—	—	6.9	—	5.1		1.7	0.9	0.02	1.9
800	303B	88.8	—	—	—	2.6	1.9		3.7	0.7	0.05	2.1

CHAPTER 18

Vielzeuf, D. and Holloway, J.R. (1988) Experimental determination of the fluid-absent melting reactions in the pelitic system. Consequences for crustal differentiation. *Contributions to Mineralogy and Petrology*, **98**, 257–276.

By the late 1970s and early 1980s it had become apparent that a great many granitic magmas were generated through partial melting of metasedimentary rocks, at granulite-facies conditions. This finding was expressed, e.g. in the works of Chappell and White (1974), which is included in this volume, White and Chappell (1977) and Fyfe (1973a,b). Many lines of evidence pointed to the fact that the magmas concerned were far from saturated with H_2O (e.g. Clemens, 1984) and were most likely to have been generated under fluid-absent conditions (Clemens and Vielzeuf, 1987). Theoretical treatments of fluid-absent partial melting had been given by Eggler (1973), while Thompson (1988) showed how such reactions specifically applied to crustal rocks.

However, Vielzeuf and Holloway (1988) published what is arguably the first and most complete experimental investigation of the fluid-absent melting behaviour of a crustal rock – a Na-enriched metapelite. The composition they chose to work on is not particularly representative of pelites in general. The rock also contained a very small amount of chlorite and some staurolite which, through dehydration, may have caused initial melting to have been fluid-present and to have resulted in the melt proportions being higher than for pure fluid-absent conditions. Also, Conrad *et al.*, Le Breton and Thompson, Pickering and Johnston and Rutter and Wyllie all published fluid-absent partial melting studies in the same year. Nevertheless, we offer Vielzeuf and Holloway as the landmark. The experiments for this work were carried out in 1985, in Holloway's experimental laboratory at Arizona State University, so the publication date is somewhat misleading.

What Vielzeuf and Holloway produced was a sweeping treatment of the subsolidus equilibria, the experiments (with analyses of all solid and melt phases and modal proportions), a consideration of the progression of reactions, models for both fluid-present and fluid-absent partial melting in the pelitic to semipelitic system, what amount to pseudosections for these at $XMg = 0.5$, a generalized T–XH_2O section and a discussion of the applications of the work to the geological evolution of the Hercynian Pyrenees, as well as the possible tectonothermal consequences of thermal buffering by such melting reactions. These ideas led directly to other interesting works such as that of Vielzeuf *et al.* (1990).

The way in which the experimental work was carried out and the treatment of the data represent exemplary models of how to do this kind of work. They have been emulated by many other researchers. Other fine examples include Conrad *et al.* (1988), Rushmer (1991), Patiño Douce and Beard (1995) and Rapp and Watson (1995). The broad scope and the great many spin-off works that the Vielzeuf and Holloway paper has produced justify its inclusion here.

Contrib Mineral Petrol (1988) 98:257–276

Contributions to
Mineralogy and
Petrology
© Springer-Verlag 1988

Experimental determination of the fluid-absent melting relations in the pelitic system

Consequences for crustal differentiation

Daniel Vielzeuf[1] and John R. Holloway[2]

[1] Département de Géologie, UA 10, 5, rue Kessler, F-63038 Clermont-Ferrand, France
[2] Departments of Chemistry and Geology, Arizona State University, Tempe-AZ 85287, USA

Abstract. In order to provide additional constraints on models for partial melting of common metasediments, we have studied experimentally the melting of a natural metapelite under fluid-absent conditions. The starting composition contains quartz, plagioclase, biotite, muscovite, garnet, staurolite, and kyanite. Experiments were done in a half-inch piston-cylinder apparatus at 7, 10, and 12 kbar and at temperatures ranging from 750° to 1250° C. The following reactions account for the mineralogical changes observed at 10 kbar between 750° and 1250° C: Bi + Als + Pl + Q = L + Gt + (Kf), Ky = Sill, Gt + Als = Sp + Q, Gt = L + Sp + Q, and Sp + Q = L + Als.

The compositions of the phases (at $T > 875°$ C) were determined using an energy-dispersive system on a scanning electron microscope. The relative proportions of melt and crystals were calculated by mass balance and by processing images from the SEM. These constraints, together with other available experimental data, are used to propose a series of $P-T$, $T-XH_2O$, and liquidus diagrams which represent a model for the fluid-present and fluid-absent melting of metapelites in the range 2–20 kbar and 600°–1250° C.

We demonstrate that, even under fluid-absent conditions, a large proportion ($\approx 40\%$) of S-type granitic liquid is produced within a narrow temperature range (850°–875° C), as a result of the reaction Bi + Als + Pl + Q = L + Gt(+/−Kf). Such liquids, or at least some proportion of them, are likely to segregate from the source, leaving behind a residue composed of quartz, garnet, sillimanite, plagioclase, representing a characteristic assemblage of aluminous granulites.

The production of a large amount of melt at around 850° C also has the important effect of buffering the temperature of metamorphism. In a restitic, recycled, lower crust undergoing further metamorphism, temperature may reach values close to 1000° C due to the absence of this buffering effect. Partial melting is the main process leading to intracontinental differentiation. We discuss the crustal cross-section exposed in the North Pyrenean Zone in the context of our experiments and modelling.

1 Introduction

Several models, based on experimental data and the geometrical analysis of phase relationships, have recently been

Offprint requests to: D. Vielzeuf

proposed for the anatexis of pelitic rocks (Abbott and Clarke 1979; Clemens and Wall 1981; Thompson 1982; Grant 1985). Most of these studies emphasize the effects of water on the melting processes and the generation and evolution of water-undersaturated granitic melts (Clemens 1984). However, we are still ignorant of (i) the actual compositions of the melts formed at various conditions, (ii) their change in composition with increasing degree of melting, and (iii) the proportion of melt formed as a function of temperature. In order to provide some constraints, the experimental melting of a metapelite has been undertaken between 7–12 kbar and 750–1250° C. Clemens (1984) showed that the majority of granitic magmas were initially water-undersaturated, indicating either that a fluid phase with a $a H_2O \ll 1$ was present during melting or that the melting reactions were fluid-absent. In the absence of free water, melting depends upon water from hydrates like muscovite or biotite. Such amounts of water are usually not sufficient to saturate a melt formed by the breakdown of the hydrates and the water released during this process is dissolved in the melt without formation of a vapor phase (Burnham 1967). This is what has been called "fluid-absent melting" (Burnham 1967; Clemens 1984), "dehydration melting" (Thompson 1982), or "vapor-absent melting" (Grant 1985) (see also Robertson and Wyllie 1971). The experimental work reported here has been performed under fluid-absent conditions because it is believed that such conditions prevail in the lower part of the crust. The purpose of these experiments was to determine

(i) the reactions involved in the partial melting of the lower crust,
(ii) the compositions of the liquids at various temperatures,
(iii) the mineralogical composition of the residuum,
(iv) the proportion of melt as a function of temperature in order to determine the temperature necessary to generate enough granitic liquid to form a mobile magma.

We then use these constraints in constructing a $T-XH_2O$ diagram and a model for both the water-saturated and the fluid-absent melting of pelites between 2 and 20 kbar. Finally, we explore the consequences of this model for the metamorphic structure of the crust and for intracontinental differentiation.

2 Experimental and analytical techniques

Most of the experiments were performed in a 0.5 inch (1.27 cm) diameter non-end-loaded piston-cylinder apparatus (Patera and

Table 1. Bulk composition of the rock starting material (average of atomic absorption analysis and EDS microprobe analyses of a glass; water content determined by thermogravimetric analysis), and representative microprobe analyses of the constituent minerals

	Gt27	St33	Bi20	Mu21	Pl40	Chl	Bulk comp.	av. clay[a]
SiO_2	37.52	27.30	35.72	46.41	59.88	24.55	64.35	64.84
Al_2O_3	20.76	51.00	20.50	36.47	25.55	23.01	18.13	17.86
FeO	31.58	14.35	18.29	0.95	0	23.38	6.26	6.35
MgO	3.83	1.93	10.65	0.88	0	15.94	2.44	2.66
CaO	5.82	0.12	0	0	6.42	0.01	1.52	1.85
Na_2O	0.	0	0.43	1.25	8.35	0.59	1.66	1.93
K_2O	0	0	8.38	9.13	0	0.32	2.56	3.64
TiO_2	0.23	0.64	1.78	0.8	0	0.12	0.82	0.86
MnO	0.29	0	0	0	0	1.08	0.09	
H_2O^b	0	4.65	4.25	4.11	0	11.00	2.15	
Total (anh.)	100.03	95.35	95.75	95.89	100.19	89.00	97.85	100
XMg	0.179	0.194	0.509	0.624		0.548		
Si	5.976	3.907	5.336	6.091	2.663	5.093		
Al	3.896	8.600	3.609	5.641	1.339	5.626		
Fe	4.206	1.717	2.284	0.104	0	4.056		
Mg	0.910	0.413	2.371	0.172	0	4.927		
Ca	0.993	0.018	0	0	0.306	0.002		
Na	0	0	0.123	0.317	0.720	0.239		
K	0	0	1.596	1.529	0	0.085		
Ti	0.027	0.069	0.199	0.078	0	0.018		
Mn	0.039	0	0	0	0	0.190		
nb. ox.	24	23	22	22	8	28		

[a] Normalized to 100% anhydrous (Shaw, 1956).
[b] By difference to 100%

Holloway 1978). The solid media assemblies were all salt below 950° C and salt + pyrex glass at higher temperatures, with a 6 mm ID graphite tube furnace and powdered pyrex surrounding the capsule. Temperatures were controlled and read by $W-Re_{26}/W-Re_5$ thermocouples and are believed to be precise to within +/−5° C. No correction was made for the effect of pressure on the EMF output of the thermocouples. Temperature gradient measurements were made (Esperança and Holloway 1986; Jakobsson and Holloway 1986) and a gradient of 14° C found from the normal thermocouple position to the center of the capsule. Thus an overall gradient of 28° C is estimated for the length of the capsule. Therefore, reported temperatures are believed to be accurate to within +/−20° C. Precision from run to run is estimated to be better than +/−5° C. Reported pressures were not corrected for the effects of friction because Esperança and Holloway (1986) found recorded pressures to be within 0.5 kbar of the published position for the Fa + Q = Fs reaction (Bohlen et al. 1980). Sealed capsules (5 mm in diameter and 7 mm in length) were of gold below 1000° C and $Pd_{40}-Ag_{60}$ at higher temperatures. They were filled with 100 to 200 milligrams of powdered rock starting material (< 5 microns) which had been dried overnight at 150° C and stored in a vacuum dessicator over magnesium perchlorate. Experiments were done by heating directly to run temperature using the hot piston-out technique. None of the reactions were reversed.

All quenched run products were identified by optical petrographic techniques and X-ray powder diffraction. Polished thin sections were made and examined by scanning electron microscopy. The SEM photographs were made in electron backscatter mode. Elemental analyses were done using an energy-dispersive analyser on the SEM. Natural minerals and a glass were used as standards. Results were normalized to 100%.

3 Starting material and subsolidus equilibria

A pelitic rock from the Cariño gneisses was used as the starting material for this experimental study. This rock is from the Cabo Ortegal complex (Galicia, NW Spain) which consists of three major rock types: (1) ultramafic rocks (peridotites and pyroxenites), (2) mafic rocks, and (3) rocks of sedimentary origin. All these underwent high-grade metamorphism ranging from high-pressure amphibolite and granulite facies to the eclogite facies (Vogel 1967; Engels 1972; Arps et al. 1977). In the "amphibolite-facies zone" the pelitic rocks (Cariño gneisses) which are non-migmatitic, are composed of quartz, plagioclase (An_{30}), kyanite, muscovite, biotite, garnet, +/−staurolite, +/−secondary chlorite. In the "granulite facies zone", the pelitic rocks (Chimparra gneisses) display the same paragenesis, except for the absence of staurolite, suggesting that these gneisses crossed the staurolite-out isograd. It is noteworthy that in a few places, incipient anatexis can be observed (Arps et al. 1977). In the "eclogite-facies zone", the associated felsic rocks are strongly migmatitic in character. Quartz, muscovite, and plagioclase formed mobilized pegmatitic fractions while garnet, kyanite, and biotite remained behind as "infusible residues" (den Tex et al. 1972). Note that, except for the presence or absence of staurolite and the migmatitic character, there is no difference in the pelitic parageneses between the different zones. The subdivision into different facies is based on typomorphic parageneses in the mafic rocks only (den Tex et al. 1972). Mafic and ultramafic rocks form up to 75% of the surface area of the catazonal Cabo Ortegal Complex and the close association with pelitic rocks has been interpreted as a volcano-detritic sequence of pelagic sediments with submarine basic lavas or tuffs (den Tex et al. 1972).

The Cariño gneisses were selected as starting materials for the melting studies at high pressure because they display a high pressure paragenesis, at equilibrium, diagnostic of temperatures just below the onset of partial melting. Furthermore, they show coexistence of muscovite and biotite which are two minerals important for the understanding of fluid-absent partial melting of pelitic rocks. The chemical composition of the rock which has been used is given in Table 1 together with the compositions of the major minerals. The modal proportions were determined by point counting (≈ 4000 points). The water content of the sample was determined by thermo-gravimetric analysis as 2.15 wt% total, of which

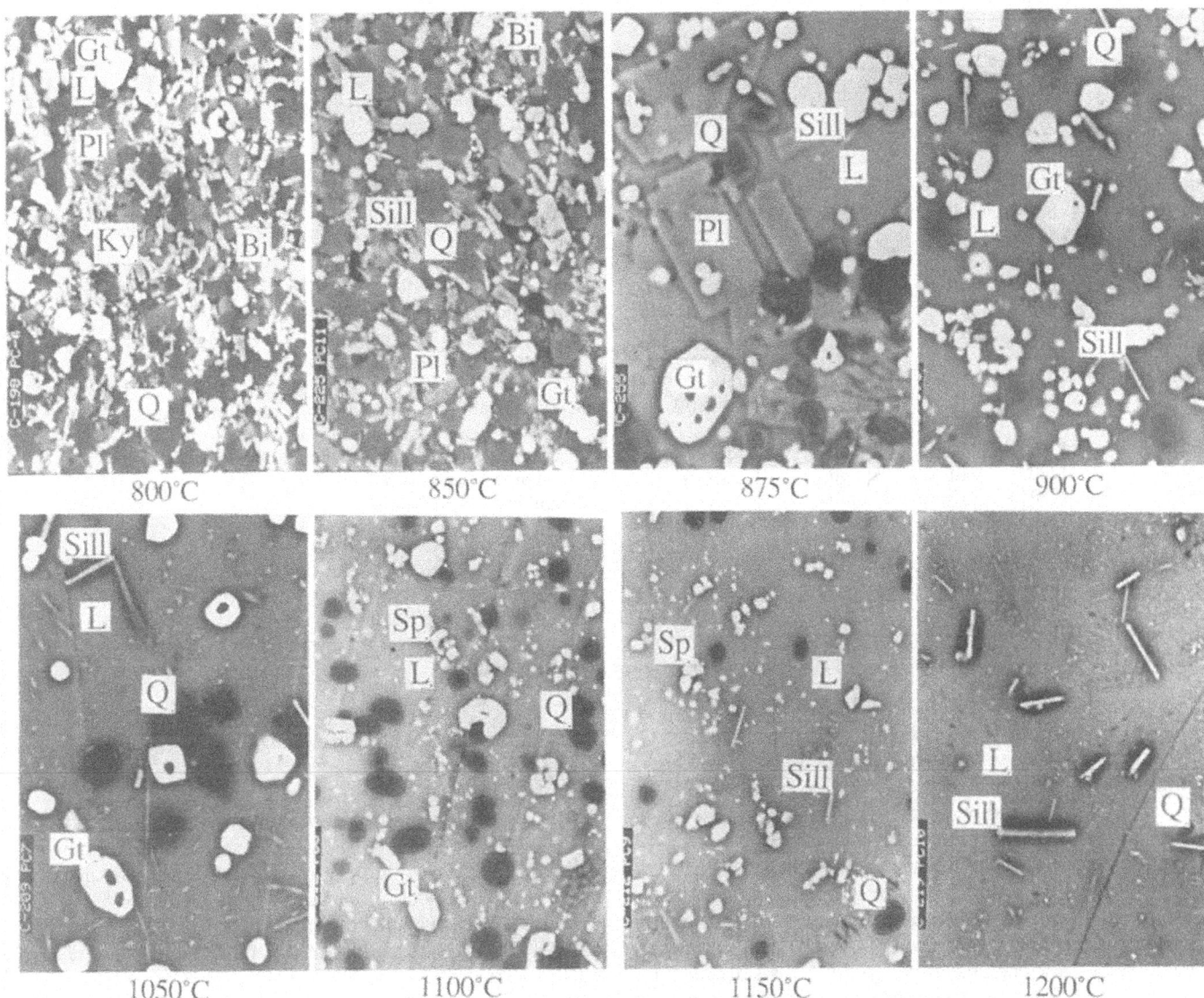

Fig. 1. SEM backscatter photographs of some runs at 10 kbar. One centimeter = 15 μ

0.29% was found to be adsorbed water. The chemical composition of this rock is very close to the average of clays as determined by Shaw (1956) except for K_2O which is significantly lower (cf. Table 1).

The physical conditions of metamorphism for the Cariño gneisses can be estimated from geothermobarometers based on mineral solid-solutions. The garnet-biotite geothermometer (Thompson 1976; Holdaway and Lee 1977; Ferry and Spear 1978) yields temperatures ranging from 640° to 680° C at 10 kbar. The $P-T$ diagram for the assemblage garnet, biotite, kyanite, quartz, muscovite, vapor in the system KFMASH constructed by Spear and Selverstone (1983), provides an estimate of 640° C and 6 kbar, plotting outside the kyanite field. Perchuk et al. (1981) proposed a grid to determine the P and T of crystallization of Bi, Gt, Sill, Mu, and Q assemblages. In the case of the Cariño gneisses this gives 610° C and 6.5 kbar. The plagioclase – biotite – garnet – muscovite geobarometer (Ghent and Stout 1981) yields a pressure of 8.5 kbar for a temperature of 660° C while a pressure of 10.5 kbar is determined by using the garnet – plagioclase – kyanite – quartz geobarometer (Newton and Haselton 1981). Considering that this last geobarometer is one of the most reliable in rocks of this type (Newton 1983), and that this rock crystallized in the field of kyanite, a pressure of 9.5+/−1 kbar and a temperature of 660+/−20° C are inferred for the conditions of crystallization of the Cariño gneisses. These are close to the conditions determined by den Tex et al. (1972) on the basis of experimental equilibria.

4 Results

4.1 Description of the run products

The results of the experiments are shown in the SEM images of Fig. 1, and in Table 2. It is known that in the range 4–10 kbar, partial melting of the muscovite, quartz, alkalifeldspar assemblage occurs at temperatures below 700° C (Thompson and Algor 1977). Our lowest temperature experiment was 750° C and muscovite was not observed in any of our run products. We also did not observe staurolite and it is assumed that staurolite disappeared in the subsolidus area. At 10 kbar, and from 750° to 860° C, the phase assemblage is composed of quartz, plagioclase, biotite, garnet, aluminum – silicate (kyanite and/or sillimanite), and glass. On the SEM images, it can be seen that the liquid forms a film around the crystals and is concentrated at triple junctions. The proportion of liquid is small, certainly less than 10%. The size of the crystals is about 5 μ. Between 832° and 850° C, kyanite reacted to produce distinct sillimanite needles. At higher temperatures (between 850° and 862° C) the proportion of liquid increases dramatically; this important change corresponds to the disappearance of biotite. The resulting assemblage is composed of quartz, pla-

Table 2. Experimental results

Run number	$T °$ C	P kbar	Duration	Assemblage	Remarks
PC21	750°	10	7 days	L Gt Ky Q Pl Bi	disappearance of muscovite and staurolite below 750° C
PC04	800°	10	7 days	as above	progressive increase of garnet and decrease of biotite and kyanite
PC05	832°	10	6 days	as above	
PC11	850°	10	7 days	L Gt Ky — Sill Q Pl Bi	appearance of sillimanite needles
PC18	862°	10	6 days	L Gt Sill Q Pl	disappearance of biotite and large increase in the proportion of melt
PC17	875°	10	5.5 days	L Gt Sill Q Pl	
PC20	887°	10	5 days	L Gt Sill Q Pl	
PC03	900°	10	4 days	L Gt Sill Q (Pl)	large increase in sillimanite, very small amount of plagioclase
PC02	950°	10	24 hrs	L Gt Sill Q	disappearance of plagioclase
PC06	1000°	10	24 hrs	as above	progressive decrease of garnet
PC07	1050°	10	24 hrs	as above	
PC08	1100°	10	2 hrs	L Gt Sp Q	crystallization of spinel, increase of quartz, decrease of garnet, disappearance of sillimanite
PC09	1150°	10	30 min	L Sp Sill Q	disappearance of garnet. Sillimanite reappears
PC10	1200°	10	20 min	L Sill (Q)	disappearance of spinel and large decrease of quartz
PC01	1250°	10	30 min	L	liquidus
PC16	950°	12	15 hrs	L Gt Ky (Sill) Q	rare needles of sillimanite
PC15	975°	12	14 hrs	as above	
PC27	850°	7	7 days	L Gt Bi Sill Q Pl	
PC24	875°	7	7 days	L Gt Sill Q Pl	disappearance of biotite

(in all the runs below the liquidus, kyanite persisted metastably in the field of sillimanite)

gioclase, aluminum — silicate, garnet, and glass. At 875° C the sample was accidently quenched after 18 h (a brief high temperature excursion occurred) and then rerun at 875° C for 5 days. This procedure produced large euhedral crystals of garnet (10 μ), quartz (5 μ), and plagioclase (20 microns), and the nucleation of numerous small garnets. At 900° C, the proportion of plagioclase is very small and probably close to the Pl-out curve. In some of these runs, it is likely that garnet did not equilibrate completely with the melt. The cores of these crystals are relicts from the starting garnets separated from the newly-formed rims by regions of small glass inclusions. The differences in the chemical compositions of the cores and rims will be discussed later. Between 900° and 1050° C, there is a large interval in temperature over which no mineralogical changes occur; quartz, garnet, aluminum — silicate, and glass coexist. Metastable kyanite persists together with stable sillimanite. An important boundary is crossed above 1050° C with the appearance of spinel and disappearance of sillimanite; the assemblage is then composed of quartz, garnet, spinel, and glass. In the 1050–1100° C interval, the amount of garnet decreases while the modal proportion of quartz increases strongly. At 1150° C there is no more garnet, spinel is abundant, and the phase assemblage is composed of quartz, spinel, sillimanite, and glass.

Interestingly, aluminium — silicate (sillimanite or mullite?) re-appears. At 1200° C spinel is absent and the glass coexists with sillimanite and quartz. The liquidus is reached between 1200° and 1250° C.

4.2 Composition of the phases

Spinel. Spinels were analysed in two runs at 1100° and 1150° C (Table 3); they belong to the spinel — hercynite series. Their XMg (Mg/Mg + Fe) increases with temperature (0.51 at 1100° C and 0.61 at 1150° C), it is greater than

XMg in the glass (0.38) but not significantly different from XMg in the garnet (0.48 at 1100° C).

Garnet. In most cases garnet did not reach equilibrium with the glass, as shown by the presence of relict cores and newly-formed rims. The compositions of the cores are usually close to the composition of the starting garnet (= 6% CaO and XMg = 0.18) while the compositions of the rims are always poorer in calcium, and far more magnesian. XMg in the garnet increases from 0.40 at 875° C to 0.48 at 1100° C. In a previous paper (Vielzeuf 1983), and on the basis of coexisting spinel, glass, and garnet in a xenolith, it has been argued that, at very high temperature, XMg Sp > XMg Gt > XMg Gl. This apparently contradicts the data given by Clemens and Wall (1981) who observed that XMg Gl > XMg Gt, and Grant (1985) who concluded that the situation described by Vielzeuf (1983) did not represent an "approach to equilibrium". It must be stressed that the present study gives an answer to this apparent contradiction. Above 950° C XMg Gt is greater than XMg Gl, on the contrary at 950° C and at lower temperatures XMg Gt is less than XMg Gl. This is in agreement with other experimental studies (Green 1977; Ellis 1986). According to the present experiments, the reversal of the Fe — Mg partitioning between garnet and glass would occur between 950° and 1050° C at 10 kbar.

Plagioclase. This mineral has been analysed only in one experiment at 875° C; it is slightly more albitic than the starting plagioclase (Ab$_{73}$ compared to Ab$_{70}$). The coexistence of garnet, plagioclase, quartz, aluminum-silicate allows the use of the Gt — Pl — Als — Q geobarometer (Newton and Haselton 1981). This barometer gave a pressure of 16 kbar at 875° C with sillimanite and 15 kbar with kyanite (instead of 10 kbar). This descrepancy is probably due to (1) difficulties in analysing small crystals (< 10 microns) in the experimental charges, and (2) disequilibrium between plagioclase and liquid (Johannes 1978); hence the compositions of the minerals listed in Tables 3 a, b, c should be judged accordingly.

Glass. Careful attention has been paid to analysing the glasses: they were analysed at different times, with different sets of standards including a glass made of the starting material CO821. The results for runs between 875° and 1200° C are listed in Table 4. All of them plot in the adamellite field. From low to high temperatures, SiO_2 decreases regularly up to 1100° C, and then increases slightly or remains constant up to 1250° C in connection with the crystallization of spinel. Al_2O_3 and MgO increase regularly throughout the studied temperature interval, while FeO shows an abrupt increasing step between 1000 and 1050° C, perhaps in connection with the reversal of the Fe−Mg partitioning between glass and garnet. CaO concentration is low at 875° C where a significant amount of plagioclase is observed, a pronounced increase can be noted between 875° and 900° C, then the calcium content of the liquid remains remarkably constant until the liquidus is reached. Na_2O and K_2O decrease regularly from low to high temperatures. The evolution of the XMg in the liquid is interesting: at first it decreases up to 950° C then remains constant up to 1100° C (if we except the value for the run performed at 1000° C), and finally increases until the liquidus is reached. This can be ascribed to the crystallization/dissolution of garnet, and the crystallization of spinel together with the partitioning of Fe−Mg between these phases and the glass. In his review of this paper, Grant plotted the analyses of the glasses on an isocon diagram (Grant 1986a) and noted that Ti, Mg, and Fe change coherently from 875° to 900° to 1000° C, but that the FeO value for 950° C should be more like 1.9 than 2.9. As a consequence, the XMg value of 0.53 at 1000° C may be less suspect than the value of 0.43 obtained for the glass at 950° C.

The variation of the composition of the liquids as a function of temperature is also shown in the AKF diagram (Fig. 2). This diagram clearly shows that the major change in the analytical data on the glasses between 1000° and 1050° C is largely due to major dissolution of garnet in the liquid.

4.3 Proportion of melt

It is possible to calculate the modal proportions of the phases in the run products using a mass balance approach based on a least squares method. Some of the calculated values have been checked by image-processing of SEM photographs (Table 5). No suitable means has been found to calculate the modal proportions below 875° C and, as a result, the estimates at low temperatures in Fig. 3 are approximate. Only one estimate of the proportion of melt (at 1050° C) is inconsistent with the others (Table 5). The interpolated line between these points results from the assumption that a reaction promotes a sudden increase of the proportion of melt while the dissolution of phases (eg. garnet, sillimanite in the 900°–1050° C interval) produces a smooth, incremental increase in melt proportion. In Fig. 3, three main stages of melting can be distinguished:
(1) the first one (speculative) occurs below 800° C and corresponds to the melting of muscovite; the proportion of melt is small and did not exceed 10–15%;

(2) the range 850–900° C is marked by a dramatic increase in the proportion of melt changing from about 10% to 50–60%. This step corresponds to the melting of biotite and is followed by a "plateau" along which garnet and sillimanite are progressively dissolved by the liquid;
(3) the third (and last) stage is represented by the melting of garnet and spinel at very high temperatures (above 1100° C).

Concerning the other phases, the following points can be made (Table 5). Up to 1050° C, the proportion of sillimanite is constant; this mineral disappears at 1100° C and crystallizes again at 1150° C. Quartz increases markedly at 1100° C in connection with the first crystallization of spinel, the decrease of garnet, and disappearance of sillimanite. In the starting material, garnet represents 2% of the mode; at 875° it represents 24%. Thus, a large amount of garnet is produced in the range 600–875° C. This proportion decreases regularly up to 1000° C and abruptly above this temperature.

5 Modelling the partial melting of metapelites and metagraywackes

The partial melting processes can be modelled using the same principles which are commonly used in the subsolidus region. However, complications arise from the fact that a melt may be regarded as an "ultimate solid-solution" changing rapidly in composition as a function of P, T, aH$_2$O. These changes are likely to produce some modifications in the chemographic relationships. The articles published by Thompson (1982) and Grant (1985) represent important contributions to the understanding of phase equilibria in partial melting of pelitic rocks. In this paper we will emphasize some less developed aspects in these previous studies. Our purpose is to construct internally consistent $P−T$, $T−X$H$_2$O, and liquidus diagrams representing three complementary ways of viewing the partial melting processes. This model is based on available experiments and natural observations; but it should be kept in mind that such models are hypothetical because they are also based on some assumptions. Even so, they are good support for reasoning, represent efficient ways of emphasizing the critical aspects of the partial melting processes and have predictive power. There is no doubt that some of the depicted phase relations will have to be changed as more information and constraints become available.

5.1 Subsolidus equilibria

One of the first crucial points is to determine the subsolidus phase relationships in the high-grade region which will be overlapped by melting. The phase relationships among aluminum−silicate (Als), muscovite (mu), K-feldspar (Kf), biotite (Bi), garnet (Gt), and cordierite (Cd) (+quartz (Q) and vapour (V)) in the system Al_2O_3 (A)−K_2O (K)−FeO (F) or MgO (M) (+SiO_2 (S) and H_2O (H)) *(case 1)* have been studied by Thompson (1976, 1982), Vielzeuf and Boivin (1984), and Vielzeuf (1984). In the following model, phengitic substitution in the muscovite (Mu$_{ss}$) is taken into consideration leading to the replacement of the degenerate reaction (Mu+Q=Kf+Als+V by one of the following: Mu$_{ss}$+Q=Cd+Kf+Als+V, Mu$_{ss}$+Q=Bi+Kf+Als+V, and Mu$_{ss}$+Q=Gt+Kf+Als+V, depending upon the pres-

Table 3. EDS microprobe analyses of garnet, plagioclase, and spinel in different runs at 10 kbar. Structural formulae on the basis of 24, 8, and 32 oxygens respectively. Analyses are normalized to 100%. **a** – garnets

Gt	Start. mat.	1100°C				1050°C			950°C				900°C				875°C		
		Gt19c	Gt20r	Gt21c	Gt22r	Gt26	Gt27	Gt28	Gt83c	Gt84r	Gt85c	Gt86r	Gt90c	Gt92	Gt93c	Gt94r	Gt38c	Gt39r	Gt40c
SiO_2	37.52	39.1	42.0	36.9	37.0	39.1	40.6	39.7	38.4	42.5	38.1	40.3	40.1	42.0	38.8	42.2	38.5	42.3	37.7
Al_2O_3	20.76	21.6	22.7	20.6	21.6	22.0	22.1	21.7	19.6	21.3	19.4	20.8	19.9	21.5	20.3	20.8	21.1	21.6	20.8
FeO	31.58	28.0	21.6	32.3	27.1	25.1	23.8	24.4	31.6	24.5	32.3	28.4	29.5	23.9	29.9	24.6	30.1	24.1	30.8
MgO	3.83	7.1	11.4	5.3	10.9	12.3	11.8	12.5	5.0	9.0	4.1	5.9	4.8	9.4	4.6	9.5	4.4	9.2	4.2
CaO	5.82	4.2	1.6	4.0	2.2	1.1	1.1	1.0	4.5	2.4	6.0	4.6	5.6	3.1	5.6	2.6	5.5	2.9	4.8
TiO_2	0.23		0.3		0.3		0.2	0.4		0.2									
MnO_2	0.29		0.3	0.8	0.8	0.4	0.4	0.4	0.8						0.8	0.2	0.4		
Total	100.03																		
24 ox.																			
Si	5.976	6.06	6.23	5.90	5.74	5.95	6.10	6.00	6.11	6.40	6.09	6.25	6.28	6.32	6.13	6.37	6.06	6.35	6.06
Al	3.896	3.94	3.96	3.88	3.95	3.93	3.92	3.87	3.67	3.78	3.65	3.80	3.68	3.82	3.77	3.71	3.93	3.83	3.93
Fe	4.206	3.63	2.68	4.31	3.52	3.19	2.99	3.09	4.20	3.09	4.32	3.68	3.86	3.01	3.95	3.11	3.97	3.03	4.13
Mg	0.910	1.64	2.52	1.26	2.52	2.79	2.65	2.81	1.18	2.03	0.97	1.37	1.12	2.12	1.09	2.14	1.03	2.05	1.01
Ca	0.993	0.70	0.26	0.68	0.36	0.17	0.18	0.16	0.77	0.39	1.03	0.76	0.95	0.50	0.94	0.43	0.93	0.46	0.83
Ti	0.027		0.04		0.04		0.02	0.04		0.03									
Mn	0.039		0.04	0.12	0.10	0.05	0.05	0.05	0.10						0.10	0.03	0.05		
XMg	0.18	0.31	0.48	0.23	0.42	0.47	0.47	0.48	0.22	0.40	0.18	0.27	0.23	0.41	0.22	0.41	0.21	0.40	0.20
Alm.	0.684	0.608	0.487	0.677	0.540	0.514	0.510	0.505	0.671	0.558	0.683	0.634	0.651	0.535	0.650	0.545	0.664	0.546	0.692
Pyr.	0.148	0.275	0.458	0.197	0.388	0.450	0.451	0.460	0.189	0.367	0.154	0.235	0.189	0.376	0.179	0.375	0.173	0.370	0.169
Gro.	0.161	0.117	0.047	0.107	0.056	0.028	0.031	0.026	0.124	0.070	0.163	0.131	0.159	0.089	0.154	0.075	0.155	0.083	0.139
Spe.	0.006		0.008	0.019	0.016	0.008	0.008	0.008	0.017	0.005					0.017	0.005	0.008		

c: core; r: rim

b – plagioclases

	Pl Start. mat.	875° C		
		Pl42	Pl43	Pl44
SiO₂	59.9	64.4	65.6	64.4
Al₂O₃	25.5	23.2	22.4	23.3
CaO	6.4	4.0	3.6	4.0
Na₂O	8.3	7.5	7.6	7.5
K₂O	0	0.8	0.8	0.7
Si	2.66	2.83	2.88	2.83
Al	1.34	1.20	1.16	1.21
Ca	0.31	0.19	0.17	0.19
Na	0.72	0.64	0.64	0.64
K	0	0.05	0.04	0.04
Ab	0.701	0.735	0.753	0.736
An	0.298	0.213	0.197	0.217
Or	0	0.053	0.053	0.047

Structural formulae on the basis of 8 ox. Normalized to 100%

c – Spinels

	1150° C				1100° C	
	Sp10	Sp11	Sp12	Sp13	Sp17	Sp18
Al₂O₃	67.5	67.3	67.4	67.0	63.8	65.3
FeO	17.1	17.0	17.2	17.7	22.9	21.6
MgO	15.4	15.8	15.4	15.2	13.3	13.1
Al	16.26	16.20	16.25	16.20	15.85	16.09
Fe	2.92	2.90	2.94	3.04	4.03	3.77
Mg	4.69	4.80	4.69	4.65	4.19	4.09
XMg	0.62	0.62	0.61	0.60	0.51	0.52

Structural formulae on the basis of 32 ox. Normalized to 100%

sure (see Thompson 1982 for a careful discussion of this problem).

At higher temperatures, the phase relationships among Als, Kf, Bi, Opx, Gt, Cd (Q, V) in the KFMASH system (*case 2*) were modelled by Vielzeuf (1980a) and Vielzeuf and Boivin (1984). In a more recent study (Vielzeuf 1984), aluminous biotite (Bi$_{ss}$) was used instead of a biotite in the annite−phlogopite series (see Holdaway 1980) changing the degenerate reaction Bi + Q = Opx + Kf + V into one of the following reactions: Bi$_{ss}$ + Q = Cd + Opx + Kf + V, Bi$_{ss}$ + Q = Opx + Als + Kf + V, or Bi$_{ss}$ + Q = Opx + Kf + Gt + V, depending upon the pressure and the Mg/Fe ratio in the system.

A third multisystem of great interest at very high temperatures is the one involving the phases Als, Cd, Opx, Sp, Gt (+Q) in the FMAS system (*case 3*). These relations were studied by Vielzeuf (1983) and discussed by Grant (1985) and Hensen (1986). The modifications proposed by Hensen (1986) concern the slope of the reaction Sp + Q = Opx + Als located in the high temperature portion of the diagram which is metastable with respect to melting reactions in our experiments; this aspect will not be discussed in this paper.

It is interesting to note that, in each of these three systems, the high-temperature region is obliterated by some melting reactions representing successively (and as a first approximation) the muscovite melting stage, the biotite melting stage, and the garnet melting stage. Now the problem is to determine what subsolidus reactions will be interrupted by melting. As a very good approximation this important change occurs close to the (1) Kf + Q + V = L reaction in the KFMASH system or at lower temperature, and close to the Ab + Kf + Q + V = L reaction in the KNFMASH system. However, the reactions under consideration are not only dependent on P and T but also on aH₂O in the system. Most importantly, the majority of the reactions present in the subsolidus systems involve Fe−Mg solid solutions. Variations in the XMg of the system may affect the location of the divariant fields by several tens of degrees and/or several kilobars. Also, it has been demonstrated that Fe−Mg substitution can shift reaction boundaries in a grid sufficiently to cause a topological inversion (Vielzeuf 1983; Vielzeuf and Boivin 1984). Later work (Montel et al. 1986; Hensen 1986) has confirmed the utility of this approach for pelitic systems. Such inversion occurs in cases 2 and 3 and, as a result, reaction (1) Kf + Q + V = L will not necessarily intersect the same reactions in the pure Fe and pure Mg systems. Considerable attention has been given to these aspects by Thompson (1982) and Grant (1985). Thompson emphasized the effects of a change in the relative position of reaction (1) in a given grid and Grant dealt with the effects of reversals of the geometrical relationships on the melting reactions. A method of studying the effects of Fe−Mg solid solutions on the geometrical relations has been proposed in a previous paper (Vielzeuf and Boivin 1984) and these aspects will not be emphasized here. Based on the fact that shales and graywackes have remarkably constant XMg (atomic Mg/Mg + Fe) (between 0.4 and 0.5; Blatt et al. 1972), we will construct a model for a fixed and average value of XMg close to 0.5. For clarity the divariant bands in the diagrams will be represented as lines on Figs. 5 and 7.

The following reactions are important for positioning the grid in $P - T$ space:

Table 4. Composition of the liquids as a function of temperature at 10 kbar

Wt% Oxide	Anhydrous liquid compositions T° C									c 8 Kbar
	1250	1200	1150	1100	1050	1000	950	900	875	800
SiO_2	65.8	64.8	65.6	65.5	66.6	69.3	69.3	71.3	73.2	74.7
TiO_2	0.8	1.0	0.9	1.1	1.0	0.6	0.6	0.6	0.3	0.1
Al_2O_3	18.6	18.2	17.5	17.6	17.7	17.2	17.0	16.1	16.1	16.4
FeO	6.4	6.8	6.5	5.5	5.1	2.4	2.9	1.9	1.5	2.2
MgO	2.5	2.5	2.2	2.1	2.0	1.6	1.2	1.3	1.0	1.2
CaO	1.6	1.6	1.7	1.9	1.7	1.7	1.8	1.6	0.4	2.0
Na_2O	1.7	2.2	2.6	2.7	2.8	3.5	2.8	3.0	3.1	0.2
K_2O	2.6	3.1	3.1	3.6	3.1	3.7	4.2	4.2	4.4	2.9
XMg	0.41	0.39	0.38	0.41	0.42	0.53	0.43	0.55	0.54	0.50
H_2O[a]	2.15[b]	2.33	2.65	2.96	2.83	3.43	3.36	3.50	3.74	9.61[d]

[a] Calculated from the proportion of melt and water content in the starting material (except for [b] and [d])
[b] Determined by thermo-gravimetric analysis
[c] Composition of the liquid in equilibrium with Bi, Sill, Q, Gt, and V at 8 Kbar and 800° C
[d] Calculated using Burnham's model (1979)

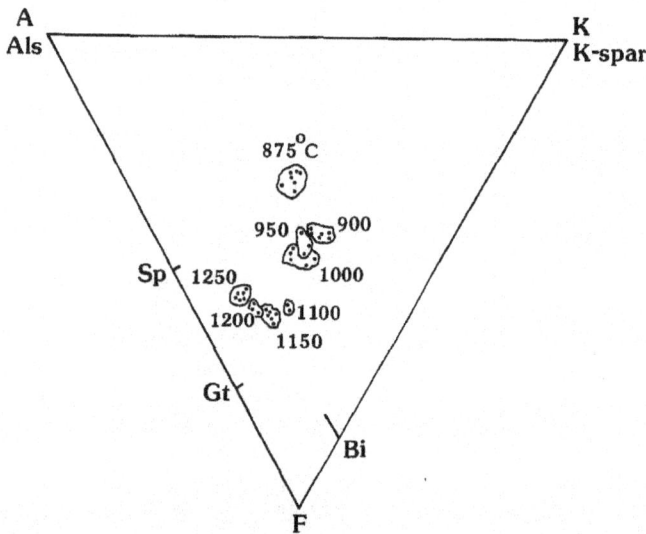

Fig. 2. Composition of liquids as a function of temperature in the AKF diagram (mole %). $A = Al_2O_3 - (K_2O + Na_2O + CaO)$; $K = K_2O$; $F = FeO + MgO$

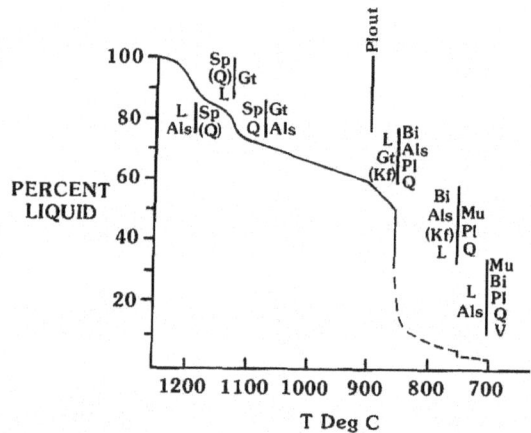

Fig. 3. Weight percent liquid versus T, estimated by mass balance, and inferred melting reactions

(2) $Mu + Q = Kf + Als + V$,
(14) $Cd = Als + Gt + Q + V$,
(39) $Bi_{ss} + Als + Q = Gt + Mu_{ss} + V$,
(24) $Bi_{ss} + Als + Q = Cd + Kf + V$,
(10) $Gt + Q + V = Cd + Opx$,
(6) $Bi + Q = Kf + Opx + V$.

The univariant reaction (2) $Mu + Q = Kf + Als + V$ has been investigated in several experimental studies, and a summary of the results is given by Helgeson et al. (1978). Since Fe or Mg preferentially enter muscovite, the addition of Fe or Mg must increase the stability field of muscovite until saturation, which corresponds to the crystallization of cordierite, biotite or garnet, depending on the pressure (Thompson 1982). Theoretical and experimental data on the melting reactions of the muscovite – quartz assemblage (i.e. at the intersection of reactions (2) and (3) $Mu + Kf + Q + V = L$) (Figs. 4 and 5) are given by Lambert et al. (1969), Storre and Karotke (1972), Storre (1972), Thompson and Algor (1977), Thompson (1982).

$PH_2O - T$ conditions for the iron end-member of reaction (24) have been determined by Holdaway and Lee (1977): 640° C at 2 kbar and 710° C at 2.7 kbar. A $P - X$Fe $-$ Mg diagram for the Fe $-$ Mg analogue of this reaction has been contructed from these data and natural Kd values. Hoffer (1976) has performed hydrothermal experiments on this reaction using natural minerals ($CdFe_{34}$, $BiFe_{50}$) and his results average about 0.8 kbar lower than Holdaway and Lee's extrapolation. Reaction (24) is terminated at high temperature by the intersection with the melting reaction (28) $Bi_{ss} + Als + Kf + Q + V = L$ at $T = 720°$ C and $P = 3.7$ kbar. Reaction (28) was studied at 7 and 10 kbar ($Ps = PH_2O$) by Hoffer (1976). The first melt is generated at 703° C at 7 kbar and 685° C at 10 kbar. Beyond the invariant point [Cd, Gt, Opx], reaction (28) becomes metastable and is replaced by (25) $Mu_{ss} + Bi_{ss} + Kf + Q + V = L$ (see Fig. 5).

The reaction (14) $Cd = Als + Gt + Q + V$ and its Fe $-$ Mg divariant field have been studied by numerous authors; a summary of these results will be found in Lonker (1981). From an extensive review of the literature, Newton (1983) considers that cordierite is usually not stable at pressures greater than 6 kbar in most common metapelites. A similar conclusion has been independently reached from the geo-

Table 5. Modal proportion of minerals and glass calculated by mass balance (Wt%). Results of the image processing are indicated in the brackets. The modal proportion of the starting material was determined by combining point-counting and mass balance calculations

Wt%	1250	1200	1150	1100	1050	1000	950	900	875	850
L	100	92	81	72	75	63	64	62	57	
Sill	–	3	1	–	5	6	6	6	5	
Q	–	5	12	18	10	13	11	11	9	
Sp	–	–	6 (4)	9 (11)[a]	–	–	–	–	–	
Gt	–	–	–	1	10 (8)	18 (12)	19 (20)	21 (15)	24 (21)	(9)
Pl	–	–	–	–	–	–	–	–	5	

Starting material: Ky 7, Q 39, Gt 2, Pl 19, Mu 9, Bi 21, St 1, Chl 1, Op 0.2

[a] Spinel and garnet cannot be separated by image processing.

barometric study of the granulites found in some Hercynian massifs where the disappearance of cordierite can be observed as a function of depth (Vielzeuf 1984).

The mineral pair cordierite − orthopyroxene, whose stability field is bounded by the reaction (10) Gt+Q+V= Cd+Opx, has long been considered as typical of the pyroxene hornfels facies (Turner and Verhoogen 1960). According to de Waard (1965) its breakdown marks the boundary between pyroxene contact metamorphism and pyroxene granulite facies. However this association is also present in high grade metamorphic rocks which crystallized or recrystallized in low pressure regional metamorphism or plutono − metamorphism (Bard 1969; Vielzeuf 1980a; Harris and Holland 1984). We believe that in most cases, for a average Fe − Mg ratio in the rock, a pressure of crystallization less than 3 kbar can be expected for rocks with cordierite − orthopyroxene assemblages.

From the experimental data on reactions 2, 14, and 24, it can be calculated that the position of the iron end-member of reaction (39) Bi_{ss}+Als+Q=Gt+Mu_{ss}+V is almost independent of temperature and is located between 6.5 and 7 kbar (Vielzeuf 1984). On the other hand, calculations using the standard thermodynamic properties of these minerals indicate much lower pressures for this reaction (<1 kbar according to Spear and Selverstone (1983); 2 kbar at 600° C and a much steeper slope according to Perchuk et al. (1981)). However, the errors associated with the thermodynamic properties of these Fe minerals are so large that there is no certainty in the results. The situation is complicated by the fact that ΔH, ΔS, and ΔV of this reaction are very small and a small error in any one of these will have a drastic effect on the calculations. Finally it must be stressed that the multivariant field of reaction (39) has to be very wide because the association of biotite, aluminum − silicate, quartz, garnet, muscovite is observed at pressures as low as 4 kbar in the stability field of andalusite (Novak and Holdaway 1981) and as high as 28 kbar in the stability field of kyanite and pyrope + quartz (Chopin 1984).

According to Eugster and Wones (1962) and Rutherford (1969), the iron end-member of reaction (6) Bi+Q=Kf+Opx+V would be at about 720° C at 2 kbar. On the other hand, if we consider that the Fe − Mg solid-solutions in biotite and orthopyroxene are close to ideal, the $P-T$ locations of the Mg and Fe end-members should not be very different since the Mg − Fe distribution coefficient between biotite and orthopyroxene is only slightly greater than 1 (1.2 according to Clemens (1984); 1.1 using the data from the granulites in the Pyrenees (Vielzeuf 1984)) with

XMgBi > XMgOpx (see also Grant 1981 p. 1135). This prediction of a narrow divariant field conflicts with the reversal reported by Wones and Dodge (1966) at about 840° C and 0.5 kbar. Helgeson et al. (1978) suggested that these very high temperatures for the Mg end-member could be a result of Al/Si disorder in the phlogopite formed in Wones and Dodge's experiments. The univariant equilibrium curve generated from retrieved thermodynamic data by Helgeson et al. (1978) is located 150° C–200° C below the curve generated from Wones and Dodge's reversal. On the other hand, Clemens et al. (1987) have shown that both natural and synthetic biotites are nearly completely Al − Si disordered at T>600° C but suggest that the synthetic Fe-rich micas may show significant degrees of tetrahedral order. This might explain the experimental observations. The curve calculated by Clemens et al. (1987) for the Mg end-member of reaction (6) Bi Q=Opx Kf V has been used in Fig. 5.

The addition of excess alumina to the system will increase the stability field of biotite until saturation and crystallization of one of the following phases occurs: cordierite, garnet, or sillimanite (see Fig. 5). Reaction (6) terminates in melting a few degrees below the Kf+Q+V=L reaction at the invariant point I_2 (Luth 1967; Wendlandt 1981; Bohlen et al. 1983).

5.2 Construction of the model

5.2.1 Melting at $PH_2O=P$total

In predicting phase relationships, it is useful to start from simple systems and then to study the influence of additional components rather than attempt to immediately derive the phase relations in the full system. This is because the reactions in the simplest system represent limiting conditions for the complex systems.

For the study of the Q- and V-saturated melting of metapelites and metagraywackes, the simplest limiting reaction is (1) Kf+Q+V=L in the KASH system (where KAlSi$_3$O$_8$ is the only independent component). This reaction has been located in the $P-T$ space by Tuttle and Bowen (1958), Luth et al. (1964), Shaw (1963), and Lambert et al. (1969). Its displacement as a function of aH$_2$O has been experimentally investigated and calculated by Clemens (1981) and Bohlen et al. (1983).

The addition of Al$_2$O$_3$ in excess of normative feldspar will necessarily increase the stability of the liquid and reaction (1) will move towards lower temperature until the system is saturated in muscovite or aluminum − silicate, de-

266

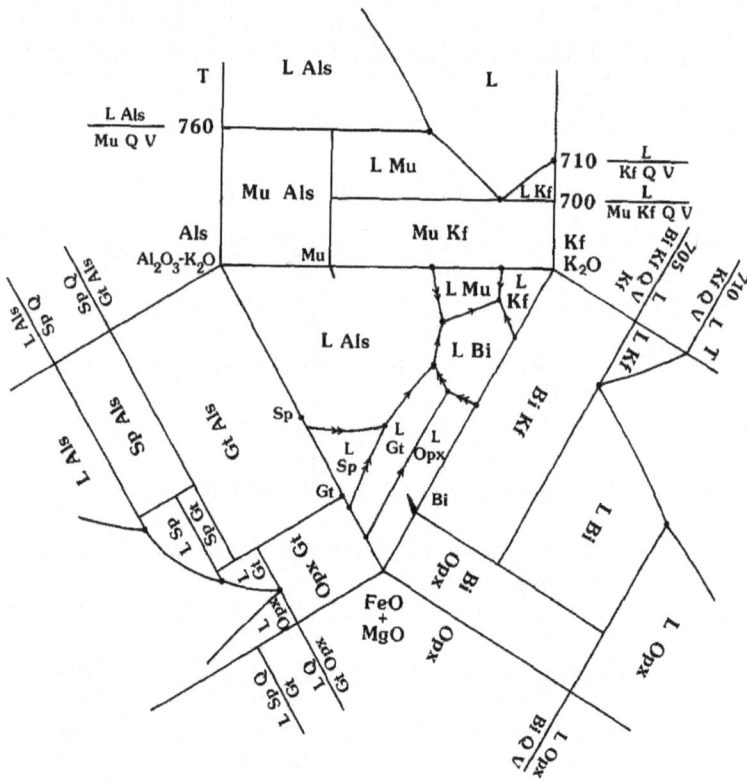

Fig. 4. Pseudo-binary $T-X$ diagrams in the systems $Al_2SiO_5 - KAlSi_3O_8$, $KAlSi_3O_8 - (Fe, Mg) SiO_3$, and $(Fe, Mg) SiO_3 - Al_2SiO_5$ (with excess SiO_2 and H_2O), and derivation of a pseudo-ternary liquidus diagram in the AKF system at 10 kbar. Not to scale. Reactions in the $Al_2SiO_5 - (Fe, Mg) SiO_3$ system occur at much higher temperature than those in the two other subsystems (see Fig. 5)

pending upon pressure. The reactions (3) Kf+Mu+Q+V=L and (5) Kf+Als+Q+V=L are thus analogues of reaction (1) in a pseudobinary system $Al_2SiO_5 - KAlSi_3O_8$ ($+SiO_2$ and H_2O in excess) (Fig. 4). The relations of all curves around the invariant point [I$_1$] (Fig. 5) were derived schematically by Lambert et al. (1969) and determined experimentally by Storre and Karotke (1972) and Storre (1972). According to these studies, the invariant point [I$_1$] would be located near 6 kbar and 730° C.

The addition of FeO or MgO which preferentially enter into the liquid, will also move reaction (1) Kf+Q+V=L towards lower temperatures until the system is saturated with biotite or orthopyroxene depending upon pressure and temperature. Reactions (7) Kf+Bi+Q+V=L and (9) Kf+Opx+Q+V=L are thus analogues of reaction (1) in the pseudo-binary system ($KAlSi_3O_8 - (FeO$ or $MgO)SiO_2$ (SiO_2, H_2O in excess) (Fig. 4). What is not known is if the temperature effect resulting from the addition of FeO is greater than that resulting from the addition of MgO or excess Al_2O_3. This question is related to the relative solubilities of these oxides in the liquid. The relations involving the phases Kf−Q−V−Opx−Bi−L are shown in Fig. 5 – invariant point [I$_2$] (Luth 1967). Some experimental data are available concerning the reaction phlogopite+quartz+vapor=enstatite+liquid (Wones and Dodge 1977; Bohlen et al. 1983). Up to 10 kbar, these studies indicate that this reaction has a negative slope and occurs only a few degrees above the Kf+Q+V=L curve. If we trust these experiments, we must conclude that, at high pressure, the assemblage Phl−Q−V melts at lower temperature than the assemblage Mu−Q−V. Since the effect of adding Fe to the system is likely to extend the stability field of liquid and orthopyroxene, the Bi−Q−V assemblage will melt at even lower temperatures. This is contrary to all available natural observations. The fact that phlogopite and quartz never disappeared in the experiments of Bohlen et al.

(1983), and the likely stabilizing effect on biotite of other components (e.g. Ti) may be invoked to explain this discrepancy. We have ascribed a nearly vertical slope to the reaction (8) Bi+Q+V=L+Opx in the temperature range 800°–850° C in Fig. 5. This is in good agreement with the preliminary report of experiments on the Mg end-member of that reaction by Peterson and Newton (1987).

The last pseudobinary system of interest in this model is the $FeSiO_3$ or $MgSiO_3 - Al_2SiO_5$ system (SiO_2, H_2O in excess) (Fig. 4). Very few experimental data are available. The assemblage enstatite+quartz+vapor melts congruently at about 1000° C and 20 kbar (Kushiro and Yoder 1969). The addition of FeO to the system should lower the temperature of melting. On the other hand, the stability of almandine at elevated pressures and temperatures has been studied by Keesmann et al. (1971) who showed that almandine melts incongruently to hercynite+quartz+liquid at 10 kbar. At pressures between about 12 and 20 kbar the products of incongruent melting are hercynite+liquid only, and at still higher pressures almandine melts congruently. In this paper, and on the basis of the liquid compositions given by Keesmann et al. (1971), we will assume that spinel+quartz melt incongruently to liquid+aluminum−silicate. Using these data, the hypothetical pseudobinary system (Fe, Mg)SiO$_3$ − Al$_2$SiO$_5$ is represented in Fig. 4 for a pressure above the stability limit of cordierite. At one atmosphere, Fe and Mg cordierite melt incongruently to produce mullite+liquid+/−tridymite (Schairer and Yagi 1952; Schreyer and Yoder 1959). This reaction has been extrapolated towards higher pressures, considering that the temperature of incongruent melting is lowered by pressure.

The addition of both FeO−MgO and excess Al_2O_3 transforms the one-component limiting system $KAlSi_3O_8$ into a 3 component system $Al_2SiO_5 - (Fe, Mg)SiO_3 - KAlSi_3O_8$ ($+$excess SiO_2 and H_2O) and still increases the stabil-

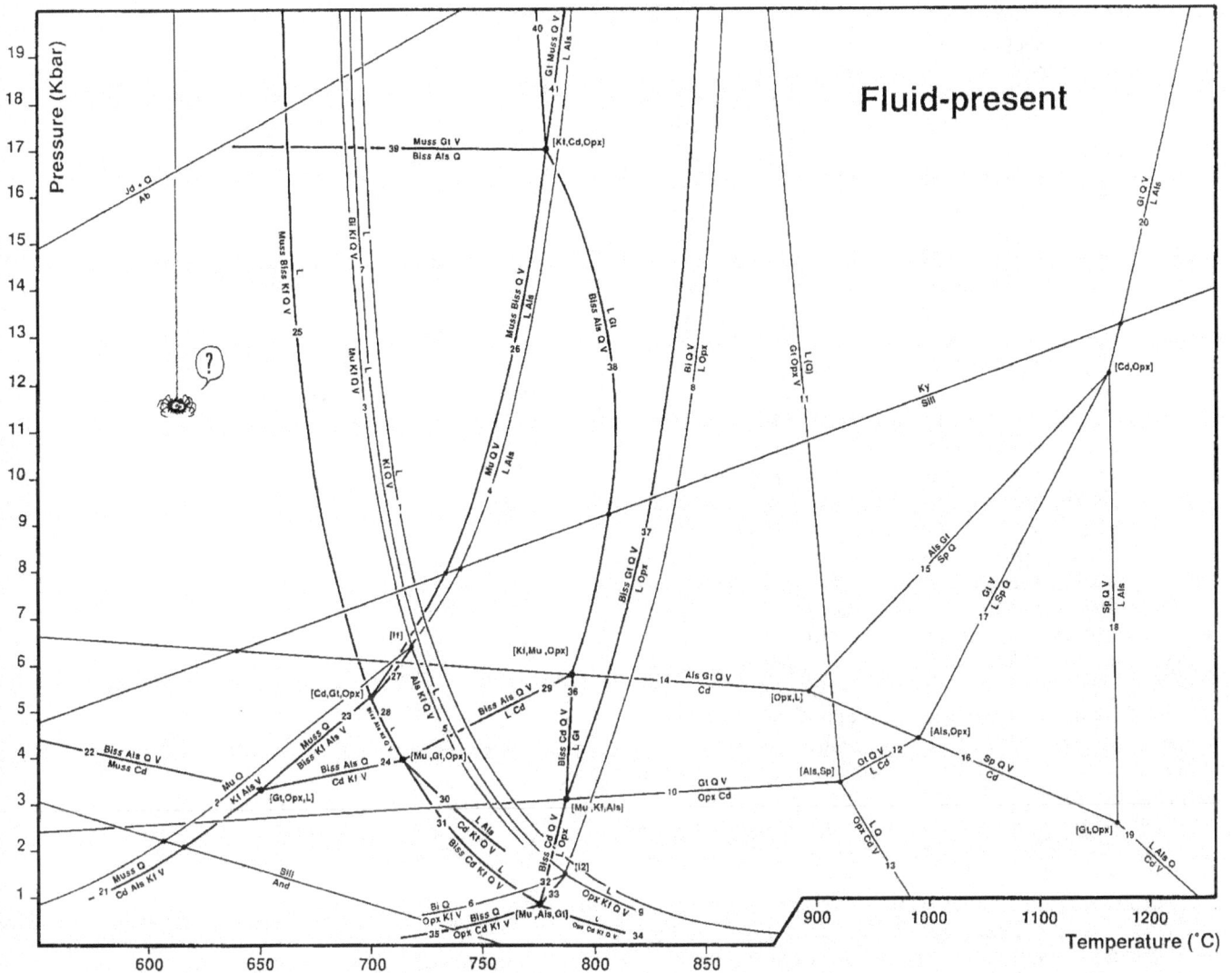

Fig. 5. Pressure-temperature diagram for the fluid-present (pure H_2O) and quartz saturated partial melting of metapelites and metagray-wackes (XMg close to 0.5). Reaction 1 refers to the pseudo-unary system $KAlSi_3O_8$ with the phases Kf, Q, V, L. Reactions 2–5 around the invariant point [I_1] refer to the pseudo-binary system $KAlSi_3O_8 - Al_2SiO_5$, with the phases Kf, Mu, Als, L, Q, V. Reactions 6–9 around [I_2] refer to the pseudo-binary system $KAlSi_3O_8 - (Fe, Mg) SiO_3$, with the phases Kf, Bi, Als, L, Q, V. Reactions 10–20 around the invariant points [Als, Sp], [Opx, L], [Als, Opx], [Gt, Opx], and [Cd, Opx] refer to the pseudo-binary system $(Fe, Mg) SiO_3 - Al_2SiO_5$, with the phases Gt, Als, Sp, Cd, Opx, L, Q, V. Note the refractory character of this subsystem. Reactions 21 to 41 (*heavy lines*) around the invariant points [Gt, Opx, L], [Cd, Gt, Opx], [Mu, Gt, Opx], [Mu, Als, Gt], [Mu, Kf, Als], [Mu, Kf, Opx], and [Kf, Cd, Opx] refer to the pseudo-ternary system $Al_2SiO_5 - KAlSi_3O_8 - (Fe, Mg) SiO_3$ with the phases Als, Mu_{ss}, Kf, Bi_{ss}, Opx, Gt, Cd, L, Q, and V. Note that the temperature scale changes above 900° C. The following reactions are shown by number only: (*27*) $Mu_{ss} + Q = L + Als + Kf$; (*33*) $Bi_{ss} + Q + V = L + Opx + Kf$; (*40*) $Bi_{ss} + Mu_{ss} + Q + V = L + Gt$

ity field of the liquid until saturation in these two extra components is reached. In this multisystem the phases involved are aluminum—silicate, muscovite (ss), K-feldspar, biotite (ss), orthopyroxene, garnet, cordierite, spinel, and liquid (+quartz and vapor). In this compositional triangle, assumptions have to be made about the chemographic position of the liquid. In this respect one of the questions is to which side of the extension of the join biotite (ss) – orthopyroxene the liquid composition lies. Considering (i) that in the absence of sillimanite (which is often the case when orthopyroxene is present) biotites are not aluminum-rich and (ii) that orthopyroxenes can be significantly alumi-nous, we consider that the composition of the liquid is lo-cated on the sillimanite side of the extended join Opx – Bi_{ss}. Grant (1985) chose the other case. When the liquid is lo-cated on the sillimanite side of the join Bi_{ss} – Opx, as in

Fig. 5, the reaction (8) $Bi + Q + V = L + Opx$ is located at higher T than the reaction (37) $Bi_{ss} + Gt + Q + V = L + Opx$ (excess Al_2O_3 in the system decreases the stability field of $Bi + Q + V$). When the liquid is located on the Kf side of the join Bi_{ss} – Opx, the reaction (8) is located at lower T than the $Bi_{ss} + Q + V = Gt + Opx + L$ (excess Al_2O_3 in the system increases the stability field of $Bi + Q + V$). This prob-lem is related to the relative activities of Al_2O_3 in biotite and liquid. Selecting the first possibility, we keep the sym-metry with the relations involving muscovite around I_1 and [Cd, Gt, Opx]. The liquid composition R15-B reported by Hoffer and Grant (1980) supports this choice though the alternative chemography cannot be completely ruled out. On the basis of his experiments, Seifert (1976) concluded that the quartz, K-feldspar, cordierite, vapor assemblage melts congruently. According to Grant (1985) the simplest interpretation is that a thermal barrier exists generating the relations shown in his Fig. 3.6 indicating that an aluminous minimum melt can be produced at low pressure. For the sake of simplicity, this aspect has been neglected in our model (Fig. 5). The breakdown of sillimanite, biotite (ss), and quartz in the presence of albite, to melt + cordierite was determined experimentally at conditions of $PH_2O = P$ total by Hoffer (1978). This reaction is the analogue of reaction (29) $Bi_{ss} + Als + Q + V = L + Cd$ with the addition of Na to the system. Hoffer found that the reaction curve has a steep negative slope (nearly vertical) at temperatures of about 650° C. In our $P-T$ diagram we drew a positive slope for this reaction because of the location of the reactions (24) $Bi_{ss} + Als + Q = Cd + Kf + V$ and (38) $Bi_{ss} + Als + Q + V = L + Gt$. This is the same relationship as shown in Grant (1985) Fig. 3.10 for the system KFASH. Some hydrothermal experiments were performed on reaction (38) at 4, 6, and 8 kbar in an internally heated pressure vessel (Vielzeuf 1980b), starting with a mixture of natural minerals: biotite (XMg 0.50), sillimanite, K-feldspar, garnet (XMg 0.35), and quartz. The QFM buffer was used to control the oxygen fugacity. These experiments show that the Bi_{ss} – Sill assemblage is still stable at 8 kbar–800° C. The composition of the liquid in equilibrium with Bi_{ss}, Sill, Q, Gt, and V at 8 kbar and 800° C is given in Table 4. Thus, reaction (38) must be located at temperatures as high as 800° C, with a nearly vertical slope if we compare it to its subsolidus equivalent $Bi_{ss} + Sill + Q = Gt + Kf + V$ (Phillips 1980; Vielzeuf 1984). It seems possible that Hoffer mapped the solidus in the system Ab + Q with excess alumina and addition of FeO and MgO, rather than the reaction $Bi_{ss} + Ab + Sill + Q + V = L + Cd$ itself. Concerning the reaction $Bi + Ab + Q + Cd + V = L + Opx$, it seems unlikely that cordierite and orthopyroxene of average compositions will coexist at pressures as high as 7 kbar as is implied by Hoffer and Grant (1980). Finally, due to the uncertainty on the curvature of reaction (38) $Bi_{ss} + Als + Q + V = L + Gt$ and the fact that this reaction and the reaction (26) $Mu_{ss} + Bi_{ss} + Q + V = L + Als$ have very steep slopes, the uncertainty in the pressure location of the invariant point [Kf, Cd, Opx] is very important.

The $P-T$ diagram of Fig. 5 has been constructed on the basis of these data and assumptions; it represents an hypothetical model for the water saturated melting of pelites and graywackes between 2 and 20 kbar and 600°–1250° C. Our model differs from those proposed by Thompson (1982) and Grant (1985) in the following points:
– the choice in some chemographic relationships differ, namely: the relations between biotite – orthopyroxene – liquid are different from those used by Grant (1985); cordierite is assumed to be hydrous (see Thompson 1982, Fig. 4); reaction 39 $Mu_{ss} + Gt + V = Bi_{ss} + Als + Q$ is considered to be vapor present (see Thompson 1982, Fig. 4; Grant 1985, Figs. 3.10 and 3.11);
– reactions 10 $Opx + Cd = Gt + Q + V$ and 35 $Bi_{ss} + Q = Opx + Cd + Kf + V$ do not intersect in the subsolidus domain (see Thompson 1982, Fig. 4);
– this model is made for a single, average Fe – Mg ratio (XMg ≈ 0.50), and is thus less general but perhaps more applicable to common natural rocks;
– the mutual relationships between simple (AKSH, AFMSH) and more complex (KFMASH) systems are shown allowing the construction of consistent $P-T$, liquidus AKF, and $T-X$H$_2$O diagrams;

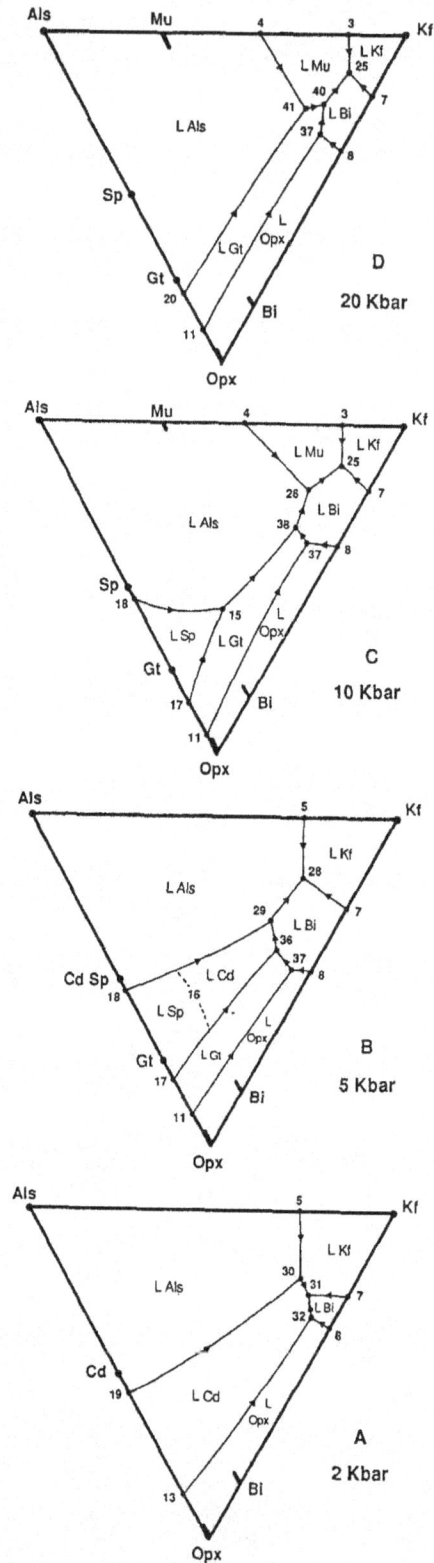

Fig. 6. Isobaric liquidus diagrams at $PH_2O = 2, 5, 10$, and 20 kbar derived from *Fig. 5*

– the diagram covers the complete temperature range (600°–1250° C) relevant to melting in pelites and graywackes.

As shown by Eggler (1973), the simultaneous use of $P-T$ and liquidus diagrams is a powerful tool to depict partial melting processes (see also Grant 1985). Four

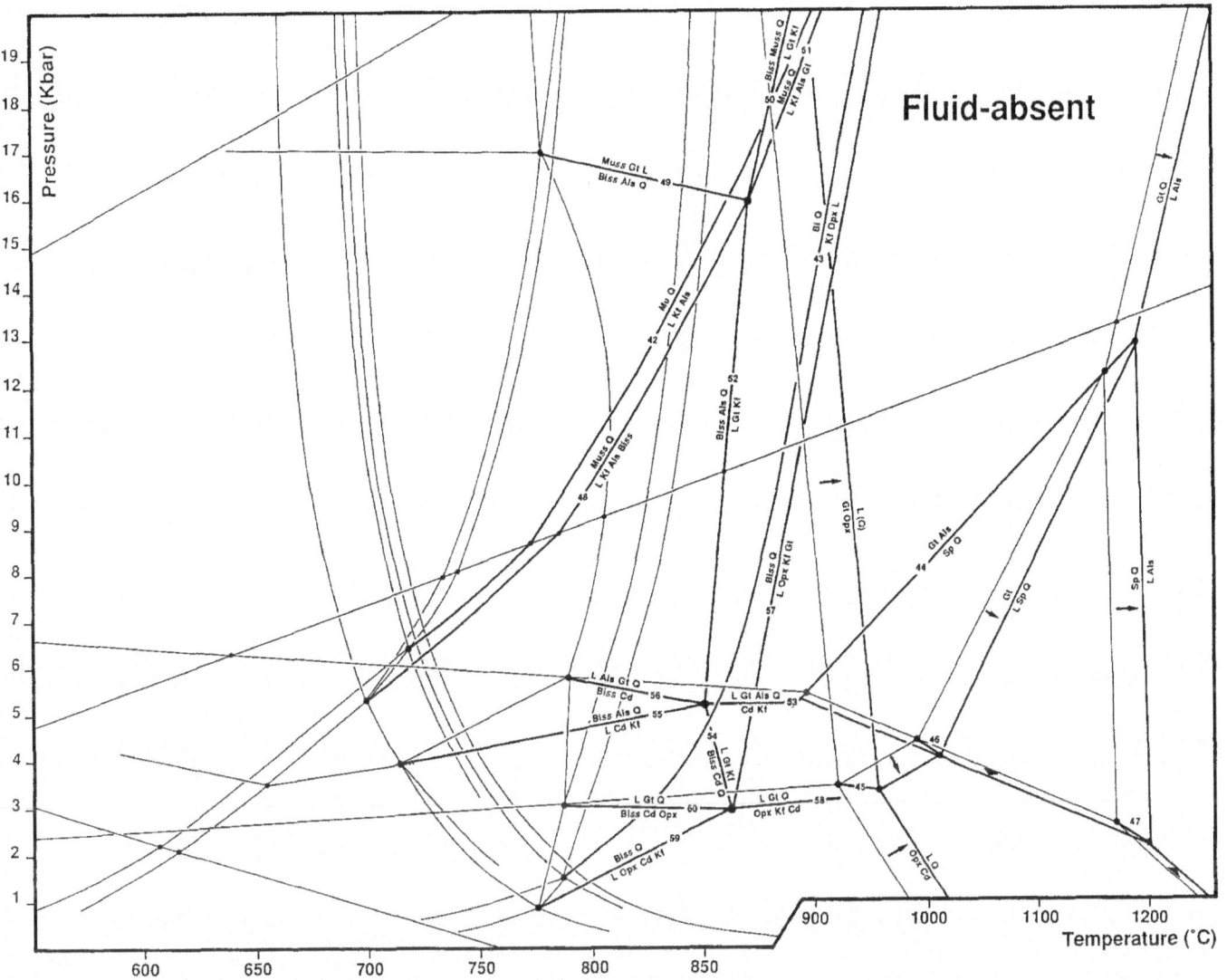

Fig. 7. $P-T$ diagram for the fluid-absent and quartz-saturated partial melting of metapelites and metagraywackes (XMg close to 0.5). Some reactions from *Fig. 5* are also shown (*light lines*). *Arrows* indicate the direction of displacement of the reactions involving only melt as a potential hydrous phase, in response to lowering the activity of water. The following reactions are shown by number only: (45) Opx + Cd = Gt + L; (46) Cd + Gt = Sp + L; (47) Cd = L + Als + Sp

isobaric, polythermal, liquidus diagrams derived from Fig. 5 are displayed in Fig. 6 in order to show the possible evolution of liquid compositions.

The addition of an albitic component will shift the curves involving a liquid towards lower temperatures (approximately 100° C) until an albitic plagioclase appears on the low temperature side of the reaction. The further addition of an anorthitic component will reduce this effect. The influence of these two components is thoroughly discussed by Thompson and Algor (1977), Thompson and Tracy (1979), and Grant (1985). Other components such as Ti are likely to have effects that are important but difficult to quantify. However, following Grant (1985 p. 116), it seems that "...the system KFMASH remains most useful as far as understanding partial melting processes in pelitic rocks is concerned".

5.2.2 Fluid-absent melting

Fluid-absent melting has been considered by Yoder and Kushiro (1969), Robertson and Wyllie (1971), Maaløe and

Wyllie (1975), Thompson (1982), and Grant (1985), among others. General geometric relations have been discussed by Eggler (1973) and Eggler and Holloway (1977). From the model constructed under fluid-present conditions at aH$_2$O = 1, it is possible to derive a model for fluid-absent conditions (Fig. 7) using Schreinemakers' principles.

Fluid-present curves, around an invariant point, can be displaced by lowering aH$_2$O. This situation can be obtained by externally buffering the water activity or, in other words, by diluting the pure H$_2$O with another fluid-species, possibly CO$_2$ (Fyfe et al. 1978). A fictive line can be drawn along which a given invariant point in the water-saturated system evolves in response to lowered activity of water. This line which could be considered as a composite vector of substitution of CO$_2$ − H$_2$O in the fluid-phase, has no physical reality in the pure H$_2$O, fluid-present situation. In nature, we expect that metamorphism or melting will occur under either fluid-present of fluid-absent conditions and thus that fluid-present and fluid-absent conditions will not overlap in real situations. This is one reason why fluid-present and fluid-absent reactions have not been drawn on a single diagram.

Under fluid-absent conditions, in the domain of melting, the activity of water in the system can be buffered by the presence of an assemblage involving a crystalline hydrate and its breakdown products. Under such conditions, the only reactions that can occur are fluid-absent providing that the melt is able to dissolve more water than is present in the crystalline hydrates.

The melting reaction obtained in a fluid-absent situation and the line defined by the migration of the invariant point (in a mixed-fluid system) will coincide if the diluting species in the fluid-phase is insoluble in all the other phases, and in the melt in particular.

When fluid-absent reactions intersect, they are likely to generate additional invariant points of "higher order" (Vielzeuf 1983); in this respect, the effects on the geometrical phase-relations of changing the composition of the fluid-phase are comparable to what has already been described for Fe−Mg substitutions (Vielzeuf 1983; Vielzeuf and Boivin 1984). The shift in the positions of the reactions due to a change in composition of the fluid-phase can be sufficient to reverse the grid by crossing an invariant point of higher order.

The various fluid-present reactions shown in Fig. 5 (for pure water) may have a different status in the fluid-absent situation. Those involving a crystalline hydrate are replaced by fluid-absent reactions involving additional phases. It is interesting to note that, due to chemographic relations, some phases present on the low-temperature side of a fluid-present curve may be on the high-temperature side of the corresponding fluid-absent reaction (e.g. Bi_{ss} in reactions 26 and 48; and Gt in reactions 37 and 57). On the contrary, reactions involving only melt as a potential hydrous phase (e.g. reactions 11, 12, 13, 17–20) will still exist under fluid-absent conditions but will be displaced towards higher temperatures as a function of reduced activity of water. Note that there is a significant uncertainty on the T location of these reactions in Fig. 7. Reactions involving no hydrous phases (e.g. (15) Als + Gt = Sp + Q) are insensitive to the presence or absence of a fluid-phase. In this model, cordierite is considered as a hydrous mineral but can be anhydrous also, at high temperatures. The model presented in Fig. 7 is derived from the fluid-present model and the few available experiments in the fluid-absent domain.

Data for the reaction (42) Mu + Q = L + Kf + Als were obtained by Storre (1972) between 7 and 20 kbar. Reaction (48) Mu_{ss} + Q = Bi_{ss} + Als + Kf + L is the analogue of reaction (42) in the system with FeO and MgO. From the present experiments and those presented by Le Breton and Thompson (in prep.), it is shown that reaction (52) Bi_{ss} + Als + Q = L + Gt + Kf is located at about 860° C and has a nearly vertical slope. Reactions (52) and (48) intersect and generate an invariant point. If we consider that reaction (49) has a very small slope, not very different from that of reaction (39) Bi_{ss} + Als + Q = Mu_{ss} + Gt + V (Thompson (1982) and Grant (1985) consider that reaction (39) is fluid-absent and degenerate), then we have, with this invariant point, an indirect indication of the location of reaction (39) with respect to pressure.

The melting of phlogopite−quartz under fluid-absent conditions has been determined from 5 to 20 kbar by Bohlen et al. (1983). At 10 kbar, the temperature of melting is close to 900° C. Peterson and Newton (1987) found a much lower temperature of about 815° C implying a gap of only 25° C between the fluid-present and the fluid-absent

reactions. This latter result is incompatible with the location of the reaction Bi_{ss} + Als + Q = L + Gt + Kf at about 860° C. This apparent discrepancy may be due to the effects of other components (e.g. excess Al, Fe, Ti) in the system. The modifications of the phase-relations involving quartz, K-feldspar, phlogopite, enstatite, liquid, and vapor proposed by Grant (1986b) and in particular the existence of a thermal divide, are not taken into consideration in our Figs. 5 and 7 because they are based on the experimental results of Wendlandt (1981) which are difficult to interpret.

6 Interpretation of the experiments at 10 kbar

The experiments at 10 kbar can be interpreted in the light of the above model. Since no experiments were done below 750° C, the interpretation of the breakdown of chlorite, staurolite, and muscovite is hypothetical. The following reactions could be responsible for the disappearance of chlorite and staurolite respectively:

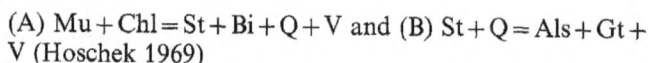

(A) Mu + Chl = St + Bi + Q + V and (B) St + Q = Als + Gt + V (Hoschek 1969)

However it is unlikely that equilibrium was reached during the rapid increase in temperature in our experiments.

In the absence of K-feldspar in the starting material (excess aluminum−silicate on the high pressure-low temperature side of the Mu + Q = Kf + Als + V reaction) the eutectic melting reaction Mu + Bi + Kf + Pl + Q + V = L did not occur. A small amount of free water was available in the starting material (adsorbed water plus that released by the breakdown of chlorite, staurolite, and part of the muscovite). Thus, we believe that the following reaction:

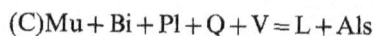

(C) Mu + Bi + Pl + Q + V = L + Als

took place until disappearance of the vapor phase, and that the first drop of liquid was saturated in water. A mass balance approach indicates that the amount of liquid which can be produced by reaction (C) is close to 4 Wt%. This reaction is able to consume all the free water and thus it represents the transition between a fluid-present and a fluid-absent situation.

The disappearance of muscovite is probably a result of the fluid-absent reaction:

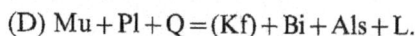

(D) Mu + Pl + Q = (Kf) + Bi + Als + L.

Mass balance calculations show that reaction (D) produces only a tiny amount of liquid ($\approx$1%) and also some K-feldspar which could possibly enter as a solid-solution in the plagioclase or dissolve in the liquid already present. From these results, we conclude that under fluid-absent conditions the muscovite melting stage of metapelites generates only a small amount of nearly water-saturated liquid. However, the proportion of melt is strongly dependent on the proportion of muscovite in the starting material.

The next step corresponds to the "biotite melting stage" marked by the reaction

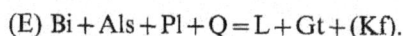

(E) Bi + Als + Pl + Q = L + Gt + (Kf).

The mass balance approach indicates that, contrary to what is suggested by previous models (including ours), K-feldspar is not required as a product of such reactions as (E). The appearance of K-feldspar depends upon the amount of biotite present in the starting rock (Thompson, pers. com.). These calculations indicate also that a large amount

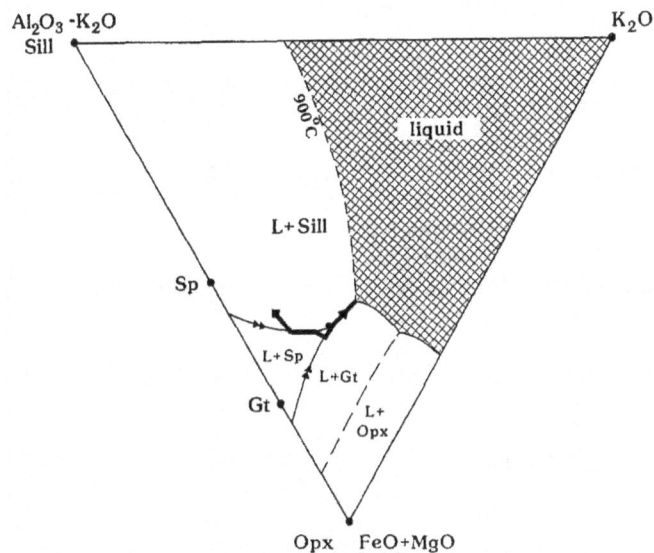

Fig. 8. Phase relations at high temperature and 10 kbar in the pseudo-ternary AKF diagram. The heavy line shows the path followed by the liquid. A schematic liquid field is shown for $T=900°$ C. Compare with *Fig. 2*

(≈ 45–50%) of water-undersaturated melt can be generated by this reaction. This amount depends on the modal proportion of biotite in the starting material. In order to confirm this important result, calculations were done based on modelling of fluid-absent melting in the simplified system Q$-$Or$-$Ab$-$H$_2$O (Clemens and Vielzeuf 1987). The amounts of melt predicted in this model are consistent with those produced in these experiments.

Above 875° C, the evolution of the modal proportions and of the liquid composition can be interpreted in terms of selective dissolution of plagioclase and garnet in the liquid and crystallization of a small amount of sillimanite and quartz. The intervention of the reaction Pl+Q=L is ruled out because, in our runs, the modal proportion of quartz increased instead of decreasing.

From this point, the experimental data can be interpreted in the light of the path shown on the ternary liquidus diagram in Fig. 8. With increasing temperature, and particularly above 1000° C, the liquid dissolves a large proportion of the garnet (and a small amount of aluminum$-$silicate) until the boundary of the Gt$-$Als assemblage is reached (between 1050° and 1100° C). This limit corresponds to the peritectoid reaction Gt+Als=Sp+Q. This observation is in agreement with the experimental data reported by Bohlen et al. (1986) on the Fe end-member of that reaction. Interestingly, the Fe$-$Mg divariant field of this reaction is probably narrow since the Fe$-$Mg partitioning between spinel and garnet is small (Vielzeuf 1983). The peritectoid reaction Gt+Als=Sp+Q stops when all the sillimanite is consumed. Thereafter the liquid evolves along the Gt$-$Sp cotectic which corresponds to the peritectic reaction Gt=L+Sp+(Q) until all the garnet is consumed. Then the liquid leaves this peritectic and dissolves spinel (and quartz) until a second peritectic reaction is reached: Sp+(Q)=L+Als. It is interesting to note that the aluminum$-$silicate (sillimanite or mullite) disappears when Sp and Q crystallize and re-appears when the Sp$-$Q assemblage breaks down. Above 1200° C all the spinel disappears and the liquid dissolves the aluminum$-$silicate until the liquidus is reached at about 1250° C.

For a granitic liquid composition at a given pressure, the temperature at which the liquid coexists with quartz and plagioclase of a certain composition places close constraints on the H$_2$O content of the melt (Nekvasil and Burnham 1987). For instance, the 875° C run products include quartz and An$_{30}$ plagioclase. For a granitic melt with the 875° C composition to coexist with those phases at 10 kbar and 875° C requires a melt water content of about 4 Wt.% H$_2$O (based on the Burnham-Nekvasil [1986] model). This is in very close agreement with the 3.7 Wt.% calculated by mass balance (Table 3). These H$_2$O contents are equivalent to H$_2$O activities of about 0.3, illustrating the highly H$_2$O-undersaturated condition existing just above the biotite-out temperature.

The important question is, how wide is the T interval over which biotite melts? The actual width will depend in part on the melting loop caused by Mg$-$Fe partitioning between the phases, and in part on the reaction geometry with respect to H$_2$O. The initial H$_2$O content of the rock is lower (2.15%) than the H$_2$O content of the melt (minimum of 3.7%). This would place the bulk composition to the left (away from the H$_2$O end-member) of any peritectic or eutectic involving biotite in diagrams such as shown by Eggler and Holloway (1977, Fig. 2). Such a position would result in no melting interval for the simple system case considered by Eggler and Holloway (1977). In the system studied here, we conclude that the melting interval will be due solely to the solid solutions, and that it will be small, probably less than 20° C.

$T$$-$$XH_2$O diagram – In all crustal geological systems undergoing partial fusion, water is an important limiting factor in determining the amount of melt formed at a given temperature. Figure 9 displays a T-composition section for a pseudo-binary join A$-$H$_2$O (Yoder and Kushiro 1969; Eggler 1973; Eggler and Holloway 1977; Whitney 1975) where A is a composition in the Al$_2$O$_3$$-K_2O/K_2$O/FeO+ MgO triangle. Quartz is in excess and it has been considered that biotite is in excess with respect to muscovite in the reaction Mu+Bi+Q+V+(Pl)=L+Als, sillimanite in excess with respect to biotite in the reaction Bi+Als+(Pl)+ Q=L+Gt+Kf, and garnet in excess with respect to sillimanite in the reaction Gt+Als=Sp+Q. Temperature coordinates are given for a system in which plagioclase coexists with the other phases below 900° C. This diagram was constructed by recording the phases coexisting along the A$-$$XH_2$O join at various temperatures. It illustrates the mutual relationships between excess fluid-present and fluid-absent reactions. The dashed line schematically shows the path of the liquid for the bulk composition studied. This evolution should be compared to what is shown in the $P$$-$$T$ and liquidus diagrams. It can be seen that below 860° C the water content of the liquid is buffered by the presence of hydrous minerals. This ceases to be the case at higher temperatures. $P$$-$$T$, $T$$-$$XH_2$O, and liquidus diagrams represent three complementary useful ways to understand the different aspects of the melting processes.

7 Discussion

7.1 Magma production

The experimental observations lead to the following considerations: melts below 860° C and corresponding to the mus-

272

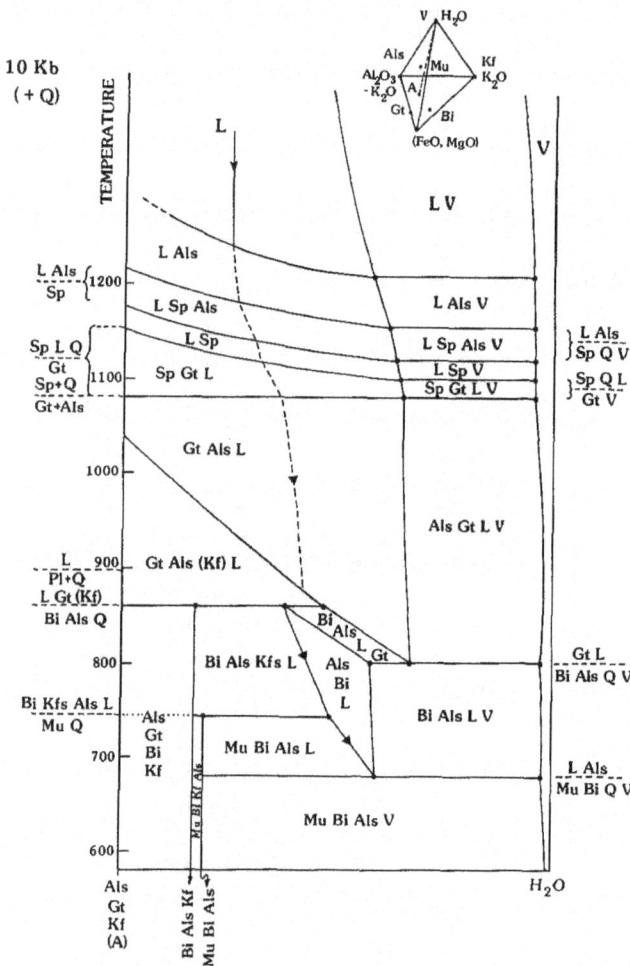

Fig. 9. $T-X$ section for the pseudo-binary join $A-H_2O$ at 10 kbar, where A is a composition in the $Al_2O_3 - K_2O/K_2O/FeO + MgO$ triangle. Quartz is in excess and temperature coordinates are given for a system in which plagioclase coexists with the other phases below 900° C

Such liquid fractions (45–50%) are above the critical melt fraction determined by van der Molen and Paterson (1979) and thus will form a mobile, buoyant magma as soon as biotite disappears. The residual minerals are quartz, garnet, sillimanite, and plagioclase (or K-feldspar). Some of these minerals may be carried up with the magma as a restite fraction (e.g. White and Chappel 1983) or remain in the source region with some residual liquid, where they would form characteristic aluminous granulites. In the lower crust, temperatures as high as 860° C seem reasonable, in particular during the major thermal events which occur during the late stages of evolution of orogenic belts (Albarède 1976; Couturié and Kornprobst 1977; England and Richardson 1977; England and Thompson 1986; see also Pin and Vielzeuf, 1983 for the Hercynian belt, and Frey et al., 1974 for the Alps).

The production of a large amount of melt around 850° C, even under such limiting conditions as the fluid-absent situation, has the important consequence of buffering the temperature of metamorphism in the area of melting. The temperature of metamorphism will not exceed 850° C until the process of partial melting is completed. Melting of 50% of the crust must require a large amount of energy which will rarely, if ever, be available. As a result, temperatures of metamorphism in excess of 850° C will rarely be exceeded in a metapelitic crust undergoing a first thermal event. On the contrary, in the case of a recycled crust in which partial melt had already been extracted in a previous cycle, the temperature of metamorphism would no longer be buffered and temperatures close to 1000° C could be reached if there were sufficient heat supply. In this case, and depending on the composition of the pelitic rocks (restites whose composition is a function of the pressure during the first partial melting event), mineral assemblages such as spinel + quartz, sapphirine + quartz, orthopyroxene + sillimanite, and osumilite may appear. If this is true, these assemblages should be more common in the regional metamorphic terranes belonging to old, recycled cratons. A review of the literature indicates that this is the case. Such assemblages have been described in the old basement of Antarctica (Ellis 1980; Grew 1982a); Eastern Siberia (Karsakov et al. 1975); India (Grew 1982b); Hoggar, Algeria (Ouzegane 1981), Namaqualand, South Africa (Waters 1986); Rogaland, Norway (Maijer et al. 1977) and Central Labrador, Canada (Morse and Talley 1971).

The buffering effect of metamorphism by partial melting processes appears general, and the absence of such a thermal buffer is a resonable explanation for the occurrence of these unusual, high-temperature parageneses.

In a paper dealing with osumilite − sapphirine − quartz granulites from Enderby Land, Ellis (1980) proposed that crustal rocks in other granulitic terranes may have experienced such extreme conditions of metamorphism also. Ellis further proposed that uplift at high temperature could have eradicated petrographic evidence for this early $P-T$ history. He concluded that the $P-T$ path of cooling and uplift may be more important than the extreme conditions of metamorphism in explaining why such high-temperature granulites are not exposed elsewhere in the world. As an example, Ellis considered the Bohemian massif belonging to the Hercynian belt of median Europe, and suggested that such extreme $P-T$ mineral assemblages from an earlier part of the rock history were obliterated by a later thermal event. Even if metamorphic conditions changed from high-

covite melting stage are water-rich and are generated in small amount (<10%) in most common metapelites. As suggested by Burnham (1967) and Clemens (1984), at 5 kbar, these nearly H_2O-saturated melts will probably form veins, pockets and small plutons unable to move from their site of generation. The migmatitic zones developed in metapelites close to the muscovite-out isograd probably have such an origin. At 10 kbar, such liquids are less likely to be H_2O-saturated since the solubility of H_2O in melt increases with P (Goranson 1931; Burnham 1979). As a result they might rise some distance, but still not very far.

One of the most important results of this study is the fact that, at 7 and 10 kbar, the breakdown of the biotite + sillimanite + plagioclase + quartz assemblage produces a large amount of liquid within a narrow range of temperature (850°–875° C). This conclusion can be extended to lower pressures since (i) the proportion of melt produced from a given starting material and under fluid-absent conditions increases as pressure decreases (Clemens and Vielzeuf 1987), and (ii) the reaction $Bi_{ss} + Als + Pl + Q = L + Gt + (Kf)$ is nearly vertical in $P-T$ space. Such liquids are strongly water-undersaturated; they have water contents of about 4% while a water-saturated liquid of that composition would have a water content of about 12% at 10 kbar.

P to low-P and high-T (Jakeš 1969; Marchand 1974; Lasnier 1977; see also Pin and Vielzeuf 1973 for a summary), early granulitic assemblages belonging to the high-pressure granulite facies are widespread and well preserved in the Hercynian belt. Pelitic granulites here are composed of kyanite + garnet + rutile + K-feldspar + plagioclase + quartz +/− biotite and geobarometry yields pressure estimates up to 21 kbar (Vielzeuf 1984), higher than those recorded by Ellis in Enderby Land. Furthermore, there is no evidence of very-high-temperature metamorphism in the Hercynian belt. Thus, we do not believe that the conditions of formation of the high-temperature granulites are common to all mountain belts. Rather we consider that the presence of a recycled crust is required.

7.2 An application to the Hercynian in the Pyrenees

An example of the process of continental differentiation (which is not meant to be generalized) is provided by the North Pyrenean Zone. During Alpine times (40–100 Ma), this zone underwent major tectonic processes (crustal thinning related to transcurrent movements and followed by a succession of compressive stages) which allowed the outcropping of various parts of the lithosphere from the upper mantle (Lherzolites) to the upper crustal levels (Vielzeuf and Kornprobst 1984). Thus, it is possible to reconstitute a cross-section of the crust as it was left after the Hercynian orogeny (300–400 Ma) (Vielzeuf 1984). It is believed that an important thermal anomaly, occurring around 310 Ma in response to the extension following collisional processes (Pin and Vielzeuf 1983), is responsible for a complete (re-)structuring of the crust. The following processes could be involved:

– Following compressive stages, the onset of crustal extension may allow upwelling of the asthenosphere accompanied by the intrusion of mafic magmas (now observed as layered complexes) into the lower crust. These processes induced fluid-absent melting of the lower crust. Fourty to sixty percent of crustal melts were generated at temperatures buffered around 850° C (the biotite melting stage). These H₂O-undersaturated melts, or at least some proportion of them, were able to segregate, leaving behind a residue composed of garnet, sillimanite, plagioclase, quartz, rutile and residual liquid. On the basis of mass balance, some granulitic paragneisses found in the North Pyrenean Zone are interpreted as metapelites from which up to 40% of granitic melt was extracted (Vielzeuf 1980b). Following Leterrier (1972) and Debon (1975) it is believed that, in some cases, crustally and mantle derived liquids are able to interact and contaminate each other to generate I-type granites.

– Closer to the surface, the fluid-present melting of metapelites at about 2–3 kbar and 650° C (the muscovite melting stage) generated a thick layer of migmatites unable to migrate because of their nearly H₂O-saturated character.

– The upper levels of the crust underwent metamorphism within a high temperature gradient and they are now characterized, in the field, by a rapid succession of isograds parallel to that for anatexis.

– The stratified mafic complexes from the lower crust in this region yield an age of 315 Ma (Postaire 1982; Respaut and Lancelot 1983) while the last granites were emplaced 280 Ma ago (Michard-Vitrac et al. 1980). As a result, it is believed that the process of crustal differentiation in the

Pyrenees was promoted by partial melting, and lasted 20 to 40 Ma.

Acknowledgements. This study was undertaken while DV was at Arizona State University on sabbatical leave from CNRS, Clermont-Ferrand. Funding for this was provided principally through a NATO research fellowship and CNRS. Experimental work was supported by NSF grant EAR 8407742-01 to JRH. Part of the work was also funded by the INSU through ATP "Transfert" contracts 1558 and 1592 to DV. SEM facilities were provided by the Center for Solid State Science at ASU. Image processing was done at the Physics laboratory (LERM) of the University of Clermont II with the help of D. Lafon. DV also thanks the inhabitants of Depths of the Earth Inc. for providing a stimulating and amicable work environment. Tracy Rushmer gave advice and assistance with piston-cylinder techniques. Dave Joyce is particularly thanked for animated discussions of phase relations in partial molten systems. John Clemens suggested constructive revisions to the various versions of the manuscript. Rapid, thorough, and critical reviews were provided by Jim Grant and Alan Thompson.

References

Abbott RN Jr., Clarke DB (1979) Hypothetical liquidus relationships in the subsystem Al₂O₃−FeO−MgO projected from quartz, alkali feldspar and plagioclase for a(H₂O)<1. Can Mineral 17:549–560

Albarède F (1976) Thermal models of post-tectonic decompression as examplified by the Haut-Allier granulites (Massif Central, France). Bull Soc Geol Fr 18:1023–1031

Arps CES, van Calsteren C, Hilgen JD, Kuijper RP, den Tex E (1977) Mafic and related complexes in Galicia: an excursion guide. Leidse Geol Meded 51:3–94

Bard JP (1969) Le métamorphisme régional progressif des Sierras d'Aracena en Andalousie occidentale (Espagne). Sa place dans le segment hercynien sub-ibérique. Thèse d'Etat, Montpellier, 398p

Blatt H, Middleton G, Murray R (1972) Origin of Sedimentary rocks. Prentice-Hall Inc., 634 p

Bohlen SR, Boettcher AL, Wall VJ, Clemens JD (1983) Stability of phlogopite-quartz and sanidine-quartz: a model for melting in the lower crust. Contrib Mineral Petrol 83:270–277

Bohlen SR, Dollase WA, Wall VJ (1986) Calibration and applications of spinel equilibria in the system FeO−Al₂O₃−SiO₂. J Petrol 27:1143–1156

Bohlen SR, Essene EJ, Boettcher AL (1980) Reinvestigation and application of olivine−quartz−orthopyroxene barometry. Earth Planet Sci Lett 47:1–10

Burnham CW (1967) Hydrothermal fluids at the magmatic stage. In: Barnes HL (ed) Geochemistry of hydrothermal ore deposits. Holt, Reinhart and Winston, New York, pp 38–76

Burnham CW (1979) The importance of volatile constituents. In: Yoder HS (ed) The evolution of the igneous rocks (Fiftieth anniversary perspectives), Princeton University Press, Princeton, pp 439–482

Burnham CW, Nekvasil H (1986) Equilibrium properties of granite pegmatite magmas. Am Mineral 71:239–263

Chopin C (1984) Coesite and pure pyrope in high-grade blueschists of the Western Alps: a first record and some consequences. Contrib Mineral Petrol 86:107–118

Clemens JD (1981) The origin and evolution of some peraluminous acid magmas (experimental, geochemical and petrological investigations). Unpubl. Ph. D. Thesis, Monash University, Australia, 577 p

Clemens JD (1984) Water contents of intermediate to silicic magmas. Lithos 11:213–287

Clemens JD, Circone S, Navrotsky A, McMillan PF, Smith BK, Wall VS (1987) Phlogopite: new calorimetric data and the effect of stacking disorder on thermodynamic properties. Geochim Cosmochim Acta 51:2569–2578

Clemens JD, Vielzeuf D (1987) Constraints on melting and magma production in the crust. Earth Planet Sci Lett 86:287–306

Clemens JD, Wall VJ (1981) Crystallization and origin of some peraluminous (S-type) granitic magmas. Can Mineral 19:111–132

Couturie JP, Kornprobst J (1977) Une interprétation géodynamique de l'évolution polyphasée des assemblages des granulites dans les chaînes bético-rifaines et le Massif Central Français. CR Somm Soc Geol Fr 5:289–291

Debon F (1975) Les massifs granitoïdes à structure concentrique de Cauterets-Panticosa (Pyrénées Occidentales) et leurs ensembles. Sci de la Terre, Mem. n° 33, Nancy, 420 p

den Tex E, Engels JP, Vogel DE (1972) A high-pressure intermediate-temperature facies series in the Precambrian at Cabo Ortegal (Northwest Spain). 24th Int Geol Cong, 1972 Section 2:64–73

Eggler DH (1973) Principles of melting of hydrous phases in silicate melt. Carnegie Inst Wash Yrbk 72:491–495

Eggler DH, Holloway JR (1977) Partial Melting of peridotite in the presence of H_2O and CO_2: principles and review. Magma Genesis, Oregon Dept Geol Min Ind Bull 96:15–36

Ellis DJ (1980) Osumilite − sapphirine − quartz granulites from Enderby Land, Antarctica: $P − T$ conditions of metamorphism, implications for garnet − cordierite equilibria and the evolution of the deep crust. Contrib Mineral Petrol 74:201–210

Ellis DJ (1986) Garnet − liquid $Fe^{2+} −$ Mg equilibria and implications for the beginning of melting in the crust and subduction zones. Am J Sci 286:765–791

Engels JP (1972) The catazonal polymetamorphic rocks of Cabo Ortegal (NW Spain), a structural and petrofabric study. Leidse Geol Med 48:83–133

England PC, Richardson SW (1977) The influence of erosion upon the mineral facies of rocks from different metamorphic environments. J Geol Soc Lond 134:201–213

England PC, Thompson AB (1986) Some thermal and tectonic models for crustal melting in continental collision zones. In: Coward MP, Ries AC (eds) Collision tectonics. Geol Soc Spec Pub 19:83–94

Esperança S, Holloway JR (1986) The origin of the high-K latites from Camp Creek, Arizona: constraints from experiments with variable fO_2 and aH_2O. Contrib Mineral Petrol 93:504–512

Eugster JP, Wones DR (1962) Stability relations of the ferruginous biotite, annite. J Petrol 3:82–125

Ferry JM, Spear FS (1978) Experimental calibration of the partitioning of Fe and Mg between biotite and garnet. Contrib Mineral Petrol 66:113–117

Frey M, Hunziker JC, Franck W, Bocquet J, Dal Piaz GV, Jager E, Niggli E (1974) Alpine metamorphism of the Alps. A review. Schweiz Mineral Petrogr Mitt 54, 2/3:247–291

Fyfe WS, Price NJ, Thompson AB (1978) Fluids in the Earth's Crust. Developments in Geochemistry, 1, Elsevier, Amsterdam, 383 p

Ghent ED, Stout MZ (1981) Geobarometry and geothermometry of plagioclase − biotite − garnet − muscovite assemblages. Contrib Mineral Petrol 76:92–97

Goranson RW (1931) The solubility of water in granite magmas. Am J Sci 22:481–502

Grant JA (1981) Orthoamphibole and orthopyroxene relations in high-grade metamorphism of pelitic rocks. Am J Sci 281:1127–1143

Grant JA (1985) Phase equilibria in partial melting of pelitic rocks. In: Migmatites, Ashworth JR (ed) Glasgow, Blackie and Son, pp 86–144

Grant JA (1986a) The isocon diagram − A simple solution to Gresens' equation for metasomatic alteration. Econ Geol 81:1976–1982

Grant JA (1986b) Quartz − phlogopite − liquid equilibria and origins of charnockites. Am Mineral 71:1071–1075

Green TH (1977) Garnet in silicic liquids and its possible use as a $P − T$ indicator. Contrib Mineral Petrol 65:59–67

Grew ES (1982a) Osumilite in the sapphirine-quartz terrane of Enderby Land, Antarctica: Implications for osumilite petrogenesis in the granulite facies. Am Mineral 67:762–787

Grew ES (1982a) Sapphirine, kornerupine, and sillimanite + orthopyroxene in the charnockitic region of south India. J Geol Soc India 23, 10:469–505

Harris NBW, Holland TJB (1984) The significance of cordierite-hypersthene assemblages from the Beitbridge region of the Central Limpopo belt; evidence for rapid decompression in the Archaean? Am Mineral 69:1036–1049

Helgeson HC, Delany JM, Nesbitt HW, Bird DK (1978) Summary and critique of the thermodynamic properties of rock-forming minerals. Am J Sci 278-A:1–229

Hensen BJ (1986) Theoretical phase relations involving cordierite and garnet revisited: the influence of oxygen fugacity on the stability of sapphirine and spinel in the system Mg − Fe − Al − Si − O. Contrib Mineral Petrol 92:362–367

Hoffer E (1976) The reaction sillimanite + biotite + quartz = cordierite + K-feldspar + H_2O and partial melting in the system $K_2O − FeO − MgO − Al_2O_3 − SiO_2 − H_2O$. Contrib Mineral Petrol 55:127–130

Hoffer E (1978) Melting reactions in aluminous metapelites: stability limits of biotite + sillimanite + quartz in the presence of albite. Neues Jahrb Mineral Monatsh 9:396–407

Hoffer E, Grant JA (1980) Experimental investigation of the formation of cordierite-orthopyroxene parageneses in pelitic rocks. Contrib Mineral Petrol 73:15–22

Holdaway MJ (1980) Chemical formulae and activity models for biotite, muscovite and chlorite applicable to pelitic metamorphic rocks. Am Mineral 65:711–719

Holdaway MJ, Lee SM (1977) Fe − Mg cordierite stability in high-grade pelitic rocks based on experimental, theoretical and natural observations. Contrib Mineral Petrol 63:175–198

Hoschek G (1969) The stability of staurolite and chloritoid and their significance in metamorphism of pelitic rocks. Contrib Mineral Petrol 22:208–232

Jakobsson S, Holloway JR (1986) Crystal-liquid experiments in the presence of a C − O − H fluid buffered by graphite + iron + wustite: experimental method and near liquidus relations in basanite. J Volcanol Geotherm Res 29:265–291

Jakeš P (1969) Retrogressive changes of granulite-facies rocks − an example from the Bohemian Massif. Spec Publ Geol Soc, Australia, 2:367–374

Johannes W (1978) Melting of plagioclase in the system Ab − An − H_2O and Qz − Ab − An − H_2O at $PH_2O = 5$ kbar, an equilibrium problem. Contrib Mineral Petrol 66:295–303

Karsakov LP, Shuldiner VI, Lennikov AM (1975) Granulite complex of the eastern part of the Stanovoy fold province and the Chogar facies of depth. (in Russian), Izvest Akad Nauk SSSR Ser Geol 5:47–61

Keesman I, Matthes S, Schreyer W, Seifert F (1971) Stability of almandine in the system $FeO − (Fe_2O_3) − Al_2O_3 − SiO_2 − (H_2O)$ at elevated pressures. Contrib Mineral Petrol 31:132–144

Kushiro I, Yoder HS (1969) Melting of forsterite and enstatite at high pressure under hydrous conditions. Carnegie Inst Washington, Ann Rept Dir Geophys Lab 1967–68:153–161

Lambert IB, Robertson JK, Wyllie PJ (1969) Melting reactions in the system $KAlSi_3O_8 − SiO_2 − H_2O$ to 18.5 kilobars. Am J Sci 267:609–626

Lasnier B (1977) Persistance d'une série granulitique au cœur du Massif Central français (Haut Allier). Les termes basiques, ultrabasiques et carbonatés. Thèse d'Etat, Nantes, 351 p

Le Breton N, Thompson AB (in prep.) Fluid-absent (dehydration) melting of biotite in metapelites in the early stages of crustal anatexis

Leterrier J (1972) Etude pétrographique et géochimique du massif granitique de Quérigut (Ariège). Sci de la Terre, Mem Fr, 23, 320 p

Lonker SW (1981) The $P − T − X$ relations of the cordierite − garnet − sillimanite − quartz equilibrium. Am J Sci 281:1056–1090

Luth WC (1967) Studies in the system $KAlSiO_4 - Mg_2SiO_4 - SiO_2 - H_2O$: I Inferred phase relations and petrologic application. J Petrol 8:372–416

Luth WC, Jahns RH, Tuttle OF (1964) The granite system at pressures of 4 to 10 kilobars. J Geophys Res 9:759–773

Maaløe S, Wyllie PJ (1975) Water content of a granite magma deduced from the sequence of crystallization determined experimentally with water-undersaturated conditions. Contrib Mineral Petrol 52:175–191

Maijer C, Jansen JBH, Wevers J, Poorter RPE (1977) – Osumilite, a mineral new to Norway. Norsk geologisk Tidsskrift 57:187–188

Marchand J (1974) Persistance d'une série granulitique au cœur du Massif Central français – Haut Allier. Les Termes acides. Thèse 3ème cycle, Nantes, 207 p

Michard-Vitrac A, Albarède F, Dupuis C, Taylor HPJ (1980) The genesis of Variscan (Hercynian) plutonic rocks: Inferences from Sr, Pb and O studies on the Maladeta igneous complex, central Pyrenees (Spain). Contrib Mineral Petrol 72:57–72

Montel JM, Weber C, Pichavant M (1986) Biotite – sillimanite – spinel assemblages in high-grade metamorphic rocks: occurrences, chemographic analysis and thermobarometric interest. Bull Mineral 109:555–573

Morse SA, Talley JH (1971) Sapphirine reactions in deep-seated granulites near Wilson lake, Central Labrador, Canada. Earth Planet Sci Lett 10:325–328

Nekvasil H, Burnham CW (1987) The calculated individual effects of pressure and water content on phase equilibria in the granite system. In: BO Mysen (ed) Magmatic Processes: Physicochemical Principles. Geochem Soc Spec pub 1:433–445

Newton RC (1983) Geobarometry of high grade metamorphic rocks. Am J Sci 283A:1–28

Newton RC, Haselton HT (1981) Thermodynamics of the garnet – plagioclase – Al_2SiO_5-quartz geobarometer. In: RC Newton, A Navrotsky, BJ Wood (Eds) Thermodynamics of minerals and melts, Springer, Berlin Heidelberg New York, 129–145

Novak JM, Holdaway MJ (1981) Metamorphic petrology, mineral equilibria, and polymetamorphism in the Augusta quadrangle, south central Maine. Am Mineral 66:51–69

Ouzegane K (1981) Le métamorphisme polyphasé granulitique de la région de Tamanrasset (Hoggar Central). Thèse 3ème cycle, Paris VI

Patera ES, Holloway JR (1982) Experimental determination of the spinel-garnet boundary in a Martian mantle composition. J Geophys Res [Suppl] 87:A31–A36

Perchuk LL, Podlesskii KK, Aranovich L YA (1981) Calculation of thermodynamic properties of end-member minerals from natural parageneses. In: RC Newton, A Navrotsky, BJ Wood (eds) Thermodynamics of minerals and melts, Springer, New York, pp 111–129

Peterson JW, Newton RC (1987) Reversed biotite + quartz melting reactions. EOS 68, 16:451

Phillips GN (1980) Water activity changes across an amphibolite – granulite facies transition, Broken Hill, Australia. Contrib Mineral Petrol 75:377–386

Pin C, Vielzeuf D (1983) Granulites and related rocks in Variscan median Europe: a dualistic interpretation. Tectonophysics 93:47–74

Postaire B (1982) Systématique Pb commun et U – Pb sur zircons. Application aux roches de haut grade métamorphique impliquées dans la chaîne hercynienne (Europe de l'Ouest) et aux granulites de Laponie (Finlande). Thèse 3ème cycle, Rennes, 71 p

Respaut JP, Lancelot JR (1983) Datation de la mise en place synmétamorphe de la charnockite d'Ansignan (massif de l'Agly) par la méthode U – Pb sur zircons et monazites. N Jhb Miner Abh 147:21–34

Robertson JK, Wyllie PJ (1971) Rock-water systems with special reference to the water-deficient region. Am J Sci 271:252–277

Rutherford MJ (1969) An experimental determination of iron biotite – alkali feldspar equilibria. J Petrol 10:381–408

Schairer JF, Yagi K (1952) The system $FeO - Al_2O_3 - SiO_2$. Am J Sci (Bowen vol) 471–512

Schreyer W, Yoder HS (1959) Stability of Mg-cordierite. Bull Geol Soc Amer 70:1672 (abstract)

Seifert F (1976) Stability of the assemblage cordierite + K feldspar + Quartz. Contrib Mineral Petrol 51:179–185

Shaw DM (1956) Geochemistry of pelitic rocks, III. Bull Geol Soc Am 67:919–934

Shaw HR (1963) The four-phase curve sanidine – quartz – liquid – gas between 500 and 4000 bars. Am Mineral 48:883–896

Spear FS, Selverstone J (1983) Quantitative $P - T$ paths from zoned minerals: theory and tectonic applications. Contrib Mineral Petrol 83:348–357

Storre B (1972) Dry melting of muscovite + quartz in the range Ps = 7 kb to Ps = 20 kb. Contrib Mineral Petrol 37:87–89

Storre B, Karotke E (1972) Experimental data on melting reactions of muscovite + quartz in the system $K_2O - Al_2O_3 - SiO_2 - H_2O$ to 20 kb water pressure. Contrib Mineral Petrol 36:343–345

Thompson AB (1976) Mineral reactions in pelitic rocks: I Prediction of $P-T-X$ (Fe Mg) phase relations. II Calculation of some $P-T-X$ (Fe–Mg) phase relations. Am J Sci 276:401–454

Thompson AB (1982) Dehydration melting of pelitic rocks and the generation of H_2O-undersaturated granitic liquids. Am J Sci 282:1567–1595

Thompson AB, Algor JR (1977) Model systems for anatexis of pelitic rocks. I Theory of melting relations in the system $KAlO_2 - NaAlO_2 - Al_2O_3 - H_2O$. Contrib Mineral Petrol 3:247–269

Thompson AB, Tracy RJ (1979) Model systems for anatexis of pelitic rocks. II Facies series melting and reactions in the system $CaO - KAlO_2 - NaAlO_2 - Al_2O_3 - SiO_2 - H_2O$. Contrib Mineral Petrol 70:429–438

Turner FJ, Verhoogen J (1960) Igneous and metamorphic petrology. McGraw-Hill book company, Inc New-York Toronto London, 694 pp

Tuttle OF, Bowen NL (1958) Origin of granite in the light of experimental studies in the system $NaAlSi_3O_8 - KAlSi_3O_8 - SiO_2 - H_2O$. Geol Soc Amer Mem 74:153 p

van Der Molen I, Paterson MS (1979) Experimental deformation of partially-melted granite. Contrib Mineral Petrol 70:229–318

Vielzeuf D (1980a) Orthopyroxene and cordierite secondary assemblages in the granulitic paragneisses from Lherz and Saleix (French Pyrenees). Bull Minéral 103:66–78

Vielzeuf D (1980b) Pétrologie des écailles granulitiques de la région de Lherz (Ariège-Zone Nord-Pyrénéenne). Introduction à l'étude expérimentale de l'association grenat (Alm-Pyr) –feldspath potassique. Thèse 3ème cycle, Clermont-Ferrand, 219 p

Vielzeuf D (1983) The spinel and quartz associations in high grade xenoliths from Tallante (SE Spain) and their potential use in geothermometry and barometry. Contrib Mineral Petrol 82:301–311

Vielzeuf D (1984) Relations de phases dans le faciès granulite et implications géodynamiques. L'exemple des granulites des Pyrénées. Thèse Doctorat d'Etat, Clermont-Ferrand, 288 p

Vielzeuf D, Boivin P (1984) An algorithm for the construction of petrogenetic grids – Application to some equilibria in granulitic paragneisses. Am J Sci 284:760–791

Vielzeuf D, Kornprobst J (1984) Crustal splitting and the emplacement of the pyrenean lherzolites and granulites. Earth Planet Sci Lett 67:87–96

Vogel DE (1967) Petrology of an eclogite and pyrigarnite-bearing polymetamorphic rock complex at Cabo Ortegal, NW Spain. Leidse Geol Med 40:121–213

Waard D de (1965) A proposed subdivision of the granulite facies. Am J Sci 263:455–461

Waters DJ (1986) Metamorphic history of sapphirine-bearing and related magnesian gneisses from Namaqualand, South Africa. J Petrol 27:541–565

Wendlandt RF (1981) Inflence of CO_2 on melting of model granulite facies assemblages: a model for the genesis of charnockites. Am Mineral 66:1164–1174

White AJR, Chappell BW (1983) Granitoid types and their distribution in the Lachlan fold belt, Southeastern Australia. Geol Soc Am Mem 19:21–34

Whitney JA (1975) The effects of pressure, temperature, and $X\mathrm{H_2O}$ on phase assemblage in four synthetic rock compositions. J Geol 83:1–31

Wones DR, Dodge FCW (1966) On the stability of phlogopite. Geol Soc Am Spec Pap 101:242 (abstr)

Wones DR, Dodge FCW (1977) The stability of phlogopite in the presence of quartz and diopside. In: DG Fraser (ed) Thermodynamics in Geology. D Reidel, Dordrecht, 229–247

Yoder HS Jr, Kushiro I (1969) Melting of a hydrous phase: phlogopite. Am J Sci 267A:558–582

Received July 20, 1987 / Accepted December 14, 1987

Editorial responsibility: J. Hoefs

CHAPTER 19

Bea, F. (1996) Residence of REE, Y, Th and U in granites and crustal protoliths; implications for the chemistry of crustal melts. *Journal of Petrology*, **37**, 521–552.

Petrology has often suffered from attempts to apply a unifying philosophy to embrace both (essentially liquid) mafic magmas and (essentially solid) granitic "migmas" (Hughes, 1982).

The Granite Controversy (see, for example, the papers of Read, 1948, and Bowen 1948, included in this volume) ended with the prevalence of magmatist over transformist ideas. The resulting wide acceptance that most granitic rocks had crystallized from magmas was a major advance in understanding granite genesis. However, it also had some undesired consequences. Chief among these undesirable consequences was the implicit assumption that geochemical tools successfully employed to investigate the evolution of basaltic magmas could be used equally as well for investigating granitic magmas. Magmas were magmas, were they not?

During the 1970s and 1980s, the distribution of the Rare Earth Elements (*REE*) between groups of potentially related rocks became increasingly popular as a petrogenetic tool (Haskin, 1979). The systematic decrease in ionic radius from La^{3+} to Lu^{3+}, the fact that almost 100% of the Eu occurs as Eu^{2+}, with an ionic radius similar to Sr^{2+}, and the variable Ce^{3+}/Ce^{4+}, depending on the oxygen fugacity of the system, make whole-rock, chondrite-normalized *REE* patterns easily interpretable in terms of the proportions of melted versus residual minerals, for mantle-derived magmatic systems (but see Haskin, 1990). After successes in studying terrestrial basalts and lunar materials, the geochemistry of *REE* was embraced enthusiastically by granite petrologists. However, the success and utility of the method here was highly questionable. This is because, more often than not, *REE*-based models of granite genesis contradicted other lines of evidence.

It is clear that trace elements that we wish to use in geochemical modelling of petrological processes must reside in major minerals. Though not universally the case, the trace elements found in accessory minerals usually form essential structural components. As a consequence, their concentrations in partial melts are not governed by crystal-melt distribution coefficients but by solubility relations and solution kinetics. Moreover, a fraction of the crystals of accessory minerals will be present as inclusions inside crystals of the major minerals. This fraction can remain physically isolated from the melt, thus preventing equilibration, or may be entrained into the melt along with host restitic crystals. With fractionation, the latter process will potentially produce crystal concentrations with trace-element contents higher than expected from solubility relations.

During the 1980s, Watson and Harrison (1984) and Watson (1988) showed which of the *REE* reside, at least partially, in accessory minerals. Inspired by these works, Bea undertook a systematic study to determine the concentrations of *REE*, Y, Th and U in the major and accessory minerals of common crustal rocks, using LA-ICP-MS and EMP as analytical techniques. The results are described in this 1996 paper.

This investigation revealed that, in contrast to mafic rocks where the *REE* mostly reside in major minerals, the budget of these elements in granitic rocks is essentially controlled by the accessory mineral assemblage, the nature of which depends critically on the aluminosity of the magma. The major finding was that these elements are unsuitable for use in geochemical modelling of granitic rocks by means of equilibrium-based trace-element fractionation equations. In counterpoint, however, Bea realized that these elements are still very useful petrogenetic tools because their distribution reflects the behaviour of accessory minerals and some key major minerals such as garnets, feldspars, and amphiboles. Thus, data on the concentrations of these elements can provide valuable information about the conditions of partial melting, melt segregation and crystallization of granite magmas in different crustal regimes.

Bea (1996) is included as a landmark because it liberated granite petrologists from the fetters of basalt-conditioned attitudes toward the use of trace elements in petrogenetic studies and because it demonstrated how these elements could still be used to good effect. It showed us the problem and also gave us the solution.

JOURNAL OF PETROLOGY | VOLUME 37 | NUMBER 3 | PAGES 521–552 | 1996

F. BEA

DEPARTMENT OF MINERALOGY AND PETROLOGY, CAMPUS FUENTENUEVA, UNIVERSITY OF GRANADA, 18002 GRANADA, SPAIN

Residence of REE, Y, Th and U in Granites and Crustal Protoliths; Implications for the Chemistry of Crustal Melts

A systematic study with laser ablation–ICP-MS, scanning electron microscopy and electron microprobe revealed that ~70–95 wt % of REE (except Eu), Y, Th and U in granite rocks and crustal protoliths reside within REEYThU-rich accessories whose nature, composition and associations change with the rock aluminosity. The accessory assemblage of peraluminous granites, migmatites and high-grade rocks is composed of monazite, xenotime (in low-Ca varieties), apatite, zircon, Th-orthosilicate, uraninite and betafite–pyrochlore. Metaluminous granites have allanite, sphene, apatite, zircon, monazite and Th-orthosilicate. Peralkaline granites have aeschinite, fergusonite, samarskite, bastnaesite, fluocerite, allanite, sphene, zircon, monazite, xenotime and Th-orthosilicate. Granulite-grade garnets are enriched in Nd and Sm by no less than one order of magnitude with respect to amphibolite-grade garnets. Granulite-grade feldspars are also enriched in LREE with respect to amphibolite-grade feldspars. Accessories cause non-Henrian behaviour of REE, Y, Th and U during melt–solid partitioning. Because elevated fractions of monazite, xenotime and zircon in common migmatites are included within major minerals, their behaviour during anatexis is controlled by that of their host. Settling curves calculated for a convecting magma show that accessories are too small to settle appreciably, being separated from the melt as inclusions within larger minerals. Biotite has the greatest tendency to include accessories, thereby indirectly controlling the geochemistry of REE, Y, Th and U. We conclude that REE, Y, Th and U are unsuitable for petrogenetical modelling of granitoids through equilibrium-based trace-element fractionation equations.

KEY WORDS: *accessory minerals; geochemical modelling; granitoids; REE, Y, Th, U*

INTRODUCTION

The geochemistry of rare earth elements (REE), yttrium, thorium and uranium forms the basis for many important methods of igneous petrogenesis. In particular, REE are probably the most extensively used elements for petrogenetic modelling, as each major mineral participating in the genesis of a rock is supposed to leave its own recognizable signature on the REE pattern of that rock (e.g. Haskin, 1979, 1990; McKay, 1989). REE-based modelling was first applied to mantle-related systems, where it has found undeniable success. It was then adopted for crustal systems, despite the fact that crustal rocks always contain REE-, Y-, Th-, U-rich accessory minerals (hereafter called REEYThU-rich minerals), which usually account for an elevated fraction of REE, Y, Th and U contents in bulk rock (e.g. Gromet & Silver, 1983; Sawka, 1988; Bea *et al.*, 1994a, 1994b) and may therefore disturb or even completely mask the effects produced by major minerals during melting and crystallization (Miller & Mittlefehdlt, 1982; Yurimoto *et al.*, 1990).

Since the recognition that REEYThU-rich accessories may play an important role in controlling the geochemistry of crustal melts (e.g. Watson & Harrison, 1984; Watson, 1988; Watt & Harley, 1993), a considerable amount of work has been done in an attempt to understand their effects. However, this effort has been almost exclusively focused on three minerals: zircon, monazite and apatite. Nevertheless, the variety of REEYThU-rich accessories in granite rocks is neither limited to these three minerals nor are they always the main REE, Y, Th and U carriers. Perusal of available descriptions of granite rocks as well as our routine scanning electron microscope (SEM) studies on Uralian and Iberian granitoids revealed a plethora of REEYThU-rich minerals whose nature, grain-size, textural position, and hence their presumable influence on the behaviour of REE, Y, Th and U during melting and crystallization, change notably from one rock type to another.

This paper presents the results of a systematic study—by scanning electron microscope (SEM), electron microprobe and laser ablation–inductively coupled plasma mass spectrometry (LA–ICP-MS)—on the residence of REE, Y, Th and U in a variety of samples of granitoids, migmatites and granulites from Iberia, Ivrea–Verbano, the Urals, Kola and Transbaikalia. The research was carried out with three objectives: (1) to study the REE, Y, Th and U composition of major minerals in different geological environments; (2) to learn the nature, abundance, composition, grain-size distribution, textural relationships and associations of REEYThU-rich accessories; (3) to determine the relative contributions of major and accessory minerals to REE, Y, Th and U budgets in granitoids and common crustal protoliths.

Based on these results, we address the behaviour of REE, Y, Th and U in granite melts, the role of accessories during crustal melting, melt segregation and crystallization, the effects of pressure on the REE composition of restitic minerals and hence of melts, the influence of accessories on the contrasting geochemistry of REE, Y, Th and U in different granite types, and the mobility of these elements during post-magmatic stages. We then examine the question of whether REE-based petrogenetic modelling of granite rocks—understood as a method for inferring the behaviour of major minerals during melt and crystallization—makes sense or not.

SAMPLES AND METHODS

Samples

About 150 samples from 38 different geological units were selected for this work (Table 1). Slides for thin sections (optical, SEM, microprobe and LA–ICP-MS analyses) were cut from each sample. Crushing of 5–10 kg of rock to a grain-size of <5 mm was done in a crusher with hardened steel jaws adjusted to an output size of 5 mm. Powders for chemical analysis were obtained by grinding about 50 g per sample of crushed rock in a tungsten carbide jar until grain-size was <25 μm. This system produced no detectable contamination with REE, Y, U or Th.

Whole-rock analysis

REE, Y, Th and U determinations were done by inductively coupled-plasma mass-spectrometry (ICP-MS) after $HNO_3 + HF$ digestion of 0·1000 g of sample powder in a Teflon-lined vessel for 150 min at high temperature and pressure, evaporation to dryness, and subsequent dissolution in 100 ml of 4 vol. % HNO_3. Measurements were carried out in triplicate with a PE Sciex ELAN-5000 spectrometer using Rh and Re as internal standards. Coefficients of variation calculated by dissolution and subsequent analysis of 10 replicates of powdered sample were better than ±3 rel. % and ±8 rel. % for analyte concentrations of 50 and 5 p.p.m., respectively.

Textural and modal analysis

Thin sections were studied by optical and scanning electron microscopy (SEM) with back-scattered electron (BSE) and energy dispersive X-ray microanalysis (EDAX). Modal counting was done on selected samples, using four thin sections per sample. The mass fraction of major minerals was calculated by mass balance using average microprobe data of major minerals and whole-rock chemical composition. The modal proportion of REEYThU-rich accessories was estimated with the SEM in two steps, first calculating the overall modal abundance of all bright minerals (REEYThU-accessories, sulphides, Fe–Ti oxides, barite, etc.) by image analyses, and then estimating the relative abundance of each species by identification with EDAX of every bright grain present in the section and recording its apparent diameter. The textural position (inclusion, between two grains, in a triple junction, etc.) and the nature of the surrounding minerals were also recorded. Values of the ratio $\Sigma_1^{i=n} X_i C_i / C_{\text{whole-rock}}$ were between 0·8 and 1·15 for most elements, where X_i is the modal fraction on each phase, and C_i and $C_{\text{whole-rock}}$ are the concentration of a given element in each phase and in the whole rock, respectively.

REE, Y, U and Th analysis of minerals

Major minerals and accessory grains with a grain-size >100 μm were analysed by LA–ICP-MS. Smaller accessory grains were analysed by electron microprobe.

LA–ICP-MS analyses were performed in the Perkin Elmer ICP-MS Applications Laboratory at Überlingen (Germany) with a prototype UV laser ablation system coupled to a high-sensitivity PE Sciex ELAN-6000 ICP-MS spectrometer. The excimer SOPRA Ne:Xe:HCl gas laser was set at

Table 1: Studied geological units

Geological unit	Region	Main features
Differentiated two-mica granites and leucogranite		
Pedrobernardo	Central Iberia	ASI ~ 1·27. Accessory dumortierite
Albuquerque	Central Iberia	ASI ~ 1·25. Cordierite-bearing facies. Accessory andalusite and dumortierite. Related to U ore deposits
Trujillo	Central Iberia	ASI ~ 1·24. Accessory cordiente and tourmaline
Jalama	Central Iberia	ASI ~ 1·24. Sillimanite-bearing facies
Cabeza de Araya	Central Iberia	ASI ~ 1·21. Cordierite- and garnet-bearing facies
Boquerones	Central Iberia	ASI ~ 1·27. Accessory andalusite and cordierite
Ronda leucogranites	SW Iberia	ASI ~ 1·1–1·24. High B contents. Abundant tourmaline and cordierite
Murzinka	Central Urals	ASI ~ 1·1. Accessory garnet in aplites
Uvildy-Ilmen	Central Urals	ASI ~ 1·05–1·13. Accessory garnet
Autochthonous anatectic leucogranites		
Almohalla	Central Iberia	ASI ~ 1·18. Garnet-rich leucogranite, with accessory muscovite and biotite
Tarayuela	Central Iberia	ASI ~ 1·26. Cordierite leucogranite with accessory garnet
Peña Negra	Central Iberia	ASI ~ 1·24. Cordierite leucogranite with accessory biotite
Peraluminous adamellites, granodiorites and tonalites		
Hoyos-Gredos	Central Iberia	ASI ~ 1·24. Biotite–cordierite granodiorites and adamellites
Alberche	Central Iberia	ASI ~ 1·18. Biotite adamellites, with accessory cordierite
Dzyabyk	Central Urals	ASI ~ 1·14. Two-mica adamellites and biotite granodiorites
Cheliabinsk	Central Urals	ASI ~ 1·14. Biotite granodiorites with muscovite-bearing facies
Elanchink	Central Urals	ASI ~ 1·08. Biotite adamellites
Subaluminous granites, granodiorites, tonalites and diorites		
Verkisest	Central Urals	ASI ~ 1·02. Abundant magmatic epidote and titanite. K-poor facies
Shartash	Central Urals	ASI ~ 1·03. Na > K
Shabry	Central Urals	ASI ~ 1·05. Na ~ K
Elanchick	Central Urals	ASI ~ 1·03. Na ≤ K
Stepninsk	Central Urals	ASI ~ 1·04. K-rich facies
Tagil-Chernolstochinsk	Central Urals	ASI ~ 1·01. Amphibole leucodiorites with no biotite
Khabarny	South Urals	Ophiolitic complex with plagiogranites
Peralkaline granites		
Galiñeiro	N Iberia	Very high REE, Zr, Y, Nb, Th contents. Mineralized facies
Barcarrota	SW Iberia	Riebeckite–aegirine granites
Kelvy	Kola	Riebeckite–aegirine granites
Kharitonovo	Transbaikalia	Riebeckite–aegirine granites
Magnitogorsk	Central Urals	Ferrohastingsite–riebeckite granites
Migmatites		
Peña Negra	Central Iberia	Cordierite–sillimanite–biotite migmatites with occasional accessory garnet
Los Villares	SE Iberia	Cordierite–garnet–sillimanite–biotite
Murmansk block	Kola	Garnet–sillimanite–biotite
Murzinka	Central Urals	Garnet–biotite–sillimanite
Osinovsk	Central Urals	Garnet–biotite–sillimanite–cordierite
Granulites		
Strona	Ivrea–Verbano	Garnet–sillimanite–Kfsp
Murmansk block	Kola	Opx–garnet–kyanite–sillimanite
Ronda	SE Iberia	Garnet–sillimanite–Kfsp
Cabo Ortegal	NW Iberia	Garnet–kyanite–Kfsp

523

JOURNAL OF PETROLOGY | VOLUME 37 | NUMBER 3 | JUNE 1996

3·8 bars gas pressure. Wavelength was 308 nm, repetition rate 5 Hz, pulse length 30 ns and pulse energy 100 mJ. Data were acquired for 50 s, starting acquisition 10 s after ablation was initiated. The diameter of the laser beam and gas pressure were fixed so as to produce craters with a diameter of ~80 μm in diameter and 40–60 μm in depth, depending on the ablated mineral. Under these conditions, crystals smaller than 100 μm cannot be reliably analysed. Isobaric interferences were automatically corrected by the TotalQuant III$^{©}$ software implemented in the ELAN-6000, which also allows the response factors for all the elements to be adjusted by calibrating the system with an external standard with just a few elements covering the mass range. Calibration was done using the NBS-612 glass, which contains ~40 p.p.m. of most trace elements, as external standard. Concentration values were later refined using silicon as internal standard, previously determined by electron microprobe on the same minerals. In these conditions, detection limits for REE, Y, Th and U were better than 10 p.p.b. Coefficients of variation (CV) obtained by measuring five replicates on a single grain of astrophyllite, which appeared to be unzoned and free of inclusions, were ±4·2 rel.% for 45·6 p.p.m. La, ±18·9 rel.% for 0·57 p.p.m. Y, ±6·4 rel.% for 8·63 p.p.m. U, ±17·5 rel.% for 0·28 p.p.m. Th, ±4·4 rel.% for 4·71 p.p.m. Eu, and ±11·3 rel.% for 0·62 p.p.m. Yb. For elements with concentration higher than 100 p.p.m., CV are normally in the range of ±1–3 rel.%.

Electron microprobe mineral analysis

Major-element, REE, Y, Th and U analyses of minerals were obtained by wavelength dispersive analyses with an ARL electron microprobe operated with PROBE software using as standards the four synthetic glasses described by Drake & Weill (1972)—REE1 Chgo/2, REE2 Chgo/3, REE3 Chgo/4 and REE4 Chgo/5—and the following minerals: Albite Amelia Chgo/18, Mn-Hortonolite Chgo/35, Monazite SPI/32 and Zirconia SPI/47. Accelerating voltage was 20 kV and beam current was 20 nA. Interelemental interferences were suppressed by peak-overlap corrections (Roeder, 1985). Detection limits were better than 0·10–0·05% depending on the element and mineral analysed. Coefficients of variation were close to ±4 rel.% for 1 wt% concentration and ±10 rel.% for 0·25 wt% concentration.

REE, Th AND U COMPOSITION OF MAJOR MINERALS

Micas

REE, Y, Th and U concentrations in micas are very low (Table 2), close to or lower than detection limits (~0·01 p.p.m.) in most cases. Early LA–ICP-MS work on muscovites gave somewhat higher levels (Bea et al., 1994a), but it is now recognized that they were probably produced by microinclusions of monazite and xenotime within the ablated spot. Micas from U-rich granites may have several p.p.m. U, especially in hydrothermalized facies, which suggests that uranium is absorbed as uranyl ions, probably in the interlayer position.

Amphiboles and pyroxenes

Orthopyroxenes also have negligible contents of REE, Y, U and Th (Table 3). Amphiboles and clinopyroxenes have appreciable REE and Y, but low Th and U contents (Table 3). Chondrite-normalized REE patterns of Ca-amphibole (Fig. 1) increase progressively from La to Pr and then decrease smoothly from Nd to Lu with no Eu anomaly or a small negative one, more intense in amphiboles coexisting with primary epidote (e.g. analysis 1 in Table 3). Partition coefficients for coexisting Ca-amphibole–Ca-clinopyroxene pairs are $D_{U,Th}^{amp/cpx} \sim 1$ and $D_{REE,Y}^{amp/cpxx} \sim 2$–4. REE patterns of Ca-clinopyroxene are analogous to those of Ca-amphiboles, but at lower concentrations (Fig. 1), and have very small negative or positive Eu anomalies [see also Mazzuchelli et al. (1992)].

Na-amphiboles are richer in La–Sm and Tm–Lu than Ca-amphiboles (Table 3). Chondrite-normalized REE patterns decrease uniformly from La to Dy and then increase from Er to Lu, with $Er_N/Lu_N \sim 0.3$–0·6 (Fig. 1). Partition coefficients for Na-amphibole–Na-clinopyroxene pairs are $D_{U,Th}^{Na-amp/Na-cpx} \sim 2$–5, $D_{LREE}^{Na-amp/Na-cpx} \sim 6$–10, $D_{Eu-Er,Y}^{Na-amp/Na-cpx} \sim 4$–6 and $D_{Tm-Lu}^{Na-amp/Na-cpx} \sim 1$–3. REE patterns of Na-clinopyroxene are parallel to those of Na-amphiboles from La to Ho, but the enrichment from Er to Lu is more pronounced (Fig. 1).

K-feldspar and plagioclase

Feldspars have moderate to low LREE contents, and low to very low HREE, Y, Th and U contents (Table 4). Partition coefficients for coexisting plagioclase–K-feldspar pairs are $D_{U,Th,Y}^{plag/Kfsp} \sim 1$, $D_{LREE}^{plag/Kfsp} \sim 1$–5, $D_{Eu}^{plag/Kfsp} \sim 0.3$–3 and $D_{HREE}^{plag/Kfspx} \sim 1$. Chondrite-normalized REE patterns of

Table 2: Selected LA–ICP-MS analyses of primary micas (results are in p.p.m.)

	Biotite						Muscovite					
	1	2	3	4	5	6	1	2	3	4	5	6
Y	0·05	0·25	0·01	0·08	0·09	0·09	0·19	0·01	0·16	0·01	0·06	0·01
U	0·32	0·00	0·00	3·41	0·17	0·01	0·77	0·07	0·16	0·00	0·63	0·21
Th	0·04	0·00	0·10	0·00	0·00	0·21	0·11	0·00	0·03	0·04	0·09	0·00
La	0·11	1·17	0·03	0·03	0·22	0·50	0·34	0·00	0·06	0·04	0·06	0·00
Ce	0·11	0·17	0·00	0·00	0·31	0·50	0·76	0·00	0·15	0·05	0·07	0·02
Pr	0·00	0·11	0·00	0·00	0·02	0·08	0·09	0·04	0·03	0·01	0·03	0·00
Nd	0·00	0·25	0·00	0·00	0·11	0·11	0·00	0·06	0·03	0·01	0·00	0·00
Sm	0·00	0·00	0·00	0·00	0·04	0·04	0·00	0·00	0·03	0·01	0·00	0·00
Eu	0·00	0·04	0·00	0·00	0·11	0·04	0·10	0·00	0·12	0·00	0·00	0·06
Gd	0·00	0·00	0·00	0·00	0·06	0·03	0·03	0·00	0·04	0·02	0·05	0·00
Tb	0·00	0·00	0·00	0·00	0·01	0·00	0·07	0·00	0·00	0·00	0·00	0·00
Dy	0·00	0·00	0·00	0·00	0·05	0·04	0·10	0·00	0·00	0·04	0·00	0·00
Ho	0·00	0·00	0·00	0·00	0·01	0·01	0·03	0·00	0·00	0·01	0·00	0·00
Er	0·00	0·00	0·00	0·00	0·02	0·02	0·11	0·00	0·00	0·02	0·00	0·00
Tm	0·00	0·01	0·00	0·00	0·01	0·00	0·02	0·00	0·00	0·00	0·00	0·00
Yb	0·00	0·00	0·00	0·00	0·01	0·02	0·24	0·00	0·00	0·02	0·00	0·00
Lu	0·00	0·00	0·00	0·00	0·00	0·00	0·00	0·00	0·00	0·00	0·00	0·00
Eu/Eu*	—	—	—	—	6·87	3·53	—	—	10·6	0·00	—	—
La_N/Yb_N	—	—	—	—	14·8	16·8	0·96	—	—	1·35	—	—

Biotites: 1, migmatite (Peña Negra); 2, subaluminous granodiorite (Vierkisest); 3, peraluminous adamellite (Hoyos); 4, U-rich two-mica leucogranite (Albuquerque); 5, subaluminous tonalite (Shabry); 6, peralkaline granite (Magnitogorsk). Muscovites: 1, U-rich two-mica leucogranite (Albuquerque); 2, two-mica granite (Pedrobernardo); 3, peraluminous adamellite (Hoyos); 4, migmatite (Peña Negra); 5, migmatite (Ronda); 6, Be-rich pegmatite (Murzinka).

K-feldspar (Fig. 1) show a variable but always strong LREE–HREE fractionation, usually with intense positive Eu anomaly. REE patterns of plagioclase (Fig. 1) are similar but slightly more enriched in LREE. The intensity of the Eu anomaly in both feldspars is highly variable. Maximum positive anomalies (Eu/Eu* ~25–40) occur in feldspars from migmatites (see analyses 4 in Table 4), whereas the smallest Eu anomalies, even negative ones, appear in feldspars coexisting with magmatic epidote (analyses 5 in Table 4) and in feldspars from extremely differentiated leucocratic segregates from peraluminous granites (Bea et al., 1994a). K-feldspars and plagioclases from granulite facies metapelites are remarkably enriched in LREE, with $La_N \geq 50$ and $La_N/Sm_N \sim 10$ [e.g. analyses 3 in Table 4; see also ion-probe data of Reid (1990) and Watt & Harley (1993)].

Garnet

Among major minerals, garnet is the richest in Y and HREE (Table 5). The composition of garnet in crustal protoliths shows a marked dependence on whether they come from amphibolite or granulite facies.

Garnets from amphibolite facies are rich in HREE and Y, but have near-zero LREE, Th and U contents [Table 5; see also data of Harris et al. (1992)]. Chondrite-normalized REE patterns (Fig. 2) are steeply positive from La to Dy–Er, with Sm_N/Gd_N <0·6 and either no Eu anomaly or a small negative one (Eu/Eu* ~0·4–1·4), and then are either flat or decrease smoothly to Lu. Amphibolite-grade garnets are usually zoned, with the HREE and Y either coupled with Mn (Hickmott et al., 1987; Hickmott & Shimizu, 1990) or decoupled from major components (Lanzirotti, 1995; Delima et al., 1995).

Garnets from granulite facies have somewhat lower Y and Dy–Lu, much higher Nd and Sm, higher Sm/Gd ratios (Sm_N/Gd_N >0·6), a more pronounced Eu negative anomaly (Eu/Eu* ~0·01–0·2), and a flatter HREE chondrite-normalized pattern than amphibolite-grade garnets [Fig. 2; Table 5; compare also ion-probe data of Reid (1990) and Watt & Harley (1993) with those of Harris et al.

525

JOURNAL OF PETROLOGY | VOLUME 37 | NUMBER 3 | JUNE 1996

Table 3: Selected LA–ICP-MS analyses of amphiboles, clinopyroxenes and orthopyroxenes (results are in p.p.m.)

	Amphibole						Cpx				Opx	
	1	2	3	4	5	6	1	2	3	4	1	2
Y	17·9	19·3	8·69	4·53	9·30	10·2	4·10	2·71	0·37	2·65	0·34	0·28
U	1·76	0·07	0·10	0·14	0·02	0·51	0·18	0·09	0·16	0·13	0·03	0·05
Th	0·25	0·01	0·00	0·00	0·00	0·53	0·00	0·04	0·00	0·10	0·00	0·00
La	8·70	6·85	2·81	1·24	3·82	29·4	1·19	0·20	0·24	3·34	0·03	0·02
Ce	39·4	35·3	12·0	6·72	17·4	83·9	3·16	0·96	0·93	7·43	0·02	0·08
Pr	8·00	7·46	1·93	1·22	2·66	11·8	0·49	0·20	0·18	1·02	0·00	0·01
Nd	39·9	35·1	9·13	6·84	12·5	40·5	2·24	1·29	0·84	4·64	0·00	0·02
Sm	11·5	11·3	2·61	2·01	2·74	9·27	0·74	0·47	0·33	1·16	0·00	0·00
Eu	1·44	3·43	0·88	0·68	0·86	2·77	0·24	0·21	0·10	0·41	0·00	0·00
Gd	9·48	8·47	2·46	1·59	2·43	7·99	0·65	0·56	0·34	1·16	0·00	0·01
Tb	1·53	1·34	0·29	0·23	0·48	1·12	0·10	0·09	0·05	0·18	0·00	0·00
Dy	7·23	7·07	1·25	1·05	2·32	5·13	0·62	0·53	0·32	0·85	0·00	0·00
Ho	1·43	1·37	0·30	0·22	0·51	1·12	0·12	0·11	0·05	0·20	0·00	0·00
Er	3·70	4·11	0·83	0·42	1·36	3·06	0·33	0·27	0·11	0·67	0·00	0·00
Tm	0·51	0·61	0·12	0·06	0·20	0·58	0·05	0·04	0·01	0·24	0·00	0·00
Yb	2·90	3·63	0·78	0·25	1·24	5·00	0·30	0·23	0·08	2·89	0·00	0·00
Lu	0·45	0·54	0·15	0·04	0·20	1·04	0·05	0·03	0·01	0·65	0·00	0·00
Eu/Eu*	0·42	1·07	1·08	1·16	1·02	0·98	1·06	1·25	0·91	1·08	—	—
La$_N$/Yb$_N$	2·02	1·31	2·43	3·35	2·08	3·97	2·68	0·59	2·02	0·78	—	—

Amphiboles: 1, granodiorite (Vierkisest); 2, tonalite (Vierkisest); 3, leucodiorite (Chernoistochinsk); 4, tonalite (Shabry); 5, tonalite (Shabry); 6, riebeckite, peralkaline sienite (Kharitonovo). Clinopyroxenes: 1, leucodiorite (Chernoistochinsk); 2, diorite (Chernoistochinsk); 3, gabbro–norite (Khabarny); 4, aegirine, peralkaline granite (Kharitonovo). Orthopyroxenes: 1, granulite (Murmansk); 2, charnockite (Kola).

(1992)]. In some granulitic garnets from Ivrea–Verbano stronalites we also detected significant La and Ce contents (see Fig. 2) with La$_N$/Sm$_N$ ~0·4–0·8. This feature is discussed below.

Garnets from granites usually have an 'amphibolitic' composition. Some crystals occasionally have a 'granulitic' core, probably relict in origin, surrounded by an 'amphibolitic' rim which seems to be equilibrated with the melt at low pressure [analyses 4c and 4r in Table 5, for example; see also data of Sevigny (1993)].

Cordierite

The contents of REE, Y, Th and U in cordierite range from low or very low to zero (Table 5). Chondrite-normalized REE patterns (Fig. 2) show a moderate decrease from La to Sm, a small negative Eu anomaly and variable Gd–Lu profiles with either a negative or positive slope.

Epidote

Magmatic epidote has high LREE and U contents, and moderate U, HREE and Y contents (Table 6). Primary epidote grains are usually rimmed by a film of secondary LREE carbonates (Fig. 3) such as bastnaesite or parisite, produced by reaction with late- or post-magmatic fluids. This causes the REE concentrations of epidote crystals from the same thin section to vary by a factor of 50. Chondrite-normalized REE patterns (Fig. 4) display a steep decrease from La to Sm, a moderate to strong positive Eu anomaly, and a moderate decrease from Gd to Lu. Remarkably, the partitioning of Eu into epidote is much more intense than into feldspars ($D_{Eu}^{epidote/Kfsp}$ ~20–30; $D_{Eu}^{epidote/plagioclase}$ ~8–20).

Tourmaline

Tourmaline has low REE, Y, Th and U contents (Table 6). Chondrite-normalized REE patterns are

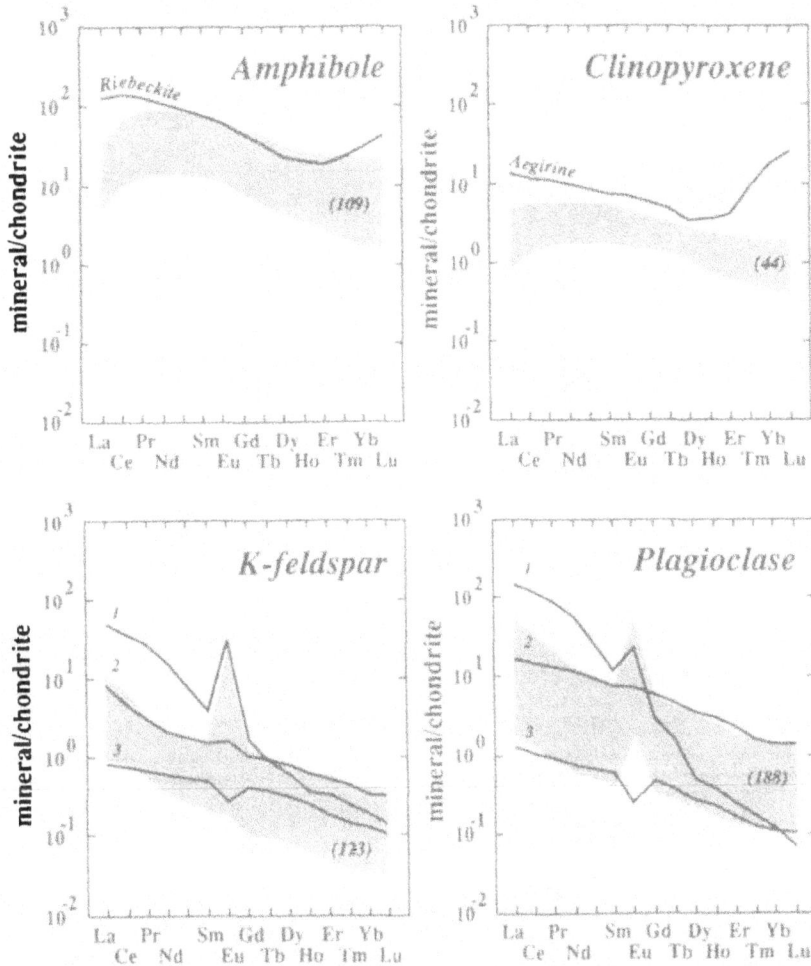

Fig. 1. Chondrite-normalized REE patterns of amphiboles, clinopyroxenes, K-feldspar and plagioclase. Shaded areas represent fields occupied by analysed specimens; the number of analyses represented is shown in parentheses. For K-feldspar and plagioclase, lines 1, 2 and 3 are representative examples of feldspars from a granulite (note the LREE enrichment), feldspars coexisting with primary epidote (note the small positive Eu anomaly) and feldspar from highly fractionated peraluminous rocks (note the low REE contents and negative Eu anomaly), respectively.

so variable (Fig. 2) that at present it is not possible to make any reasonable generalizations.

Al₂O₅ polymorphs

The concentration of REE, Y, Yh and U in all analysed grains of andalusite, sillimanite and kyanite was always lower than the sensitivity of the LA–ICP-MS system.

REEYThU-RICH ACCESSORY MINERALS

We identified about 26 species of REEYThU-rich minerals in common granite rocks and crustal protoliths. Twenty species—monazite, cheralite, xenotime, huttonite, thorite, allanite, cerianite, ura-

ninite, betafite, pyrochlore, brannerite, uranosferite, bastnaesite, parisite, loparite, samarskite, aeschinite, fergusonite, zirkelite and fluocerite—have at least one REE, or Y, or Th, or U as an essential structural component. Six species have at least one of these elements as an occasional abundant impurity: up to a few percent in zircon, up to a few thousand p.p.m. in apatite and sphene, and up to a few hundred p.p.m. in baddeleyite, rutile and fluorite. In the sections below we describe those aspects of the occurrence, composition and textural relationships of REEYThU-rich minerals which are relevant to understanding the geochemistry of these elements in granite rocks. These descriptions only reflect the characteristics of REE-, Th- and U-rich accessories such as they appear in common rocks,

527

JOURNAL OF PETROLOGY | VOLUME 37 | NUMBER 3 | JUNE 1996

Table 4: Selected LA–ICP-MS analyses of K-feldspar and plagioclase (results are in p.p.m.)

	K-feldspar						Plagioclase					
	1	2	3	4	5	6	1	2	3	4	5	6
Y	0·13	1·25	0·44	0·16	0·40	0·39	0·03	0·08	0·31	0·43	0·65	0·56
U	0·08	10·2	0·37	0·32	0·23	0·37	0·02	0·07	0·23	0·43	0·10	0·42
Th	0·18	0·01	0·10	0·01	0·09	0·08	0·11	0·03	0·11	0·10	0·03	0·09
La	0·19	1·78	10·5	1·39	1·86	0·78	0·64	2·20	34·2	7·33	4·55	1·47
Ce	0·44	2·81	20·8	1·63	2·23	1·71	1·40	3·57	70·8	5·60	9·33	2·62
Pr	0·06	0·28	2·40	0·12	0·28	0·21	0·17	0·40	8·4	0·49	1·19	0·20
Nd	0·27	0·82	6·57	0·29	0·97	0·47	0·50	0·81	25·4	1·38	3·99	0·52
Sm	0·07	0·10	0·59	0·08	0·23	0·06	0·07	0·15	1·77	0·18	1·13	0·12
Eu	0·01	0·66	1·68	0·89	0·09	0·28	0·01	0·23	1·25	1·44	0·29	1·30
Gd	0·08	0·14	0·33	0·08	0·21	0·07	0·07	0·13	0·59	0·15	0·98	0·12
Tb	0·01	0·03	0·03	0·01	0·03	0·01	0·01	0·02	0·08	0·02	0·16	0·02
Dy	0·08	0·13	0·15	0·06	0·19	0·05	0·07	0·09	0·12	0·10	1·01	0·14
Ho	0·01	0·03	0·02	0·01	0·03	0·01	0·01	0·02	0·02	0·02	0·16	0·03
Er	0·03	0·07	0·05	0·02	0·08	0·03	0·04	0·04	0·04	0·06	0·26	0·07
Tm	0·00	0·02	0·01	0·00	0·01	0·00	0·00	0·01	0·00	0·01	0·03	0·01
Yb	0·02	0·07	0·03	0·02	0·05	0·02	0·02	0·03	0·02	0·06	0·12	0·06
Lu	0·00	0·02	0·00	0·00	0·01	0·00	0·01	0·00	0·00	0·01	0·02	0·01
Eu/Eu*	0·41	17·1	11·6	34·0	1·25	13·2	0·44	5·00	3·74	26·8	0·84	33·1
La$_N$/Yb$_N$	6·41	17·1	236	46·9	25·1	26·3	21·6	49·4	11·54	82·4	25·5	16·5

1, Aplite (Pedrobernardo); 2, phenocryst in porphyritic U-rich granite (Albuquerque); 3, granulite (Strona); 4, migmatite (Peña Negra); 5, epidote-bearing granite (Vierkisest); 6, tonalite (Shabry).

not in pegmatites, metasomatites, or highly alkaline types.

Minerals with REE, Y, U or Th as essential structural components

Monazite (LREEPO$_4$) and cheralite [ThCa(PO$_4$)$_2$]

Monazite is the most widespread LREE-rich accessory mineral in granitoids, where it appears as isolated minute crystals with a diameter usually smaller than 50–60 μm, which have a great tendency to be included within biotite (Fig. 5), garnet (Fig. 6) and apatite (Fig. 7). Monazite is more abundant in silicic peraluminous types, but may be found in all types of granitoids, regardless of their silica content and degree of alumina saturation. However, when crystallization conditions were extraordinarily favourable for the formation of allanite, such as in granites with magmatic epidote, monazite may be very scarce or absent.

The composition of monazite (Table 7) varies widely owing to substitutions among the rare earths, the effects of solid solution with cheralite [ThCa(PO$_4$)$_2$] and huttonite (ThSiO$_4$) molecules, and partial replacement of Th by U (e.g. Vlasov,

1966, p. 285; Förster, 1993; Wark & Miller, 1993; Förster & Tischendorf, 1994; Förster & Rhede, 1995). Monazite is a selective mineral for LREE. Chondrite-normalized REE patterns are almost flat from La to Pr and then decrease smoothly from Pr to Yb and Lu, with a strong negative Eu anomaly (Fig. 8). Nd-rich monazites are not uncommon, especially in granulites (analysis 3 in Table 7). Primary monazite has high concentrations of thorium (ThO$_2$ ~5–27 wt % in unmineralized granites) and moderate concentrations of uranium (UO$_2$ ~0·2–3 wt %), except in U-rich granites where the situation is reversed (e.g. analysis 2, Table 7). Most thorium is incorporated into the monazite lattice through the cheralitic substitution Th^{4+} + Ca^{2+} ⇌ 2LREE^{3+}; it seems, in fact, that there is a complete solubility between LREEPO$_4$ and ThCa(PO$_4$)$_2$. The percentage of cheralitic component is usually in the range of 6–18%, but may rise to 50–70% in allanite-rich granites. Single grains of monazite are often zoned, with roughly concentric patterns from a cheralite-rich nucleus to monazite-rich rims predominating. The solubility of ThSiO$_4$ is more limited, usually <5%, even in monazites coexisting with Th-orthosilicate, although it may occasionally

Table 5: Selected LA–ICP-MS analyses of garnets and cordierites from metapelitic rocks and granites (results are in p.p.m.)

	Garnet								Cordierite			
	1	2	3	4	5 c	5 r	6 c	6 r	1	2	3	4
Y	167	94·3	58·6	25·9	248	135	45·1	187	0·07	0·55	0·37	0·05
U	0·11	0·21	0·03	0·12	0·08	0·00	0·03	0·18	0·05	0·36	0·16	0·01
Th	0·09	0·10	0·01	0·09	0·00	0·00	0·04	0·07	0·05	0·31	0·19	0·07
La	0·03	0·16	0·97	1·75	0·01	0·04	0·43	0·03	0·08	0·77	0·47	0·11
Ce	0·14	0·17	3·03	4·62	0·04	0·03	0·63	0·00	0·21	1·72	1·09	0·12
Pr	0·05	0·07	0·51	0·61	0·01	0·01	0·07	0·02	0·03	0·20	0·15	0·00
Nd	1·95	1·01	3·13	3·29	0·08	0·14	0·49	0·17	0·12	0·89	0·53	0·00
Sm	6·26	5·37	5·57	5·75	0·06	0·09	2·81	0·32	0·03	0·14	0·15	0·00
Eu	0·01	0·11	0·13	0·11	0·15	0·11	0·21	0 07	0·01	0·05	0·05	0·00
Gd	8·17	8·89	7·69	8·22	5·45	4·69	7·39	3·70	0·05	0·17	0·21	0·00
Tb	1·74	2·03	1·41	2·15	4·20	3·81	2·07	2·34	0·01	0·04	0·04	0·00
Dy	14·7	15·6	10·2	20·2	60·1	44·3	13·1	33·9	0·05	0·30	0·21	0·00
Ho	3·45	4·30	2·23	6·41	21·9	10·8	2·96	14·6	0·01	0·09	0·04	0·00
Er	11·4	15·2	7·13	21·1	75·4	22·0	7·35	68·7	0·03	0·27	0·11	0·00
Tm	1·60	2·63	1·26	3·58	12·5	2·40	1·14	13·6	0·01	0·06	0·01	0·00
Yb	8·93	17·8	8·36	21·5	76·7	11·4	7·10	100	0·04	0·40	0·07	0·00
Lu	1·09	2·43	1·45	3·12	10·2	1·04	1·12	15·3	0·01	0·08	0·01	0·00
Eu/Eu*	0·00	0·05	0·06	0·05	0·80	0·52	0·14	0·20	0·79	0·99	0·86	—
Sm$_W$/Gd$_N$	1·01	0·80	0·96	0·92	0·01	0·01	0·03	0·50	0·11			

In the case of garnet, c and r mean core and rim, respectively. Garnet: 1–4, garnet granulites (stronalites, kinzigite formation, Ivrea zone); 5, garnet amphibolite (NW Iberia); 6, granite (Peña Negra). Cordierite: 1, migmatite (Peña Negra); 2, retrograde crystal after garnet (Peña Negra); 3, leucogranite (Peña Negra); 4, leucogranite (Ronda).

reach 30–40% in some (but not all) monazite grains from a given thin section.

We could not find any meaningful correlation between granite typology and the chemistry of monazite. Certainly, monazites from metaluminous granites tend to be rich in cheralitic component and those from Th-rich granites may be rich in huttonitic component, but the compositions of grains from the same thin section usually vary so much [see also Wark & Miller (1993)] that it seems almost impossible to sketch general tendencies.

Secondary monazites appear always hydrated (rabdofan?) and characteristically have very low Th and U contents (analysis 12, Table 7). Most but not all secondary monazites we analysed also have high Y contents.

Allanite

Allanite is, after monazite, the most important LREE carrier in granitoids. Primary allanite may be found in all granite types except in the most peraluminous [aluminium saturation index (ASI) > 1·2]

phosphorus-rich varieties, and is especially abundant in granites with magmatic epidote. Contrary to a common belief among petrographers, primary allanite and monazite may coexist in equilibrium (Fig. 9). Secondary allanite is very common, even in the most peraluminous granites (Fig. 7).

Primary allanite usually appears as idiomorphic or subidiomorphic grains, often metamictic, with a diameter ranging from several tens of microns up to a few millimetres. They occur either alone or included within larger crystals, generally epidote or amphibole (Fig. 10).

The chemical composition of allanite is very complex and may vary across a wide range [see Vlasov (1966, pp. 302–306)]. In granites, however, the compositional range of allanite is more restricted [Table 8; see also Petrik *et al.* (1995)]. The ΣREE concentration ranges between 14 and 18 wt % with a clear predominance of LREE and a strong negative Eu anomaly (Fig. 8). Primary allanites always contain some Th (ThO$_2$ ~0·5–3 wt %), Zr (ZrO$_2$ ~0–2 wt %) and P (P$_2$O$_5$ ~0–0·2 wt %). The concentration of these elements is positively correlated

JOURNAL OF PETROLOGY | VOLUME 37 | NUMBER 3 | JUNE 1996

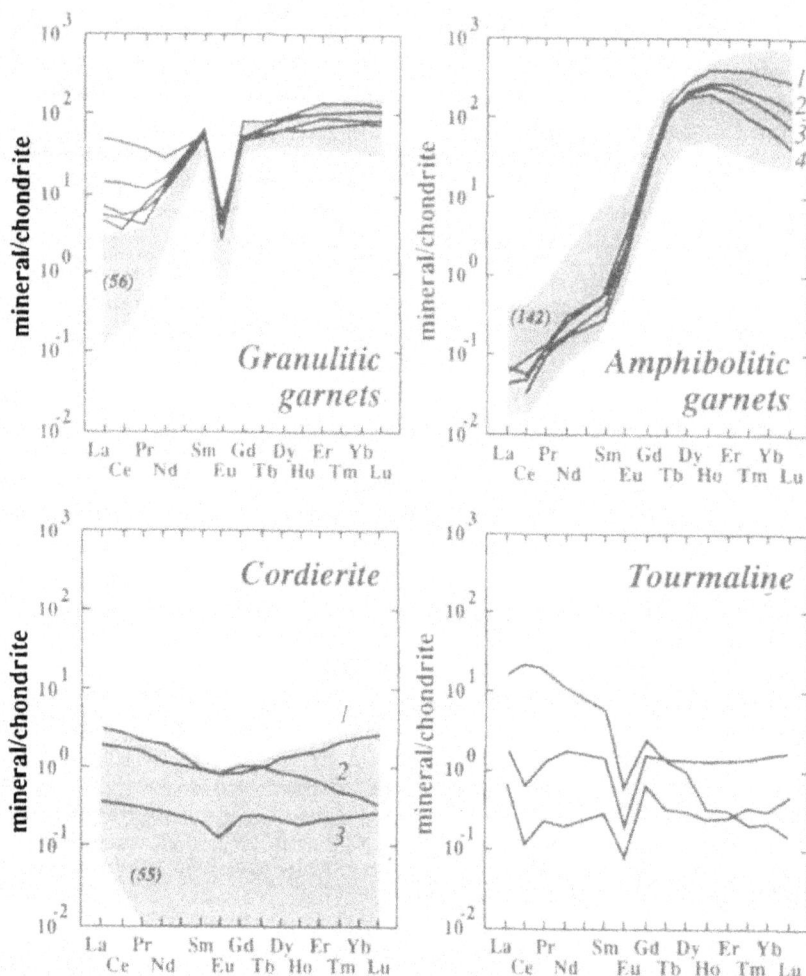

Fig. 2. Chondrite-normalized REE patterns of granulite-grade and amphibolite-grade garnets, cordierite and tourmaline. Shaded areas represent fields occupied by analysed specimens; the number of analyses represented is shown in parentheses. For granulite-grade garnets, continuous lines represent abnormally La–Ce-rich specimens from Ivrea–Verbano stronalites. For cordierites, 1 represents a retrograde crystal after garnet, 2 is an idiomorphic crystal from a granite and 3 comes from a migmatite.

with that of silica (Chesner & Ettlinger, 1989). U and Y contents are usually low and do not show any significant correlation with other elements.

Xenotime [(Y,HREE)PO₄] and intermediate xenotime–zircon phases

Xenotime is isostructural with zircon, coffinite and thorite (Vlasov, 1966, p. 230) and appears to form solid solution series with these minerals, above all with zircon. For convenience and arbitrarily, we have named all members of the xenotime–zircon solid solution series that have ZrO_2 <12 wt% xenotime, all those with P_2O_5 <6 wt% zircon, and all members between these intermediate xenotime–zircon phases (IXZP). As IXZP always occur closely

associated with xenotime, they will be described in conjunction with it.

Primary xenotime and IXZP are characteristic of peraluminous leucogranites, but they also occur in peralkaline granites. Both minerals appear either as isolated small grains with a diameter usually smaller than 50 μm or, more often, as complex intergrowths with zircon (Fig. 11).

The theoretical chemical composition of xenotime is YPO_4, but natural specimens in granitoids always contain 5–15 wt% of HREE [Table 9; see also Wark & Miller (1993)]. Besides the $ZrSiO_4$ component, primary xenotime also has ~0·5–2 wt% of LREE, 0·5–2 wt% of UO_2, 0–2 wt% of ThO_2 and appreciable amounts of FeO and CaO. Chondrite-normalized REE patterns (Fig. 8) show a steep increase from La to Sm, a deep negative europium anomaly,

BEA | RESIDENCE OF REE, Y, Th AND U IN GRANITES

Table 6: Selected LA–ICP-MS analyses of tourmalines and epidotes (results are in p.p.m.)

	Tourmaline						Epidote					
	1	2	3	4	5	6	1	2	3	4	5	6
Y	0·11	0·06	0·07.	4·30	1·02	0·01	3·77	6·03	9·56	7·19	3·55	1·81
U	0·24	1·04	0·12	0·00	0·00	0·07	7·53	0·00	0·88	4·29	7·29	1·83
Th	0·20	0·00	0·07	1·56	0·00	0·00	0·37	0·71	0·04	0·60	0·70	0·52
La	0·16	0·25	0·41	0·95	1·24	3·87	53·5	97·1	2·99	20·7	13·0	15·0
Ce	0·07	0·63	0·39	3·21	2·09	13·0	61·3	129·	9·07	48·6	24·3	23·2
Pr	0·02	0·06	0·12	0·68	0·24	1·65	5·26	10·1	1·54	5·90	2·94	2·46
Nd	0·09	0·20	0·75	2·94	0·86	4·80	13·5	21·4	6·19	20·0	12·5	7·14
Sm	0·04	0·00	0·23	0·94	0·13	0·82	2·23	2·71	1·64	3·74	1·96	0·91
Eu	0·01	0·00	0·01	0·08	0·00	0·04	2·70	5·40	0·69	2·75	0·88	1·06
Gd	0·13	0·00	0·32	1·62	0·26	0·49	1·56	1·51	1·13	3·26	1·93	0·67
Tb	0·01	0·00	0·05	0·25	0·04	0·05	0·20	0·23	0·18	0·46	0·24	0·09
Dy	0·07	0·00	0·26	1·52	0·32	0·22	0·91	1·37	0·98	2·53	1·54	0·63
Ho	0·01	0·00	0·07	0·32	0·07	0·02	0·20	0·26	0·20	0·51	0·30	0·14
Er	0·04	0·00	0·18	0·91	0·23	0·05	0·51	0·69	0·46	1·31	0·72	0·37
Tm	0·01	0·00	0·03	0·16	0·04	0·01	0·06	0·10	0·06	0·22	0·10	0·06
Yb	0·04	0·00	0·21	0·85	0·26	0·03	0·41	0·58	0·36	1·37	0·54	0·28
Lu	0·01	0·00	0·07	0·14	0·05	0·00	0·06	0·09	0·05	0·20	0·09	0·04
Eu/Eu*	0·42	—	0·11	0·20	0·00	0·19	4·43	8·17	1·55	2·41	1·38	4·15
La$_N$/Yb$_N$	2·53	—	1·32	0·75	3·22	87·0	88·1	113	5·61	10·2	16·2	36·1

Tourmaline: 1 and 4 leucogranites (Ronda); 2, U-rich granite (Albuquerque); 3 and 5, migmatites (Peña Negra); 6, granite (Trujillo). Epidote: 1–3, granodiorites (Vierkisest); 4, granite (Vierkisest); 5 and 6, granodiorites (Shartash).

and then another increase from Gd to Yb and Lu, not as abrupt as before. In some rare cases, LREE rise to 3 wt% and produce REE patterns which appear as if some monazitic component could also be dissolved in xenotime (Fig. 8). Additional support for this idea comes from the fact that Th increases together with LREE.

Intermediate xenotime–zircon phases have lower REE contents than pure xenotime, and the predominance of HREE over LREE is also less pronounced (Fig. 8). In addition, they are richer in Al_2O_3, FeO and CaO, which increase with the zircon component.

Thorium orthosilicate (ThSiO$_4$) minerals: huttonite and thorite

Very low but still significant amounts of thorium orthosilicate minerals—the monoclinic huttonite and the tetragonal thorite—are scattered through all types of granitoids. They appear as minute grains included in major (Fig. 12) and accessory minerals, above all zircon (Fig. 13) [see also Rubin *et al.* (1989)]. Because the huttonite and thorite can be differentiated with SEM only by the morphology of

the sections, we could not establish the relative proportion of these two minerals, although the fact that $ThSiO_4$ minerals in metaluminous granites are often metamictic, whereas in peraluminous granites they appear remarkably unaltered, could perhaps indicate that they consist of thorite and huttonite, respectively (Bayer, 1969).

$ThSiO_4$ minerals from granites always contain several weight percent of REE (Table 10), with a strong predominance of HREE over LREE and deep negative Eu anomalies (Fig. 8). Some P-rich varieties having high concentrations of LREE are probably intermediate monazite–huttonite phases (analysis 3, Table 10). Uranium is usually in the range of 1–10 wt%, although grains with subequal contents of Th and U (uranothorite?) are frequent.

Uraninite (UO$_2$)

Uraninite is a common accessory in peraluminous leucogranites, where it appears as very small grains (diameter usually <5 μm) included either in major or accessory minerals, especially zircon (Fig. 14). Uraninite contains up to 20 wt% ThO_2 and may also have up to 5 wt% of Y_2O_3 and 1 wt% of REE,

JOURNAL OF PETROLOGY | VOLUME 37 | NUMBER 3 | JUNE 1996

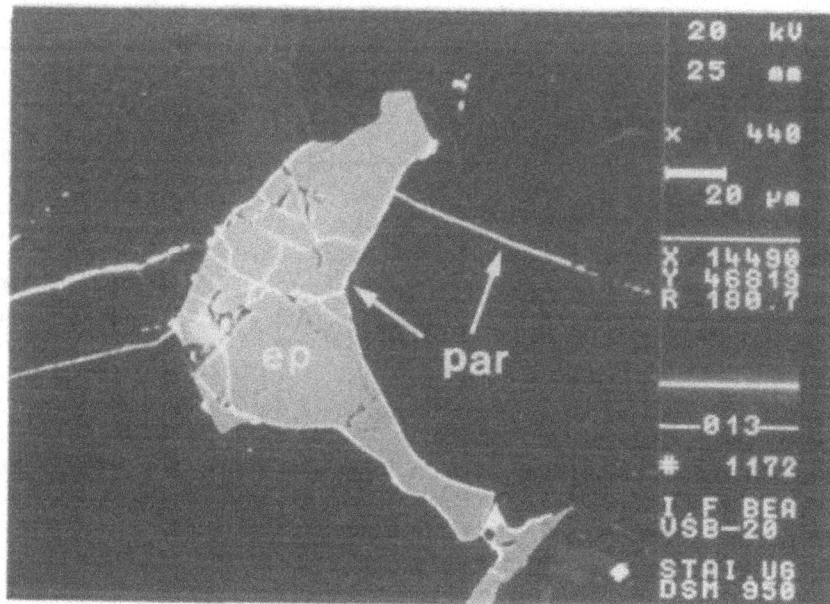

Fig. 3. Back-scattered electron (BSE) SEM image of a primary epidote crystal rimmed by REE-carbonate (parisite?). (Note how REE-carbonate fills radiating cracks.) Metaluminous granodiorite, Verkisest batholith.

with a clear predominance of HREE (Table 10; Fig. 15).

Complex uranium minerals

Rocks containing uraninite and xenotime usually contain other primary U minerals as well. The most common species are minerals from the pyrochlore–betafite $[(U,Ca)(Nb,Ta,Ti)_3O_9.nH_2O]$ group. A U–Bi mineral, probably uranosferite $[BiUO_3(OH)_2]$ and brannerite (UTi_2O_6) are also occasionally found. All these minerals contain variable amounts of Th and REE, and are generally HREE selective (Table 10; Fig. 15).

Aeschinite, samarskite, fergusonite, loparite

Complex minerals of Nb, Ti, Y, U and REE are common in peralkaline granitoids, usually forming complex aggregates of phases with a highly variable composition (Montero, 1995) (Table 10; Fig. 15). The most abundant species are aeschinite (low U and Y, high LREE), fergusonite (moderate U, Y and LREE, high HREE) and samarskite (high U, Y and HREE, low LREE). Loparite is one of the richest minerals in La–Ce (see analysis 11, Table 10) although it appears to be limited to highly peralkaline rocks.

Bastnaesite and parisite

Primary bastnaesite is found only in REE-rich per-

alkaline granitoids. Secondary REE carbonates, probably parisite and/or bastnaesite, are fairly common in metaluminous granites, where they appear as a product of the decomposition of allanite and epidote (Figs 3 and 10). These minerals are extremely selective for LREE and contain negligible amounts of Th and U.

Cerianite (CeO_2)

The oxide of tetravalent Ce is found in some hydro-thermalized granites filling cracks and veins. As cerianite is usually related to monazites with a deep negative Ce anomaly (Fig. 8; analysis 1 in Table 7), it would seem to be produced through the oxidation and subsequent leaching of Ce^{4+} from monazite caused by hydrothermal fluids. The concentration of Y, Th, U and REE other than Ce in cerianite is below microprobe sensitivity.

Zirkelite $[(Zr,Ca,Ti,Fe,Mg,REE,U,Th)_3O_5]$

Zirkelite is a rare mineral which appears in low-silica granitoids together with thorite and badde-leyite, usually as very small inclusions in zircon (Fig. 13). Its rarity and small grain-size have not permitted us to obtain reliable analyses, although semi-quantitative EDAX analyses with SEM indicate $UO_2 \sim 3$–7 wt %, and the predominance of HREE over LREE, with $Dy_2O_3 \sim 0.5$ wt %, $Er_2O_3 \sim 0.7$ wt % and $Yb_2O_3 \sim 0.9$ wt %.

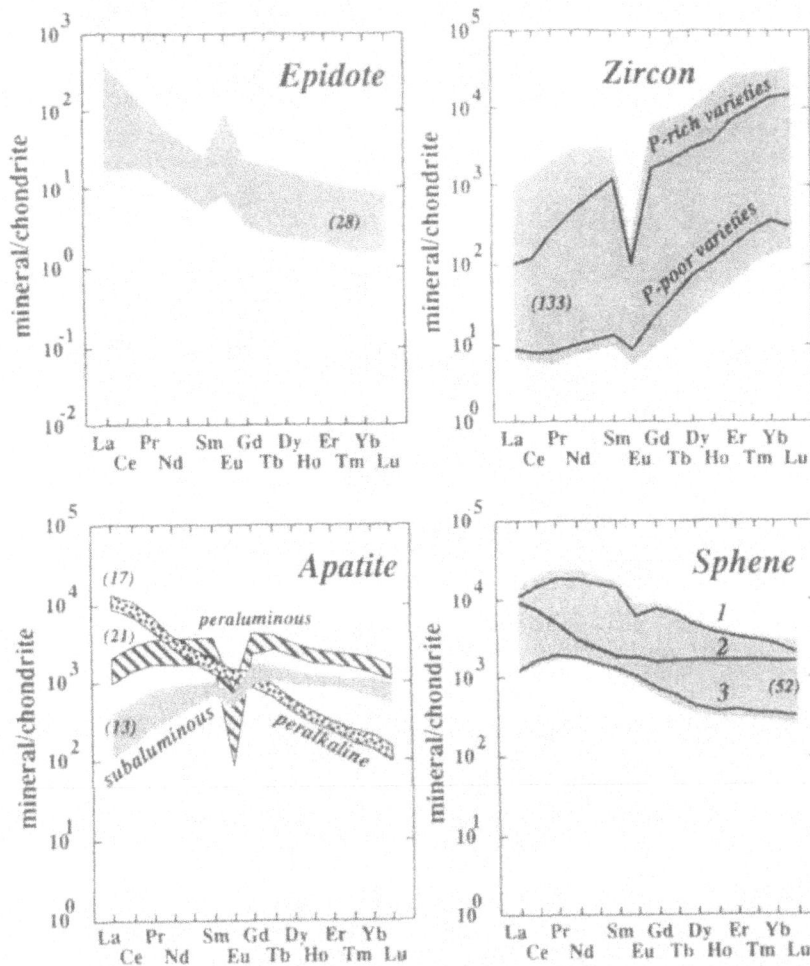

Fig. 4. Chondrite-normalized REE patterns of epidote, zircon, apatite and sphene. Shaded areas represent fields occupied by analysed specimens; the number of analyses represented is shown in parentheses. For sphene, 1 and 3 represent extreme cases of sphenes from metaluminous granitoids (note the decreased intensity of the Eu anomaly as ΣREE decreases; 2 represents the characteristic pattern of sphenes from peralkaline granites.

Fluocerite [(La,Ce)$_3$F]

Fluocerite is an occasional mineral in peralkaline granitoids (Imaoka & Nakashima, 1994; Montero, 1995), where it normally appears as a late mineral growing in interstitial positions. It contains about 18–20 wt % La, 33–35 wt % Ce, 1·5–2 wt % Pr, 6–6·5 wt % Nd, 0·4–0·5 wt % Sm, 0·4–0·6 wt % Gd and <0·1 wt % of the other REE. Th content may be as high as 2 wt %. U and Y are usually very low.

Minerals with REE, Y, Th or U as abundant impurities

Zircon (ZrSiO$_4$)

Zircon is the most ubiquitous accessory mineral in granitoids, having been extensively studied owing to its many applications in isotopic dating and as a petrogenetic indicator (Pupin, 1980; Krasnobayev, 1986; Aleinikoff & Stoeser, 1989; Heaman & Parrish, 1991). Zircon appears as grains with ϕ_{max} usually between 60 and 1 μm, which frequently contain micron-sized inclusions of thorite (Fig. 13) and uraninite (Fig. 14), and may form complex intergrowths with xenotime (Fig. 11) and, less commonly, with monazite (Fig. 16).

The chemical composition of zircon varies within a broad range owing to replacement of Zr^{4+} by Hf^{4+}, Th^{4+} and U^{4+}, as well as to solid solution with xenotime, which introduces elevated amounts of Y and HREE in the zircon lattice [Table 11; see also Heaman *et al.* (1990), Pupin (1992), Rub *et al.* (1994) and Barbey *et al.* (1995)]. The concentrations of Th and U are highly variable, even among grains from the same thin section. ThO_2 is rarely greater

JOURNAL OF PETROLOGY | VOLUME 37 | NUMBER 3 | JUNE 1996

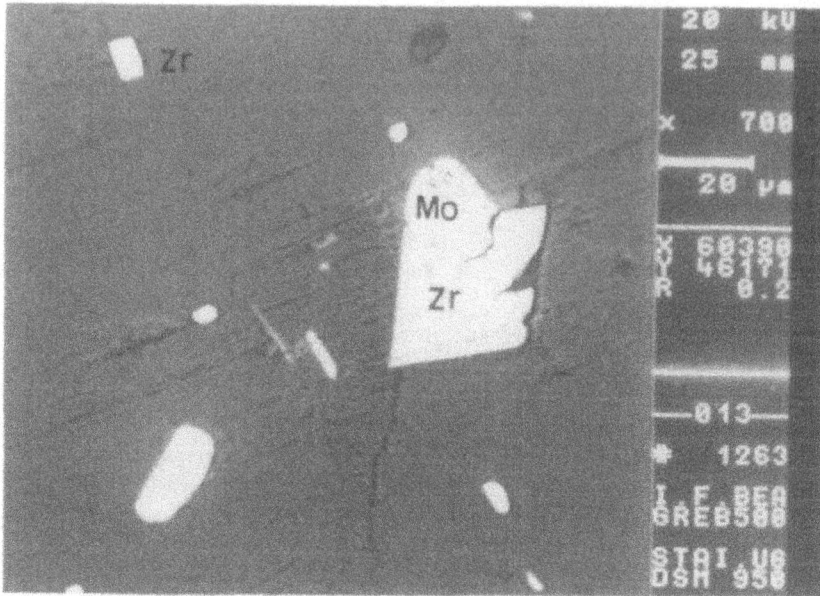

Fig. 5. BSE-SEM image of a biotite crystal from a peraluminous granite. [Note the great abundance and small grain-size of zircon (Zr) and monazite (Mo and unlabelled) inclusions.] Pedrobernardo granite.

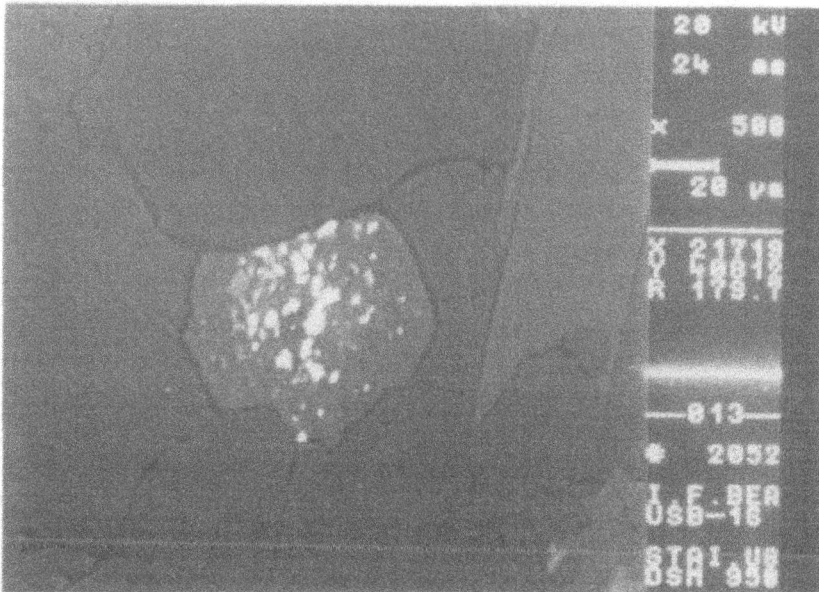

Fig. 6. BSE-SEM image of monazite inclusions in a garnet from a peraluminous leucogranite. (Note how monazite is selectively included in garnet.) Murzinka granite.

than 0·75 wt %, and LA–ICP-MS data show that values between 100 and 200 p.p.m. are common (Table 11). UO_2 may occasionally reach 10 wt %, but normal concentrations are between 10 and 100 p.p.m. U (Table 7). The concentration of Y varies from a few p.p.m. to 6 wt %, and increases with increasing phosphorus contents (Tourrette et al.,

1991). ΣHREE are in the range of 0–2 wt % and also increase with phosphorus, which suggests that solid solution with xenotime is the main mechanism of Y and REE incorporation into the zircon lattice. Chondrite-normalized REE patterns (Fig. 4) show La_N/Sm_N in the range 1–5, a moderate negative Eu anomaly and a progressive increase in HREE, with

534

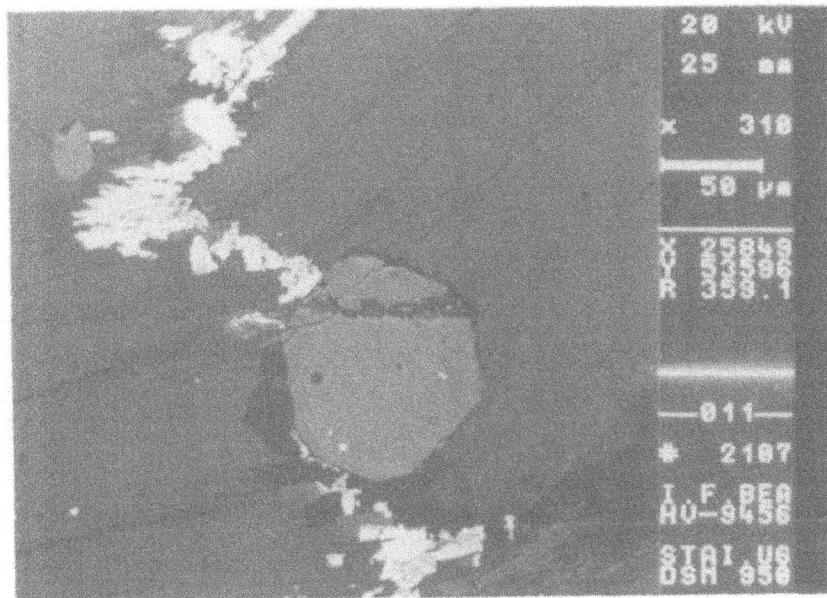

Fig. 7. BSE-SEM image of secondary allanite that developed along a fracture in biotite. The pseudohexagonal large crystal is apatite. White small grains included in apatite and biotite are monazite. Hoyos granodiorite.

$Gd_N/Yb_N \sim 5$–10 and occasionally $Yb_N/Lu_N > 1$. Sc is a common trace element with a concentration in zircons from granites rarely higher than some tens of p.p.m. (Heaman *et al.*, 1990), although we have found one case (Albuquerque granite, Iberia) with zircons having up to 10 wt% Sc_2O_3 (analysis 11, Table 11).

Apatite [$Ca_5(PO_4)_3(OH,F,Cl)$]

Apatite is a well-known abundant accessory mineral in granite rocks, especially in peraluminous granodiorites and monzogranites, where it may reach 1% modal abundance. Apatite appears in grains of variable size and morphology, from small (~ 30–5 μm), needle-like crystals, more characteristic of I- and A-type granites, to large (~ 500–50 μm), roughly equidimensional crystals typical of strongly peraluminous S-type granites. Apatite crystals usually contain small inclusions of monazite (Fig. 7).

The concentration of REE, Y, Th and U in apatite is highly variable (Roeder *et al.*, 1987) and shows marked differences according to the rock's aluminosity (Table 12; Fig. 4). Apatites from peraluminous rocks are the richest in Y, U, Th and HREE, and have flat chondrite-normalized REE patterns ($La_N/Lu_N \sim 1$) with a strong negative Eu anomaly ($Eu/Eu^* \sim 0.1$). Apatites from metaluminous granites have less REE, Y, Th and U, and their REE patterns show a positive slope from La to Sm, a small negative Eu anomaly ($Eu/Eu^* \sim 0.7$),

and are almost flat from Gd to Lu, with $La_N/Lu_N \sim 0.1$–0.4 (Fig. 4). Apatites from peralkaline rocks have the lowest Y, REE, Th and U contents, but are the richest in LREE, have no Eu anomaly, and show REE patterns with a steep negative slope from La to Lu ($La_N/Lu_N \sim 50$–100).

Sphene [$CaTi(SiO_4)(O,OH,F)$]

Primary sphene is a widespread accessory in metaluminous and peralkaline granites, where it usually appears as idiomorphic or subidiomorphic crystals with a grain-size in the range of 0.05–5 mm. Primary sphenes always contain a few thousand p.p.m. REE and Y, and a few hundred p.p.m. Th (Table 13). The concentration of U, however, is highly variable, from near zero to ~ 500 p.p.m. Chondrite-normalized REE patterns of sphene (Fig. 4) from metaluminous granites increase from La to Pr–Nd, have a small negative Eu anomaly (less intense the less REE the crystal has), decrease from Gd to Dy–Ho, and are almost flat from Er to Lu. Sphenes from peralkaline rocks, in contrast, have REE patterns with a negative slope from La to Sm, no Eu anomaly, and are almost flat from Gd to Lu.

Baddeleyite, fluorite and rutile

These minerals have been reported to contain somewhat elevated concentrations of REE, Y, Th and/or U (Vlasov, 1966; Heaman & Parrish, 1991).

535

JOURNAL OF PETROLOGY | VOLUME 37 | NUMBER 3 | JUNE 1996

Table 7: Selected microprobe analyses of monazite crystals (results expressed in percent)

	1	2	3	4	5	6	7	8	9	10	11	12
SiO_2	0·00	0·00	4·74	2·50	0·40	0·00	0·00	0·08	0·00	0·00	0·00	0·92
ZrO_2	0·01	0·15	0·00	0·17	0·16	0·05	0·30	0·00	0·26	0·28	0·10	0·34
Al_2O_3	0·00	0·99	0·00	0·00	0·00	0·05	0·00	0·00	0·00	0·00	0·15	0·62
FeO	0·00	0·30	0·01	0·00	0·00	0·63	0·21	0·00	0·05	0·00	0·79	0·15
MgO	0·00	0·00	0·00	0·00	0·00	0·00	0·00	0·00	0·00	0·00	0·00	0·00
CaO	2·79	2·15	0·41	1·28	1·02	1·58	2·86	2·11	2·11	1·48	1·18	1·75
Na_2O	0·00	0·02	0·00	0·00	0·00	0·00	0·05	0·00	0·00	0·00	0·00	0·11
P_2O_5	34·05	28·96	21·07	23·51	30·04	30·62	28·22	24·80	28·70	28·42	28·51	25·65
Y_2O_3	2·88	2·86	0·96	2·93	3·62	1·61	2·36	3·25	0·04	2·22	0·33	7·67
ThO_2	10·58	4·78	27·31	18·64	8·54	9·22	9·98	10·01	11·75	8·94	5·42	0·00
UO_2	0·18	13·78	0·81	1·46	1·58	1·18	6·47	1·35	0·90	0·79	1·14	0·48
La_2O_3	13·62	7·83	6·43	9·18	8·29	10·27	8·15	8·77	9·31	11·37	6·49	6·36
Ce_2O_3	2·42	21·01	18·88	23·34	25·79	26·47	24·03	25·26	27·20	28·15	29·36	20·92
Pr_2O_3	5·50	2·84	2·81	3·72	3·74	3·07	3·57	3·65	3·97	4·20	3·22	2·73
Nd_2O_3	18·34	9·76	13·59	8·51	11·20	11·48	10·46	14·99	12·63	10·17	16·80	9·58
Sm_2O_3	2·15	2·63	1·15	2·21	3·70	2·55	2·27	3·58	2·07	2·69	3·04	1·87
Eu_2O_3	0·14	0·22	0·18	0·24	0·14	0·08	0·05	0·07	0·15	0·24	0·19	0·01
Gd_2O_3	0·59	0·66	0·49	1·40	0·89	0·83	0·39	1·67	1·12	0·68	1·82	2·11
Dy_2O_3	0·27	0·42	0·46	0·64	0·59	0·37	0·15	0·91	0·38	0·13	0·31	2·42
Er_2O_3	0·06	0·02	0·18	0·07	0·08	0·15	0·08	0·16	0·14	0·07	0·16	0·77
Yb_2O_3	0·03	0·01	0·15	0·09	0·03	0·11	0·07	0·03	0·12	0·07	0·12	0·72
Lu_2O_3	0·00	0·00	0·03	0·01	0·00	0·02	0·01	0·00	0·01	0·01	0·02	0·20
Total	93·61*	99·39	99·66	99·90	99·81	100·34	99·68	100·69	100·91	99·71	99·14	85·38*

* Contains appreciable H_2O.
1, Cerianite-bearing granite (Ilmen); 2, U-rich peraluminous leucogranite (Albuquerque); 3, metapelitic granulite (Strona); 4, garnet leucogranite (Murzinka); 5, B-rich leucogranite (Ronda); 6, two-mica granite (Pedrobernardo); 7 and 8, allanite-bearing granite (Magnitogorsk); 9, allanite-bearing granodiorite (Sirostan); 10, xenotime-bearing peraluminous leucogranite (Ronda); 11, xenotime-bearing migmatite (Peña Negra); 12, secondary monazite (Galiñeiro).

However, in all the samples we studied these elements are below microprobe sensitivity. On the other hand, the small grain-size has not allowed us to perform reliable LA–ICP-MS analyses. Some laser shots at crystals slightly smaller than the laser beam—and therefore with some contribution from the surrounding crystals—indicate that these three minerals are LREE selective and the concentrations of REE and Y are lower than a few tens of p.p.m. Analysis of a baddeleyite concentrate from Kovdor (Kola) also shows a minor enrichment in HREE, producing a U-shaped REE pattern, very similar to that described by Reischmann *et al.* (1995) in baddeleyite concentrates from South Africa. The U and Th concentrations are probably a few p.p.m., except in rutile, where U is probably in the range of 20–100 p.p.m.

FRACTIONAL CONTRIBUTION OF EACH MINERAL TO WHOLE-ROCK REE, Y, Th AND U CONCENTRATIONS

The nature and, in some cases, the composition of REEYThU-rich accessories changes systematically with the rock bulk-chemistry, above all with aluminosity. Therefore, to obtain a general picture of the fractional contribution of major and accessory minerals to the bulk-rock REE budget for the whole granite spectrum, we performed a mass-balance study of 12 plutons from Iberia and the Urals with an average ASI ranging from 1·27 to 0·78. The following features stand out.

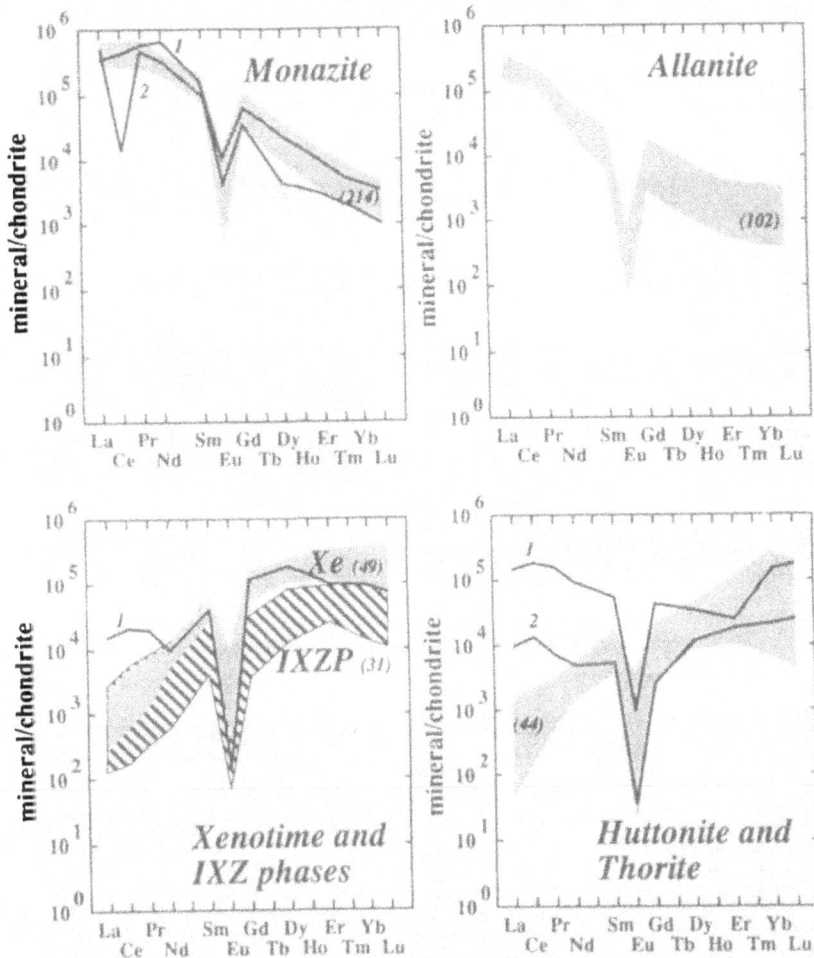

Fig. 8. Chondrite-normalized REE patterns of monazite, allanite, xenotime, intermediate xenotime–zircon phases (IXZP) and Th-orthosilicate minerals. Shaded areas represent fields occupied by analysed specimens; the number of analyses represented is shown in parentheses. Monazite: 1, Nd-rich variety; 2, Ce-depleted monazite associated with secondary cerianite (see text). Xenotime: 1, intermediate xenotime–monazite phase. Th-orthosilicates: 1 and 2, LREE- and P-rich varieties, probably monazite–huttonite solid solution phases.

Distribution of LREE (Fig. 17)

In peraluminous granites, ~90–95 wt % of bulk-rock LREE contents reside within accessories (80–85 wt % in monazite, the rest in apatite) and the remaining 5–10 wt % in feldspars. In metaluminous granites, the fraction of LREE residing within accessories decreases to ~70 wt % (50–60 wt % in allanite, the rest in sphene, apatite, monazite and REE-carbonates). Amphibole accounts for ~15–25 wt % LREE, whereas the remaining 5–10 wt % is in feldspars and, when present, epidote. In peralkaline granites, the fraction of LREE residing within accessories increases again up to 80–90 wt %. Allanite is still a substantial LREE reservoir, although bastnaesite, fluocerite and aeschinite may also be very important.

Distribution of Eu (Fig. 17)

In peraluminous granites, plagioclase and K-feldspar contain similar fractions of total Eu, and together they account for ~90 wt % Eu. Apatite plus monazite accounts for ~5–7 wt % Eu and micas could have ~1–2 wt % Eu. In metaluminous granites, plagioclase is the most important Eu reservoir and together with K-feldspar accounts for ~70–80 wt % of total Eu. Amphibole may account for ~5–15 wt % and the rest is in allanite, sphene, apatite and monazite. When present, primary epidote may also account for a significant fraction of total Eu. In peralkaline granites, ~75 wt % Eu is in feldspars, another 10 wt % in alkali amphiboles, and the rest is distributed in allanite, epidote, sphene, monazite, bastnaesite, niobotantalates, etc.

JOURNAL OF PETROLOGY | VOLUME 37 | NUMBER 3 | JUNE 1996

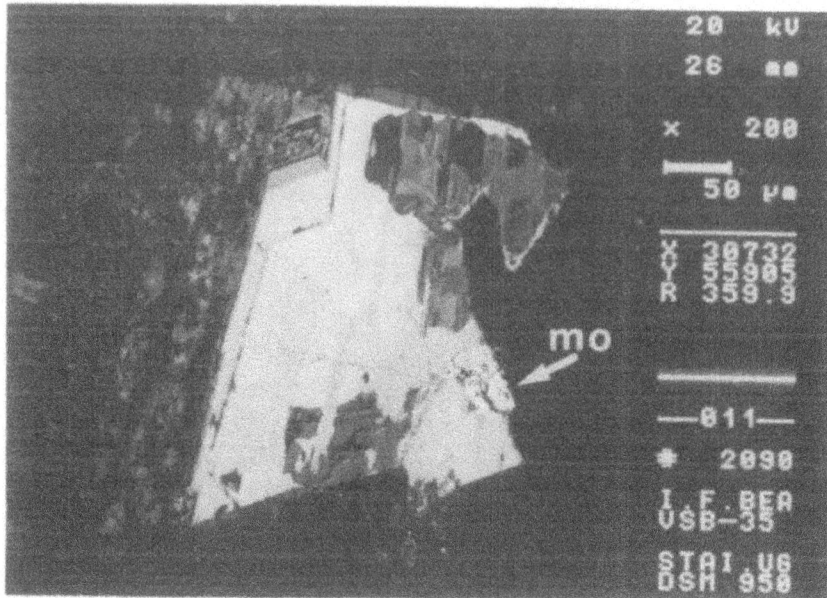

Fig. 9. BSE-SEM image of an idiomorphic-zoned allanite including a cheralitic monazite crystal, apparently in equilibrium (see text). Metaluminous plagiogranite, Ackermanovsky complex, Khabarny.

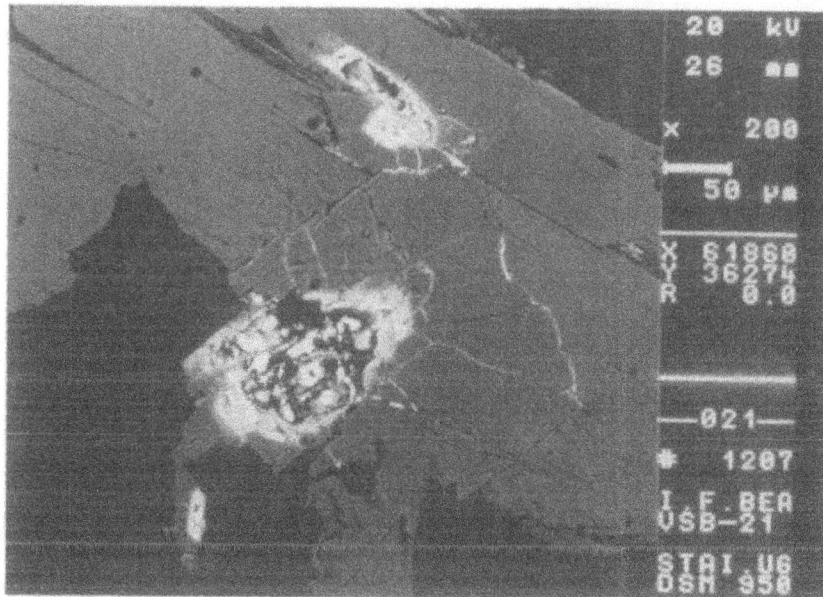

Fig. 10. Metamictic allanite crystals included in amphibole. [Note secondary REE-carbonates filling cracks (BSE-SEM image).]

Distribution of HREE and Y (Fig. 17)

In peraluminous granites, no less than 95 wt% of total HREE reside within accessories. In low-Ca varieties, the fraction of total REE that resides within xenotime is ~30–50 wt%, in apatite ~15–25 wt%, in zircon ~15–20 wt%, in monazite ~5–10 wt%, and in Th-orthosilicate ~5–10 wt%. The disappearance of xenotime in high-Ca peraluminous granites reinforces the roles of zircon (~35–40 wt% of total HREE), monazite (~20 wt%) and apatite (20–25 wt%). In metaluminous granites, amphibole may contain as much as 30–35 wt% of total HREE, whereas zircon accounts for ~30–50 wt%, allanite

Table 8: Selected analyses of allanite crystals (results expressed in percent)

	1	2	3	4	5	6	7	8	9	10	11
SiO_2	32·95	37·28	34·15	30·14	36·43	38·21	42·66	42·35	33·43	32·10	31·88
ZrO_2	0·06	0·08	0·02	0·00	0·28	0·76	1·64	1·45	0·07	0·13	0·00
Al_2O_3	14·86	12·14	16·16	14·02	15·67	16·04	13·56	13·36	11·01	10·74	13·68
FeO	18·22	17·53	17·50	17·07	10·87	10·83	9·09	8·91	16·23	17·07	13·51
MgO	1·20	0·02	0·75	0·24	1·03	1·00	0·67	0·84	0·10	0·01	1·22
CaO	10·39	9·31	9·86	9·09	10·95	10·15	9·87	9·34	8·37	8·78	10·14
Na_2O	0·12	0·08	0·14	0·22	0·19	0·23	0·56	0·50	0·19	0·22	0·06
P_2O_5	0·08	0·02	0·04	0·03	0·04	0·16	0·12	0·14	0·00	0·01	0·09
Y_2O_3	0·34	0·22	0·37	0·19	0·52	0·24	0·365	0·229	0·147	0·236	0·271
ThO_2	0·72	0·64	0·56	0·66	1·31	1·67	1·79	2·05	1·05	1·31	0·96
UO_2	0·00	0·00	0·00	0·00	0·00	0·00	0·22	0·011	4·38	0·101	0·093
La_2O_3	4·47	5·06	4·75	8·94	4·84	4·87	4·87	3·97	5·72	4·630	4·67
Ce_2O_3	9·50	9·82	8·97	14·13	10·01	9·76	8·53	8·79	12·27	14·73	13·96
Pr_2O_3	1·28	1·32	1·18	1·65	1·21	1·25	0·90	1·28	1·40	1·71	1·61
Nd_2O_3	2·68	2·74	2·45	0·46	3·16	2·17	1·30	1·88	4·07	4·82	5·06
Sm_2O_3	0·43	0·43	0·43	0·08	0·47	0·36	0·27	0·269	0·230	0·300	0·640
Eu_2O_3	0·00	0·00	0·00	0·00	0·00	0·00	0·00	0·001	0·001	0·003	0·001
Gd_2O_3	0·25	0·36	0·09	0·07	0·37	0·17	0·14	0·169	0·130	0·174	0·430
Tb_2O_3	n.d.	n.d.	n.d.	n.d.	n.d.	n.d.	n.d.	0·021	0·016	0·023	0·055
Dy_2O_3	0·19	0·23	0·03	0·07	0·10	0·06	0·11	0·101	0·085	0·130	0·210
Ho_2O_3	n.d.	n.d.	n.d.	n.d.	n.d.	n.d.	n.d.	0·020	0·014	0·020	0·033
Er_2O_3	0·07	0·07	0·01	0·03	0·04	0·01	0·05	0·048	0·036	0·053	0·073
Tm_2O_3	n.d.	n.d.	n.d.	n.d.	n.d.	n.d.	n.d.	0·006	0·004	0·006	0·008
Yb_2O_3	0·05	0·06	0·01	0·02	0·03	0·01	0·03	0·031	0·020	0·030	0·032
Lu_2O_3	0·00	0·01	0·00	0·00	0·00	0·00	0·00	0·004	0·003	0·003	0·004
Total	97·86	97·42	97·47	97·11	97·52	97·95	96·75	95·76	98·97	97·34	98·69

*Contains appreciable H_2O.
REE, Y, Th and U were determined by LA–ICP-MS in samples 8–11. All the other data were obtained by electron microprobe. 1, 2 and 10, granodiorite (Vierkisest); 3 and 4, granite (Magnitogorsk); 5 and 11, plagiogranite (Khabarny); 6, leucogranite (Shartash); 7, 8 and 9, tonalite (Sirostan). n.d., not determined.

for ∼5–15 wt %, sphene for ∼5–10 wt %, Th-orthosilicate for ∼2–5 wt %, and apatite plus monazite for the remaining 5–10 wt %. In peralkaline granites, the fraction of HREE in major minerals decreases down to 5 wt %. Niobotantalates contain ∼40–60 wt % HREE, xenotime has ∼20–30 wt %, and the rest is in zircon and thorite. Yttrium shows basically the same distribution pattern as HREE.

Distribution of Th (Fig. 18)

In peraluminous granites, the fraction of Th residing within major minerals is very low, ∼5 wt %. Monazite, with ∼65–80 wt %, and Th-orthosilicate, with ∼20–30 wt %, are the most important Th reservoirs, whereas zircon, xenotime and apatite account for ∼1–2 wt % each. In metaluminous granites, amphibole contains ∼15–20 wt % of total Th, feldspars ∼5 wt %, allanite ∼10–40 wt %, monazite ∼15–70 wt %, Th-orthosilicate ∼10–25 wt %, sphene and zircon ∼2–10 wt %, and the rest is in apatite. In peralkaline granites, the fraction of Th contained in major minerals decreases down to 5–15 wt %, whereas the roles of Th-orthosilicate (∼20–40 wt %) and niobotantalates (∼1–10 wt %) increase. Monazite contains ∼20–40 wt % Th, allanite ∼5–20 wt %, sphene ∼5 wt %, and the rest is in zircon and apatite.

JOURNAL OF PETROLOGY | VOLUME 37 | NUMBER 3 | JUNE 1996

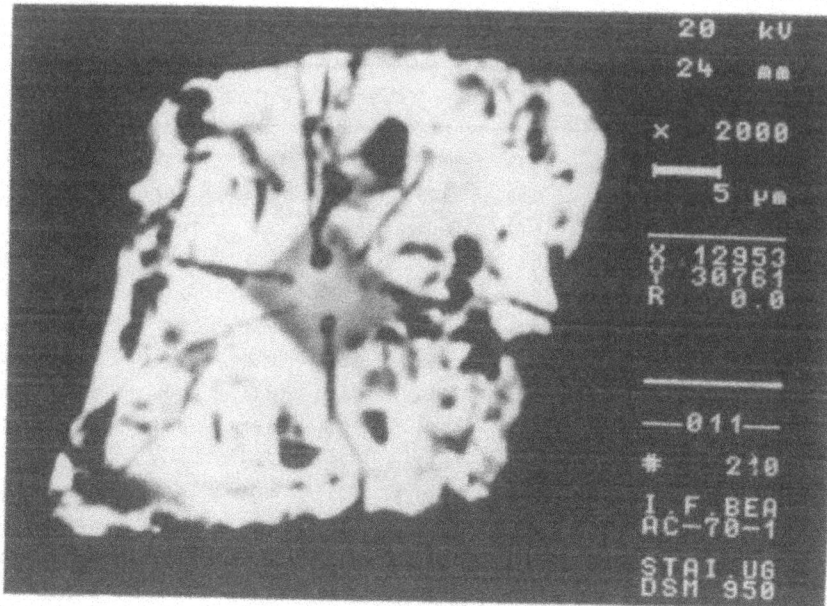

Fig. 11. BSE-SEM image of a xenotime–zircon intergrowth. The idiomorphic nucleus has an external part composed of zircon and a core comprising IXZP. The external part is mostly composed of xenotime, with alternating thin layers of IXZP. Ronda leucogranites.

Distribution of U (Fig. 18)

The fraction of U contained in major minerals is always <5 wt %. In low-Ca peraluminous granites, U resides mostly in xenotime (~50–60 wt %), followed by uraninite, monazite ~5–20 wt %, betafite–pyrochlore and zircon. In high-Ca peraluminous granites, zircon and xenotime contain ~85–90 wt % of total U, whereas Th-orthosilicate and uraninite account for 5–10 wt % U. In metaluminous granites, zircon accounts for 50–60 wt % U, allanite for 10–20 wt % U, apatite plus xenotime for 15–30 wt % and Th-orthosilicates for 5–10 wt % U. In peralkaline granites, niobotantalates, especially samarskite, may become the most important U reservoirs.

DISCUSSION

Effects of accessories on the geochemistry of REE, Y, Th and U

Effects derived from the high fraction of REE, Y, Th and U contained in accessories

The fractionation of minute amounts of accessories will dramatically influence the partitioning of REE, Y, Th and U between melt and solids, simply because accessories contain most of these elements. This study on the contribution of major and accessory minerals, despite the unavoidable analytical inaccuracies and the arbitrariness of sample selection, unequivocally confirms that major minerals play a very subordinate role with respect to that of accessories (Gromet & Silver, 1983), especially in peraluminous systems. At the same time, as REE, Y, Th and U are essential structural components in at least one accessory phase of every crystallizing granite and melting protolith, they do not generally obey Henry's law during melt–solid partitioning, their concentrations in partial melts not being ruled by crystal–melt distribution coefficients but by solubility relations and dissolution kinetics (Rapp & Watson, 1986). These latter, in their turn, depend strongly on the volatile content and bulk-composition of the system (Watt & Harley, 1993).

Effects of textural position during partial melting

The textural position and small grain-size of REEYThU-rich accessories further complicates the partitioning of REE, Y, Th and U during melt segregation. The behaviour of accessory minerals during partial melting depends on whether they are placed at major-phase grain boundaries or are included within major minerals (Watson *et al.*, 1989). In the first case, accessories are available for the melt and so react with it. In the second case, accessories may remain physically isolated from the melt, thus preventing any reaction, or may be entrained as inclusions if major minerals are incorporated into the melt as restitic crystals (Watt & Harley, 1993; Bea *et al.*, 1994*b*). Based on facial energy considerations and experimentation, Watson

Table 9: Selected microprobe analyses of xenotime and intermediate xenotime–zircon (IXZ) crystals (results expressed in percent)

	1	2	3	4	5	6	7	8	9	10	11	12
SiO_2	1·11	0·31	0·22	0·31	0·75	1·02	1·00	3·20	8·37	16·00	9·65	11·52
ZrO_2	0·00	0·11	0·13	0·16	0·24	0·29	0·41	4·95	11·08	23·91	28·29	38·45
HfO_2	0·03	0·12	0·05	0·00	0·00	0·00	0·00	0·02	0·32	1·05	0·28	0·93
Al_2O_3	0·00	0·00	0·00	0·00	0·00	0·00	0·00	0·78	0·66	3·86	2·17	2·00
FeO	0·26	0·13	0·36	0·61	0·00	0·10	0·00	2·31	0·30	3·16	1·59	1·35
MgO	0·00	0·00	0·00	0·04	0·00	0·00	0·00	0·02	0·01	0·11	0·03	0·07
CaO	0·10	0·14	0·11	0·12	0·15	0·98	0·13	0·80	0·29	0·90	1·64	1·61
Na_2O	0·00	0·05	0·01	0·00	0·02	0·05	0·00	0·08	0·09	0·32	0·05	0·06
P_2O_5	31·84	33·01	33·34	32·88	31·85	29·02	30·75	25·18	25·90	20·87	23·31	19·28
Y_2O_3	46·91	50·07	48·97	44·82	47·17	44·84	47·20	45·51	36·50	4·86	22·36	15·30
ThO_2	0·85	0·18	0·12	0·17	0·69	0·00	1·15	1·73	1·62	16·77	0·00	0·15
UO_2	1·90	0·00	0·00	0·00	2·27	9·44	2·02	1·94	0·59	1·25	2·10	1·81
La_2O_3	0·02	0·03	0·03	0·01	0·03	0·02	0·06	0·04	0·42	0·07	0·07	0·00
Ce_2O_3	0·14	0·15	0·16	0·13	0·16	0·07	0·17	0·26	1·55	0·80	0·11	0·03
Pr_2O_3	0·05	0·04	0·03	0·00	0·05	0·01	0·04	0·05	0·21	0·12	0·04	0·01
Nd_2O_3	0·42	0·38	0·36	0·43	0·39	0·36	0·56	0·43	0·54	0·61	0·35	0·19
Sm_2O_3	0·63	0·63	0·56	0·65	1·08	0·68	0·95	0·86	0·68	0·29	0·42	0·39
Eu_2O_3	0·06	0·00	0·00	0·00	0·00	0·00	0·00	0·00	0·00	0·00	0·00	0·00
Gd_2O_3	1·86	1·91	1·84	1·60	3·33	1·94	2·84	2·34	2·61	0·53	0·83	0·63
Dy_2O_3	4·53	5·31	5·11	4·49	6·42	5·31	5·64	5·27	5·16	0·95	2·38	1·88
Er_2O_3	4·41	4·32	4·61	5·23	2·95	3·04	3·26	2·74	1·84	0·55	2·00	1·33
Yb_2O_3	4·30	2·71	3·96	6·32	2·36	2·78	1·74	1·75	1·65	0·30	2·01	1·57
Lu_2O_3	0·65	0·29	0·61	1·04	0·30	0·55	0·18	0·20	0·21	0·03	0·38	0·24
Total	100·07	99·89	100·58	99·01	100·11	100·50	98·10	100·46	100·60	97·35	100·06	98·80

1, Peraluminous granodiorite (Hoyos); 2, migmatite leucosome (Peña Negra); 3, cordierite leucogranite (Boquerones); 4, two-mica granite (Pedrobernardo); 5; B-rich leucogranite (Ronda); 6, U-rich leucogranite (Albuquerque); 7, peraluminous leucogranite (Trujillo); 8 and 9, leucogranite (Murzinka); 10–12, IXZ from Albuquerque leucogranites.

et al. (1989) concluded that, although included accessories are common in anatexites and high-grade metamorphic rocks, larger accessories representing a significant mass fraction tend to be located at major-phase grain boundaries and must therefore be involved in crustal melting. Our observations on peraluminous migmatites and high-grade gneisses from Iberia agree with that idea but emphasize the role of grain-size. Certainly, in the case of apatite grains—almost equidimensional, with a diameter between 1000 and 10 μm—the mass fraction located at major-mineral boundaries is ~80 wt%, in excellent agreement with Watson *et al.* (1989). However, this is not the situation for monazite, xenotime and zircon, because their grain-size is significantly smaller, usually <30 μm. Results of modal counting with SEM repeatedly indicate that the mass fraction of monazite, xenotime and zircon located at major-phase grain boundaries is <20 wt%, whereas >70 wt% of the mass of these minerals is included within biotite, which thus physically controls the behaviour of REE, Y, Th and U during melt segregation.

Effects of grain-size during fractional crystallization

The vertical distribution of crystals in a convecting magma chamber may be estimated using Bartlett's equation [1969, equation (24)]:

$$\ln Np_2/Np_1 = \left[\frac{(\delta_c - \delta)g}{18\eta K} \left(\frac{1700vK^{1/3}}{\alpha_T g(T_1 - T_2)} \right) \right] d_c^2 \quad (1)$$

where Np_2 and Np_1 are the mineral particle popu-

JOURNAL OF PETROLOGY | VOLUME 37 | NUMBER 3 | JUNE 1996

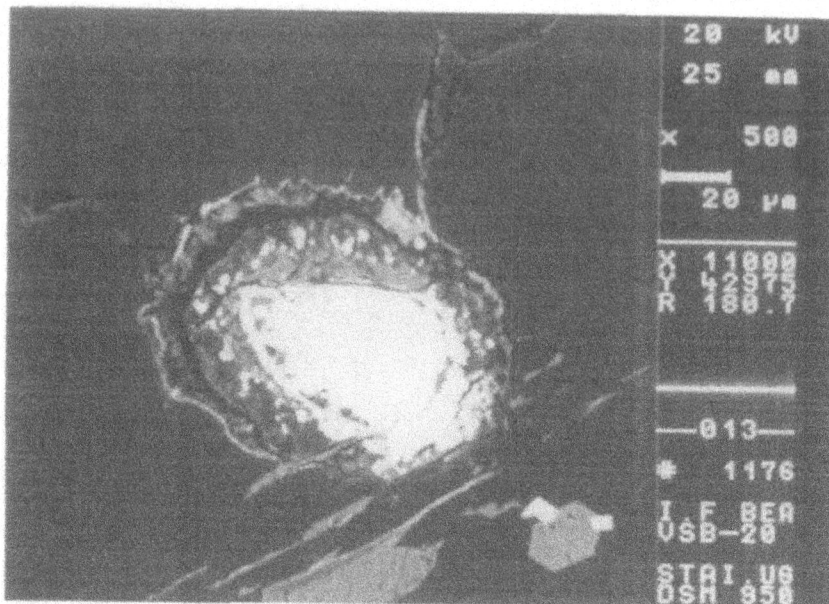

Fig. 12. BSE-SEM image of a radiation-damage structure around a metamictic Th-orthosilicate grain, probably thorite, included in amphibole. The small pseudo-hexagonal grain in the lower right part is an apatite half-including two zircon grains. Sirostan tonalites.

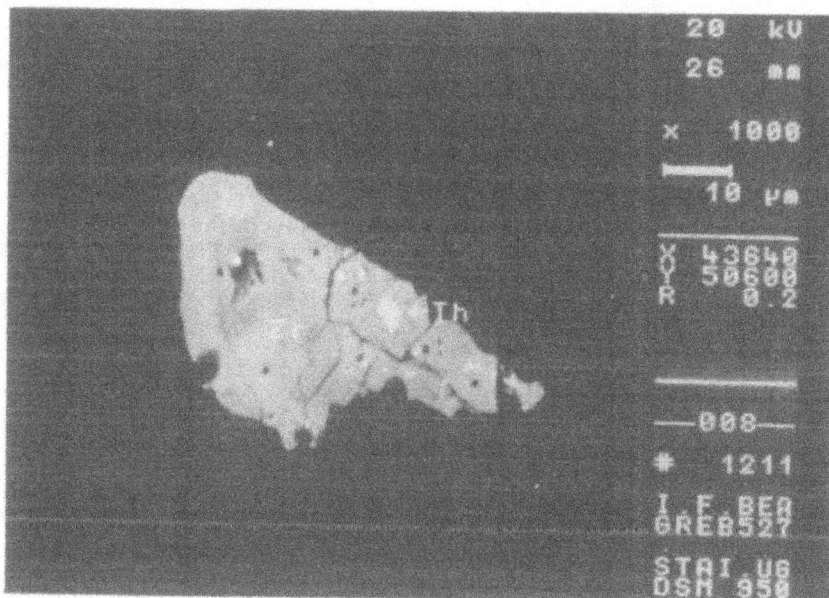

Fig. 13. BSE-SEM image of a xenomorphic zircon grain with microinclusions of Th-orthosilicate, baddeleyite and zirkelite. Burguillos del Cerro diorite.

lation densities in the magma near the ceiling and the floor, respectively; δ_c and δ are the densities of crystal and melt, respectively; g is the gravity acceleration; K is the magma thermal diffusivity; η is the magma viscosity; ν is the kinematic viscosity; $T_1 - T_2$ represents the temperature difference between floor and ceiling; α_T is the thermal expansion coefficient; and d_c is the crystal diameter.

Figure 19 shows settling curves as a function of the diameter of settling minerals, calculated from equation (1) for crystallizing granite melts with a density of 2·5 g/cm^3, viscosity from 10^5 to 10^8 poises,

542

L314

Table 10: Selected microprobe analyses of other REEYThU-rich accessories (results expressed in percent)

	1	2	3	4	5	6	7	8	9	10	11
SiO_2	13·26	17·66	12·60	0·28	0·58	2·14	0·10	0·39	0·14	0·61	0·00
ZrO_2	0·10	0·00	0·12	0·01	0·01	0·25	0·31	0·71	0·81	0·89	0·00
Nb_2O_5	0·41	0·31	0·05	0·66	0·48	54·21	29·22	4·90	24·59	25·60	12·28
Ta_2O_5	0·05	0·03	0·00	0·21	0·17	3·17	7·61	0·31	0·82	1·07	0·19
$FeO^{tot.}$	1·93	7·53	1·24	0·97	2·35	2·61	4·39	3·90	7·13	8·54	0·02
TiO_2	0·00	0·03	0·00	0·00	0·06	9·18	17·07	17·41	3·59	5·79	30·51
CaO	1·48	0·56	4·51	0·00	0·01	3·99	4·29	5·63	6·48	1·39	2·37
Na_2O	0·02	0·14	0·05	0·03	0·04	0·00	0·21	0·19	1·99	0·05	7·14
P_2O_5	3·05	0·55	10·07	0·21	0·63	0·13	0·08	0·01	0·01	0·04	0·00
Y_2O_3	0·84	3·14	3·57	0·18	0·49	3·5	0·62	2·35	8·34	12·97	0·10
ThO_2	76·44	55·56	38·78	17·51	11·22	2·14	5·69	2·59	3·14	2·98	0·58
UO_2	1·46	7·37	1·11	78·93	83·36	7·28	22·80	2·41	5·76	13·33	0·06
La_2O_3	0·02	0·00	3·99	0·00	0·01	0·06	0·12	12·33	2·19	0·50	17·08
Ce_2O_3	0·06	0·01	12·70	0·01	0·03	0·14	0·24	22·43	6·48	0·90	24·39
Pr_2O_3	0·02	0·01	1·65	0·00	0·01	0·02	0·03	3·08	1·33	0·41	1·68
Nd_2O_3	0·10	0·07	4·82	0·03	0·04	0·06	0·12	11·90	6·44	3·08	2·57
Sm_2O_3	0·09	0·11	0·90	0·08	0·08	0·03	0·03	2·62	3·07	2·46	0·10
Eu_2O_3	0·00	0·01	0·01	0·00	0·01	0·00	0·00	0·25	0·39	0·30	0·01
Gd_2O_3	0·15	0·10	0·93	0·11	0·14	0·06	0·03	2·41	4·16	4·29	0·04
Dy_2O_3	0·21	0·49	0·96	0·16	0·19	0·16	0·04	2·53	7·19	7·74	0·00
Er_2O_3	0·16	1·00	0·45	0·13	0·17	0·21	0·04	1·25	4·46	5·08	0·00
Yb_2O_3	0·11	4·63	0·40	0·09	0·14	0·34	0·07	0·55	2·08	2·57	0·00
Lu_2O_3	0·01	0·61	0·07	0·00	0·04	0·06	0·01	0·05	0·19	0·23	0·00
Total	99·97	99·82	98·98	99·60	100·26	89·74	93·02	100·20	99·78	100·82	99·02

Th-orthosilicates: 1, peraluminous granodiorite (Hoyos); 2, peralkaline granite (Galiñeiro); 3, phosphothorite, peralkaline granite (Galiñeiro). Uraninite: 4, cordierite-bearing granodiorite (Gredos); 5, U-rich leucogranite (Albuquerque). Pyrochlore: 6, peralkaline granite (Keivy). Betafite: 7, peraluminous granodiorite (Hoyos). Aeschynite: 8, peralkaline granite (Kharitonovo). Fergusonite: 9, peralkaline granite (Galiñeiro). Samarskite: 10, peralkaline granite (Galiñeiro). Loparite: 11, peralkaline syenite (Kola).

mineral densities from 4 to 6 g/cm^3 and T_1-T_2 of 50°C. As the dependence of T_1-T_2 is rather weak (Bartlett, 1969), the above parameter set will represent the situation in most crystallizing granites. Figure 19 also shows a curve representing the relative mass fraction of REEYThU-rich accessories as a function of grain-size calculated by averaging modal counting on granites mentioned in Fig. 17. It is evident that even grains of 150 μm with a density as high as 6 g/cm^3 cannot settle appreciably within a melt with a viscosity as low as 10^5 poises. Unmodified uraninite crystals may have a density of 10·3 g/cm^3, but their diameter is usually so small that gravity settling is also impossible. We therefore conclude that the fate of early crystallized REEYThU-rich accessories is to remain in sus-pension within the melt until a crystallizing major mineral includes them.

The reasons why accessories are selectively included within a few major minerals are not well understood yet, but must surely be diverse. In rocks crystallized from a melt, heterogeneous nucleation—either of the major mineral on an early-crystallized accessory or vice versa—coupled with crystallization related to local saturation adjacent to a growing phenocryst (Bacon, 1989) seem to be the most probable mechanisms. In metasedimentary rocks, detrital inheritance and the neoformation of acces-sories in some preferential places during sedi-mentation and diagenesis, complicated later by recrystallization during metamorphism, are the key factors. It is clear in both cases, however, that

JOURNAL OF PETROLOGY | VOLUME 37 | NUMBER 3 | JUNE 1996

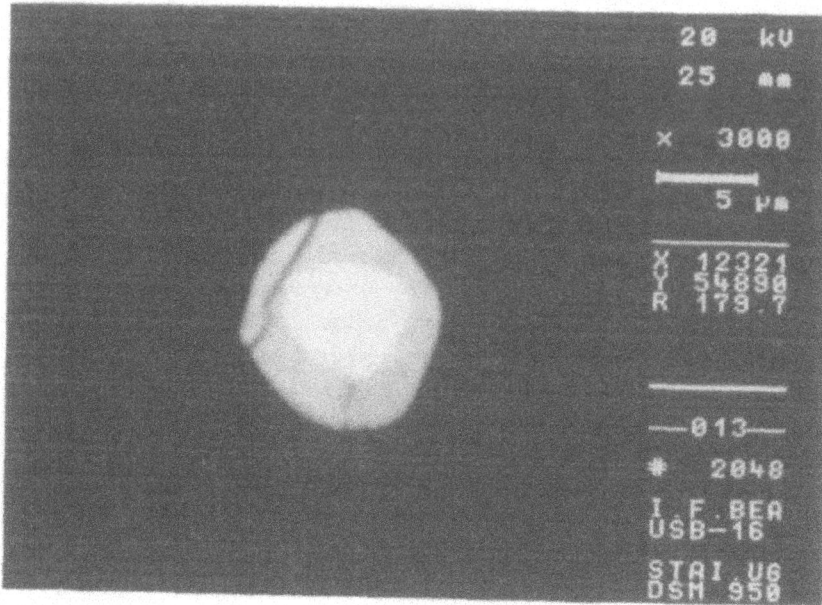

Fig. 14. Zircon growing on a uraninite crystal (BSE-SEM image). Murzinka granite, the Urals.

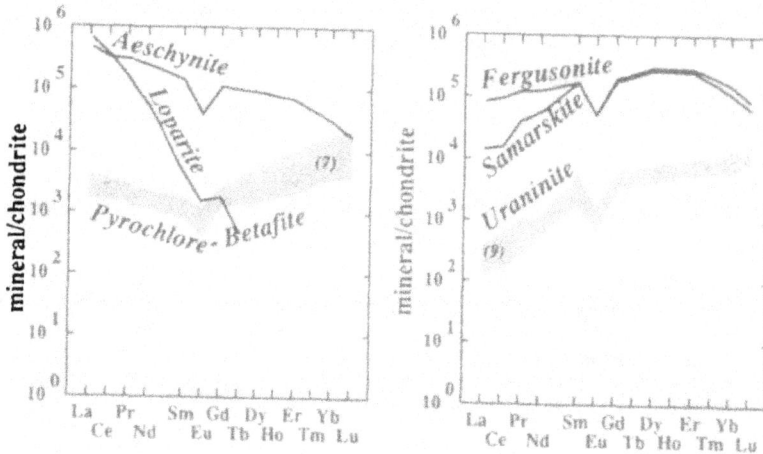

Fig. 15. Chondrite-normalized REE patterns of REEYThU-rich niobotantalates, uraninite and pyrochlore–betafite minerals.

selective inclusion of REEYThU-rich accessories in a given major mineral confers on it an indirect but still significant control over the geochemistry of REE, Y, Th and U, which is almost impossible to quantitatively model with present knowledge.

Contrasting geochemistry of REE, Y, Th and U in subaluminous vs peraluminous granites

The geochemical behaviour of REE, Y, Th and U in differentiated granites changes with aluminium saturation. In general, the higher the aluminium saturation index [ASI = mol. $Al_2O_3/(CaO + Na_2O + K_2O)$], the stronger the depletion in REE

(except Eu), Y, Th and U in leucocratic differentiates (Bea, 1993). A good example, which may have important consequences for the understanding of heat production in the crust, is the contrasting vertical distribution of REE, Y, Th and U (and K) in vertically zoned plutons: in I-type granitoids the concentrations of REE, Y, Th and U increase from the least to the most differentiated facies, accumulating upwards (Sawka & Chappell, 1988), whereas in S-type granitoids, in contrast, the concentrations of REE, Y, Th and U decrease from the least to the most differentiated facies, accumulating downwards (Bea *et al.*, 1994*a*). These differences are due to variations in the nature of the major and accessory

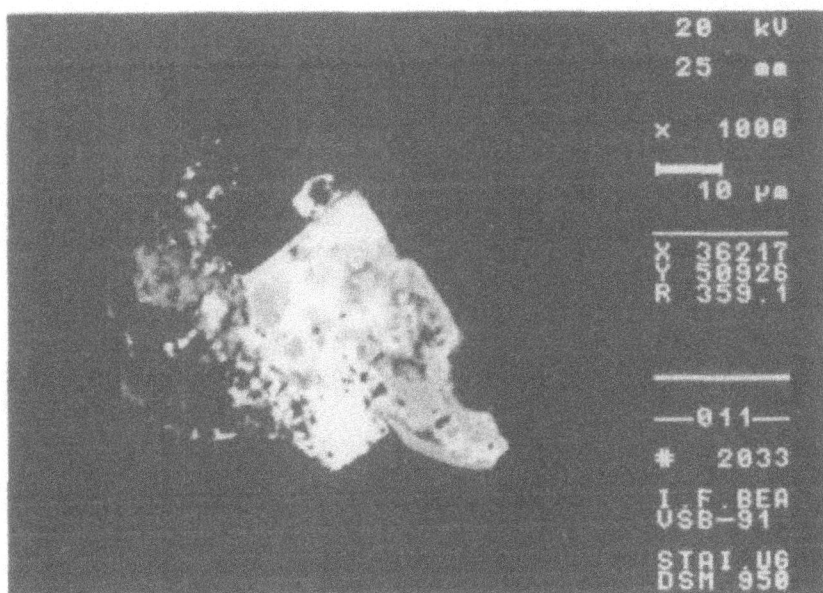

Fig. 16. BSE-SEM image of a complex monazite (white)–zircon (grey) intergrowth.

minerals as a consequence of changes in the aluminium saturation, the key factor being the progressive replacement of allanite and sphene by monazite and xenotime with increasing ASI and decreasing CaO, which seems to occur in the following way.

Peraluminous granites have higher phosphorus contents than metaluminous granites with the same silica content, owing to the enhancement of apatite solubility at high ASI values (London, 1992; Bea *et al.*, 1992; Pichavant *et al.*, 1992; Wolf & London, 1994). As the solubility of monazite and xenotime remains very low regardless of aluminosity (Rapp & Watson, 1986; Wolf & London, 1995), phosphorus-rich peraluminous melts become saturated in monazite and xenotime at low REE concentrations. Precipitated REE-phosphates are then effectively removed from the melt as inclusions in biotite, also an early crystallized mineral, thus producing a strong depletion of REE, Y, Th and U in residual melts. In metaluminous melts, however, the scarcity of phosphate anions makes the saturation in REE-phosphates occur at higher REE concentrations than in peraluminous melts, and allanite and sphene become the dominant REEYThU-carriers. As these minerals are by far less effective than monazite and xenotime at removing these elements from the melt, and they do not usually precipitate massively at the beginning of crystallization, bulk crystal–melt distribution coefficients should remain at $D_{REE,Y,Th,U}^{crystal/melt} < 1$ during most of the crystallization

history, thus producing somewhat enriched differentiates.

Post-magmatic mobility of REE, Y, Th and U

REE, Y, Th and U are highly mobile during post-magmatic stages. SEM images revealed that (1) primary epidote is usually rimmed by a film of LREE carbonates which also fill cracks radiating from the epidote grain (Fig. 3), (2) primary allanite is also the source of secondary migrating LREE carbonates (Fig. 10), and (3) secondary allanite forms easily from monazite along fractures in micas (Fig. 7). The mobility of REE, Y, Th and U is related to the fact that many accessories are metamictic and are surrounded by radiation damage structures (see Fig. 12) capable of channelling the migration of REE, Y, Th and U. It is important to emphasize that characteristic accessories from peraluminous granites (monazite, xenotime, huttonite) are inherently more stable than accessories from metaluminous and peralkaline rocks (allanite, thorite).

Effects of the pressure on the composition of major minerals in crustal protoliths

The REEYThU composition of major minerals in amphibolite-grade metasediments and orthogneisses is very similar to that of peraluminous granites. Granulite-grade garnets, however, are noticeably enriched in Nd and Sm and have a larger Eu negative anomaly (Fig. 2). Granulite-grade feldspars

JOURNAL OF PETROLOGY | VOLUME 37 | NUMBER 3 | JUNE 1996

Table 11: Selected analyses of zircon crystals

	1	2	3	4	5	6	7	8	9	10	11*
SiO$_2$	33·35	34·13	35·04	33·23	34·31	32·38	30·96	31·78	31·30	30·68	28·43
ZrO$_2$	64·77	62·69	61·97	61·05	63·26	63·03	53·93	60·65	56·76	58·00	53·63
HfO$_2$	1·97	2·17	1·14	2·29	0·78	1·73	2·31	1·73	1·89	1·35	1·75
Al$_2$O$_3$	0·02	0·00	0·00	0·00	0·00	0·40	0·00	0·09	0·01	0·00	0·29
FeO	0·50	0·09	0·83	0·01	0·05	0·18	0·00	0·14	0·04	1·08	0·44
MgO	0·03	0·00	0·05	0·00	0·00	0·02	0·00	0·00	0·00	0·02	0·02
CaO	0·00	0·01	0·16	0·01	0·00	0·17	0·01	0·02	0·02	0·01	0·06
Na$_2$O	0·00	0·07	0·07	0·00	0·07	0·00	0·00	0·01	0·03	0·10	0·12
P$_2$O$_5$	0·01	0·06	0·00	0·53	0·34	1·11	1·79	1·71	2·79	3·16	2·48
					percent						
Y	77·0	199	139	1228	0·70	0·64	2·38	1·74	2·52	3·31	0·79
Th	159	105	259	1011	0·21	0·00	0·66	0·10	0·24	0·08	0·00
U	12·2	64·5	28·1	104	0·00	0·11	5·67	0·88	2·82	0·00	0·22
La	2·60	5·57	1·65	9·52	0·00	0·01	0·01	0·00	0·00	0·01	0·02
Ce	7·75	13·	6·37	26·9	0·03	0·01	0·05	0·01	0·02	0·01	0·03
Pr	1·13	2·84	0·50	4·66	0·01	0·00	0·01	0·00	0·00	0·00	0·01
Nd	6·11	13·3	3·34	29·5	0·03	0·01	0·05	0·01	0·03	0·01	0·05
Sm	1·43	2·48	2·45	15·6	0·03	0·03	0·02	0·01	0·03	0·01	0·03
Eu	0·32	0·63	1·15	4·02	0·01	0·00	0·00	0·00	0·00	0·00	0·00
Gd	1·94	3·74	8·90	42·4	0·04	0·06	0·10	0·04	0·12	0·03	0·05
Tb	0·48	0·81	4·41	14·7	n.d.	n.d.	n.d.	n.d.	n.d.	n.d.	n.d.
Dy	5·42	9·04	47·3	157	0·09	0·07	0·16	0·13	0·22	0·20	0·09
Ho	2·23	3·75	19·0	64·4	n.d.	n.d.	n.d.	n.d.	n.d.	n.d.	n.d.
Er	9·96	20·1	82·5	357	0·08	0·16	0·29	0·22	0·36	0·43	0·11
Tm	2·84	4·05	20·4	73·0	n.d.	n.d.	n.d.	n.d.	n.d.	n.d.	n.d.
Yb	27·8	30·9	188	510	0·17	0·18	0·28	0·37	0·44	0·48	0·13
Lu	9·18	3·73	27·1	69·3	0·04	0·02	0·04	0·10	0·06	0·08	0·04

*Contains 10·05% Sc$_2$O$_3$.

REE, Y, Th and U were determined by LA–ICP-MS in samples 1–4 and are expressed in p.p.m. All the other data were obtained by electron microprobe and are expressed in percent. 1, Peraluminous leucogranite (Boquerones); 2, quartz syenite (Magnitogorsk); 3, plagiogranite (Chemolstochinsk); 4, peralkaline granite (Galiñeiro); 5, peralkaline granite (Magnitogorsk); 6, granodiorite (Vierkisast); 7, kinzigite (Ronda); 8, cordierite-bearing granodiorite (Hoyos); 9, xenotime-rich leucogranite (Ronda); 10, xenotime-rich migmatite leucosome (Peña Negra); 11, xenotime-rich granite (Albuquerque). n.d., not determined.

also appear enriched in LREE (Fig. 1), although not as much as garnets. As no changes in LREE composition of major minerals have been found related to increasing temperature (e.g. REE patterns in garnets from biotite gneisses and sanidine-bearing peraluminous dacites are very similar), the increasing pressure seems, in principle, to be mainly responsible for the above-mentioned LREE enrichment in granulite-grade feldspars and garnets. Additional evidence about the role of increasing pressure may be gathered from Harris *et al.* (1992),

whose ion-probe data on plagioclase and garnet from a kyanite schist and a sillimanite gneiss also revealed consistently high values of LREE in minerals from the higher-pressure rock.

Much systematic work, not only geochemical but also experimental and crystallographic, is still needed to discover the extension of this phenomenon and the reasons for its occurrence. However, we tentatively suggest that one potentially important mechanism may be the destabilization of monazite by reaction with garnet, which basically consists in

Table 12: Selected LA–ICP-MS analyses of apatite crystals

	Peraluminous						Metaluminous			Peralkaline		
	1	2	3	4	5	6	1	2	3	1	2	3
Y	1806	2467	1486	984	1736	1598	632	951	1024	257	317	284
U	41·1	551	394	0·51	828	129	11·1	35·9	88·7	4·3	15·3	18·1
Th	9·86	14·9	0·61	0·93	81·4	40·7	0·70	2·45	3·17	12·7	46·1	18·0
La	480	630	329	297	444	463	26·2	25·9	60·4	2721	2755	2829
Ce	1452	1709	1156	1181	1632	1659	103	96·6	177	4710	4692	6930
Pr	308	291	198	212	294	295	24·6	31·1	43·6	466	536	656
Nd	1633	1537	865	769	1412	1269	166	214	300	1220	1643	1701
Sm	591	589	324	264	506	456	102	124	135	189	264	235
Eu	4·69	23·7	12·5	14·6	10·2	8·85	28·4	36·5	44·8	58·8	77·5	84·4
Gd	541	769	359	274	707	652	178	243	221	185	262	207
Tb	100	132	59·2	45·3	126	105	36·3	50·5	44·8	22·5	33·1	26·0
Dy	575	742	322	216	616	569	235	289	320	100	135	107
Ho	110	135	59·8	40·2	109	103	51·9	63·5	74·8	18·3	23·6	18·6
Er	302	362	148	94·5	288	272	153	196	197	43·3	50·5	46·8
Tm	43·6	51·9	20·6	13·6	41·7	41·9	20·3	29·1	29·8	4·71	5·53	5·77
Yb	285	298	116	87·5	269	218	122	152	178	24·1	22·7	34·1
Lu	33·9	39·4	13·9	10·5	41·7	27·2	17·1	24·5	28·4	2·49	2·79	3·37
Eu/Eu*	0·03	0·11	0·11	0·17	0·05	0·05	0·64	0·64	0·79	0·96	0·90	1·17
La$_N$/Yb$_N$	1·14	1·43	1·91	2·29	1·11	1·43	0·14	0·11	0·23	76·2	81·9	55·9

From peraluminous granites: 1, cordierite-bearing granodiorite (Hoyos); 2, migmatite (Peña Negra); 3, two-mica granite (Pedro-bernardo); 4, leucogranite (Trujillo); 5, U-rich leucogranite (Albuquerque); 6, garnet-bearing granite (Murzinka). From metaluminous granites: 1 and 2, granodiorite (Vierkisest); 3, tonalite (Sirostan). From peralkaline granites: 1–3, granite (Keivy).

Table 13: Selected LA–ICP-MS analyses of primary sphene crystals (results are in p.p.m.)

	1	2	3	4	5	6	7	8	9	10	11	12
Y	1150	1612	2127	1100	217	732	100	245	376	2239	1301	1251
U	0·00	10·5	69·8	19·2	391	157	226	32·3	21·9	65·2	1471	439
Th	166	171	163	161	142	213	219	46·9	110	87·7	3268	278
La	1109	1752	2394	1189	3153	1087	1488	281	561	2135	3547	2942
Ce	3596	6102	9273	4315	8157	3337	4786	987	1861	7552	7908	2819
Pr	790	1256	1778	819	1656	608·	845	182	341	1628	973	452
Nd	3240	5279	7453	3484	6839	2519	3525	831	1427	7773	3017	1384
Sm	788	1367	1994	830	2475	510	707	185	291	2084	569	292
Eu	204	240	300	212	398	140	185	57·5	85·8	346	138	99·8
Gd	606	1107	1390	654	1844	391	534	140	229	1539	621	314
Tb	89·5	134	202	89·8	259	55·1	78·0	19·7	30·7	238	74·0	62·2
Dy	458	698	1003	451	1241	290	417	102	165	1184	415	398
Ho	88·5	130	176	82·8	205	58·1	82·0	19·3	32·2	215	85·2	90·5
Er	251	352	452	227	535	160	240	57·8	89·6	564	266	306
Tm	39·5	51·4	63·4	34·23	71·3	23·8	36·5	8·46	13·7	72·5	41·7	47·8
Yb	265·	336	390	230	410	162	234	57·0	89·4	418	296	310
Lu	36·8	41·4	47·8	32·2	46·4	21·7	31·2	7·9	11·5	40·0	40·4	33·9
Eu/Eu*	0·90	0·60	0·55	0·88	0·57	0·96	0·92	1·09	1·02	0·59	0·71	1·01
La$_N$/Yb$_N$	2·82	3·52	4·14	3·49	5·19	4·53	4·29	3·33	4·23	3·45	8·09	6·40

1–6, epidote-bearing metaluminous granites and granodiorites (Vierkisest); 7, metaluminous tonalite (Sirostan); 8 and 9, metaluminous diorite (Burguillos del Cerro); 10, 11 and 12, peralkaline granite (Keivy).

JOURNAL OF PETROLOGY | VOLUME 37 | NUMBER 3 | JUNE 1996

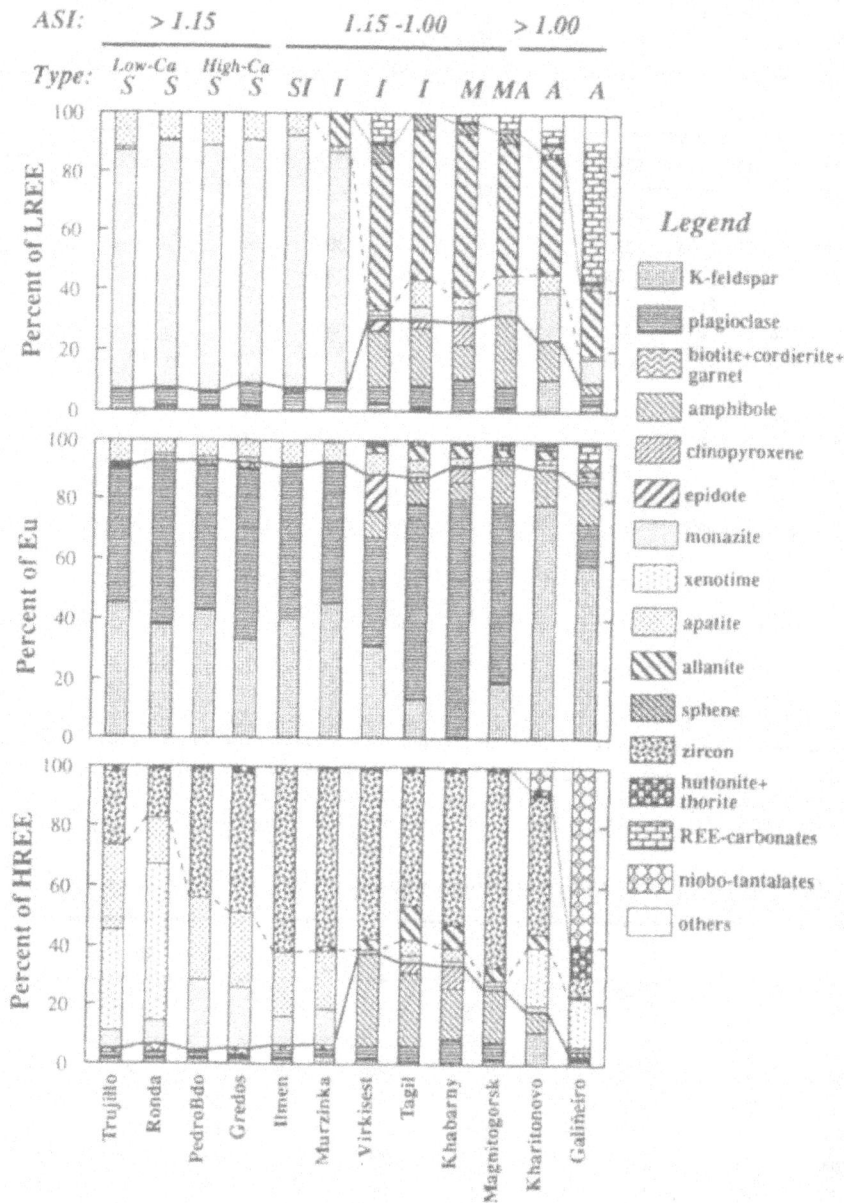

Fig. 17. Fractional contribution of each mineral to the whole-rock LREE, Eu and HREE budgets in selected granite plutons. Continuous line separates the contribution of major from accessory minerals. Dashed line separates the contribution of accessory phosphates from silicates. Dotted line separates accessory silicates from niobotantalates and carbonates. (See discussion in text.)

the exchange of Ca (from garnet) by LREE (from monazite). In a current study on the amphibolite–granulite transition in the Ivrea–Verbano zone (Bea et al., in preparation), we detected that an elevated proportion of monazite grains included within granulite-grade garnets are partially or totally pseudomorphed to a mixture of apatite and cheralite-rich monazite, whereas the garnet around the inclusion is enriched in LREE. Those garnet analyses with abnormally high La–Ce contents reported in Table 5 and Fig. 2 were probably caused by the ablation of such zones.

If pressure-related differences in the REE composition of major minerals are confirmed with more systematic studies, their role as a potential source for differences in the REE chemistry of melts equili-

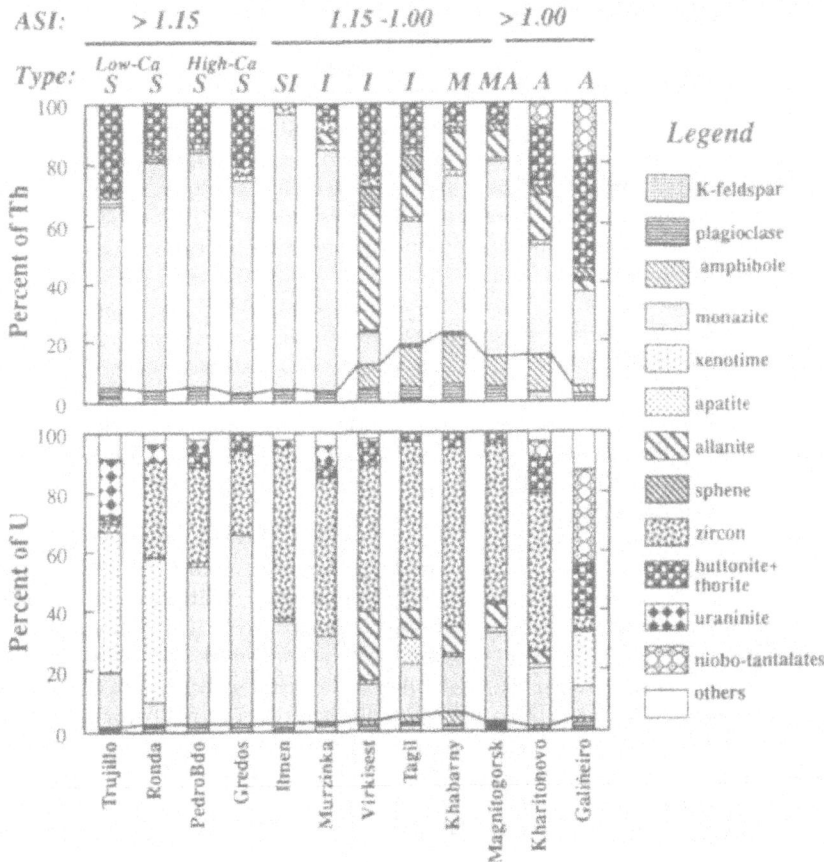

Fig. 18. Fractional contribution of each mineral to the whole-rock Th and U budgets in selected granite plutons. Continuous line separates the contribution of major from accessory minerals. Dashed line separates the contribution of accessory phosphates from silicates. Dotted line separates accessory silicates from niobotantalates and carbonates. (See discussion in text.)

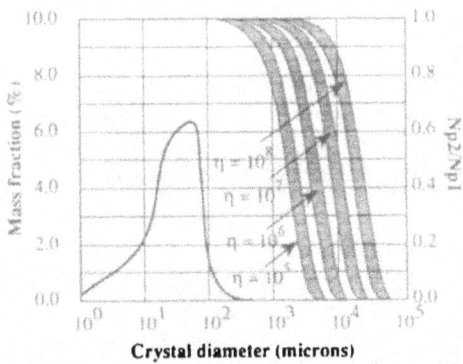

Fig. 19. Settling curves as a function of the diameter of settling minerals calculated with the Bartlett (1969) equation for density intervals between 4 and 6 g/cm³. The right vertical axis represents the ratio between the mineral particle population densities in the magma near the ceiling (Np_2) and the floor (Np_1). The curve in the left part of the diagram represents the mass fraction of accessories as a function of their diameter. (Note how even 150-μm grains with a density as high as 6 g/cm³ cannot settle appreciably within a melt with a viscosity as low as 10^5 poises.)

brated at amphibolite- and granulite-grade conditions, respectively, should therefore be considered as a subject for further investigation.

Can REE be used in geochemical modelling of granite rocks?

The objective of trace-element-based petrogenetic modelling of igneous rocks is to infer the behaviour of minerals during melting and crystallization by comparing actual trace-element distribution patterns with theoretical models generally calculated from fractionation equations based on the laws of chemical equilibrium (Allègre & Minster, 1978; Hanson, 1989). To be useful in petrogenetical modelling, a given trace element must be essentially contained within major minerals, obey Raoult–Henry's law, and not move appreciably during post-magmatic stages. It is evident that in the case of granite rocks, REE (except Eu), Y, Th and U do not satisfy any of these conditions and simply cannot therefore be used for modelling the genesis of granitoids in that manner.

CONCLUSIONS

The LREE, HREE, Y and Th fractions that reside within major minerals are as low as 5–10 wt% in peraluminous and peralkaline granites, but may rise to 20–30 wt% in amphibole-rich metaluminous granites. Eu is always essentially contained within feldspars, although primary epidote may also contain a significant proportion. The fraction of U that resides within major minerals is always <5 wt%.

The geochemistry of REE (except Eu), Y, Th and U in granite rocks and crustal protoliths is essentially controlled by the behaviour of REEYThU-rich accessory minerals, whose nature, associations and composition change with rock aluminosity. The accessory assemblage of peraluminous granites is composed of monazite, xenotime (restricted to low-Ca varieties), apatite, zircon, Th-orthosilicate (huttonite?), uraninite and betafite–pyrochlore minerals. Metaluminous granites have allanite, sphene, apatite, minor monazite, zircon and Th-orthosilicate (thorite?). Peralkaline granites have the same accessories as metaluminous granites, but also contain niobotantalates (aeschinite, samarskite, fergusonite, occasionally loparite), batnaesite, fluocerite and xenotime.

Migmatites and high-grade rocks contain the same accessory assemblage and similar REE, Y, Th and U distribution patterns among minerals as granites with similar aluminosity. Only garnet, owing to its higher abundance, plays a much more important role here, especially for Y and HREE. Compared with the same minerals from amphibolite-grade metapelites, granulitic feldspars are enriched in LREE, above all in La and Ce. In the same way, granulitic garnets have Nd–Sm contents higher by one order of magnitude than amphibolitic garnets and a precipitous negative Eu anomaly.

In common peraluminous migmatites, the mass fraction of monazite, xenotime and zircon included within biotite is very high. Whether these inclusions are available for the melt during anatexis or stay within their host—either in restites or entrained within restitic crystals suspended in the melt—depends completely on the behaviour of biotite. During crystallization, the small grain-size of newly formed accessories makes their separation from the melt physically impossible until a growing major mineral includes them. Biotite, which has near-zero contents of REE, Y, Th and U, is the mineral which shows the greatest tendency to include REEYThU-

rich accessories, probably owing to the combined effects of local saturation adjacent to growing biotite crystals and heterogeneous nucleation.

REE, Y, Th and U are not suitable for geochemical modelling of granitoids by means of equilibrium-based trace-element fractionation equations, but are still useful petrogenetic tools. Apart from the obvious importance of Th and U as heat-producing elements, and REE in isotopic systems such as Sm–Nd, Lu–Hf or La–Ba, the geochemistry of REE, Y, Th and U reflects the behaviour of accessories and some key major minerals such as garnet, feldspars and amphibole, and may therefore give valuable information about the conditions of partial melting, melt segregation and crystallization of granite magmas in different crustal regimes.

ACKNOWLEDGEMENTS

I acknowledge with many thanks help from the following people and institutions: P. G. Montero, for her assistance with LA–ICP-MS analyses and fruitful comments and criticism—she also provided me with samples from Galiñeiro; G. B. Fershtater, for his invaluable help during field-work in the Urals and many hours of passionate discussion on granites; G. Garuti and F. Zaccharini, for introducing me to the geology of the Ivrea–Verbano zone and their help with microprobe analyses at Modena University; Isabel and Alicia, for their patience during many hours of tedious studies with the SEM; A. Acosta and L. G. Menéndez, for providing samples from Ronda and Albuquerque; C. Laurin, for her help and patience in improving the original Spanglish manuscript. Revisions made by Gordon Watt, David London and Simon Harley are gratefully acknowledged. Perkin Elmer allowed free use of a new UV-LA probe prototype for ICP-MS analyses. This work has been financially supported by the Spanish Interministry Commission for Science and Technology (CICYT), Projects AMB93-0535 and AMB94-1420.

REFERENCES

Aleinikoff, J. N. & Stoeser, D. B., 1989. Contrasting zircon morphology and U–Pb systematics in peralkaline and metaluminous post-orogenic granite complexes of the Arabian Shield, Kingdom of Saudi Arabia. *Chemical Geology* **79**, 241–258.

Allègre, C. J. & Minster, J. F., 1978. Quantitative models of trace element behavior in magmatic processes. *Earth and Planetary Science Letters* **38**, 1–25.

Bacon, C. R., 1989. Crystallization of accessory phases in magmas by local saturation adjacent to phenocrysts. *Geochimica et Cosmochimica Acta* **53**, 1055–1066.

Barbey, P., Alle, P., Brouand, M. & Albarede, F., 1995. Rare-earth patterns in zircons from the Manaslu granite and Tibetan Slab migmatites (Himalaya): insights in the origin and evolution of a crustally-derived granite magma. *Chemical Geology* 125, 1–17.

Bartlett, R. W., 1969. Magma convection, temperature distribution, and differentiation. *American Journal of Science* 267, 1067–1082.

Bayer, G., 1969. Thorium A. Crystal chemistry. In: Wedepohl, K. H. (ed.) *Handbook of Geochemistry*. Heidelberg: Springer-Verlag.

Bea, F., 1993. Aluminosity-dependent fractionation patterns in differentiated granite–leucogranite systems. *EOS* 74(16), 343.

Bea, F., Fershtater, G. & Corretgé, L. G., 1992. The geochemistry of phosphorus in granite rocks and the effect of aluminium. *Lithos* 29, 43–56.

Bea, F., Pereira, M. D., Corretgé, L. G. & Fershtater, G. B., 1994a. Differentiation of strongly peraluminous, perphosphorous granites. The Pedrobernardo pluton, central Spain. *Geochimica et Cosmochimica Acta* 58, 2609–2628.

Bea, F., Pereira, M. D. & Stroh, A., 1994b. Mineral/leucosome trace-element partitioning in a peraluminous migmatite (a laser ablation–ICP-MS study). *Chemical Geology* 117, 291–312.

Chesner, C. A. & Ettlinger, A. D., 1989. Composition of volcanic allanite from Toba Tuffs, Sumatra, Indonesia. *American Mineralogist* 74, 750–758.

Delima, E. S., Vannucci, R., Bottazzi, P. & Ottolini, L., 1995. Reconnaissance study of trace element zonation in garnet from the Central Structural Domain, Northeastern Brazil: an example of polymetamorphic growth. *Journal of South American Earth Sciences* 8, 315–324.

Drake, M. J. & Weill, D. F., 1972. New rare earth element standards for electron microprobe analysis. *Chemical Geology* 10, 179–181.

Förster, H. J., 1993. Th-Y-REE-bearing accessory minerals in Hercynian granites of the Erzgebirge (Ore Mountains), Germany. *Geological Society of America, Abstracts with Programs* 25, 42–43.

Förster, H. J. & Rhede, D., 1995. Extreme compositional variability in granitic monazites. In: Brown, M. & Piccoli, P. M. (eds) *Third Hutton Symposium on the Origin of Granites and Related Rocks*, College Park, MD. *US Geological Survey Circular* 1129, 54–55.

Förster, H. J. & Tischendorf, G., 1994. Evolution of the Hercynian granite magmatism in the Erzegebirge metallogenic provinces. *Mineralogical Magazine* 58A, 284–285.

Gromet, L. P. & Silver, L. T., 1983. Rare earth element distribution among minerals in a granodiorite and their petrogenetic implications. *Geochimica et Cosmochimica Acta* 47, 925–940.

Hanson, G. N., 1989. An approach to trace element modeling using a simple igneous system as an example. In: Lipin, B. R. & McKay, G. A. (eds) *Geochemistry and Mineralogy of REE*. *Mineralogical Society of America, Reviews in Mineralogy* 21, 79–97.

Harris, N. B. W., Gravestock, P. & Inger, S., 1992. Ion-microprobe determinations of trace-element concentration in garnets from anatectic assemblages. *Chemical Geology* 100, 41–49.

Haskin, L. A., 1979. On rare-earth element behavior in igneous rocks. In: Ahrens, L. H. (ed.) *On Rare-Earth Element Behavior in Igneous Rocks*. Oxford: Pergamon, pp. 175–189.

Haskin, L. A., 1990. PREEconceptions pREEvent pREEcise pREEdictions. *Geochimica et Cosmochimica Acta* 54, 2353–2361.

Heaman, L. & Parrish, R., 1991. U–Pb geochronology of accessory minerals. In: Heaman, L. & Ludden, J. N. (eds) *Short Course Handbook on Applications of Radiogenic Isotope Systems to Problems in Geology*. Toronto, Ont.: Mineralogical Association of Canada, pp. 59–102.

Heaman, L. M., Bowis, R. & Crocket, J., 1990. The chemical composition of igneous zircon suites: implications for geochemical tracer studies. *Geochimica et Cosmochimica Acta* 54, 1597–1608.

Hickmott, D. D. & Shimizu, N., 1990. Trace element zoning in garnet from the Kwoiek Area, British Columbia: disequilibrium partitioning during garnet growth? *Contributions to Mineralogy and Petrology* 104, 619–630.

Hickmott, D. D., Shimizu, N., Spear, F. S. & Selverstone, J., 1987. Trace-element zoning in a metamorphic garnet. *Geology* 15, 573–576.

Imaoka, T. & Nakashima, K., 1994. Fluocerite in a peralkaline rhyolite dyke from Cape Ashizuri, Shikoku-Island, Southwest Japan. *Neues Jahrbuch für Mineralogie, Monatshefte* 12, 529–539.

Krasnobayev, A. A., 1986. *Zircon kak indikator geologicheskij prozessov*. Moscow: Nauka.

Lanzirotti, A., 1995. Yttrium zoning in metamorphic garnets. *Geochimica et Cosmochimica Acta* 59, 4105–4110.

London, D., 1992. Phosphorus in S-type magmas: the P_2O_5 content of feldspars from peraluminous granites, pegmatites, and rhyolites. *American Mineralogist* 77, 126–145.

Mazzuchelli, M., Rivalenti, G., Vannucci, R., Botazzi, P., Ottolini, L., Hoffmann, A. W. & Parenti, M., 1992. Primary positive Eu anomaly in clinopyroxenes of low-crust gabbroic rocks. *Geochimica et Cosmochimica Acta* 56, 2363–2370.

McKay, G. A., 1989. Partitioning of rare earth elements between major silicate minerals and basaltic melts. In: Lipin, B. R. & McKay, G. A. (eds) *Geochemistry and Mineralogy of REE*. *Mineralogical Society of America, Reviews in Mineralogy* 21, 45–77.

Miller, C. F. & Mittlefehdlt, D. W., 1982. Light rare earth element depletion in felsic magmas. *Geology* 10, 129–133.

Montero, P. G., 1995. Accumulation of REE and HFS elements in peralkaline granitoids: the Galiñeiro pluton, NW Spain. In: Brown, M. & Piccoli, P. M (eds) *Third Hutton Symposium on the Origin of Granites and Related Rocks*, College Park, MD. *US Geological Survey Circular* 1129, 98–99.

Petrik, I., Broska, I., Lipka, J. & Siman, P., 1995. Granitoid allanite-(Ce): substitution relations, redox conditions and REE distributions (on an example of I-type granitoids, Western Carpathians, Slovakia). *Geologica Carpathica* 46, 79–94.

Pichavant, M., Montel, J. M. & Richard, L. R., 1992. Apatite solubility in peraluminous liquids: experimental data and an extension of the Harrison–Watson model. *Geochimica et Cosmochimica Acta* 56, 3855–3861.

Pupin, J. P., 1980. Zircon and granite petrology. *Contributions to Mineralogy and Petrology* 73, 207–220.

Pupin, J. P., 1992. Les zircons des granites océaniques et continentaux: couplage typologie–géochimie des élements en traces. *Bulletin Société géologique du France* 163, 495–507.

Rapp, R. P. & Watson, E. B., 1986. Monazite solubility and dissolution kinetics: implications for the thorium and light rare earth chemistry of felsic magmas. *Contributions to Mineralogy and Petrology* 94, 304–316.

Reid, M. R., 1990. Ionprobe investigation of rare earth elements distribution and partial melting of metasedimentary granulites. In: Vielzeuf, D. & Vidal, P. (eds) *Granulites and Crustal Evolution*. Dordrecht: Kluwer Academic, pp. 506–522.

Reischmann, T., Brugmann, G. E., Jochum, K. P. & Todt, W. A., 1995. Trace element and isotopic composition of baddeleyite. *Mineralogy and Petrology* 53, 155–164.

Roeder, P. L., 1985. Electron-microprobe analysis of minerals for rare-earth elements: use of calculated peak-overlap corrections. *Canadian Mineralogist* 23, 263–371.

Roeder, P. L., MacArthur, D., Ma, X. P. & Palmer, G. R., 1987. Cathodoluminescence and microprobe study of rare-earth elements in apatite. *American Mineralogist* 72, 801–811.

Rub, V. G., Rub, A. K. & Salmin, Y. P., 1994. On the peculiarities of REE and some trace elements in the zircons of ore-bearing granites. *Geokhimiya* 11, 1577–1590.

Rubin, J., Henry, C. D. & Price, J. G., 1989. Hydrothermal zircons and zircon overgrowths, Sierra Blanca Peaks, Texas. *American Mineralogist* 74, 865–869.

Sawka, W. N., 1988. REE and trace element variations in accessory minerals and hornblende from the strongly zoned McMurry Meadows Pluton, California. *Transactions of the Royal Society of Edinburgh: Earth Sciences* 79, 157–168.

Sawka, W. N. & Chappell, B. W., 1988. Fractionation of uranium, thorium and rare earth elements in a vertically zoned granodiorite: implications for heat production distribution in the Sierra Nevada batholith, California, U.S.A. *Geochimica et Cosmochimica Acta* 52, 1131–1144.

Sevigny, J. H., 1993. Monazite controlled Sm/Nd fractionation in leucogranites: an ion microprobe study of garnet phenocrysts. *Geochimica et Cosmochimica Acta* 57, 4095–4102.

Tourrette, T. Z. L., Burnett, D. S. & Bacon, C. R., 1991. Uranium and minor-element partitioning in Fe–Ti oxides and zircon from partially melted granodiorite, Crater Lake, Oregon. *Geochimica et Cosmochimica Acta* 55, 457–470.

Vlasov, K. A., 1966. *Geochemistry and Mineralogy of Rare Elements and Genetic Types of their Deposits.* Jerusalem: Israel Program for Scientific Translation.

Wark, D. A. & Miller, C. F., 1993. Accessory mineral behavior during differentiation of a granite suite: monazite, xenotime, and zircon in the Sweetwater Wash pluton, southeastern California. *Chemical Geology* 110, 49–67.

Watson, E. B., 1988. The role of accessory minerals in granitoid geochemistry. In: *First Hutton Meeting on the Origin of Granite and Related Rocks, Vol. 1.* Edinburgh: Royal Society of Edinburgh, pp. 19–20.

Watson, E. B. & Harrison, T. M., 1984. Accessory minerals and the geochemical evolution of crustal magmatic systems: a summary and prospectus of experimental approaches. *Physics of the Earth and Planetary Interiors* 35, 19–30.

Watson, E. B., Vicenzi, E. P. & Rapp, R. P., 1989. Inclusion/host relations involving accessory minerals in high-grade metamorphic and anatectic rocks. *Contributions to Mineralogy and Petrology* 101, 220–231.

Watt, G. R. & Harley, S. L., 1993. Accessory phase controls on the geochemistry of crustal melts and restites produced during water-undersaturated partial melting. *Contributions to Mineralogy and Petrology* 114, 550–556.

Wolf, M. B. & London, D., 1994. Apatite dissolution into peraluminous haplogranitic melts: an experimental study of solubilities and mechanisms. *Geochimica et Cosmochimica Acta* 58, 4127–4146.

Wolf, M. B. & London, D., 1995. Incongruent dissolution of REE- and Sr-rich apatite in peraluminous granitic liquids: differential apatite, monazite, and xenotime solubility during anatexis. *American Mineralogist* 80, 765–775.

Yurimoto, H., Duke, E. F., Papike, J. J. & Shearer, C. K., 1990. Are discontinuous chondrite-normalized REE patterns in pegmatite granite systems the results of monazite fractionation? *Geochimica et Cosmochimica Acta* 54, 2141–2145.

RECEIVED JUNE 20, 1995
REVISED TYPESCRIPT ACCEPTED DECEMBER 8, 1995

JOURNAL OF PETROLOGY | VOLUME 37 | NUMBER 6 | PAGE 1601 | 1996

Corrigendum

F. Bea, 1996. Residence of REE, Y, Th and U in Granites and Crustal Protoliths; Implications for the Chemistry of Crustal Melts. *Journal of Petrology* **37**, 521–552.

Owing to an error in a computer program, some of the LA–ICP–MS data for yttrium presented in Tables 3, 4, 5, 6, 12 and 13 of Bea (1996) are incorrect. The following changes should be made:

Table no.	Analysis no.	Published value	Correct value
3	Amphibole 1	17·9	39·4
3	Amphibole 2	19·3	42·4
3	Amphibole 4	4·53	9·97
3	Amphibole 5	9·30	20·4
3	Amphibole 6	10·2	22·4
3	Cpx 3	0·37	0·81
3	Cpx 4	2·65	5·83
4	K-feldspar 5	0·40	0·88
4	K-feldspar 6	0·39	0·86
4	Plagioclase 1	0·03	0·07
4	Plagioclase 2	0·08	0·17
5	Garnet 4	25·9	51·8
5	Garnet 5c	248	546
5	Garnet 5r	135	297
5	Garnet 6c	45·1	99·2
5	Garnet 4	187	411
5	Cordierite 1	0·07	0·15
5	Cordierite 2	0·55	1·21
5	Cordierite 3	0·37	0·81
5	Cordierite 4	0·05	0·11
6	Tourmaline 4	4·30	9·46
6	Tourmaline 5	1·02	2·24
6	Tourmaline 6	0·01	0·02
6	Epidote 4	7·19	15·82
6	Epidote 5	3·55	7·81
6	Epidote 6	1·81	3·98
12	Peraluminous 1	1806	3612
12	Peraluminous 5	1736	3472
12	Metaluminous 1	632	1264
12	Metaluminous 2	951	1902
12	Metaluminous 3	1024	2048
12	Peralkaline 1	257	514
12	Peralkaline 2	317	634
12	Peralkaline 3	284	568
13	1	1150	2300
13	2	1612	3224
13	3	2127	4254
13	4	1100	2200
13	5	217	434
13	6	732	1610
13	7	100	2200
13	8	245	539
13	9	376	827
13	10	2239	4479
13	11	1301	2602
13	12	1251	2602

REFERENCE

Bea, F., 1996. Residence of REE, Y, Th and U in granites and crustal protoliths; implications for the chemistry of crustal melts. *Journal of Petrology* **37**(3), 521–552.

CHAPTER 20

Burnham, C.W. and Ohmoto, H. (1980) Late-stage processes of felsic magmatism. *Mining Geology Special Issue*, **8**, 1–11.

Neither of this volume's editors can claim any distinction in the fields of economic geology and ore genesis. Nevertheless, this is a subject area of considerable importance to humanity. We have therefore chosen to include Burnham and Ohmoto (1980) as one of our landmark papers. It was originally presented on January 29, 1979, at an international symposium on 'Granitic Magmatism and Related Mineralization', held in Tokyo, Japan, and subsequently published in *Mining Geology* Special Issue No 8, published by the Society of Mining Geologists of Japan (now The Society of Resource Geology). The journal is now known as *Resource Geology* and remains somewhat obscure. The Burnham and Ohmoto paper was the first in that volume of 245 pages, perhaps signalling the esteem in which it was held at the time.

Although there are spatial relationships between ore deposits and granites, the genetic connections are not always obvious. Nevertheless, there are clear associations between very felsic S-type granites and major tin deposits, and between some porphyry copper/molybdenum deposits and their host I-type granitic rocks. Commonly a cooling igneous intrusion just provides the heat engine for the initiation and maintenance of a hydrothermal system, and the wall rocks are the real sources of the metals. However, the granitic magmas themselves can and do dissolve a range of metals, as well as H_2O, chlorine and sulphur. As many granite-related ore deposits are composed of metal sulphides, there is an important question concerning the availability of S in granitic magmas, at the late stages of crystallization, when ore metals need to be transported as sulphur-bearing complexes. We chose this paper as a landmark because, as far as we are aware, it was the first to point out the relationships between the oxidation state of a granitic magma (largely inherited from its protolith), the form in which sulphur is held in the silicate solution, and the location and availability of that sulphur to form metal sulphides, precipitated by fluids at the late-magmatic stages. Though it does not appear in the abstract, the important implication of the analysis presented in this paper is that reduced granitic magmas, such as typical S-types, are unlikely to be associated with magmatic sulphide mineral deposits, but that some oxidized I-types are far more liable to have this association.

MINING GEOLOGY SPECIAL ISSUE, No 8, p. 1-11, 1980

TOKYO, JAPAN

Late-Stage Processes of Felsic Magmatism

C. Wayne BURNHAM* and Hiroshi OHMOTO*

Abstract: The formation of hydrothermal ore deposits, whether of the geothermal, vein, skarn, or porphyry type, is dependent largely upon processes that operate during the late stages of felsic magmatism. The effectiveness of these processes depends, in turn, upon numerous factors, principal among which are: (1) magma composition, including metal, sulfur, chloride, and especially H_2O contents; and (2) geologic environment, including depth, of magma emplacement. The importance of H_2O content lies in its control of the stage of crystallization, at a given depth, when an aqueous phase (hydrothermal fluid) is evolved by second boiling, which in turn determines the extent of wallrock fracturing and subsequent hydrothermal activity.

Second boiling releases large amounts of mechanical energy ($P\Delta V_r$) which, for $X_{w}^{m} > 0.24$ (> 2 wt. %), may produce explosive volcanism at P (load pressure) < 0.5 kbar or extensive fracturing at P < 2 kbar. It also produces an aqueous phase into which chlorides and sulfur are strongly partitioned. Chlorides exist predominantly as neutral complexes with alkalies, iron (and other chalcophile metals), hydrogen, and calcium, whereas sulfur exists as both H_2S and SO_2. The metal carrying capacity of these aqueous fluids is directly dependent upon their chloride content and the SO_2/H_2S fugacity ratio. This latter ratio is dependent, in turn upon fo_2, and may exceed 100 by diffusive loss of H_2.

Escape of these chloride- and sulfur-bearing fluids into cooler wallrocks results mainly in potassium and hydrogen metasomatism of aluminosilicates (potassic and phyllic alteration) and Fe-silicate metasomatism of carbonates (skarns). Cooling of these fluids causes precipitation of metal sulfides (also anhydrite) from the chloride complexes, mainly through hydrolysis of SO_2 to H_2S and H_2SO_4. Also, the HCl produced in this hydrolysis promotes additional phyllic alteration.

Magmatic fluids that are less than approximately 0.3 molal in ΣCl are not effective ore-forming fluids. They may acquire greater effectiveness, however, by condensation at low pressures and temperatures.

INTRODUCTION

Several decades of active exploration for porphyry-type copper, molybdenum, and tin deposits by perhaps thousands of geologists have firmly established two important but perplexing facts: (1) not all felsic porphyry stocks are hydrothermally altered, and (2) not all hydrothermally altered stocks or their wallrocks are economically mineralized. Why? Unfortunately, there does not appear to be a simple answer to this question. There are good reasons to believe, however, that the answer will be forthcoming when the relevant magmatic and hydrothermal processes, and the factors that control them, are better understood. In this context, relevant processes include not only those processes that operate in the near-surface environment, but also the numerous others that operate at deeper levels in the earth to yield hydrothermal ore-forming fluids when emplaced in the near-surface environment.

The geological and mineralogical features that are common to virtually all porphyry copper deposits have been appropriately characterized by GUSTAFSON and HUNT (1975) as a basic theme, a theme that is developed through orchestration of numerous processes whose operation is essential to the formation of each deposit. They also regarded the differences in geological and mineralogical features, from one deposit to another, as variations on the basic theme, variations that may reflect operation of additional processes or differing intensities of basic-theme processes. Whether these variations are caused by one or the other of these factors, they may or may not be critical to the formation of ore in a given deposit.

The processes that constitute both the basic theme and the variations on this theme, whether in a porphyry copper or other magmatic-hydrothermal system, naturally fall into three groups according to the environmental factors (pressure, temperature, phase assemblage, geologic and tectonic settings) that dominate their control. The first group includes those processes that traditionally have been referred to as **orthomagmatic processes**, which involve silicate melts; it is enlarged here to include some highly important processes by which hydrous, metal-bearing magmas are generated from different kinds of source rocks in different tectonic regimes, as well as the more commonly considered processes of magma emplacement and crystallization. These **orthomagmatic processes** and the factors that control them historically have been more the province of igneous petrologists than of students

---

* Department of Geosciences, The Pennsylvania State University, University Park, Pa., 16802, U.S.A. (大本洋)

Keywords: Felsic magmatism, Hydrous magma, Orthomagmatic processes, Hydrothermal processes, Porphyry-type deposits

2   *C. W. BURNHAM, H. OHMOTO*

of ore deposits. Thus, although many of them are moderately well understood in terms of their roles in the formation of igneous rocks, these processes have not generally been given adequate consideration in the construction of genetic models for the formation of associated hydrothermal ore deposits.

At the other end of the process continuum lies a group of strictly **hydrothermal processes** that involve aqueous fluids and solid phases (wallrocks and hydrothermal minerals). These **hydrothermal processes** can be further subdivided into two groups: (1) those that involve fluids of magmatic origin, and (2) those that involve waters of extraneous origin (meteoric water or seawater), whose only direct connection with magmas may be in the thermal energy that drives their circulation. Much has been learned about these processes, especially in the last ten years, but much more must be learned before our understanding of them is adequate to effectively guide exploration for the ore deposits they produce. The difficulties in understanding this group of processes arise in large part from the very large number, and extreme variability, of the environmental factors that affect them.

In the middle ground between this group and the **orthomagmatic processes** mentioned above lies an ill-defined and poorly-understood group of **transitional processes** that are not readily distinguishable from **orthomagmatic processes** at high temperatures and generally higher pressures and from strictly **hydrothermal processes** at lower temperatures and generally lower pressures. For present purposes, the upper limit of **transitional processes** may be arbitrarily set at the stage where the operation of orthomagmatic processes leads to formation of a separate magmatic aqueous phase, and the lower limit may be even more arbitrarily set at the H_2O-saturated solidus of the magma, i.e., the temperature, for a given total pressure, below which hydrous silicate melt is not thermodynamically stable. It must be realized, however, that kinetic factors play an important role in these dynamic systems and quasi-orthomagmatic conditions may extend well into the sub-solidus region in some cases.

ORTHOMAGMATIC PROCESSES

Source Rocks and Magma Generation Processes

Orthomagmatic processes of importance in ore formation commence with the generation of magmas by partial melting of older rocks, whether in mafic oceanic crust of subduction zones, in mafic amphibolites of the lower continental crust, or in felsic metasedimentary rocks of the latter region. Owing to their relatively higher average contents of the base and precious metals, mafic source rocks tend to yield partial melts (magmas) that are enriched in these elements. On the other hand, felsic metasedimentary rocks tend to yield partial melts that are commonly enriched in tin, the alkali metals, beryllium, tantalum, niobium, and the rare earths. In both types of source rocks, however, the partial melting process constitutes an important step in the concentration of the ore elements.

Partial melting is induced by the presence of H_2O in one form or another, and the pressure-temperature regime in which melting occurs depends to a large extent on the form in which H_2O occurs. At the great pressures of the deep continental crust — and even greater pressures in subduction zones — it is probable that essentially all the H_2O in the source rocks, just prior to the onset of melting, is bound as hydroxyl in hydrous minerals and the rocks have essentially zero porosity. In mafic rocks, such as subducting oceanic crust, the principal hydrous mineral is amphibole, and in felsic metasedimentary rocks the micas predominate. In both rock types, however, the total H_2O content probably is ordinarily less than two weight percent, and this places definite constraints on the amount of melt that can be produced at a given depth (pressure) and temperature.

At depths shallower than approximately 70 km, as indicated in Figure 1, a non-porous mafic amphibolite begins to melt at temperatures between 940° and 1040°C, and the H_2O content of the melt must exceed approximately 2.7 wt %, hence in the order of 30% of the source rocks can be melted for each percent of H_2O in the original rock. At depths greater than 75 to 80 km, as in a subduction zone, amphibole is not stable at any temperature; hence, melting of mafic amphibolite can begin at temperatures as low as 660°C. However, for melting to occur at such low temperatures, the melt must contain approximately 27 wt % H_2O; therefore, the amount of melt produced initially is generally less than five percent for each percent of H_2O in the original amphibolite. Non-porous muscovite-bearing metasedimentary rocks also begin to melt at temperatures in the range of 670° to 720°C under pressures of the lower continental crust. In this case, however, the first-formed melt must contain in excess of 8.4 wt % H_2O; hence, only 10 to 12% of the source rocks can be melted for each percent of H_2O in the original rock.

The melting relationships briefly outlined above serve to illustrate three features of magma generation in the lower continental crust and in subduction zones that are of utmost importance in ore genesis. First, partial melting of amphibole or mica-bearing rocks at high pressures, even in the absence of pore fluid, produces initial melts in which the H_2O content always exceeds 2.7 wt %, regardless of the total amount of H_2O initially bound in hydrous minerals. Second, these melts range in composition from granitic to dioritic, as do the igneous rocks that are characteristically associated with magmatic/hydrothermal ore deposits. Third, the amount of initial melt formed is directly proportional to the H_2O content of the original rock and, for a geologically reasonable H_2O content of 1.0

wt.%, typically ranges from 10 to 25%.

The first feature, a minimum of 2.7 wt.% H_2O in the initial melts, is important for the following reasons: (1) it assures that the magma produced, when emplaced in the shallow crustal environments required for hydrothermal ore-forming processes to operate, will evolve a separate magmatic aqueous phase upon cooling and crystallization; (2) it enhances the solubilities of metal sulfides in the initial melts by perhaps an order of magnitude over those in anhydrous melts of the same silicate composition, through equilibrium reactions such as 2FeS (solid) + $2H_2O$ (melt) + SiO_2 (melt) = Fe_2SiO_4 (melt) + $2H_2S$ (melt), and thereby provides the hydrous magmas with high sulfur-carrying capacity; and (3) it generally leads to the crystallization of hornblende and/or biotite from the magmas upon cooling at depths greater than about two kilometers, thus providing a possible exploration guide to those intrusive igneous bodies with which hydrothermal ore-forming processes might have been associated.

The second feature, the dioritic to granitic compositions of the hydrous magmas formed, is important because most ore deposits of the magmatic/hydrothermal type are associated with intrusive igneous bodies of these compositions. Thus, hydrothermal activity is closely associated with these compositional types of igneous rocks, because hydrous minerals in the source rocks, whether micas in felsic metasedimentary rocks or amphiboles in mafic amphibolites, play a major role in the generation of hydrous magmas of intermediate to felsic composition.

The direct dependence of the amount of melt formed on the total H_2O content of a given source rock provides a mechanism for greatly enriching early-formed partial melts with certain elements, relative to their concentrations in the source rocks. For those elements that are contained in minor mineral phases of the source rocks, which dissolve completely in the early-formed melts, enrichment factors are inversely proportional to the H_2O content of the source rock. On the other hand, for those elements that are contained in major mineral solid solutions which coexist with early-formed melts, enrichment factors are dependent upon partition coefficients amongst the coexisting phases. An element such as copper, which is highly concentrated in the minor iron-rich sulfides of mafic amphibolites, for example, may undergo a five-fold or greater enrichment in early-formed melts; however, an element such as lead, which substitutes for potassium in alkali feldspars, may undergo little, if any, enrichment in melts formed from metasedimentary rocks. For those elements that undergo marked enrichment in early-formed melts, hydrous magma generation by partial melting constitutes a critical initial process in the multiplicity of processes that ultimately lead to ore formation.

Magma Emplacement

The intrusive igneous bodies with which magmatic/hydrothermal base and precious metal deposits are associated appear to have been emplaced at depths generally less than 10 kilometers; many of them reached depths as shallow as one or two kilometers, and some of them appear to have vented at the surface to produce compositionally equivalent extrusive rocks. They can thus be appropriately regarded as parts of subvolcanic systems and, as such, tend to have been emplaced passively in extensional tectonic regimes. Zones of structural weakness in the basement rocks, therefore, have commonly exerted considerable influence on the localization of these bodies and the volcanic centers with which they are commonly associated.

Geological evidence suggests that magmatic differentiation by gravitational settling of early-formed

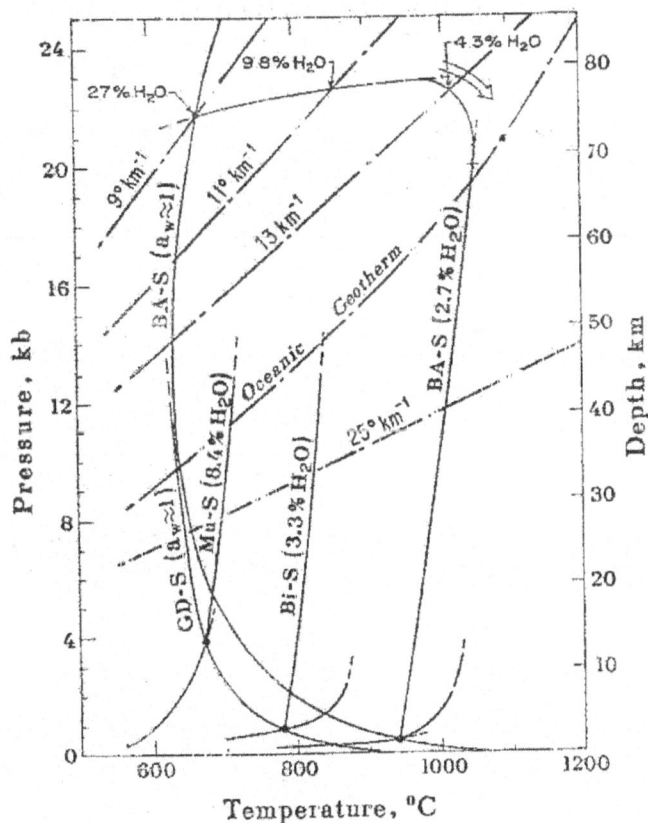

Fig. 1 Pressure-temperature projections of melting relations for muscovite- and biotite-bearing schists and gneisses, and hornblende-bearing basaltic amphibolites. The curves labeled GD-S and BA-S ($a_w \approx 1$) are the H_2O-saturated solidus for average granodiorites and basaltic amphibolites, respectively; the curves labeled Mu-S (8.4% H_2O), Bi-S (3.3% H_2O), and BA-S (2.7% H_2O) represent the divariant, fluid-absent, beginning of melting of non-porous gneisses and amphibolites. The percentage H_2O figures are the H_2O contents, in weight percent, of the first-formed melts produced by incongruent solution of muscovite, biotite, and hornblende, respectively. Adapted from Figs. 3.3 and 3.4 of BURNHAM (1979).

crystals has played a relatively minor role in the development of most magmatic/hydrothermal ore-forming systems. Differentiation during flow of a crystal-charged magma or by squeezing out ("filter pressing") of interstitial melt, on the other hand, appears to have played a more important role in some cases, especially in the formation of some ore bodies in pegmatite. What crystal settling does occur in the marginal parts of a hydrous magma body results in enrichment of the remaining melt in H_2O. Because of this increase in H_2O content, the remaining melt becomes proportionately less dense, thereby inhibiting the establishment of convective circulation. For this and other reasons, such as small size, geometry, high viscosities, and the exothermic effects of volatile separation (second boiling), convective-type circulation in magma bodies of interest here probably is not an important process in the development of magmatic-hydrothermal ore-forming systems.

Hydrous magmas of intermediate composition generally are out of chemical equilibrium with more silicic and potassic wallrocks through which they pass or in which they are emplaced. There is thus a tendency for the marginal parts of these magmas to reactively assimilate this type of wallrock, causing the melt fraction to become enriched in silica, potassium, and perhaps other elements such as tin. The process of assimilation, therefore, might be an important factor in the development of some porphyry tin systems, as in Bolivia (GRANI et al., this volume), but it is probably not an important factor in most other types of magmatic-hydrothermal ore-forming systems.

Crystallization of Hydrous Magmas

Although the nature of the source rocks and the processes of hydrous magma generation and emplacement are critically important to ore formation in magmatic/hydrothermal systems, certain processes must operate during crystallization of a magma to further concentrate the ore elements into minable deposits. Among these numerous processes, the most critical one is the evolution of a separate, H_2O-rich (aqueous) volatile phase. Were it not for this process, pegmatite ores would not form (JAHNS and BURNHAM, 1969), the fracture systems of porphyry copper/molybdenum and other stockwork deposits would not develop, contact metasomatic processes would not operate, and many of the ore-localizing structures produced by explosive volcanism would not exist.

The evolution of an aqueous phase from a given magma is controlled principally by the solubility of H_2O in the melt, which is very strongly pressure dependent, but only weakly temperature dependent. As shown in Figure 2, at a pressure of 500 bars, which is approximately the lithostatic pressure at a depth of two kilometers, the maximum solubility of H_2O in melts of dioritic to granitic composition is only 2.7 to 3.0 wt. %, whereas at a pressure of 2,000 bars (~8 km

depth) it is 6.1 to 6.4% and at 5,000 bars (~18 km depth) it is 9 to 10%. Thus, such melts that initially contained 2.0 wt. % H_2O would become saturated with this substance, during cooling, after approximately 33% crystallization at two kilometers, 73% crystallization at eight kilometers, and about 83% crystallization at 18 kilometers. Further cooling and consequent crystallization beyond these points of saturation causes H_2O to separate from the residual melt by a process commonly called second, or retrograde boiling. Eventually all the original H_2O content of the magma, except that bound structurally in hydrous minerals (0.5 to 0.8 wt. % H_2O in hornblende-biotite granodiorites) must be evolved as a separate fluid phase.

During cooling of the magmas in the foregoing examples, the minerals that generally crystallize early are plagioclase, followed by pyroxene in dioritic magmas and by alkali feldspars or quartz in granitic magmas. Their crystallization enriches the residual melt not only in H_2O, but in all other constituents not partitioned in favor of these minerals. This enrichment process by fractional crystallization is thought to be important in the production of ore-bearing granite pegmatite magmas, especially in deep-seated magma

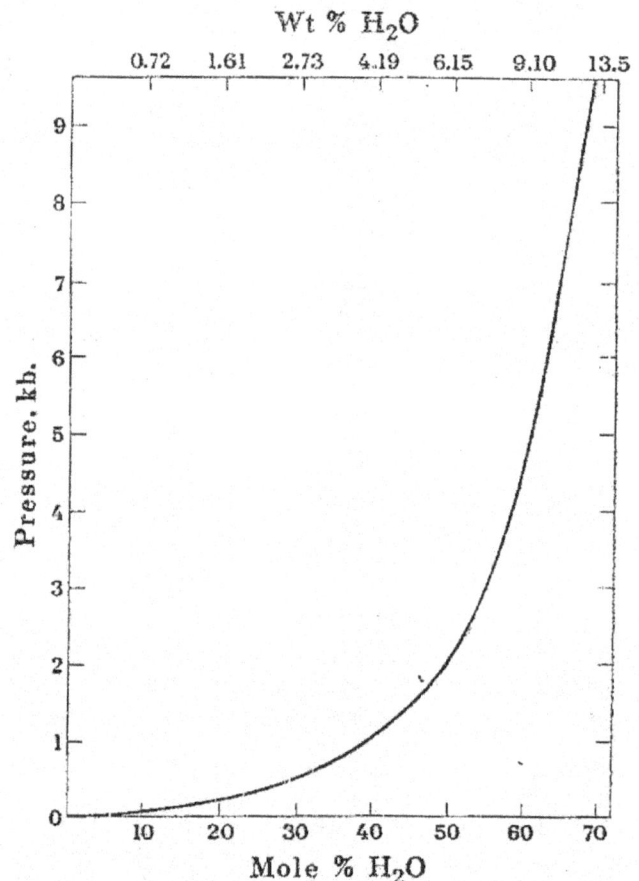

Fig. 2 The solubility of H_2O in granodioritic melts at 1,100°C. The lower abscissa is in mole % H_2O and the upper abscissa is in wt. % H_2O. Adapted from Fig. 3.1 of BURNHAM (1979).

bodies where enrichment factors in the residual melts for elements such as lithium and beryllium may be as high as ten, possibly higher. It also is thought to be important in higher-level intrusive bodies of metal sulfide-bearing magmas, where fractional crystallization concentrates the ore metals, sulfur, and chlorine in the more hydrous residual melt.

TRANSITIONAL PROCESSES

In accordance with earlier definitions, the formation of a separate magmatic volatile (aqueous) phase by second or retrograde boiling is arbitrarily chosen to distinguish those processes that operate almost entirely within condensed crystal-melt systems (orthomagmatic processes) from those that operate in coexisting condensed-phase-volatile systems, which are here called transitional processes. Chemically, orthomagmatic processes are controlled primarily by crystal-melt equilibrium, whereas transitional processes are dominated by melt-volatile (aqueous fluid) equilibrium. Also, physical orthomagmatic processes are controlled largely by magma viscosity and density contrasts between crystals and melt, whereas physical transitional processes are dominated by volume changes that accompany the second-boiling reaction, H_2O-saturated melt → crystals + volatile phase.

Physical Processes

A body of hydrous magma emplaced in colder wallrocks, whether it be a shallow-seated porphyry magma or a deeper-seated pegmatite magma, must lose heat to its surroundings, hence crystallization generally must proceed inward from the walls of the magma chamber. As a consequence of this fact and the very low diffusivity of dissolved H_2O in silicate melts, the melt in such a body of magma first becomes saturated with H_2O near the margins. The formation of this H_2O-saturated rind or carapace may, under quasi-static conditions, effectively isolate the interior of the body to the transfer of matter (except hydrogen) either in or out. These conditions are regarded as essential to the development of pegmatites (JAHNS and BURNHAM, 1969), as well as of porphyry copper/molybdenum systems and of explosive volcanism (BURNHAM, 1972, 1979).

As second boiling proceeds inside this H_2O-saturated carapace, the magma body must either expand or the internal pressure must increase, as the reaction, H_2O-saturated melt → crystals + volatile phase, takes place with an increase in volume at all crustal pressures. To a first approximation, this increase in volume is directly proportional to the H_2O content at saturation and inversely proportional to pressure. Thus, a body of H_2O-saturated pegmatite melt at 2,000 bars pressure (6.4 wt.% H_2O) will expand approximately 11% upon complete crystallization, whereas the same body saturated with H_2O at 5,000

bars (10 wt.% H_2O) will expand only 5%. Perhaps this effect of pressure provides an explanation of why some pegmatite bodies contain considerable void space (crystallization at lower pressures), in which crystals of gem minerals commonly occur, and other bodies are only moderately porous and contain no gem "pockets" (crystallization at higher pressures).

In contrast to these relatively small volume increases, a body of granodioritic melt containing 2.7 wt.% H_2O, as shown in Figure 3, will expand nearly 50% upon complete crystallization at depth of two kilometers (550 bars pressure). Furthermore, at a depth of four kilometers, this same body, now 37% crystalline at H_2O saturation, will expand more than 15%. At these shallow depths, most types of wallrocks have high rigidity and cannot accommodate such large volume changes by plastic deformation. As cooling and crystallization proceed, consequently, pressure inside the H_2O-saturated carapace must increase. Theoretically, this excess internal pressure could reach several thousand bars, but the tensile strength of the strongest wallrocks imaginable is only a few hundred bars at most. Therefore, brittle failure occurs and, because the direction of least principal stress lies essentially in the horizontal plane, expansion occurs principally in that direction and the ensuing fractures thus tend to be nearly vertical in orientation. Also, because the roofrocks are mechanically coupled to the underlying H_2O-saturated carapace through viscous interstitial melt, expansion of the magma in the horizontal plane places these roofrocks under tension. Myriads of smaller fractures therefore propagate upward to produce stockworks and the intensely shattered rocks that are so characteristic of porphyry copper/molybdenum deposits.

In essence, then, the mechanical energy released by second boiling during the inevitable cooling of a shallow-seated hydrous magma, as indicated in Figure 3, is regarded as the major cause of fracturing, which in turn is essential to the localization of ores in most magmatic-hydrothermal ore deposits. The fractures thus produced may be of the stockwork type, the larger vein type, or the even larger caldera collapse structures that accompany explosive volcanism. Whatever the type, these fractures serve as high-permeability channelways for the migration of hydrothermal ore-forming fluids, whether the fluids are of magmatic, meteoric, or seawater origin.

Chemical Processes

The generation of a magmatic aqueous phase by second boiling is accompanied by partitioning of all elements in the system such that the chemical potential or fugacity of each chemical species is the same in all phases at equilibrium. The volatile element, chlorine, which is dissolved in silicate melts mainly as chloride ion (Cl^-), is partitioned strongly toward the magmatic aqueous phase (KILINC and BURNHAM, 1972) because

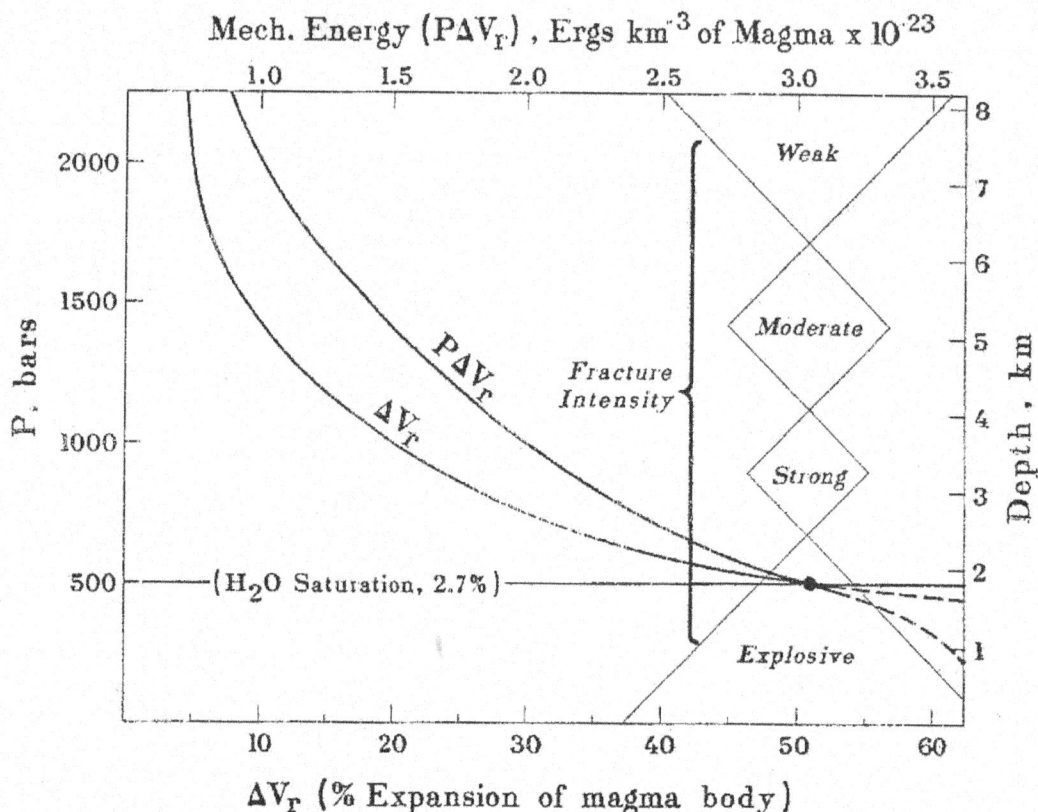

Fig. 3. The change in volume (lower abscissa) and mechanical energy released (upper abscissa) in the second-boiling reaction: H_2O-saturated melt → crystals + "vapor". Values of ΔV_r and $P\Delta V_r$ are for complete crystallization of a granodioritic magma with an initial H_2O content of 2.7wt.%. The depth of transition between "explosive" (volcanic eruption) and "strong" fracture regimes is approximate, as it depends upon the size and shape of the magma body. Curves were calculated from equations 3.10 and 3.11 of BURNHAM (1979).

(1) chloride minerals are not stable in magmas of intermediate to felsic compositions, and (2) it forms highly stable, neutral chloride complexes with hydrogen, alkali metals, alkaline earths, and heavy metals in aqueous solutions at magmatic temperatures and low to moderate pressures. Fluorine also forms stable neutral fluoride complexes in magmatic fluids, but high solubilities of fluorine in silicate melts, coupled with high thermal stabilities of minerals such as fluorite, topaz, and micas, cause fluorine to be partitioned largely toward the condensed phases. Sulfur, which is dissolved in hydrous melts principally as bisulfide ion (SH^-) (BURNHAM, 1979), also is partitioned strongly toward the magmatic aqueous phase, provided a sulfide mineral such as pyrrhotite is not stable. Carbon dioxide, dissolved sparingly in felsic melts as CO_2 and partitioned very strongly toward the volatile phase, appears to play only a minor chemical role in transitional processes at the magmatic stage.

Partition coefficients of the above named volatiles, between aqueous fluid and melt, appear to be relatively insensitive to temperature and, except for sulfur, to pressure. Because sulfur is dissolved in hydrous melts principally as SH^-, but exists in the aqueous phase as both H_2S and SO_2, its partition coefficient is sensitive to H_2O pressure (fugacity of H_2O) and to the fugacity

of oxygen (f_{O_2}). Under the f_{O_2} of a given phase assemblage, increasing pressure (f_{H_2O}, hence f_{H_2}) increases the proportion of H_2S to SO_2 in the aqueous phase, hence the partition coefficient for sulfur ($\Sigma S^v/\Sigma S^m$) is decreased. On the other hand, increasing f_{O_2} at a given pressure, as by diffusive loss of hydrogen from the system, increases the proportion of SO_2, hence the partition coefficient for total sulfur is increased. These phenomena, which arise because SO_2 is much less soluble than H_2S in hydrous magmas, are of great importance to sulfide ore formation; they provide mechanisms for generating high concentrations of both sulfur and heavy metals (almost entirely as chlorides) in the magmatic aqueous phase.

The f_{O_2} in a magma, prior to the onset of second boiling, is determined largely by the Fe^{3+}/Fe^{2+} ratio in the magma, which, in turn, is dependent to a large extent on the type of source rock from which the magma was generated. The f_{O_2} in felsic magmas that were generated by partial melting of metamorphosed igneous and volcanic rocks, as indicated in Figure 4, generally is higher than that of the QFM buffer (CARMICHAEL et al ,1974). Hence, the SO_2/H_2S fugacity ratio (and mole ratio) of fluids in equilibrium with these magmas commonly is close to unity and ranges from approximately 0.1 to 10. The f_{O_2} of felsic magmas that were generated by partial melting of carbonaceous metasediments, on the other hand, generally is lower than that of the QFM buffer (Fig. 4), and the CO_2/CH_4 fugacity and mole fraction ratios are near unity (OHMOTO and KERRICK, 1977). As a consequence, the SO_2/H_2S fugacity ratio is generally much smaller than 0.01. Aqueous fluids that separate from the former, so-called "I-type" magmas (CHAPPELL and WHITE, 1974), therefore, tend to produce sulfur-rich porphyry copper-type mineralization, whereas fluids from the latter, "S-type", magmas may yield the more sulfur-poor tin-type mineralization.

The molal concentration of chlorine in a magmatic aqueous phase is approximately 40 times that in a coexisting melt (KILINC and BURNHAM, 1972); hence, except for silica, chloride complexes commonly con-

$$(1/_{T(K)}) \times 10^3$$

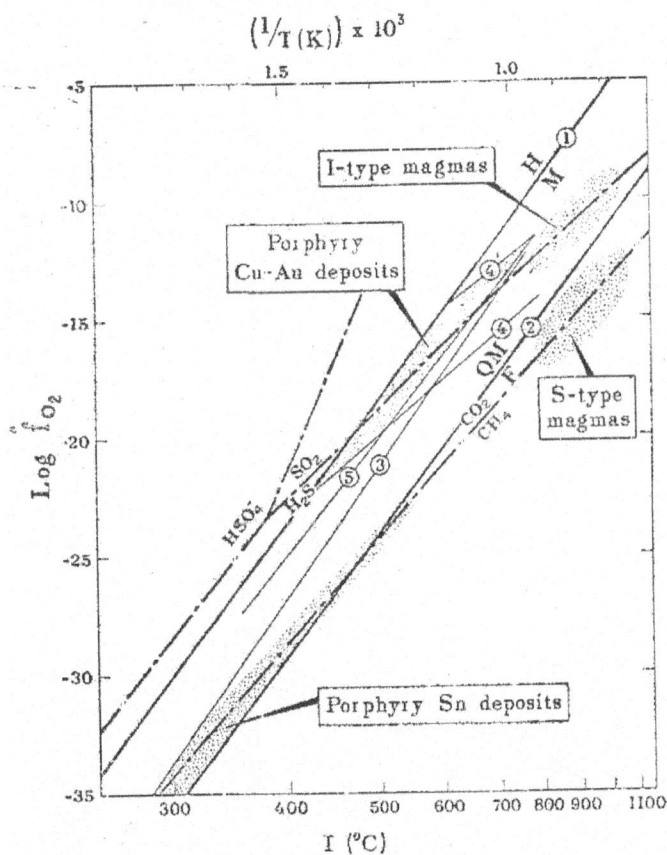

stitute the major portion of the total dissolved solutes. For this reason, the relative proportions of the various major chloride complexes and the manner in which these proportions respond to changing conditions of equilibrium are important factors in understanding such apparently diverse phenomena as mineral zoning in pegmatites and high temperature hydrothermal alteration patterns in porphyry copper deposits.

The major aqueous chloride complexes in equilibrium with typical granite melts are NaCl, KCl, and HCl, with NaCl and KCl constituting about 90% of the total, and the molal ratio of NaCl to KCl is the same as the molal ratio of sodium to potassium in the melt (BURNHAM, 1979). However, with the appearance of a mineral such as muscovite, which contains both potassium and hydrogen (OH^-), the HCl content decreases markedly, the KCl content also decreases, and the NaCl content increases to maintain the chloride stoichiometry. As a consequence, NaCl/KCl in the aqueous phase becomes greater than the sodium to potassium ratio in the melt. Thus, the relative proportions of the various chloride complexes even in this relatively simple system are complexly dependent upon both bulk melt composition and the nature of the coexisting minerals.

In a compositionally more complex system, such as a granodiorite magma, chloride complexes with calcium ($CaCl_2$) and especially iron ($FeCl_2$ and $FeCl_3$) must be included as major aqueous species ($MgCl_2$ is always a very minor constituent). The addition of these two constituents to the melt-volatile system does not affect the HCl content, for a given total chloride concentration, or the equality between NaCl/KCl and sodium/potassium in the melt, but the sum of NaCl + KCl is reduced by two or more times the molal concentration of iron and calcium. In fluids coexisting with melt, plagioclase, and magnetite at high pressures, Fe-chlorides and $CaCl_2$ are present in subequal amounts and, together, complex about 20% of the total chloride. At low pressures (500 to 1,000 bars), on the other hand, only about 5% of the chloride is complexed as $CaCl_2$, but more than 30% of the total Cl can be complexed with iron (KILINC, 1969). Under the latter conditions, then, as much Cl may be complexed with iron as with either Na or K.

Another important difference between the granite and granodiorite systems is the role of hydrous minerals. In contrast to granite systems, where precipitation of potassium-bearing micas causes NaCl/KCl in the aqueous phase to be greater than one (irrespective of sodium/potassium in the melt), early crystallization of sodium-bearing hornblende in granodioritic magmas causes NaCl/KCl to shift markedly in the opposite direction, toward values much less than one, even where the ratio of sodium to potassium in the melt is greater than one. In the presence of both hornblende and biotite, as might be expected, NaCl/KCl is close to unity, irrespective of the corresponding ratio in the

Fig. 4. Oxygen fugacity vs temperature relations for the predominant sulfur species and carbon species (broken lines) in aqueous fluids, and the stability fields of pertinent iron-bearing mineral assemblages (solid lines), all at approximately 1kb. The SO_2/HSO_4^- and HSO_4^-/H_2S boundaries are for unit activity ratios at pH=4. Mineral equilibria represented are: (1) magnetite + hematite, (2) fayalite + magnetite + quartz, (3) pyrrhotite + pyrite + magnetite, (4) biotite (38% annite) + K-feldspar + magnetite, (4′) biotite (18% annite) + K-feldspar + magnetite; and (5) anorthite + K-feldspar + pyrite + muscovite + quartz + anhydrite. Also shown (stippled) are approximate f_{O_2} - T fields for typical magnetite-bearing "I-type" and "S-type" felsic magmas, as well as for porphyry copper-gold and porphyry tin types of mineralization.

melt. Thus, the relative proportions of the major chloride complexes in the magmatic aqueous phase are strongly dependent upon melt composition, pressure, and the nature of the coexisting mineral assemblage.

The high ratio of potassium to sodium in aqueous chloride solutions coexisting with hornblende-bearing magmas provides a possible explanation for the extensive potassium metasomatism (potassic alteration) that is characteristically associated even with potassium-poor dioritic porphyry-copper porphyries. On the other hand, the relatively high concentrations of HCl in the aqueous phase, prior to the appearance of hydrous minerals, may account for the occurrence of topaz in some pegmatites and tin greisens, as well as of andalusite in some porphyry copper/molybdenum alteration haloes. Also, the very high concentrations of iron in the magmatic aqueous phase, especially at low pressures, may account not only for contact metasomatic skarn ores in carbonate wallrocks, but the overwhelming dominance of pyrite in most porphyry copper ores, as well.

Very little is known about the partitioning of most ore metals between magma and aqueous chloride solution. By analogy with manganese and zinc (HOLLAND, 1972), which are partitioned in favor of the aqueous phase by a factor of about two times the molal concentration of chloride squared, it is expected that elements more chalcophile than these two would be partitioned even more strongly toward the aqueous phase. Therefore, aqueous chloride solutions in which the fugacity of SO_2 is approximately equal to, or greater than, that of H_2S should be effective in scrubbing a magma and its immediately superjacent roof-rocks of their valuable metal content.

HYDROTHERMAL PROCESSES

With falling temperature or marked decrease in internal fluid pressures, transitional chemical processes give way to those hydrothermal processes that are dominated by crystal-volatile equilibrium. The boundary between these two regimes, which is arbitrarily defined as the H_2O-saturated solidus of the magma, may be relatively sharp in some systems and highly gradational in others. The relatively sudden brittle failure of the wallrocks in a porphyry copper system, for example, may cause part of the system to pass rather abruptly from one regime to the other. In a pegmatite system, on the other hand, both regimes may coexist in different parts of the system and "communicate" with each other through the volatile phase. Indeed, the coexistence of the two intercommunicating regimes apparently is essential to the development of mineralogic zoning in pegmatites and, hence, to the concentration of many pegmatite ores. Also, the largely closed hydrothermal circulation system established at this stage of pegmatite formation is believed to be ultimately responsible for the localiza-

tion of gem minerals, as well as minerals rich in tin, tantalum, niobium, uranium, thorium, and the rare earths.

The comparatively more rapid passage of the magmatic aqueous phase into the hydrothermal regime of a developing porphyry copper/molybdenum fracture system generally leads to conditions of gross disequilibrium between the fluids and cooler wallrocks. The nature of this disequilibrium, however, is dependent upon the initial conditions of equilibrium in the magmatic system, as well as upon the nature of the wallrocks and the extent to which temperature and pressure decrease. Aqueous chloride solutions from a high-temperature magmatic source, as discussed previously and shown in Figure 5, tend to be rich in HCl, hence they will react with feldspathic wallrocks to produce aluminum silicate (mainly andalusite or topaz) alteration, with or without biotite, at higher temperatures and muscovitic (sericitic, phyllic) altera-

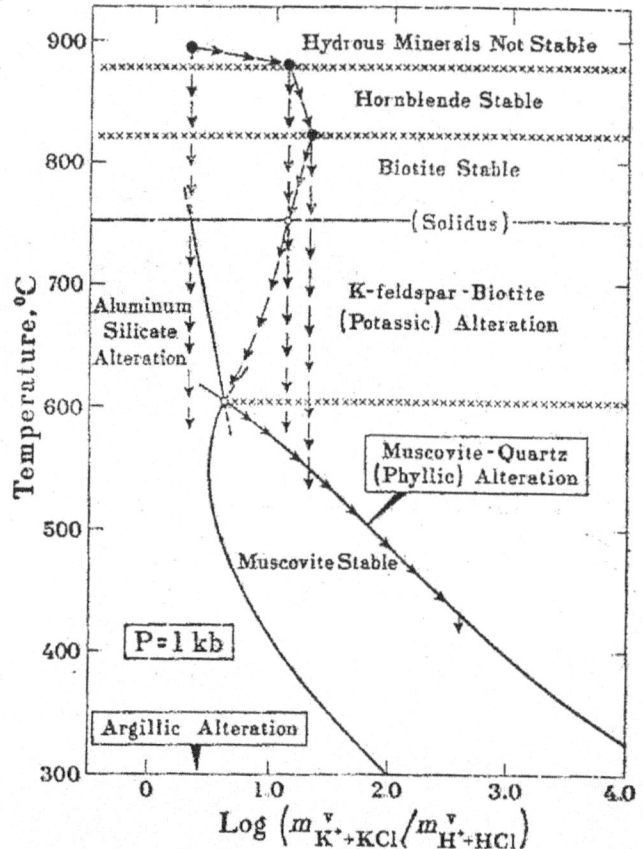

Fig. 5 $Log(m^v_{K^+ + KCl}/m^v_{H^+ + HCl})$ vs. temperature at 1 kb showing variations in m^v_{KCl}/m^v_{HCl} of magmatic aqueous chloride solutions as a function of phase assemblage, and possible non-equilibrium cooling paths of these fluids in a porphyry fracture system. Solid circles represent compositions of aqueous chloride solutions in equilibrium with granodioritic magmas, and open circles represent fluid compositions at isobaric invariant points (H_2O-saturated solidus and the muscovite + quartz ⇌ K-feldspar + aluminum silicate +"vapor" equilibrium). Modified from Fig 3.6 of BURNHAM (1979).

tion at lower temperatures. Fluids that equilibrated initially with hornblende-bearing magma, on the other hand, are enriched in KCl relative to NaCl and especially HCl, hence their interaction with non-carbonate wallrocks is largely one of fixing potassium in feldspar and biotite ("potassic alteration") by exchange with sodium and especially calcium. These exchange reactions, therefore, lower KCl/HCl in the aqueous phase, which in turn leads to entry into the stability field of muscovite ("sericitic" or phyllic alteration) near its high-temperature limit for the prevailing pressure. Thereafter, further cooling of the fluids causes KCl/HCl to increase, as potassium-rich feldspar is converted to muscovite and quartz.

The decrease in pressure and temperature that attends escape of the magmatic aqueous phase into fractured wallrocks may result in condensation of a chloride-rich liquid that consists mainly of NaCl, KCl, and Fe-chlorides. Owing to difference in density, this liquid may be segregated from the gaseous volatile phase, which has become enriched in HCl, CO_2, SO_2, and H_2S. This acid-enriched volatile phase may be responsible for part of the sericitic and argillic (clay-rich) alteration in the upper (outer), cooler parts of fracture systems.

Fluids derived from "I-type" (high fo_2) magmas, as discussed previously, may contain large quantities of SO_2 (0.1 m to >1 m), as well as $H_2S(m_{SO_2}/m_{H_2S} \approx 0.1-10)$. Because these fluids contain much smaller quantities of other reduced species, such as H_2, CO, and CH_4, molal SO_2/H_2S remains essentially constant during cooling to temperatures in the range 500° to 350°C, unless the fluids undergo condensation or reaction with large quantities of ferrous iron-bearing minerals in the wallrocks (OHMOTO and RYE, 1979). In the absence of condensation or large-scale interaction with the wallrocks, therefore, the fo_2–T paths of such fluids tend to lie near and be approximately parallel to the $SO_2/H_2S=1$ line in Figure 4. Along these cooling paths, hydrolysis of SO_2 ($4SO_2+4H_2O \rightarrow H_2S+3H_2SO_4$) results in a continuous increase in the activities, hence concentrations, of both H_2S and aqueous sulfates. The exact temperature at which this hydrolysis yields $m_{sulfates}=m_{SO_2}=m_{H_2S}$ depends upon several factors, including pH, total fluid pressure, m_{NaCl}, m_{KCl}, and m_{CaCl_2}, but for most hydrothermal fluids it lies in the range 500° to 350°C. It also may be affected by reaction with ferrous iron-bearing minerals in the wallrocks, in which the activity of H_2S is increased at the expense of SO_2 ($SO_2+6"FeO"+H_2O \rightleftharpoons H_2S+3"Fe_2O_3"$). In either case, the resultant increase in H_2S activity causes precipitation of sulfide ore minerals from the metal-chloride complexes (mainly of iron) in aqueous solution, which, in turn, produces HCl ($4FeCl_2+7H_2S+H_2SO_4 \rightarrow 4FeS_2 + 4H_2O+8HCl$). Production of HCl is further enhanced by precipitation of anhydrite ($CaCl_2+H_2SO_4 \rightarrow CaSO_4+2HCl$) from the aqueous $CaCl_2$ produced in

the earlier $K \rightleftharpoons Ca$ exchange reaction with wallrock plagioclase, as well as by direct reaction of the H_2SO_4 with the calcium-bearing minerals in the wallrocks. Thus, because the amount of HCl produced in these precipitation reactions is directly related to the total initial sulfur content of the fluids, the extent of low-temperature "acid" alteration produced by this HCl may be much greater than that produced by the HCl in the original magmatic fluid.

Fluids derived from "S-type" (low fo_2) magmas may contain as much H_2S as those derived from "I-type" magmas, but, owing to lower fo_2, they contain much smaller amounts of SO_2 and, therefore, total sulfur. As a consequence, fluids derived from "S-type" magmas generally precipitate smaller quantities of sulfides, chiefly pyrrhotite, and correspondingly larger proportions of oxides, such as cassiterite, upon cooling. The relative abundances of reduced and oxidized species in fluids derived from "S-type" magmas are generally in the order: $m_{CO_2} \simeq m_{CH_4} \simeq m_{H_2S} \gg SO_2$. As a consequence of the very high CH_4/SO_2 ratio, molal CO_2/CH_4 remains essentially constant during cooling, and the $fo_2 -T$ paths tend to lie near the $CO_2/CH_4=1$ line in Figure 4. In fact, fluid inclusions in some tin deposits have been shown by PATTERSON and OHMOTO (1976) to be rich in carbon; furthermore, CO_2 and CH_4 are present in nearly equal proportions. As in chloride-rich fluids from "I-type" magmas, precipitation of metals from "S-type" magmatic fluids, whether sulfides or oxides, also produces HCl.

In feldspathic wallrocks, the amount of HCl thus produced is controlled by hydrolysis-type reactions with the silicate mineral assemblages that commonly yield muscovite or other aluminous minerals. For a given temperature, pressure, and equilibrium assemblage, the HCl content is directly proportional to the total chloride content of the fluid. In carbonate rocks, on the other hand, HCl contents of the fluids are fixed at very low levels, irrespective of total chloride contents, by decarbonation reactions such as: $CaCO_3 + 2HCl \rightarrow CaCl_2+H_2O+CO_2$. This consumption of HCl, in turn, leads to precipitation of sulfide minerals (replacement of carbonates) by reactions such as: $ZnCl_2+H_2S \rightarrow ZnS+2HCl$. These replacement reactions commonly are accompanied, or preceded, by other carbonate replacement reactions that involve precipitation of iron-bearing silicates and oxides (mainly magnetite) to form typical skarn ores. Concomitant with these latter reactions, copious amounts of CO_2 are evolved, which, in turn, tends to inhibit further calc-silicate formation ($CaMgSi_2O_6+2CO_2 \rightleftharpoons CaMg(CO_3)_2+2SiO_2$). As a consequence, carbonate rocks eventually may be replaced by highly siliceous oxides ("jasperiod") and secondary dolomite.

Fluids emanating from a chloride-poor magmatic source are correspondingly poor in HCl, sulfur, iron, and other metals, but they are moderately richer in silica and much richer in alumina. In fact, the experi-

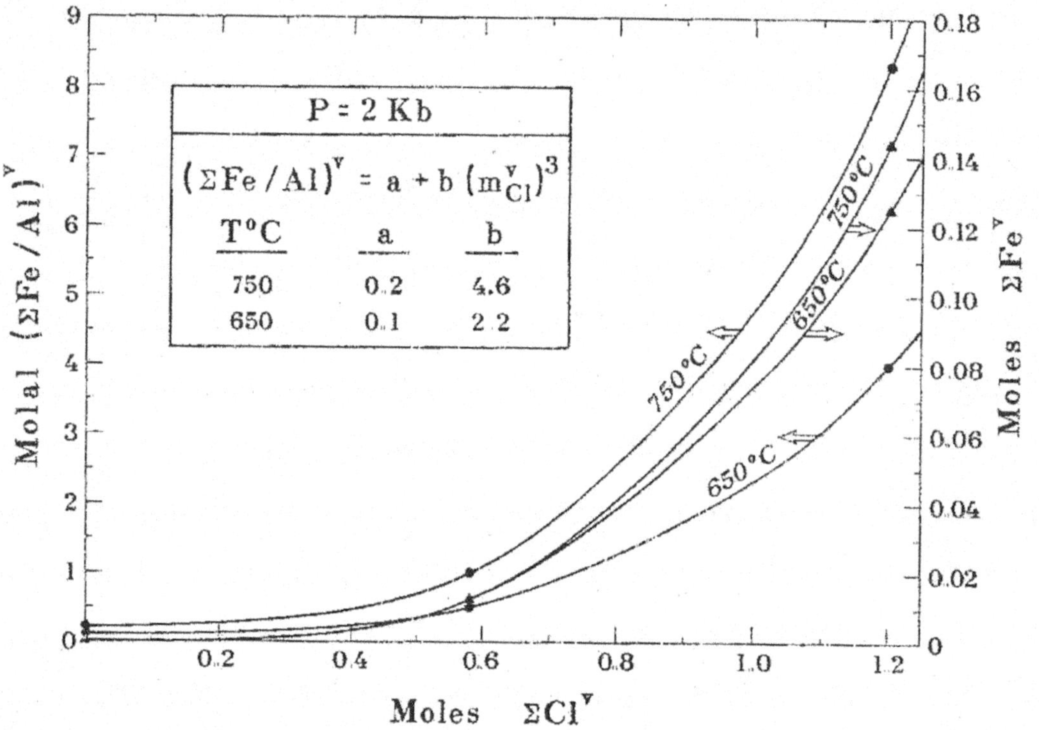

Fig. 6. The total iron to aluminum ratio [$(\Sigma Fe/Al)^v$, left ordinate] and total iron content (ΣFe^v) of aqueous fluids in equilibrium with felsic magma (750°C) and rock (650°C) at 2 kb, as a function of chloride content. Both $(\Sigma Fe/Al)^v$ and ΣFe^v vary approximately with the cube of the chloride concentration at oxygen fugacities near the upper boundary of the magnetite stability field. Data for $mCl^v = 0$ and 0.58 from BURNHAM (1967), and for $mCl^v = 1.2$ from KILINC (1969).

mental data of BURNHAM (1967) and KILINC (1969) suggest that the molal ratio of iron to aluminum, expressed as $(\Sigma Fe/Al)^v$ in Figure 6, increases almost linearly with the cube of the chloride content of the aqueous phase. As a consequence, metasomatic interactions of chloride-poor magmatic fluids with relatively pure carbonate wallrocks produce mainly calc-silicate mineral assemblages in which aluminum-rich garnet (grossularite) and magnesian pyroxene (diopside) figure more prominently than the iron-rich garnet (andradite) and pyroxene (hedenbergite) that are typically associated with skarn ores. Whether these ores are predominantly oxide or sulfide-rich, however, depends primarily on the sulfur content of the magmatic fluids.

GENERALIZATIONS

Generalizations based on limited knowledge are always easily made and commonly lead to erroneous conclusions, but they may be warranted in the present case as a means of gaining perspective of the multifarious magmatic and hydrothermal processes involved in the formation of porphyry-type ore deposits. Thus, from present understanding of these processes, as summarized above, felsic magmatism may result in extensive porphyry-type mineralization provided:

(1) The felsic magma possesses sufficient thermal energy to reach the sub-volcanic environment in a largely liquid state. Magmas that contain more than 5 wt.% H_2O generally do not posses the requisite energy.

(2) The initial H_2O content of the magma is high enough (approximately 2–4 wt.%) to result in second boiling over an appreciable depth below the roof of the magma body (in the order of 2–6 km) before the magma becomes approximately 75% crystalline. Second boiling in magmas with lower initial H_2O contents, hence at more advanced stages of crystallization, generally does not release sufficient mechanical energy to produce the extensive fracturing necessary for the egress of large quantities of mineralizing fluids from depths greater than approximately 2 km.

(3) The roofrocks of the magma chamber have sufficient strength to confine the mechanical energy released in second boiling to fracturing of the marginal parts of the porphyry stock and adjacent wallrocks, and to minor dike intrusion. Gross failure of the roofrocks, as in caldera formation, results in dissipation of potential mineralizing fluids (SILLITOE, this volume) and, hence, in a major weakening of the potential for porphyry-type mineralization. Concomitantly, however, the potential is enhanced for Kuroko-type mineralization, which appears to have formed largely from deeply circulating seawater in a submarine caldera (OHMOTO, 1978), or for Creede-type mineralization, which appears to have been produced by deeply

circulating meteoric waters in a sub-aereal caldera.

(4) The chloride content of the melt phase at the onset of second boiling is sufficient to yield a magmatic aqueous phase with the capacity to transport large quantities of metals, as chloride complexes. The chloride content of the melt required to yield porphyry copper mineralization is much higher (generally greater than 700 ppm) than that required to yield porphyry tin (cassiterite) mineralization. The amounts of metals and reduced sulfur fixed in porphyry copper deposits are much greater than in tin deposits, and large amounts of chloride are required to transport both the metals and reduced sulfur in the same fluid.

(5) The initial sulfur content of the magma, in the case of porphyry copper mineralization, is sufficient to yield an aqueous phase with a sulfur-carrying capacity, under the prevailing oxygen fugacities, adequate to eventually precipitate vast tonnages of metal sulfides (principally pyrite). The sulfur-carrying capacity of a given aqueous chloride solution is directly dependent upon the oxygen fugacity in the system, hence, fluids derived from the generally more oxidized "I-type" or "magnetite series" magmas are more effective gold and base metal-sulfide mineralizers than fluids derived from the commonly more reduced "S-type" or "ilmenite series" magmas. Chloride-rich fluids that produce magnetite-rich skarns, with only minor sulfides, probably were derived mainly from sulfur-poor magmas which, in turn, were generated either from sulfide-poor source rocks or under reducing conditions.

(6) The valuable metal content of the felsic magma is some presently unknown amount above average. The actual metal content required to yield economic mineralization depends upon the mass of magma and its crystalline products from which the metal can be extracted, the configuration of the hydrothermal plumbing system, and the efficiency of the extraction process. The mass of magma required and the configuration of the plumbing system are dependent largely upon conditions (1) to (3) above, whereas the efficiency of the extraction process is dependent primarily upon condition (4), the chloride content of the aqueous fluid. Mafic amphibolites are appropriate source rocks for porphyry copper-type magmas and pelitic metasedimentary rocks are appropriate sources of some tin- and tungsten-rich porphyries. Other porphyry magmas that are rich in tin, tungsten, and especially molybdenum perhaps were derived from granodioritic gneisses.

Acknowledgements: Support for the preparation of this article was provided by the National Sciences Foundation research grants No. EAR73-00247-A04 (C.W.B.) and EAR 76-03724 (H.O.).

REFERENCES

BURNHAM, C. W. (1967): Hydrothermal fluids at the magmatic stage. *In* Geochemistry of Hydrothermal Ore Deposits (H. L. BARNES, ed.), Holt, Rinehart and Winston, Inc., New York, 34–76.

BURNHAM, C. W. (1972): The energy of explosive volcanic eruptions. Earth and Mineral Sci., **41**, 69–70 (The Pennsylvania State University).

BURNHAM, C. W. (1979): Magmas and hydrothermal fluids. *In* Geochemistry of Hydrothermal Ore Deposits, 2nd edit. (H. L. BARNES, ed.), John Wiley and Sons, Inc., 71–136.

CARMICHAEL, I. S. E., TURNER, F. J. and VERHOOGEN, J. (1974): Igneous Petrology. McGraw Hill, Inc., 739 p.

CHAPPELL, B. W. and WHITE A. J. R. (1974): Two contrasting granite types. Pacific Geol., 8, 173–174.

GUSTAFSON, L. G and HUNT, J. P (1975): The porphyry copper deposit at El Salvador, Chile. Econ. Geol., 70, 857–912.

HOLLAND, H. D. (1972): Granites, solutions, and base metal deposits. Econ. Geol., **67**, 281–301.

JAHNS, R. H. and BURNHAM, C. W. (1969): Experimental studies of pegmatite genesis. pt. 1, a model for the derivation and crystallization of granitic pegmatites. Econ. Geol., 64, 843–864.

KILINC, I. A. (1969): Experimental metamorphism and anatexis of shales and graywackes. PhD thesis, The Pennsylvania State University, University Park, Pa 16802.

KILINC, I. A. and BURNHAM C. W. (1972): Partitioning of chloride between a silicate melt and coexisting aqueous phase from 2 to 8 kilobars. Econ. Geol., 67, 231–235.

OHMOTO, H (1978): Submarine calderas: A key to the formation of volcanogenic massive sulfide deposits? Mining Geol., **28**, 219–231.

OHMOTO, H. and KERRICK, D. M. (1977): Devolatilization equilibria in graphitic systems. Am. Jour. Sci., **277**, 1013–1044.

OHMOTO, H. and RYE, R. O. (1979): Isotopes of sulfur and carbon. *In* Geochemistry of Hydrothermal Ore Deposits, 2nd edit. (H. L. BARNES, ed.), John Wiley and Sons, Inc., 509–567.

PATTERSON, D. and OHMOTO, H. (1976): Stable isotope and fluid inclusion studies at the Renison Bell cassiterite-sulfide deposits, Western Tasmania (abst.). Intern. Conf. Stable Isotopes, New Zealand, 52.

Presented at an international symposium, "Granitic Magmatism and Related Mineralization" held in Tokyo, January 29, 1979
The manuscript received April 4, 1979; accepted August 16, 1979

References

Arndt, N. and Goldstein, S.L. (1987) Use and abuse of crust-formation ages. *Geology*, **15**, 893–895.

Balk, R. (1937) Structural behavior of igneous rocks. *Geological Society America Memoir*, **5**, 177 pp.

Barboza, S.A. and Bergantz, G.W. (1996) Dynamic model of dehydration melting motivated by a natural analogue: Applications to the Ivrea-Verbano zone, northern Italy. *Transactions of the Royal Society of Edinburg: Earth Sciences*, **87**, 23–31.

Barboza, S.A. and Bergantz, G.W. (2000) Metamorphism and anatexis in the mafic complex contact aureole, Ivrea Zone, Northern Italy. *Journal of Petrology*, **41**, 1307–1327.

Barboza, S.A., Bergantz, G.W. and Brown, M. (1999) Regional granulite facies metamorphism in the Ivrea zone: Is the Mafic Complex the smoking gun or a red herring? *Geology*, **27**, 447–450.

Bea, F. (1996a) Residence of REE, Y, Th and U in Granites and Crustal Protoliths; Implications for the Chemistry of Crustal Melts. *Journal of Petrology*, **37**, 521–552.

Bea, F. (1996b) Controls on the trace element composition of crustal melts. *Transactions of the Royal Society of Edinburg: Earth Sciences*, **87**, 33–42.

Bea, F., Montero, P. and Ortega, M. (2006) A LA-ICPMS evaluation of Zr reservoirs in common crustal rocks: implications for Zr and Hf geochemistry, and zircon-forming processes. *The Canadian Mineralogist*, **44**, 693–714.

Bea, F., Montero, P., Gonzalez Lodeiro, F. and Talavera, C. (2007) Zircon inheritance reveals exceptionally fast crustal magma generation processes in Central Iberia during the Cambro-Ordovician. *Journal of Petrology*, **48**, 2327–2339.

Bennet, V.C. and DePaolo, D.J. (1987) Proterozoic crustal history of the western United States as determined by neodymium isotopic mapping. *Geological Society of America Bulletin*, **99**, 674–685.

Bowen, N.L. (1948) The granite problem and the method of multiple prejudices. *Geological Society of America Memoir*, **28**, 79–90.

Brown, M. (2004) The mechanism of melt extraction from lower continental crust of orogens. *Transactions of the Royal Society of Edinburg: Earth Sciences*, **95**, 35–48.

Brown, M. (2007) Crustal melting and melt extraction, ascent and emplacement in orogens: mechanisms and consequences. *Journal of the Geological Society of London*, **164**, 709–730.

Brown, M. and Solar, G.S. (1999) The mechanism of ascent and emplacement of granite magma during transpression: a syntectonic granite paradigm. *Tectonophysics*, **312**, 1-33.

Bunsen, R.W. (1861) Uber die bildung des granites. *Zeitschrift de Deutchen Geologischen Gesellschaft*, **13**, 61–63.

Burnham, C.W. (1967) Hydrothermal fluids at the magmatic stage. Pp. 38–7 in: *Geochemistry of Hydrothermal Ore Deposits* (H.L. Barnes, editor). Holt, Reinhart and Winston, New York.

Burnham, C.W., and Ohmoto, H. (1980) Late-stage processes of felsic magmatism. *Mining Geology Special issue*, **8**, 1–11.

Buttner, S. and Kruhl, J.H. (1997) The evolution of a late-Variscan high T low P region: The southeastern margin of the Bohemian massif. *Geologische Rundschau*, **86**, 21–38.

Campbell, I.H. and Taylor, S.R. (1983) No water, no granites – no oceans, no continents. *Geophysical Research Letters*, **10**, 1061–1064.

Chapman, D.S. and Furlong, K.P. (1992) Thermal state of the continental lower crust. Pp. 179–199 in: *Continental Lower Crust* (D.M. Fountain, R. Arculus and R.W. Kay, editors). Elsevier. Amsterdam.

Chappell, B.W. (1984) Source rocks of I- and S-type granites in the Lachlan Fold Belt, southeastern Australia. *Philosophical Transactions of the Royal Society of London*, **A310**, 693–707.

Chappell, B.W., White, A.J.R. and Wyborn, D. (1987) The importance of residual source material (restite) in granite petrogenesis. *Journal of Petrology*, **28**, 1111–1138.

Chappell, B.W. and White, A.J.R. (1974) Two contrasting granite types. *Pacific Geology*, **8**, 173–174.

Chen, Y.D., Price, R.C., White, A.J.R. and Chappell, B.W. (1990) Mafic inclusions from the Glenbog and Blue Gum granite suites, Southeastern Australia. *Journal of Geophysical Research: Solid Earth*, **95**, 17757–17785.

Clemens, J.D. (1984) Water contents of silicic to intermediate magmas. *Lithos*, **17**, 272–287.

Clemens, J.D. (1989) The importance of residual source material (restite) in granite petrogenesis: A comment. *Journal of Petrology*, **30**, 1313–1316.

Clemens, J.D. (1998) Observations on the origins and ascent mechanisms of granitic magmas. *Journal of the Geological Society*, **155**, 843–851.

Clemens, J.D. and Mawer (1992) Granitic magma transport by fracture propagation. *Tectonophysics*, **204**, 339–360.

Clemens, J.D. and Vielzeuf, D. (1987) Constraints on melting and magma production in the crust. *Earth and Planetary Science Letters*, **86**, 287–306.

Clemens, J.D. and Wall, V.J. (1981) Origin and crystallization of some peraluminous (S-type) granitic magmas. *Canadian Mineralogist*, **19**, 111–131.

Clemens, J.D. and Watkins, J.M. (2001) The fluid regime of high-temperature metamorphism during granitoid magma genesis. *Contributions to Mineralogy and Petrology*, **140**, 600–606.

Cloos, E. (1936) Der Sierra-Nevada pluton in California. *Neues Jahrbuch für Mineralogie Geologie und Paläontologie*, **76B**, 355–450.

Conrad, W.K., Nicholls, I.A. and Wall, V.J. (1988) Water-saturated and -undersaturated melting of metaluminous and peraluminous crustal compositions at 10 kb: evidence for the origin of silicic magmas in the Taupo Volcanic Zone, New Zealand, and other occurrences. *Journal of Petrology*, **29**, 765–803.

Cook, J. and Gordon, J.E. (1964) A mechanism for the control of crack propagation in all-brittle systems. *Proceedings of the Royal Society of London*, **A282**, 508–520.

Cruden, A.R. (1998) On the emplacement of tabular granites. *Journal of the Geological Society*, **155**, 853–862.

Dallagnol, R., Lafon, J.M. and Macambira, M.J.B. (1994) Proterozoic Anorogenic Magmatism in the Central Amazonian Province, Amazonian Craton – Geochronological, Petrological and Geochemical Aspects. *Mineralogy and Petrology*, **50**, 113–138.

DePaolo, D.J. (1981) Neodymium isotopes in the Colorado Front Range and crustmantle evolution in the Proterozoic. *Nature*, **291**, 193–196.

DePaolo, D.J. (1988) Age dependence of the composition of continental crust: evidence from Nd isotopic variations in granitic rocks. *Earth and Planetary Science Letters*, **90**, 262–271.

DePaolo, D.J. and Wasserburg, G.J. (1976a) Inferences about magma sources and mantle structure from variations of $^{143}Nd/^{144}Nd$. *Geophysical Research Letters*, **3**, 743–746.

DePaolo, D.J. and Wasserburg, G.J. (1976b) Nd isotope variations and petrogenetic models. *Geophysical Research Letters*, **3**, 249–252.

Didier, J. (1973) *Granites and their Enclaves*. 393 pp. Elsevier, Amsterdam.

Eggler, D.H. (1973) Principles of melting of hydrous phases in silicate melt. *Carnegie Institute of Washington Yearbook*, **72**, 491–495.

Elsasser, W.M. (1963) Early history of the Earth. Pp. 1–29 in: *Earth Science and Meteoritics* (J. Geiss and E.D. Goldberg, editors). North-Holland, Amsterdam.

England, P.C. and Thompson, A.B. (1986) Some thermal and tectonic models for crustal melting in continental collision zones. *Geological Society Special Publication*, **19**, 83–94.

Fyfe, W.S. (1973) The granulite facies, partial melting and the Archean crust. *Philosophical Transactions of the Royal Society of London*, **A273**, 457–461.

Fyfe, W.S. (1973) The generation of batholiths. *Tectonophysics*, **17**, 273–283.

Glazner, A.F., Bartley, J.M., Coleman, D.S., Gray, W., and Taylor, R.Z. (2004) Are plutons assembled over millions of years by amalgamation from small magma chambers? *GSA Today*, **14**, 4–11.

Goranson, R.W. (1931) The solubility of water in granite magmas. *American Journal of Science*, **22**, 481–502.

Goranson, R.W. (1932) Some notes on the melting of granite. *American*

Journal of Science, **23**, 227–236.

Haederle, M. and Atherton, M.P. (2002) Shape and intrusion style of the Coastal Batholith, Peru. *Tectonophysics*, **345**, 17–28.

Hanchar, J.M. and Watson, E.B. (1993) Apatite saturation revisited: The effects of water in a metaluminous haplodacite melt. AGU 1993 Spring Meeting April 20, *EOS*, 341.

Hanchar, J.M. and Watson, E.B. (2003) Zircon saturation thermometry. Pp. 89–112 in: *Zircon* (J.M. Hanchar and P.W.O. Hoskin, editors). Reviews in Mineralogy and Geochemistry, 53, Mineralogical Society of America, Chantilly, Virginia and the Geochemical Society, St. Louis, Missouri, USA.

Harrison, T.M. and Watson, E.B. (1983) Kinetics of zircon dissolution and zirconium diffusion in granitic melts of variable water content. *Contributions to Mineralogy and Petrology*, **84**, 67–72.

Harrison, T.M. and Watson, E.B. (1984) The behavior of apatite during crustal anatexis: Equilibrium and kinetic considerations. *Geochimica et Cosmochimica Acta*, **48**, 1467–1478.

Haskin, L.A. (1979) On Rare-Earth Element Behavior in Igneous Rocks. pp. 175–189 in: *Origin and Distribution of Elements* (L.H. Ahrens, editor). Pergamon, Oxford, UK.

Haskin, L.A. (1990) PREEconceptions pREEvent pREEcise pREEdictions. *Geochimica et Cosmochimica Acta*, **54**, 2353–2361.

Henk, A., Franz, L., Teufel, S. and Oncken, O. (1997) Magmatic underplating, extension, and crustal reequilibration: Insights from a cross-section through the Ivrea Zone and Strona-Ceneri Zone, northern Italy. *Journal of Geology*, **105**, 367–377.

Hobbs, B.E. and Ord, A. (2009) The mechanics of granitoid systems and maximum entropy production rates. *Philosophical Transactions of the Royal Society of London*, **368**, 53–93.

Huang, W.L. and Wyllie, P.J. (1973) Melting relations of muscovite-granite to 35 kbar as a model for fusion of metamorphosed subducted oceanic sediments. *Contributions to Mineralogy and Petrology*, **42**, 1–14.

Huang, W-L. and Wyllie, P.J. (1975) Melting reactions in the system $NaAlSi_3O_8$–$KAlSi_3O_8$–SiO_2 to 35 kilobars, dry and with excess water. *Journal of Geology*, **83**, 737–747.

Hughes, C. J. (1982) *Igneous Petrology*. Elsevier, Amsterdam, 551 pp.

Huppert, H.E. and Sparks, R.S.J. (1988) The generation of granitic magmas by intrusion of basalt into continental crust. *Journal of Petrology*, **29**, 599–624.

Hutton, D.H.W. (1988) Granite emplacement mechanisms and tectonic controls: inferences from deformation studies. *Transactions of the Royal Society of Edinburgh: Earth Sciences*, **79**, 245–255.

Hutton, D.H.W. (1996) The 'space problem' in the emplacement of granite. *Episodes*, **19**, 114–119.

Hutton, D.H.W. (1997) Syntectonic granites and the principle of effective stress: A general solution to the space problem? *Petrology and Structural Geology*, **8**, 189–197.

Hutton, D.H.W., Dempster, T.J., Brown, P.E. and Becker, S.D. (1990) A new mechanism of granite emplacement: intrusion in active extensional shear zones. *Nature*, **343**, 452–455.

Ishihara, S. (1975) Acid magmatism and mineralization – oxidation status of granitic magma and its relation to mineralization. *Marine Science Monthly*, **7**, 756–759.

Ishihara, S. (1977) The magnetite-series and ilmenite-series granitic rocks. *Mining Geology*, **27**, 293–305.

Jahn, B.M., Wu, F.Y. and Chen, B. (2000) Massive granitoid generation in Central-Asia - Nd Isotope evidence and implication for continental growth in the Phanerozoic. *Episodes*, **23**, 82–92.

James, R.S., and Hamilton, D.L. (1969) Phase relations in the system $NaAlSi_3O_8$–$KAlSi_3O_8$–$CaAl_2Si_2O_8$–SiO_2 at 1 kilobar water vapour pressure. *Contributions to Mineralogy and Petrology*, **21**, 111–141.

Johannes, W. (1978) Melting of plagioclase in the system Ab–An–H_2O and Qz–Ab–An–H_2O at P_{H_2O} = 5 kbars, an equilibrium problem. *Contributions to Mineralogy and Petrology*, **66**, 295–303.

Knesel, K.M. and Davidson, J.P. (1999) Sr isotope systematics during melt generation by intrusion of basalt into continental crust. *Contributions to Mineralogy and Petrology*, **136**, 285–295.

LeBreton, N. and Thompson, A.B. (1988) Fluid-absent (dehydration) melting of biotite in metapelites in the early stages of crustal anatexis. *Contributions to Mineralogy and Petrology*, **99**, 226–237.

Lister, J.R. and Kerr, R.C. (1991) Fluid-mechanical models of crack propagation and their application to magma transport in dykes.

Journal of Geophysical Research-Solid Earth, **96**, 10,049–10,077.

Luth, W.C. (1969) The systems $NaAlSi_3O_8$-SiO_2 and $KAlSAi_3O_8$-SiO_2 to 20 kb and the relationship between H_2O content, P_{H2O}, and P_{total} in granitic magmas. *American Journal of Science*, **267-A**, 325–341.

Maaløe, S. and Wyllie, P.J. (1975) Water content of a granite magma deduced from the sequence of crystallization determined experimentally with water-undersaturated conditions. *Contributions to Mineralogy and Petrology*, **52**, 175–191.

McCaffrey, K.J.W. and Petford, N. (1997) Are granitic intrusions scale invariant? *Journal of the Geological Society of London*, **154**, 1–4.

McCarthy, T.C. and Patio Douce, A.E. (1997) Experimental evidence for high-temperature felsic melts formed during basaltic intrusion of the deep crust. *Geology*, **25**, 463–466.

Miller, C.F., McDowell, S.M. and Mapes, R.W. (2003) Hot and cold granites? Implications of zircon saturation temperatures and preservation of inheritance. *Geology*, **31**, 529–532.

Montel, J.M. (1993) A model for monazite/melt equilibrium and application to the generation of granite magmas. *Chemical Geology*, **110**, 127–146.

Montel, J.M., Mouchel, R. and Pichavant, M. (1988) High apatite solubilities in peraluminous melts. *Terra Cognita*, **8**, 71.

Moyen, J-F., Nedelec, A., Martin, H. and Jayananda, M. (2003) Syntectonic granite emplacement at different structural levels: the Closepet granite, South India. *Journal of Structural Geology*, **25**, 611–631.

Mrazec, M.L. (1915) Les plis diapirs et le diapirism en général. *Rumania Institute de Géologie Comptes Rendues*, **4**, 226–270.

Naney, M.T., and Swanson, S.E. (1980) The effect of Fe and Mg on crystallization in granitic systems. *American Mineralogist*, **65**, 639–653.

Nemchin, A.A., Giannini, L.M., Bodorkos, S. and Oliver, N.H. (2001) Ostwald ripening as a possible mechanism for zircon overgrowth formation during anatexis: Theoretical constraints, a numerical model, and its application to pelitic migmatites of the Tickalara Metamorphics, northwestern Australia. *Geochimica et Cosmochimica Acta*, **65**, 18.

Nicolesco, C.P. (1929) Anticlinaux diapirs sédimentaires, volcaniques et plutoniques. *Societé Géologique de France Bulletin*, **29**, 21–14.

Patio Douce, A.E., and Beard, J.S. (1995) Dehydration-melting of biotite gneiss and quartz amphibolite from 3 to 15 kbar. *Journal of Petrology*, **36**, 707–738.

Petford, N. and Gallagher, K. (2001) Partial melting of mafic (amphibolitic) lower crust by periodic influx of basaltic magma. *Earth and Planetary Science Letters*, **193**, 483–499.

Petford, N., Kerr, R.C., and Lister, J.R. (1993) Dike transport of granitic magmas. *Geology*, **21**, 843–845.

Petford, N., Lister, J.R. and Kerr, R.C. (1994) The ascent of felsic magmas in dykes. *Lithos*, **32**, 161–168.

Pickering, J.M. and Johnston, A.D. (1998) Fluid-absent melting behavior of a two-mica metapelite. *Journal of Petrology*, **39**, 1787–1804.

Pitcher, W.S. (1987) Granites and yet more granites forty years on. *Geologische Rundschau*, **76**, 51–79.

Pitcher, W.S. and Berger, A.R. (1972) *The Geology of Donegal: A Study of Granite Emplacement and Unroofing*. Wiley Interscience, New York, 435 pp.

Piwinskii, A.J. (1968) Experimental studies of igneous rock series, central Sierra Nevada batholith, California. *Journal of Geology*, **76**, 548–570.

Piwinskii, A.J. and Wyllie, P.J. (1970) Experimental studies of igneous rock series: "Felsic Body Suite" from the Needle Point pluton, Wallowa batholith, Oregon. *Journal of Geology*, **78**, 52–76.

Pollard, D.D. (1977) Derivation and evaluation of a mechanical model for sheet intrusions. *Tectonophysics*, **19**, 233–269.

Raia, F. and Spera, F.J. (1997) Simulations of crustal anatexis: Implications for the growth and differentiation of continental crust. *Journal of Geophysical Research: Solid Earth*, **102**, 22629–22648.

Ramberg, H. (1967) *Gravity, Deformation and the Earth's Crust*. Academic Press, London, 214 pp.

Rapp, R.P. and Watson, E.B. (1986) Monazite solubility and dissolution kinetics: implications for the thorium and light rare earth chemistry of felsic magmas. *Contributions to Mineralogy and Petrology*, **94**, 304–316.

Rapp, R.P. and Watson, E.B. (1995) Dehydration melting of metabasalt at

References

8−32 kbar: implications for continental growth and crust-mantle recycling. *Journal of Petrology*, **36**, 891−931.

Read, H.H. (1948) Granites and granites. *Geological Society of America Memoir*, **28**, 1−19.

Read, H.H. (1956) Granites and granites. pp. 168−193 in: *The Granite Controversy* (H.H. Read, editor). Thomas Murby & Co., London.

Reinhard, M. (1935) über Gesteinmetamorphose in den Alpen. *Auszug aus einem Vortrag in der Mijnbouwkundige Vereenining zu Delft*, 39.

Robertson, J.K. and Wyllie, P.J. (1971) Rock-water systems, with special reference to the water-deficient region. *American Journal of Science*, **271**, 252−277.

Rushmer, T. (1991) Partial melting of 2 amphibolites − contrasting experimental results under fluid-absent conditions. *Contributions to Mineralogy and Petrology*, **107**, 41−59.

Rutter, M.J. and Wyllie, P.J. (1988) Melting of vapour-absent tonalite at 10 kbar to simulate dehydration-melting in the deep crust. *Nature*, **331**, 159−160.

Scaillet, B., Holtz, F. and Pichavant, M. (1998) Phase equilibrium constraints on the viscosity of silicic magmas − 1. Volcanic-plutonic association. *Journal of Geophysical Research: Solid Earth*, **103**, 27257−27266.

Sisson, T.W., Ratajeski, K., Hankins, W.B., and Glazner, A.F. (2004) Voluminous granitic magmas from common basaltic sources. *Contributions to Mineralogy and Petrology*, **148**, 635−661.

Sleep, N.H. (1988) Tapping of melt by veins and dykes. *Journal of Geophysical Research-Solid Earth*, **93**, 10255−10272.

Solar, G.S., Pressley, R.A., Brown, M. and Tucker, R.D. (1998) Granite ascent in convergent orogenic belts: Testing a model. *Geology*, **26**, 711−714.

Thompson, A.B. (1988) Dehydration melting of crustal rocks. *Rendiconti della Societa Italiana di Mineralogia e Petrologia*, **43**, 41−60.

Thompson, A.B. and Connolly, J.A.D. (1995) Melting of the continental-crust − some thermal and petrological constraints on anatexis in continental collision zones and other tectonic settings. *Journal of Geophysical Research-Solid Earth*, **100**, 15565−15579.

Turner, J.C., Jahns, R.H., and Luth, W.C. (1975) Crystallization of alkali feldspar and quartz in the haplogranite system $NaAlSi_3O_8$-$KAlSi_3O_8$-SiO_2-H_2O at 4 kb. *Geological Society of America Bulletin*, **86**, 83−98.

Tuttle, O.F. and Bowen, N.L. (1958) Origin of granite in the light of experimental studies in the system $NaAlSi_3O_8$–$KAlSi_3O_8$–SiO_2–H_2O. *Geological Society of America Memoir*, **74**, 153 pp.

Vielzeuf, D. and Holloway, J.R. (1988) Experimental determination of the fluid-absent melting reactions in the pelitic system. Consequences for crustal differentiation. *Contributions to Mineralogy and Petrology*, **98**, 257−276.

Vielzeuf, D., Clemens, J.D., Pin, C., and Moinet, E. (1990) Granites, granulites and crustal differentiation. Pp. 59−86 in: *Granulites and Crustal Differentiation* (D. Vielzeuf, and P. Vidal, editors). Kluwer Academic Publishers, Dordrecht, The Netherlands.

Vigneresse, J.L. (1999) Should felsic magmas be considered as tectonic objects, just like faults or folds? *Journal of Structural Geology*, **21**, 1125−1130.

Voshage, H., Hofman, A.W., Mazzucchelli, M., Rivalenti, G., Sinigoi, S., Racek, I. and Demarchi, G. (1990) Isotopic evidence from the Ivrea zone for a hybrid lower crust formed by magmatic underplating. *Nature*, **347**, 731−736.

Wall, V.J., Clemens, J.D. and Clarke, D.B. (1987) Models for granitoid evolution and source compositions. *Journal of Geology*, **95**, 731−750.

Warren, R.G. and Ellis, D.J. (1996) Mantle underplating, granite tectonics, and metamorphic P-T-t paths. *Geology*, **24**, 663−666.

Watson, E.B. (1988) The role of accessory minerals in granitoid geochemistry. *Hutton Meeting*, **1**, 19−20.

Watson, E.B. (1996) Dissolution, growth and survival of zircons during crustal fusion: Kinetic principles, geological models and implications for isotopic inheritance. *Transactions of the Royal Society of Edinburg: Earth Sciences*, **87**, 43−56.

Watson, E.B. and Harrison, T.M. (1983) Zircon saturation revisited: temperature and compositional effects in a variety of crustal magma types. *Earth and Planetary Science Letters*, **64**, 295−304.

Watson, E.B. and Harrison, T.M. (1984) Accessory minerals and the geochemical evolution of crustal magmatic systems: a summary and prospectus of experimental approaches. *Physics of the Earth and Planetary Interiors*, **35**, 19−30.

Watson, E.B. and Harrison, T.M. (2005) Zircon thermometer reveals minimum melting conditions on earliest Earth. *Science*, **308**, 841−844.

Watt, G.R., Burns, I.M. and Graham, G.A. (1996) Chemical character-istics of migmatites: Accessory phase distribution and evidence for fast melt segregation rates. *Contributions to Mineralogy and Petrology*, **125**, 100−111.

Weertman, J. (1971) Theory of water-filled crevasses in glaciers applied to vertical magma transport beneath ocean ridges. *Journal of Geophysical Research-Solid Earth*, **76**, 1171−1183.

Weertman, J. and Chang, S.P. (1977) Fluid flow through a large vertical crack in the Earth's crust. *Journal of Geophysical Research-Solid Earth*, **82**, 929−932.

Weinberg, R.F. (1999) Mesoscale pervasive felsic magma migration: alternatives to dyking. *Lithos*, **46**, 393−410.

White, A.J.R. and Chappell, B.W. (1977) Ultrametamorphism and granitoid genesis. *Tectonophysics*, **43**, 7−22.

Williamson, B.J., Downes, H. and Thirlwall, M.F. (1992) The relation-ships between crustal magmatic underplating and granite genesis: an example from the Velay granite complex, Massif Central, France. *Transactions of the Royal Society of Edinburg: Earth Sciences*, **82**, 235−245.

Wilson, L. and Head (III), J.W. (1981) Ascent and eruption of basaltic magma on the Earth and moon. *Journal of Geophysical Research-Solid Earth*, **86**, 2971−3001.

Wolf, M.B. and London, D. (1993) Apatite Solubility in the Peraluminous Haplogranite System − Not Déjà Vu All Over Again. *AGU 1993 Spring Meeting April 20, EOS*, 341.

Wyborn, D., Chappell, B.W. and Johnston, R.M. (1981) Three S-Type Volcanic Suites from the Lachlan Fold Belt, Southeast Australia. *Journal of Geophysical Research*, **86**, 10335−10348.

Wyllie, P.J., Huang, W-L., Stern, C.R. and Maaløe, S. (1976) Granitic magmas: possible and impossible sources, water contents, and crystallization sequences. *Canadian Journal of Earth Sciences*, **13**, 1007−1019.

Young, D.A. (2003) *Mind Over Magma: The Story of Igneous Petrology*. Princeton University Press, Princeton, New Jersey, USA, 686 pp.

Zhao, G.C., Wilde, S.A., Cawood, P.A. and Lu, L.Z. (1999) Thermal evolution of two textural types of mafic granulites in the North China craton: evidence for both mantle plume and collisional tectonics. *Geological Magazine*, **136**, 223−240.